Rare Birds of Hampshire

Rare Birds of Hampshire

John Clark

Design and illustrations

Dan Powell

HOS

Hampshire Ornithological Society

2022

Dedicated to members of the Portsmouth Group (PG)

Pioneers of modern Hampshire and British ornithology

Thank you for generous donations which have supported the book.

In memory of Major John Welch, 1931-2020, a keen Hampshire birder.

Hampshire Biodiversity Information Centre

Ecosupport

Hampshire Ecological Services

Contents

Introduction

The aim of this book is to present a detailed review of the occurrences of rare birds in Hampshire. The species accounts cover the historical period up to 1950 and the 'modern era' of 70 years from 1951 to 2020. To qualify for inclusion in the book, a species (or recognisable subspecies) on the Hampshire List (Cox *et al.* 2019; www.hos.org.uk/home/news-recording/hampshire-birding/hampshire-bird-list) must satisfy at least one of the following criteria:

One considered by the British Birds Rarities Committee at any time since its formation in 1958.
One which has occurred in Hampshire on fewer than 200 occasions.
One which has not occurred annually in Hampshire in the 21st century (except non-natives Snow Goose, Ruddy Duck and Golden Pheasant).

Species sequence and English vernacular and scientific names follow the *Ninth Edition of the British List* (*BOU* 2017) and any subsequent changes made in *BOURC Reports* up to number 52 (*BOURC* 2021).

The species accounts have been compiled as follows. The header gives the English vernacular and scientific name and states the years in which BBRC and HFCERC/HOSRP were responsible for the assessment of records.

A status summary includes the number of records up to 1950 and in the modern era from 1951 to 2020. Terms used have a numerical value as follows:

Very rare
– fewer than 10 records ever

Rare
– less than annual

Very scarce
– up to 20 per year
or 1–10 breeding pairs

Scarce
– 21 to 200 per year
or 11–100 breeding pairs

Moderately common
– 201 to 2,000 per year
or 101 to 1,000 breeding pairs

Common
– more than 2,000 per year
or 1,000 breeding pairs

The texts start with a summary of any historic records. These rely heavily on the first county avifauna *The Birds of Hampshire and the IOW* (Kelsall & Munn 1905) but reference has also been made to *The Zoologist* and *The Field*, which are now available online, and the catalogue of the Hart collection of birds which were mostly obtained in the Christchurch and New Forest areas (Hart 1913).

Black-winged Stilts, Keyhaven, May 4th 1987

THE

ZOOLOGIST:

A

POPULAR MISCELLANY

OF

NATURAL HISTORY.

CONDUCTED BY

EDWARD NEWMAN, F.L.S., Z.S., &c.

VOLUME THE FIRST,

ILLUSTRATED BY UPWARDS OF ONE HUNDRED FIGURES.

LONDON:

JOHN VAN VOORST, PATERNOSTER ROW.

M.DCCC.XLIII.

Occurrences of 13 species not recorded in the modern era were reviewed by a sub-panel of HOSRP in 2010 (Combridge & Clark 2010). Since the review, two of them, Alpine Accentor and Common Rosefinch, have occurred in the county. Of the remaining 11, eight were retained in category B of the Hampshire List (reserved for species not recorded since 1950) and are included in the main species accounts. The remaining three were not included in category B and are included in Appendix III. The author has doubts about the validity of the first records of certain other species. Records are only accepted if at least two of the following criteria have been satisfied.

Extant specimen, or good evidence that the specimen did exist.
Dated record to year, preferably to month.
Description in a contemporary journal.
Reputable ornithologist involved.

Records of species in the modern era are described in full or summarised. The first such record is described with an account of the bird's discovery and/or a description of the bird. Many of these accounts, particularly for the period up to the 1990s, have not previously been published.

Bluethroat, male, white-spotted race, midge-catching, South Scrape, Titchfield Haven, March 29th 1998

This is followed by a chronology of subsequent records. All but the rarest species have charts showing the annual and monthly or weekly distribution of records. Species considered by BBRC for all or part of the period under consideration have a narrative giving information about each record. An informal approach has been used with first names for observers where known. Records of more numerous species such as Temminck's Stint and Pectoral Sandpiper have their records summarised. Species with birds which have returned in successive winters (mostly wildfowl and gulls) have a table with each 'individual' numbered so that their history can be followed.

The total number of records quoted for each species includes only those accepted by BBRC and/or HFCERC/HOSRP. Additional records may be mentioned including those considered by the author as 'harshly not proven' – in effect a recommendation for recirculation – and non-submitted records. In some cases, comment has been made on the totals of not proven and non-submitted records.

The species accounts are followed by various appendices as listed on the Contents Page.

Pectoral Sandpiper, South Scrape, Titchfield Haven, September 9th 2017

Acknowledgements

It has been a labour of love to compile this book. Some ten thousand records referring to over 200 species have been analysed. I hope that readers will enjoy reading the many previously unpublished notes and anecdotes which are included. Dan Powell has done a fantastic job in designing the cover and layout of the book. He has included many superb paintings and sketches, most of which illustrate his encounters with the actual birds. These are complemented by an unrivalled set of photographs which has been assembled by Alan Lewis.

Thanks are due to all the photographers whose images appear in the book: David Aitken, Matthew Barfield, Steve Bassett, Martin Bennett, Keith Betton, Jerry Bird, J.B. & S. Bottomley, Dennis Bright, John Burnett, Jen Churchill, John Clark, Trevor Codlin, Simon Colenutt, Andy Collins, Steve Copsey, Mike Creighton, Jason Crook, Tony Croucher, David Cuddon, Mike Darling, Alan & Caroline Dawson, Kit Day, Brian Dempsey, David Dimmock, John Dodds, Geoff Doré, Pete Drury, Barry Duffin, Mike Duffy, George Else, Darren Evans, John Eyre, Philip Fawkes, Geoffrey Fisher, Richard Ford, Olly Frampton, Mark Francis, Lee Fuller, Paul Gale, Steve Gantlett, Fred Gibson, Daryl Gorman, Alan Hayden, John Hilton, Chris Holt, John Humble, Simon Ingram, Richard Jacobs, Andy Johnson, Simon Johnson, Nigel Jones, Lynn Kelly, Simon M. King, Steve Lavington, Maxwell Law, Jim Lawrence, Mike Lawrence, Alan Lewis, Barry Lintott, Bob Marchant, Keith Maycock, Jeremy McClements, Keith Metcalf, Peter Milinets-Raby, Linda Miller, Marc Moody, Roger Murfitt, Pete Northcote, Jerry O'Brien, Dave Perrett, Ian Pibworth, Rob Porter, Dan Powell, Ted Pressey, Mike Read, Gareth Rees, Andy Rhodes, Louis Rumis, David Ryves, Tommy Saunders, Gordon Small, George Spraggs, Barry Stalker, Joe Stockwell, Keith Stone, Chris Tabor, Tony Taylor, Andy Thorpe, Mike Trew, Mike Trim, Philip Underwood, Simon Vale, Jason Ward, Marcus Ward, Mike Wearing, Adrian Webb, Andy Welch, Ian Wells, John Wichall, Pete Wheeler, Ian Williamson, Paul Winter, Simon Woolley and Russell Wynn.

Every effort has been made to contact photographers for permission to use their pictures and to ensure that they are correctly credited.

Keith Betton, Chairman of HOS and County Recorder, has been continually supportive of the project. Bryan Coates and John Shillitoe, HOS Sales Officer and Webmaster respectively, have expertly managed orders for the book. Andrew Hunter and Mark Brouard, of our printers Henry Ling, have ensured the quality of the final product.

Spotted Sandiper, South Scrape, Titchfield Haven, October 4th 1986. A 'tennis ball on legs'.

Red-flanked Bluetail, Sandy Point, October 19th 2010

John Cloyne has been extremely helpful with the provision of several historic references and made thorough comments on the text. Mike Chalmers, Pete Combridge, Alan Cox, Jason Crook, Nigel Jones, Marc Moody and Ed Wiseman have also provided important assistance and commented on the text. The following made comments on the text and/or provided information about particular records: Rosemary Billett, Dave Bishop, Simon Buckell, Richard Carpenter, Paul Castle, Bob Chapman, Anthony Cheke, Robert Cheke, David Christie, Rob Clay, Chris Cockburn, Trevor Codlin, Andy Collins, Barry Collins, Martin Cooper, Tim Doran, Barry Duffin, Mark Edgeller, Trevor Ellery, John Eyre, Philip Fawkes, Richard Ford, Pete Gammage, Steve Gantlett, George Green, Andy Hanby, Alan Hayden, Chris Holt, Ashley Howe, Rob Hume, Simon Ingram, Ron Johns, Andy Johnson, Steve Keen, Richard Levett, James Lidster, Bob Marchant, Peter Milinets-Raby, Richard Millington, John Newnham, John Norton, Stuart Orr, Barry Page, Andy Parfitt, Martin Pitt, Pete Potts, Dan Powell, Leo Pyke, Graham Rees, Amy Robjohns, Mick Scott, Brian Small, Graham Stephenson, David Taylor, Martin Terry, David Thelwell, Mike Toms, Jenni Tubbs, Dave Unsworth, Tracey Viney, John Walters, Ian Watts, Keith Wills, Paul Winter, Dave Wooldridge and Russell Wynn.

I have also had invaluable help from staff at several museums: Jo Hatton (Horniman Museum, London), Claire Cooper and Heather Southorn (Leicester Museum), Ross Turle (Chilcomb House, Hampshire Cultural Trust), Christine Taylor (Portsmouth Museums), Alex Bond (British Museum, Natural History) and Pam Field (Bournemouth Natural Science Society Museum). Suzanne Foster and Amber Davenport kindly arranged access to the Winchester College collection in the Science School there.

I would like to thank Roger Riddington and Stephen Menzie, past and present editors of *British Birds*, for permission to use national data and distributional information from the annual Rare Birds and Scarce Migrant Reports, Paul French, Chas Holt and Nigel Hudson (BBRC), Chris McInerny (BOURC), the BTO for the loan of David Billett's Langstone Harbour yearbooks, Keith Naylor for assistance with historical records (see historicalrarebirds.info) and David Allen for access to the London NHS library.

Finally, my thanks to Keith Betton, Mike Chalmers, Pete Combridge, Tim Doran, Rosie Powell and Graham Stephenson who carried out a last-minute proof-check. Responsibility for any errors which escaped their close attention is entirely mine!

John Clark, April 2022

About the author

John at Fatu Hivu, French Polynesia, September 2019

Born in Aldershot in 1953, John has been a life-long Hampshire resident. He was interested in many aspects of natural history from an early age. At Farnborough Grammar School, influenced by his biology teacher Ewart Jones and friends 'Spike' Millington and Trevor Price, he caught the birding bug. From 1971, he spent four years at Southampton University, completing a degree in Zoology and a PGCE. He also produced the first *Hants/Surrey Border Bird Report* which continued until 1997.

John took up a teaching post in Eastleigh in 1975 and started his twice-monthly visits to the Avon Valley, a very under-watched area at the time. He joined the Hampshire Editorial Committee in 1977 and went on to co-edit or edit the *Hampshire Bird Report* for much of the 1980s. His book *The Birds of the Hants/Surrey Border* was published in 1984. He was a major contributor to the fieldwork for the first Hampshire Tetrad Atlas in 1986–91. The results of this were included in *The Birds of Hampshire* which John co-edited with John Eyre and was published in 1993. In that year he became County Recorder, a role which he held until 2010. Recently, he edited the *Hampshire Bird Report* from 2012–2014 and re-joined the HOSRP in 2017.

Birding in Hampshire has always been John's main passion with serious recording and survey work to the fore but always with an eye on his county list. When he took early retirement from teaching in 2008, he took increasingly frequent trips abroad. He has now birded in almost 90 countries and seen over 8,600 species. In any discussion about his Hampshire list, he always mentions the eight species he has missed while birding abroad including blockers such as the Little Bittern (at Fleet Pond), Red-flanked Bluetail and Eastern Olivaceous Warbler!

John does have interests outside birding! He is a great fan of the music of Neil Young and has seen him live 41 times. He also likes Jackson Browne and Bruce Springsteen. He has been a lifelong supporter of hometown club Aldershot Town although there has rarely been much to celebrate! Support of Hampshire County Cricket Club has been almost as frustrating although they did come close to winning the county championship in 2021. He enjoys watching most other sports.

About the artwork

It has been a joy looking back through some 40 years of my Hampshire field sketchbooks and studio paintings, reliving the successful moments during bouts of twitchiness, now under control. When watching any wildlife, the drive to make marks on a page is an emotional one, a reaction to the moment, light and shapes the subject is forming. The sketches have been scanned directly from the page, appearing warts and all. The paintings were mainly completed in the studio, some taking their time before wanting to flow out of the brushes. A few of the smaller paintings may be familiar having been published in past *Hampshire Bird Reports*.

I have travelled many miles in good birding company, too many folk to mention here, but I must thank Rosie for sharing the adventure these last three decades.

Dan Powell, April 2022

Early efforts:

Red-breasted Goose, River Avon, Bisterne, January 8th 1984

Short-toed Lark Calshot, October 19th 1983

White-winged Gulls, 1984

Nature reserves and protected areas

The list shows the owners and managers of selected localities mentioned in the book. Site designations such as LNR and NNR have been omitted from the species accounts.

Beacon Hill NNR (Natural England)
Blashford Lakes NR (HIWWT)
Calshot Marshes (HCC)
Castle Bottom NNR (HCC)
Farlington Marshes (Portsmouth City Council; managed by HIWWT)
Fleet Pond LNR (Hart District Council)
Fishlake Meadows NR (HIWWT)
Hook-with-Warsash LNR (HCC)
Langstone Harbour Islands (RSPB)
Lepe Country Park (HCC)
Lower Test Marshes NR (HIWWT)
Ludshott Common (Wealden Heaths SPA; owned and managed by National Trust)
Lymington and Keyhaven Marshes LNR (HCC) and **NR** (New Forest District Council, Managed by HIWWT)
Martin Down NNR (Natural England)
New Forest (New Forest National Park Authority, managed by Forestry England)
North Solent NNR (including Needs Ore) (Natural England)
Old Winchester Hill NNR (Natural England)
Sandy Point LNR (HCC)
Sturt Pond LNR (Milford on Sea Parish Council)
Testwood Lakes NR (HIWWT)
The Kench LNR, Hayling Island (HCC)
Titchfield Haven NNR (HCC)
West Hayling LNR (Havant Borough Council; managed by RSPB)
Wildgrounds NR (Gosport Borough Council)
Winnall Moors NR (HIWWT)
Woolmer Pond and Forest (Wealden Heaths SPA; owned by MOD and managed by HIWWT)

Abbreviations

BBRC British Birds Rarities Committee
BNSS Bournemouth Natural Science Society
BoH *Birds of Hampshire* (Clark & Eyre 1993)
BOU British Ornithologists' Union
BOURC British Ornithologists' Union Records Committee
HBR Hampshire Bird Report
HCC Hampshire County Council
HCT Hampshire Cultural Trust
HFC Hampshire Field Club
HFCERC Hampshire Field Club Editorial & Records Committee
HOS Hampshire Ornithological Society
HOSRP Hampshire Ornithological Society Records Panel
HIWWT Hampshire & IOW Wildlife Trust
IOC International Ornithological Congress
IOW Isle of Wight
JNCC Joint Nature Conservation Committee
K & M *Birds of Hampshire & the Isle of Wight* (Kelsall & Munn 1905)
LNR Local Nature Reserve
NFOC New Forest Ornithologists' Club
NHS Natural Hisory Societies
NNR National Nature Reserve
RBBP Rare Breeding Birds Panel
RSPB Royal Society for the Protection of Birds
SEBR South Eastern Bird Report
SPA Special Protection Area
WCC Winchester City Council
WeBS Wetland Bird Survey
WWT Wildfowl & Wetlands Trust

Rare Birds of Hampshire – a brief history

Before the 19th century

The first known reference to a rare bird in Hampshire is that of a Hoopoe included in *Pinax rerum naturalium Brittanicarum*, an alphabetical list of flora, fauna and minerals found in England which was published by Christopher Merrett in 1666. Almost a century later, a flock of 25 Great Bustards was seen on the downs west of Middle Wallop in November 1751 although this was not put on record until much later (Chafin 1818).

Rev. Gilbert White. Gilbert White's House & The Oates Museum

The man who was to have the greatest influence on early Hampshire ornithology was Gilbert White, born in Selborne in 1720. He was ordained in 1749, and subsequently held several curacies in the vicinity of Selborne. In 1757, on the death of his father, he moved back into the family home *The Wakes* and resided there until he died in 1793. White began record keeping in his *Garden Kalendar* in the early 1750s although the first reference to birds was not made until 1758, when on April 9th he recorded the arrival of Swallows in the village. Subsequently, White corresponded with eminent naturalists such as Thomas Pennant and Daines Barrington, his letters to them forming the core of his book *The Natural History and Antiquities of Selborne*, first published in 1789.

Waxwings and green tea at Wagamamas, Whiteley, February 12th 2017

Of the species included in *Rare Birds of Hampshire*, White recorded eleven: the resident Black Grouse, which was very scarce even by that time, the summer visitors Corncrake, Wryneck and Red-backed Shrike, and rarities Hoopoe (a pair some years prior to 1767), Waxwing in 1767, Little Auk in 1771, Hooded Crow in 1773, Spotted Crake in 1774 and Dipper in 1786. His most famous 'observation' was of the six Black-winged Stilts which came down on Frensham Great Pond on April 28th 1779 although it is not clear if White saw the birds alive. Five were shot and one was allowed to go free. Frensham Great Pond was at one time partly in Hampshire but a boundary change in 1991 transferred it all to Surrey so the occurrence no longer qualifies as a Hampshire record.

19th century

The 19th century was the era of 'shotgun ornithology'. Many wealthy collectors wanted to acquire a complete collection of British birds including rarities. This resulted in a great demand for specimens, so the art of taxidermy flourished during this period. The most significant collections in Hampshire were those of Edward Hart, the Earl of Malmesbury and Winchester College.

Edward Hart

William Hart (1811–68) started a taxidermy business in Christchurch in 1834. His son Edward, born in 1847, assumed responsibility when William died, and built up an extensive skin collection which included many specimens procured locally. Edward Hart opened a museum at the Bow House, Christchurch which by 1913 had expanded to 347 cases of birds, 53 cases of mammals and an extensive collection of birds' eggs and butterflies. Hart offered the collection for sale for £4,000 in 1903 but a public appeal to raise the asking price fell short. The collection remained at the Bow House but after Hart died in 1928 it was sold to John Hall of Stafford. He was outlived by his wife, but when she died her executors donated the collection to Stowe School, Buckinghamshire, which had been attended by Hall's children. In 1950 the school asked staff at Leicester Museum to inspect the collection which by then was in poor condition. Conservation work at Leicester saved some 290 of the original 400-plus cases. In addition, some birds from the other cases were retained as cabinet specimens. Stowe School sold the collection in 1983, when it was jointly purchased by the Horniman Museum in London, the Hampshire County Council Museum Service and Leicester Museum. The Horniman Museum holds the lion's share with 254 cases including Hampshire's first Nutcracker collected in 1846, Balearic Shearwater in 1859, Roller in 1874 and Rough-legged Buzzard and Woodchat Shrike in

1875. Some cases are on display in the museum although most are in storage and can be viewed by arrangement. Photographs of all are on the Horniman Museum website. Hampshire acquired 22 cases, which initially went to the Red House Museum in Christchurch but are now housed at Chilcomb House, Winchester. These include important specimens such as Hampshire's first Cream-coloured Courser in 1845, Collared Pratincole in 1857, American Bittern in 1876 and Lesser Grey Shrike in 1900. These can be viewed on the website of the Hampshire Cultural Trust. At Leicester, 16 cases have been retained including a pair of Hen Harriers with four nestlings procured in the New Forest in 1874 along with an extensive collection of cabinet specimens. These include Hampshire's first Snowy Owl in 1848, Scops Owl in 1866, Night-heron in 1868, Purple Heron in 1875 and also a shearwater shot by Edward in May 1878 and identified by him as a Great Shearwater. This skin was examined by the author and Marc Moody in October 2021 and appears to be a Scopoli's Shearwater rather than the similar Cory's Shearwater.

Scopoli's Shearwater, Leicester Museum. Marc Moody

Sir James Harris (1746–1820) was created the Earl of Malmesbury in 1800. The family home was at Heron Court, a large estate just south of Hurn in present-day Dorset. His son, James Edward (1778–1841), who became the 2nd Earl, was the author of the 'Shooting Journals' which form the basis of the bird records from Heron Court for the first 40 years of the 19th century. During this period, a substantial collection of birds was acquired. It occupied its own room at Heron Court until the house was sold in 1950. The collection was dispersed with most specimens going to the Red House Museum and some being donated to the Bournemouth Natural Science Society. These include the 1828 White's Thrush which was shot at Heron Court and was the first for Britain; it is on display in the society's museum in Bournemouth.

Many of the specimens in the Hart and Malmesbury collections were procured in locations which were in Hampshire at the time. As a result of the 1974 local government reorganisation, parts of south-west Hampshire including Christchurch Harbour and the Avon Valley upstream to Sopley, as well as the town of Bournemouth, were transferred to Dorset. Only those specimens known to have been taken in present-day Hampshire have been included within the main species accounts in this book. Nonetheless, many others were firsts for Hampshire at the time and have not yet been admitted to the Dorset avifauna. They are all listed in Appendix V along with other significant records for the former parts of Hampshire.

William Chalkley's shop in The Square, Winchester, 1903. © Hampshire Cultural Trust

The Winchester College NHS was formed in March 1870. It became a focal point for the collection of birds and other specimens. In 1900 a booklet *Winchester College Memorial Buildings: Collection of Hampshire Birds 1863–1900* was published which contained an annotated list of the stuffed birds in the collection. These included a Black Grouse from 1882 and two Dotterels from 1893. William Chalkley (1842–1922) was a self-taught taxidermist who helped with the college collection and later turned his hobby into a business, opening a shop in The Square, Winchester in 1880 which was continued by his nephew until 1966. He formed a collection of birds and mammals of Hampshire which was displayed in the Winchester Museum, where he became Honorary Curator in 1903. Further specimens were added to the display until the 1950s when it was dismantled. Some specimens were transferred to Winchester College where they are still on display in the Science School with others from the original college collection.

Another significant character in 19th century Hampshire ornithology was Col. Peter Hawker (1786–1853). He was an ardent wildfowler and punt-gunner who was particularly active in the Keyhaven area from 1814 until his death. He saw the first White-tailed Eagle for Hampshire in 1827 and encountered several more but failed to kill one. He secured the county's first Smew in 1830 and shot many Whooper Swans which occurred in the frequent cold winters during that period. Most of the available information is contained in his diaries which were edited into two volumes by Payne-Gallwey (1893). The originals are archived in Yale University, Connecticut, USA.

'Return to Keyhaven after a day's gunning, in the winter of 1838'. Image from Hawker's diary, The Gun Inn and Hawker's Cottage can be seen in the centre of the picture. Drawing C. Varley. Engraved H. Adlard

The Hampshire Field Club and Archaeological Society was founded in 1885. The Rev. John E. Kelsall (1864–1924) was a founder member and published *A Briefly Annotated List of Birds of Hampshire and the Isle of Wight* in the *HFC Proceedings* (Kelsall 1890). This was the precursor to the first county avifauna *The Birds of Hampshire and The Isle of Wight* written by Kelsall and Philip W. Munn (1865–1946) and published in 1905. This drew heavily on the collections already mentioned, White's writings from over a century before, other publications such as Wise (1862) and a long list of correspondents. These included George Corbin (1841–1914), a cabinet maker and upholsterer resident in Ringwood. He recorded several notable birds including the White-throated Needletail which was exhibited in London in early 1880 and is now held in the Natural History Museum at Tring. K & M also relied heavily on *The Zoologist*, a monthly journal first published in 1843, although they do seem to have missed some references to Hampshire birds in *The Field*, a weekly magazine dating from 1853. The complete run of *The Zoologist* up to 1916 is now available online in the Biodiversity Heritage Library and *The Field* up to 1911 can be accessed via online subscription to the British Newspaper Library Archive.

1900–1950

The first half of the 20th century was a period when ornithological recording was limited in Hampshire. The age of collecting specimens was in decline and although egg-collecting probably reached its peak there was a secrecy surrounding the activities of many oologists. Philip Munn's *Notes on the Birds of Hampshire and the Isle of Wight* was published in the *HFC Proceedings* in two parts (Munn 1919, 1920). Apart from a paper on New Forest birds in *Ibis* (Longstaff 1926), nothing was published in the rest of the decade but during the 1930s three publications included Hampshire bird records. Dr. Frederick Haines (1864–1946) retired to Linwood in the New Forest in 1931 and became Ornithological Secretary of HFC. He compiled a narrative report *Birds in Hampshire, 1932* and systematic reports in subsequent years which were published in the *HFC Proceedings*. The Bournemouth Natural Science Society published a report on local bird observations in their annual proceedings through the decade and Ralph Whitlock (1914–1995) initiated the *South Eastern Bird Report* in 1934 which covered Hampshire, Kent, Surrey and Sussex. Editions up to 1938 included different records to those in the *HFC Proceedings*. From 1939 the *SEBR* included a reprint from the *HFC Proceedings* but the final two reports for 1946 and 1947 did not include any Hampshire records. Bird recording fell to a nadir in 1948 when there were only 22 contributors to *Birds in Hampshire* included in the *Proceedings*. In the following year Ralph Whitlock, by then HFC Ornithological Secretary, used the *Proceedings* to appeal to Hampshire observers to send their bird records to Edwin Cohen (1895–1970), newly appointed as County Recorder by HFC. Not surprisingly, these reports had a very limited content by modern standards.

The early 1950s: the emergence of the Portsmouth Group

Things were about to change! After the Second World War, Dr. Canning (Doc) Suffern (1892–1978) started weekly walks for birdwatchers around the then private Titchfield Haven. He encouraged several young birders including Tony Norris, who found the county's first Kentish Plover and Glaucous Gull there in 1948. The most significant developments were at Farlington Marshes, where several teenagers began to watch birds. In March 1951, David Billett (1932–2014) and Colin Tubbs (1937–1997) met for the first time at Farlington. Later that year, David met Doc Suffern at Titchfield Haven and their discussion encouraged him to begin the detailed recording of all birds at Farlington and in the surrounding Langstone Harbour. In 1952, George Clay, Cliff Henty, Graham (Taff) Rees and Bryan Renyard joined the Farlington watchers and it was agreed that an annual summary of their sightings would be submitted to HFC. The first, compiled by David and Graham, was sent to Edwin Cohen early in 1953. Edwin produced the annual *Birds in Hampshire* which continued to form part of the *HFC Proceedings* until 1955 although it was also available as an offprint. In 1956 and 1957 a separate Natural History Supplement was published; that for the latter year included one page of *Diptera* observations but was otherwise devoted to birds. This inevitably led to the first separate *Hampshire Bird Report* for 1958 although officially it still formed part of the *HFC Proceedings*. For the subsequent development of the *Hampshire Bird Report* into the 21st century, see Cox (2007).

Proceedings of the
HAMPSHIRE FIELD CLUB
and Archaeological Society
Vol. XXI Part II

The Hampshire Bird Report

Edited by the Records Committee
of
The Ornithological Section

M. BRYANT E. COHEN
K. H. PALMER J. STAFFORD
J. H. TAVERNER

1958

Price 5/-

Dr. Canning Suffern, David Billett and Colin Tubbs.

Bill Truckle (top two) and Jenni Tubbs

It was Edwin who invented the term 'Portsmouth Group' (PG). The regular counts of wildfowl and waders started by the PG were the model on which the future national Birds of Estuaries Enquiry and Wetland Bird Survey were based. In the next few years, new members including Michael Bryant, Ed Wiseman, Johnny Bowers, Bill Truckle and Peter Le Brocq joined the PG. At the same time, it broadened its activities to include raptors and heathland birds in the New Forest and migration watches at St. Catherine's Point in the Isle of Wight as well as other sites in Hampshire. In the period up to 1961, PG members added 12 species to the Hampshire list including three major rarities: Broad-billed Sandpiper in 1957, Little Crake in 1959 and Desert Wheatear in 1961. For more detail on the history of the Portsmouth Group, see Wiseman (2009).

The influence of the PG spread to other parts of the county and encouraged birdwatchers to keep detailed records. This led to the discovery of increasing numbers of rarities including many new birds for Hampshire.

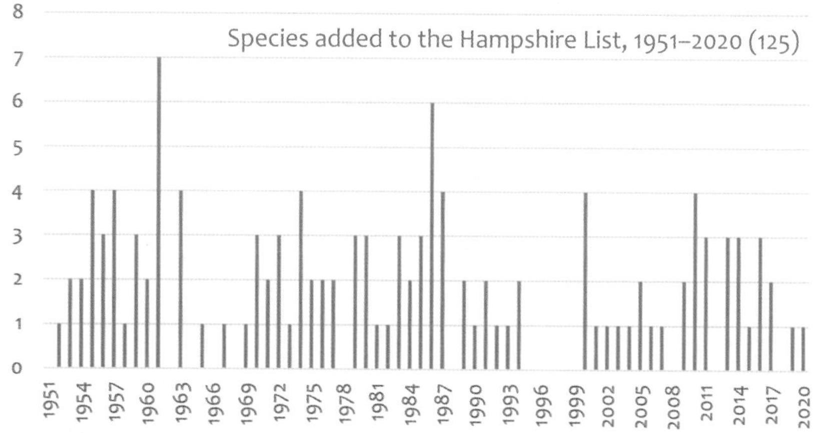

The mid-1950s and 1960s: improving coverage throughout the county

In 1956 Frank Clafton founded the Christchurch Harbour Ornithological Group, which has produced an annual report ever since. The Christchurch area had proved to be rich in rarities during the time of Edward Hart. That continued, although most of Christchurch Harbour's rarest birds have occurred since it was included in the portion of south-west Hampshire which was transferred to Dorset in April 1974. Birds recorded in the Christchurch area prior to this, including the Hart rarities, have still not been admitted to the Dorset avifauna. These are listed in Appendix V. Post-1974 records of species in Christchurch Harbour which either pre-date their first occurrence in present-day Hampshire or have never been recorded in Hampshire are listed in Appendix VI.

Coverage in other areas improved in the second half of the 1950s and through the 1960s. At Titchfield Haven, a network of birders led by Doc Suffern included Roy Dennis, now leader of the project to reintroduce the White-tailed Eagle to southern England. He saw a flock of 34 Smew there in severe weather in February 1956, a total which has never been exceeded. In subsequent years, this group expanded to include Shaun White, Martin Terry, David Price, Barry Duffin, Richard Carpenter and others. Shaun discovered a Great Reed Warbler singing there in May 1960 which stayed for a week. A rather more exciting discovery was made there on March 4th 1961 when a bird with an explosively loud song was tentatively identified as a Moustached Warbler. During the next fortnight it was heard or glimpsed by most of the leading ornithologists of the day. After much heated debate as to its identity, it was finally caught and identified as a Cetti's Warbler, the first for Britain. The controversy continued to rage in the letters pages of *British Birds* for some months. Also in the early 1960s, Martin Terry, David Price and Barry Duffin started the Gilkicker Point Group after Doc Suffern had suggested that the site could be good for migration watching. The group grew and extended its activities to other localities, producing an annual report until well into the 1970s.

Cetti's Warbler

In the Southampton area, Ken Palmer, Jim Williams, Paul Bowman, Norman Pratt, David Glue and others ensured good coverage of Dibden Bay, where the reclaimed areas were attractive to waders, and Nursling Gravel Pits. This extended to Calshot Marshes and the Fawley Oil Refinery, where Philip Fawkes and John M. Jones achieved thorough coverage in the early and mid-1970s, and the Hamble Estuary, assiduously watched by David Christie from the 1970s. The Beaulieu Estuary, including Needs Ore and The Gins (nowadays known as Blackwater) was systematically covered through the efforts of John Taverner from the late 1950s and was firmly put on the ornithological map when it was revealed that Mediterranean Gulls bred in the UK for the first time there in 1968 (Taverner 1970). The Keyhaven and Pennington Marshes area received intermittent coverage during the 1950s and early 1960s although the area's potential for attracting rare birds was already recognised with the discovery there of Lesser Yellowlegs by Bob Sharland in September 1953 and Gerald Kinsey in August 1954, a Sooty Tern by Dave Wooldridge and Doug Poole in June 1961 and a Terek Sandpiper by Ed Wiseman in May 1963. The appointment of Ed initially as a summer warden in 1966 and later in a permanent position led to much-improved coverage. This was complemented by the efforts of many birders including Jim Williams, Dave Wooldridge, Bob Dunn, Pete Hobby and later Mike and Pete Combridge, Marc Moody, Tim Parminter and Russell Wynn. For a detailed history of bird recording in the Keyhaven area, see Wiseman (2017).

Away from the New Forest, most inland areas remained poorly watched apart from the Itchen Valley, which was covered by John Taverner, John Cloyne and Winchester College NHS members, and Woolmer Forest and the South Downs at Butser Hill which were occasionally visited by PG members. In the late 1960s and early 1970s, several young birders developed under the guidance of Ewart Jones, head of Biology at Farnborough Grammar School, which led to thorough coverage of sites such as Fleet Pond, the gravel pits at Eversley and Yateley in the Blackwater Valley and the local heathlands. This group included Richard Millington and Steve Gantlett (who jointly went on to found the Bird Information Service and the journal *Birding World*), Stephen 'Spike' Millington, Graham Stephenson, Tim Doran and the author, and later Mick Scott. Among birds found at Fleet Pond were a Great Reed Warbler in May 1970 (the first of three there) and a Collared Pratincole in May 1974. Another area with improved coverage was the Test Valley, where David Thelwell, Tony Hale and Tim Norriss were particularly active. The Avon Valley, with its wintering White-fronted Geese and Bewick's Swans, and the downs in the Martin area to the west were covered by Norman Orr and John Ash during the 1950s and 1960s, the latter also studying the declining population of Red-backed Shrikes in the New Forest. The first two Red-breasted Geese for the county were found with the White-fronts at Ibsley in January 1967 and January 1969, the latter by visiting birders. Coverage improved from the mid-1970s onwards through the efforts of local birders Mike Read and Martin King and the author, who made twice-monthly visits to the area from his home in Fleet at the opposite end of the county for 40 years.

The 1970s: major rarities and the timely expansion of Hampshire's network of reserves

The 1970s began with the discovery by John Smith and David Billett of a 'mega': an adult Franklin's Gull at Farlington Marshes on February 21st 1970. This was the first for the Western Palearctic and its long stay enabled many observers to see it (Billett & Grant 1971). The decade was a good one for BBRC rarities as shown opposite although the annual totals were swelled by large numbers of Aquatic and Savi's Warblers and Red-footed Falcons. Major rarities included a Lesser Grey Shrike on Portsdown Hill in July 1973, a Yellow-billed Cuckoo at Pennington Marshes in November 1976, a Least Sandpiper at Farlington Marshes in May 1977 and a Lanceolated Warbler trapped at Damerham in September 1979; all of these were one-day birds and were not available to other birders. Dave Bishop, finder of the cuckoo, recently recalled how frustrating it was that he couldn't share his sighting with others. He had to rely on evening dissemination of the news by landline but despite a large arrival of twitchers the next morning the bird was nowhere to be seen.

A further 14 firsts for Hampshire included iconic birds such as Black Stork in 1970, Blue-winged Teal in 1972, Wilson's Phalarope and Caspian Tern in 1974 and Marsh Sandpiper in 1976 – all have been recorded again on multiple occasions.

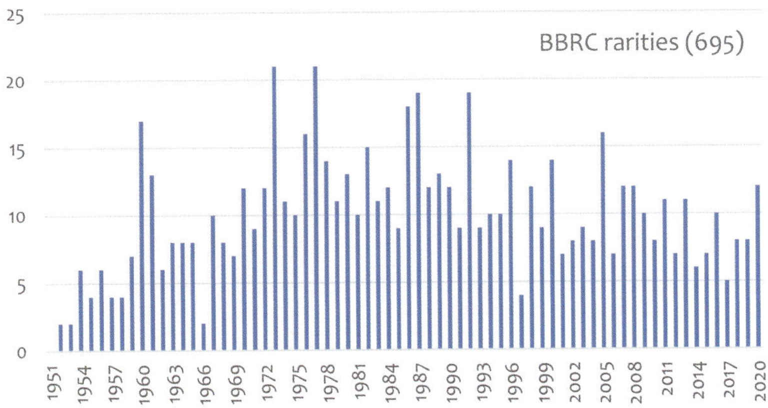

This decade also saw increased protection given to three of Hampshire's most important bird locations. Titchfield Haven was purchased by Hampshire County Council in 1972 with Barry Duffin appointed as warden. In the following year, HCC acquired Normandy Farm, Salterns Marsh and Eight Acre Pond and Ed Wiseman became full-time warden. Subsequently, the council extended their holding in the area by purchasing Pennington and Oxey Marshes in 1979 and Keyhaven Marsh in 1984. Farlington Marshes, which had been managed by the Hampshire and IOW Naturalists' Trust since 1962, officially became a nature reserve in 1974 with David Billett as full-time warden. The network of reserves managed by these two far-sighted organisations expanded in the following decades with sites such as Hook-with-Warsash, Lower Test Marshes, Sandy Point, Blashford Lakes, Testwood Lakes and Fishlake Meadows coming under their control. The RSPB purchased the islands in Langstone Harbour in 1979 while Natural England was responsible for National Nature Reserves at Martin Down, the North Solent including Needs Ore and Old Winchester Hill. Many of the staff and volunteers

Titchfield Haven

who worked at these reserves contributed much to bird conservation and recording in Hampshire and found their share of rarities. These included Bob Chapman, Adrian Clark, Chris Cockburn, Trevor Codlin, Jason Crook, Pete Durnell, Simon S. King, Richard Levett, Jess Pain, Andy Parfitt, Pete Potts, Doug Radford, Reg Thorpe and Dave Unsworth.

View across Keyhaven Marshes to Hurst Castle

Farlington Marshes

The 1980s

The 1980s started with one of the most notorious records in the annals of Hampshire ornithology. This involved the discovery of a territorial Scops Owl in Dummer in May that year. Found by Ted Raynor, news of the bird was not released immediately for fear of upsetting the village residents, but it stayed for several weeks and was eventually seen by hundreds of birders. Rumours that it was an escape subsequently surfaced although this was never proved to the satisfaction of BBRC and the record stands. 1980 also produced Hampshire's first-ever Black Kite at Emery Down in July, a species which since then has been claimed on well over 100 occasions although only 46 records have been accepted. The year also saw the discovery of Hampshire's only Killdeer by Mike Stewart and Bob Dunn at Keyhaven in late September. Fortunately, it stayed until the next day enabling many birders to see it. The decade saw the development of the Hayling Group (HG), co-ordinated by John Walters, which resulted in saturation coverage of Hayling Island particularly at migration times. This revealed the substantial numbers of migrants such as Pied Flycatcher which passed through, although firsts for the county were limited to a Subalpine Warbler found by Doug Radford in May 1984, a Great White Egret in June 1985 and a Black Brant in November 1986, the last two both found by Barry Collins. HG members were also active in other parts of the county with a Pallid Swift found by Andy Hanby at Farlington Marshes in May 1983 the undoubted highlight. That year also produced the first ever Citrine Wagtail for Hampshire, a first-winter found by Bob Chapman at Titchfield Haven which was eventually accepted by BBRC in 1997 after the diagnostic characteristics of birds of this age had been established. A long-staying male Two-barred Crossbill found by Graham Barrett at Slufters Inclosure, New Forest in March 1984 and subsequently seen by many observers was finally accepted by BBRC in 1988.

Sociable Plover and Lapwings, dropping onto the North Scrape, Titchfield Haven, October 17th 1986

The decade also featured an amazing run of firsts for Hampshire at Titchfield Haven in autumn 1986 – a Spotted Sandpiper first seen by Dan Powell on October 4th, a Sociable Plover found by Barry Duffin on October 17th and a Cattle Egret watched in flight by Pete Potts and Barry Duffin on November 6th. All three made extended stays and were seen by many observers. Three further new species for Hampshire have not been recorded again – an American Redstart found by Simon Woolley at Winchester Water Meadows on October 4th 1985, a Tundra Swan identified by the author at Ibsley on the last day of 1986 which stayed for almost a month, and a Western Black-eared Wheatear located by Geoff Farwell and Terry Jennings at Farlington Marshes on June 5th 1987 which was enjoyed by many birders thanks to the prompt release of the news. Two pipit species were added to the Hampshire List in the last years of the decade: an Olive-backed Pipit ringed by Stuart Sporne on his smallholding at Sway in November 1987 and a Red-throated found by Barry Duffin at Titchfield Haven in October 1989. Mention should also be made of the Little Bustard found by Alan Hayden at Coward's Marsh, Dorset in late December 1988; it proved to be rather mobile and pitched up in a field at Ripley, well inside post-1974 Hampshire, on New Year's Day. Although there are seven 19th and early 20th century records it remains the only occurrence in the modern era.

The 1990s

The first two years of the 1990s featured four new species for Hampshire which have not been recorded again. These were a Baillon's Crake brought into Ed and Wendy Wiseman's house at Normandy by the family cat Arthur in March 1990, a White-billed Diver found by John Garr in Oxey Creek in February 1991 (surprisingly not relocated by local observers in the area, admittedly unaware of its presence), a group of three Parrot Crossbills seen by Kevin

Rylands at Bricksbury Hill in March 1991 and a Pacific Golden Plover found by Dave Unsworth at Needs Ore which stayed for two days in June 1992. There was also a superb first-summer male Citrine Wagtail found by John Dixon at Fleet Pond in May 1993 which stayed for two days. At the time it was believed to be the first for the county although it was subsequently pre-dated by the Titchfield Haven bird mentioned above. The following year produced two new waders for the county: a Semipalmated Sandpiper identified by Paul Stancliffe at Keyhaven Lagoon in July which stayed for two days and an American Golden Plover found by Mike and Pete Combridge at nearby Lower Pennington Lane in November which moved on quickly. Both species have subsequently been recorded on several occasions. Pickings were slim for rarity hunters in the last five years of the decade with no new species added to the county list and a low of only four BBRC rarities accepted for 1997.

The knowledge of Hampshire birds increased exponentially in the second half of the 20th century. Edwin Cohen worked on his *Birds of Hampshire and the Isle of Wight* through the 1950s and constantly updated his texts as new information emerged. It covered up to the end of 1961 and was published in 1963. Edwin was working on a second edition when he died in 1970 and the work was completed by John Taverner, emerging as *A Revised List of Hampshire and Isle of Wight birds* in 1972. Bird recording in the county continued to be managed by the Hampshire Field Club Ornithological Section until 1979 when the independent Hampshire Ornithological Society was launched. This gave a new impetus to ornithology in the county and resulted in an ambitious project to map Hampshire's breeding birds at the tetrad (2 x 2 km square) level during 1986–1991. The results of this were included in the fourth county avifauna, *Birds of Hampshire*, edited by John Eyre and the present author and published in 1993.

The 2000s

Great Spotted Cuckoo, Lower Pennington Lane, April 9th 2000

The start of the new millennium produced one of the most widely appreciated rarities in Hampshire – a Great Spotted Cuckoo. It was identified by Glyn Horacek-Davis on April 2nd 2000 but promptly disappeared, much to the chagrin of the assembling crowd. Thought to have been long gone, it was seen again briefly at dusk on April 8th and relocated the next morning. It stayed for a further eight days. The year also produced the county's only Little Swift found by Jonathan Stokes at Bishop's Dyke on June 4th while leading a Naturetrek tour group. Later that year a red-letter day at Titchfield Haven on October 1st started with a Paddyfield Warbler trapped by the ringing team to be followed by a massive easterly movement of Swallows among which Dan Powell picked out an American Cliff Swallow. The Keyhaven area subsequently 'fought back' with two long-staying waders: a Stilt Sandpiper discovered by Russell Wynn in July 2002, the first of three for the county, and a Lesser Sand Plover found by Lawrie Chappell in July 2003, the only record. A surprising record of a Chinese Pond Heron found and photographed by Rob and Jen Churchill at their garden pond in East Dean in November 2004 was assumed to relate to an escape but was subsequently accepted by BOURC as the first for Britain (Cox 2017). Later in the decade, a national influx of Laughing Gulls in November 2005 produced the first for Hampshire, a long-staying adult found at Gosport by John Norton and Peter Milinets-Raby. A month later, a Radde's Warbler was found by Andrew Pink near Sherborne St. John on December 7th but was not present the next day. The following two years each produced firsts.

A Dusky Warbler was located by Andy Johnson at Sandy Point, Hayling Island on November 2nd 2006 and seen by many observers during its one day stay. A Greater Yellowlegs was found by Jason Crook and Kevin Crisp at Farlington Marshes on September 26th 2007; it was still present early the next morning but flew off before the gathered crowd could obtain good views as they were looking into the rising sun. Fortunately, a long-staying bird was available at Titchfield Haven in 2015. A chiffchaff singing an unusual song was found by Dan Houghton, Mark Rolfe and Alan Butler at New Copse Inclosure, New Forest in May 2009; it stayed for eight days and was subsequently identified as Hampshire's first Iberian Chiffchaff. Those that didn't go to see it keenly await the next occurrence! On June 1st, Simon S. King found a singing male Common Rosefinch near Godshill in the New Forest which didn't stay; this remains the only modern record of the species. Later that month, on 21st, Alan and Caroline Dawson photographed what they thought was a European Bee-eater at Needs Ore. On arriving home, they re-identified it as a Blue-cheeked (not surprisingly a county first) but it was not seen again.

Cliff Swallow, a chunky American cousin among many thousands of Swallows, River Meon, Titchfield Haven, October 1st 2000

Daurian Shrike, Grove Park, Gosport, September 11th 2010

The 2010s

The period from 2010 produced a remarkable 21 new birds for Hampshire. There were four firsts in 2010: a two-day Rustic Bunting in Russell Wynn's Brockenhurst garden in March, a Daurian Shrike found by John Norton in Gosport and a Booted Warbler located by Dave Unsworth at Needs Ore within two days of each other in September, and a long-staying and much-appreciated Red-flanked Bluetail found by Andy Johnson at Sandy Point in October. In 2011, Jason Crook found a pale-legged stint at Farlington Marshes in failing light on September 8th. His images facilitated its eventual identification as a Long-toed Stint and the record was accepted by BBRC in 2022. This was followed by a Spectacled Warbler discovered by Philip Hack at Needs Ore which stayed for two days in late October. In 2013, Bob Marchant found a Brown Shrike at Hook-with-Warsash on September 20th; he was able to share it with a steady stream of admirers during the rest of that day. Expert fieldcraft by Andy Johnson at Black Point on October 17th 2013 enabled him to pick out the calls of a Semipalmated Plover among a large flock of roosting Ringed Plovers. It remained for a further 12 days and was seen by hundreds of birders. The final rarity of that year was a wheatear near Monk Sherborne on December 7th which was not identified as a Pied Wheatear until ten days later. Sadly, after the news broke it departed overnight before anyone could see it!

The following year 2014 produced three more firsts, two of which were totally unexpected. A Short-toed Eagle had been seen at Morden Bog, Dorset on May 31st and it was the same bird that was seen by Simon Colenutt over Bishop's Dyke on June 8th and later photographed. Two days later it was seen in Ashdown Forest, East Sussex and then regularly commuted between there and the New Forest, where it was last seen on July 5th. Next, the ringing team at Titchfield Haven trapped a Pallas's Grasshopper Warbler on September 6th and finally Bryan Pinchen saw and photographed a Hume's Warbler in his garden at Pennington on November 3rd. In 2015, prompt release by Gary Howard of the news of his discovery of a stunning Eastern Black-eared Wheatear at Acres Down on June 13th enabled many birders to see it during the afternoon and evening before it left overnight. Amazingly, comparison of photographs of the bird with those of one seen in the Netherlands three days earlier confirmed that they were of the same individual.

Semipalmated Plover, hunkered with Ringed Plover and Sanderling, Eastoke Beach, Hayling Island, October 19th 2013

George Else was at Brownwich Cliffs on April 13th 2016 where he was looking for certain species of bees when he saw and photographed a bird which he did not immediately recognise. He subsequently identified it as an Alpine Accentor but despite releasing the news of his find the bird was not relocated. Prior to 2016, there had been three records of 'Siberian' type Stonechats but following the split of that taxon into Siberian and Stejneger's Stonechats they could not be ascribed to either species. However, Amy Robjohns had the good fortune to discover a stunning male Siberian Stonechat of the distinctive race *hemprichii* at Titchfield Haven on May 10th of that year where it stayed until the next day. Finally in that year, Bryan Pinchen found a juvenile Pallid Harrier at Needs Ore on November 6th which remained for seven more days and was enjoyed by many Hampshire birders. In 2017 Andy Johnson located a colour-ringed Elegant Tern in a high tide roost at Fishery Creek, Hayling Island on June 7th. Five other birders were able to see the tern before it departed. Two days later Andy saw it again briefly off Sandy Point before it settled in the tern colony at Pagham Harbour, West Sussex and remained until June 20th. The colour rings confirmed that it was an individual that was ringed in a Sandwich Tern colony in France in 2003. Later that year, a Lesser Scaup had been present at Longham Lakes, Dorset where it was last seen on October 26th. Alan Lewis thought that it might turn up at nearby Blashford Lakes and sure enough he found it there two days later. It remained in the area until November 5th and reappeared briefly in February 2018 and January 2019. Alan's excellent coverage of the gull roost at Ibsley Water paid dividends when he identified a juvenile Thayer's Gull there in late January 2018. It was regularly found by day feeding at a pig farm at Blackheath Down and continued roosting at Ibsley until early March.

Eastern Olivaceous Warbler, Farlington Marshes, September 18th 2019

Only one species was added to the county list in 2019: an Eastern Olivaceous Warbler found by Pete Gammage at Farlington Marshes on September 14th. It was subsequently seen by hundreds of birders until last seen on September 19th. Last but not least, Mark Francis located a very pale first-winter *flava*-type wagtail at Brownwich on October 5th 2020. He suspected that it might be of the recently-split Eastern Yellow Wagtail and this was confirmed the next day by reference to the excellent photographs and sonograms that were obtained.

The news was released and a small number of observers managed to see the bird before it flew to roost but unfortunately it was not located the next day. This final record illustrates the massive advances in technology and knowledge which have occurred during the 21st century; certain identification of this taxon would have been impossible in previous decades.

Postscript

Although this book covers the period up to the end of 2020, two more firsts for Hampshire were found before it went to press. A Baikal Teal was located by Olly Frampton among a large flock of Wigeon at Ripley Farm Reservoir on March 14th 2021 and seen again two days later. This East Asian duck is extremely rare in western Europe and several previous birds have been suspected of being escapes. However, this individual met all the criteria for acceptance as a wild bird by BBRC and has been accepted. Martin Cooper found a small wader at Normandy Lagoon on October 4th which he suspected was a North American peep. Simon Buckell saw the bird the next day and identified it as a Western Sandpiper. After perusal of many high-quality photographs and much debate it was eventually confirmed as that species. It stayed until October 14th. Sightings in 2021 of six other species with fewer than ten previous records are also included. But bird recording doesn't stand still, and so a line has to be drawn. Records of Hampshire's first May Grey Phalarope and first July Pink-footed Goose, as well as those of several other species included in this book, will feature in the 2021 *Hampshire Bird Report*.

Western Sandpiper, with Dunlin, Normandy Lagoon, October 9th 2021

Species accounts

Black Brant *Branta bernicla nigricans*

BBRC until 2004–05 winter; HOSRP 2005–06 winter to date

A rare winter visitor. 109 records involving a minimum of 48 individuals since 1986.

The first record for the county of this race of Brent Goose was of one found by Barry and Margaret Collins at Thorney Deeps in the Sussex part of Chichester Harbour on November 2nd 1986 which remained in the area until January 7th 1987 (not March 8th 1987 as in BoH). Barry (*in litt.*) says that it occasionally flew across to East Hayling on a rising tide but did not record dates when he saw it do this. The first definite dated record for the county is thus of one found by Richard White at Farlington Marshes on February 7th 1987 which remained until at least 12th and was no doubt the Thorney bird. This individual returned to Chichester Harbour in the following seven winters. It was regularly seen in Emsworth Channel and along the East Hayling Shore and was last recorded on March 6th 1994. It was seen again by Richard White in Langstone Harbour on January 8th 1990. Extracts from Richard's description of the bird, as taken on February 7th 1987 and submitted to BBRC, are reproduced below.

> *A flock of c. 75 Dark-bellied Brents flew on to the lake (at Farlington Marshes) and started bathing and preening. Scanning through the flock the bird was obvious due to the larger neck collar, silvery flanks and blacker breast and belly than bernicla. Its longer-necked appearance was commented on by other observers. The flock then moved to the fields where the bird was not relocated that day.*

Subsequently, there have been over 100 records but, allowing for movements between sites and birds returning in successive winters, the number of individuals has been assessed as 48.

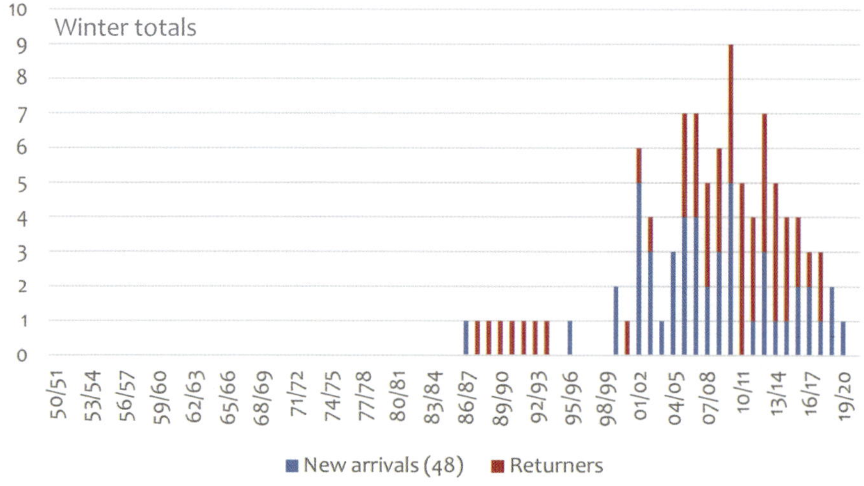

The next was one found by Dave Unsworth at Needs Ore on May 11th 1996 which stayed until May 30th. Two were seen in the 1999/2000 winter, one found by Jason Crook in Langstone Harbour on November 13th which stayed until February 18th and was considered to be the same as one at Pilsey Island, Chichester Harbour from October 8th to November 4th 1999, and another in the Sussex part of Chichester Harbour from January 13th to 22nd which was found by Jason Crook and Kevin Crisp at Black Point on February 2nd. This individual had been seen in the Sussex part of Chichester Harbour in early 1999. The Langstone Harbour bird returned in 2000/01 but was seen only in Sussex. In 2001/02 it was in Langstone Harbour from November 14th to December 1st before moving to Portsmouth Harbour where it was present from December 11th to March 7th. Around four others were in Hampshire that winter with a presumed female in the Keyhaven area from November 25th to March 29th, a presumed male there from December 16th to March 24th, another at Needs Ore from December 1st to March 16th which may also have been the bird in the Chilling/Hook-with-Warsash area from February 10th to March 23rd, and one at Weston Shore on February 25th.

Between 2002/03 and 2019/20 there were records in every winter with mostly three to six individuals present but a high of nine in 2009/10 and lows of one in 2003/04, two in 2018/19 and one in 2019/20. The recent reduction is mirrored in Sussex where none has been recorded since November 2016 (John Newnham *in litt.*). The table below shows numbers of new arrivals (left) and the totals including returning birds (right) in various areas.

Area	New	Total
Hampshire part of Chichester Harbour including grasslands and farmland along East Hayling shore	7	29
Langstone Harbour, Farlington Marshes, West Hayling shore and nearby grasslands in Farlington and on Portsea Island such as the Tangier Road goose refuge	13	32
Portsmouth Harbour and nearby grasslands at Gosport such as the HMS Sultan playing field	7	13
Hill Head – Chilling – Hook-with-Warsash area	3	3
Weston Shore	1	1
Beaulieu Estuary and nearby grasslands and farmland at Needs Ore and Park Farm	6	10
Tanners Lane	1	1
Normandy – Pennington – Keyhaven Marshes and nearby grasslands and farmland at Lymore Lane and New Lane	10	20

In addition to the first for the county which returned for seven winters, the longest-lived individual was one at Gosport from 2004/05–2009/10. All were presumed to be adults apart from a first-winter identified by Rob Hume at

Mount Lake, Keyhaven on April 12th 2010 which had been present since April 6th. That most were adults is not surprising since lone adult Black Brants among a population of Dark-bellied Brent Geese would be more likely to interbreed and produce hybrid young. Male Black Brants paired with female Dark-bellied Brent Geese and accompanied by hybrid young have been recorded on two occasions. Jason Crook reported such a pair with two hybrid young in Langstone Harbour between October 23rd 2005 and March 24th 2006 and Andy Johnson found a pair with four hybrid young in Chichester Harbour on October 27th 2011. Adult hybrids have also been recorded. An intergrade or hybrid individual was documented by Jason Crook as present in Langstone Harbour between the 1997/98 and 2005/06 winters and was accompanied by two young in 2000/01 and 2005/06 (Martin 2002, Crook 2002). Two further hybrids, assumed to be the 2005/06 offspring, were recorded in Langstone Harbour from 2006/07 to 2008/09. The only other records of adult hybrids involved singles in Chichester Harbour between October 23rd and November 19th 2014 and on February 4th 2016.

Black Brants: Sturt Pond, February 22nd 2015 and Eastney Playing Fields, January 14th 2013.
Alan Lewis and Peter Milinets-Raby

Arrival dates of 17 recorded in October are evenly spread through the month with the earliest ever in Langstone Harbour on 1st in 2002 and possibly the same at Sturt Pond, Milford on Sea on 7th. Neither of these was seen again. Another in Langstone Harbour on October 6th 2007 was assumed to be a returning bird. The chart opposite shows new arrivals continuing until February but it is likely that some of these will have been overlooked in earlier months. Of nine recorded in April, five had departed by 12th but one remained in Langstone Harbour until 30th in 2003 and another at Needs Ore from 25th to 27th in 2010 was on its own and is assumed to be a migrant. One moved east off Sandy Point on 6th in 2013 and another east off Hurst Castle two days later, presumably having wintered further west. The latest ever was the bird at Needs Ore from May 11th to 30th 1996.

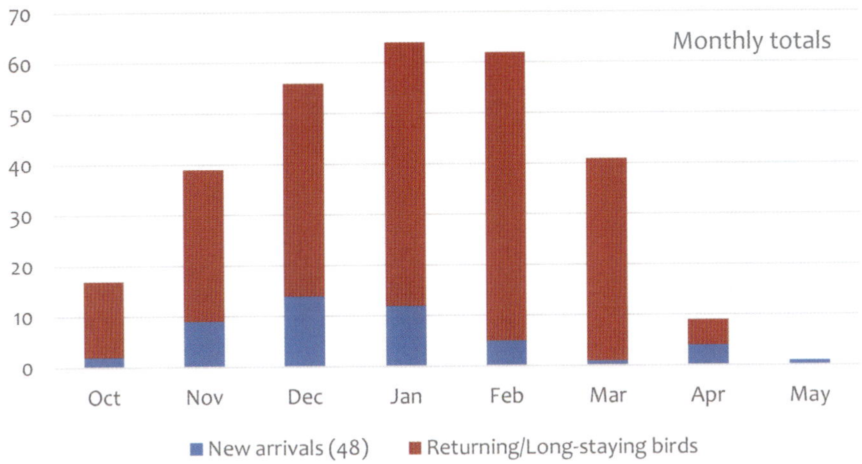

Although most are found accompanying Dark-bellied Brent Geese in intertidal areas or on adjacent grasslands and farmlands, Black Brants have twice been detected among flocks of Brents which have flown inland from the eastern harbours – at Denmead on December 3rd 2005 and presumably the same individual at Widley four days later.

A record of one at Needs Ore on May 4th 1992 was published in the *HBR* but referred to a Pale-bellied Brent Goose.

GB status: 407, 1958–2019. Stabilising after increase, mean 13 in 2010–19. (Breeds Arctic NE Siberia W to Lena delta, where overlaps with nominate race. Majority breed low Arctic from N & W Alaska, E through parts of N coast Canada. Migratory, wintering locally on Pacific coast from Alaska south to S Baja California, and coastal N China, Korean Peninsula & Japan where now rare.)

Red-breasted Goose *Branta ruficollis*

BBRC

A rare winter visitor. 14 records of presumed wild birds involving a minimum of ten individuals since 1967.

The first for the county was a first-winter found by Peter Mays feeding with around 500 White-fronted Geese in the water meadows just above Ibsley Bridge in the Avon Valley on January 18th 1967. Peter evidently told John Ash, who saw it on January 21st. It was subsequently seen by many observers up to February 19th. It was almost certainly the same individual that was present at Slimbridge, Gloucestershire from January 5th to 15th 1967.

Subsequently, a further nine individuals have occurred, two of which returned in two successive winters, as shown on the next page. At least eight other birds, most of which were ringed, are presumed to have been escapes from captivity.

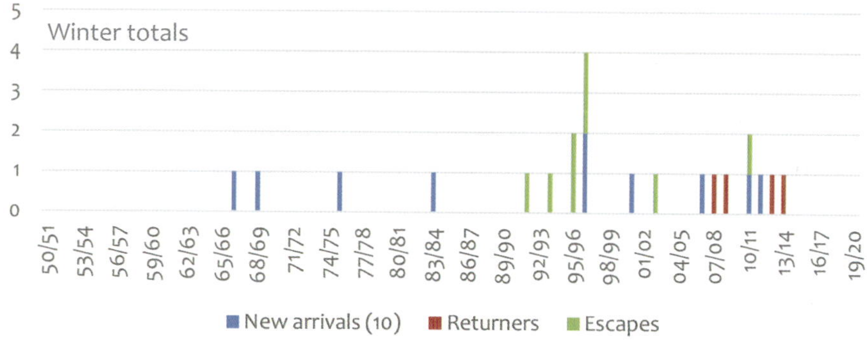

The Avon Valley's monopoly was broken on November 8th 1975 when an adult was found by Ron Huxford with Brent Geese just off the point at Farlington Marshes. It remained with Brents in the northern part of Langstone Harbour until November 23rd, when it moved to West Hayling and subsequently before the end of the year to fields at North Hayling where it often fed alone. It remained in the area until February 29th 1976. This was probably the individual present in Essex from October 12th until November 7th 1975.

The fourth for Hampshire was an adult found with a flock of 325 White-fronts at Bisterne Water Meadows by Tim Doran and the author on December 27th 1983. It was still there two days later but was at Blashford from 31st until January 4th. It was back at Bisterne on January 5th where it remained until 14th and was last reported at Blashford on 15th. Two days later presumably the same bird arrived at Slimbridge where it stayed until March 1st 1984.

A known escape was at Needs Ore between February 16th and April 26th 1992 and at Hill Head on March 29th. An adult was at Anton Lakes, Andover on March 13th 1994 and presumably the same at Needs Ore intermittently between March 17th and June 25th and at Pennington Marshes on May 26th and June 6th. Another or the same was at Needs Ore between August 19th and December 11th 1995 and at Keyhaven Marshes on September 2nd and 23rd, with two reported at Needs Ore on August 24th. One was at Needs Ore

on August 10th 1996. None of these sightings indicated if the bird(s) were ringed but, given the location (inland at Andover) and subsequent dates, it seems highly likely that they referred to escapes and they have not been included in the county total.

The fifth wild bird for the county was an adult with Brent Geese found by a visiting birder from Norfolk, Mr. A Cohen, at The Kench, Langstone Harbour on October 12th 1996 and seen by him again the next day. He submitted his report direct to BBRC but it was also found independently by Jenni and Colin Tubbs and Pete Durnell and Andy Parfitt on 15th. Pete submitted a description to BBRC in which he confirmed that the bird was unringed. It was not seen there again but *HOS Newsletter* number 73 reports the bird at Mill Rythe, Chichester Harbour on November 7th. The story now moves to Needs Ore, when on November 9th Keith Godfrey reported one with a black ring on the left leg and a yellow tag on the closed left wing. This bird evidently remained at Needs Ore until at least January 11th 1997 and was picked up dead six days later. However, at the same location Pete Potts saw one at very close range on November 22nd 1996 which left west with Brents; this was unringed and may have been the bird first seen at The Kench. One was at Keyhaven on January 11th and from February 11th to March 31st 1997. On February 17th it was confirmed by Graham Walbridge as an unringed first-winter and was no doubt the bird present at Compton, IOW from January 1st to 6th. Also during this period, an adult with a black colour ring was with Brents at Hook-with-Warsash from February 15th to April 9th, nearby at Hill Head on April 8th and at Needs Ore from April 12th to 22nd. Thus all the records for the 1996/97 winter indicate the presence of a wild adult and first-winter and two escapes which presumably were those first seen in 1995–96.

The seventh for the county was an unringed adult found independently by Dave Taylor and Jeff Ross with 200 adult Brents at Lepe on September 24th 2000. This is an early date for a Red-breasted Goose but coincided with a large overnight arrival of Brents in Langstone Harbour where Jason Crook noted an

Red-breasted Goose with Brent Geese, Needs Ore, January 19th 2014. Ian Williamson

increase from 370 on 23rd to 1,125 on 24th. It was still present the next day but on 26th was found by Jason in Langstone Harbour off the south-west side of Farlington Marshes. It was not seen again and there were no other reports from southern England that winter, so possibly the bird moved to France.

Another escape was in the Upper Itchen Valley in the 2002/03 winter, with sightings at The Grange Lake, Northington on October 12th, Avington Lake on December 27th and 28th and back at The Grange Lake on March 23rd.

The next report was an intriguing one of a first-winter at Needs Ore on December 15th 2006 which unfortunately was not documented. Meanwhile, an adult which had arrived at the Fleet, Dorset on November 4th remained in that area until 15th before transferring to Poole Harbour where it was last recorded on January 25th 2007. The next day Rosemary Cook found it feeding with Brents in a field at Lymore Lane, Keyhaven. It remained in that area until January 31st when Paul Winter saw it leave with Brents. Shortly after Mike Rafter recorded it at Park Shore and by the next day it was in fields at nearby Blackwater, Needs Ore where it stayed until February 13th. Three days later it was found by Glenn Madison, Danny Cooper and Paul Spencer grazing with Brents in Sinah Warren paddocks adjacent to Langstone Harbour. It was still there the next day when Jason Crook saw it fly into the harbour at dusk. Andy Johnson rediscovered it near Tournerbury in Chichester Harbour on 21st. It remained in the harbour until March 7th but was mostly in West Sussex territory, the only other sightings in Hampshire being on 24th and 27th.

Interestingly a Red-breasted Goose was seen and photographed at the Fleet, Dorset on February 15th 2007 which BBRC assumed was the same individual as seen in Hampshire on 13th and 16th. However, images on the Portland Bird Observatory website show that the two birds at the Fleet in November and February were different and that the Hampshire bird was the November one. What is assumed to be the same individual returned in the following winter and was first found by Pamela Read during the WeBS count at Gutner Point, East Hayling on November 10th. It remained in Chichester Harbour throughout the winter and was last seen on March 6th 2008. The following winter it was refound by Ed Wiseman at Keyhaven on October 31st and remained in the area until February 6th 2009 when it was seen by Marcus Ward at Normandy Marshes. It then moved to Chichester Harbour where it was seen by Andy Johnson on four dates between February 13th and March 6th.

The next was an adult bearing a small red ring found by Andy Johnson at Mengham Salterns in Chichester Harbour which stayed in the area from October 2nd to 4th 2010 and then was seen the next day by Marc Moody moving west off Keyhaven with ten Brents. An unringed adult, the ninth for the county, was found by John Shillitoe at Chilling on January 19th 2011. It stayed in the area between Hill Head and Hook-with-Warsash until February

6th. This is assumed to be the same bird which flew west past Dungeness, Kent on December 2nd, stayed at Pett Level, East Sussex from December 26th to January 1st and was at Sandwich Bay, Kent on February 8th. BBRC considered the Sandwich Bay bird and one at Freshwater, IOW on February 12th to be the same individual (Hudson *et al.* 2013) but the latter may have been the adult at Lower Pennington Lane on February 14th. This was initially assumed to be the Chilling bird but later proved to be bearing a red ring, which suggests it was the bird at Chichester Harbour the previous October. It remained in the area until 24th.

Red-breasted Goose and Brent Geese, Chilling, January 21st 2011

An unringed first-winter, the tenth for Hampshire, was found by Paul Winter at Pennington Marshes on February 25th 2012. It was still there the next day but on 27th was relocated by Ashley Howe at Needs Ore where it remained until March 24th. It had been first noted at Christchurch Harbour, Dorset between October 18th and 28th 2011, before pausing at the Fleet, Dorset on October 30th and 31st and arriving at the Exe Estuary, Devon on November 5th 2011 where it was last recorded on February 18th. What is assumed to be the same individual returned to southern England the following winter when it was at South Swale, Kent from October 14th to 23rd 2012 but was found

Red-breasted Goose with Brent Goose, Sturt Pond, November 16th 2013. Jason Crook

by Marc Moody at Keyhaven the next day. It had moved to Langstone Harbour on 25th when Bob Chapman found it grazing on Farlington Marshes. It remained in the area until January 12th 2013 and was also seen grazing at Tangier Road goose refuge and Southsea Cricket Ground. On 13th it moved east then west off Selsey Bill, West Sussex before settling in Chichester Harbour. It subsequently spent most of its time on Thorney Island but did venture back into Hampshire when it was recorded in Emsworth Harbour on February 2nd, at Gutner Point on 9th and off Langstone village on 19th. It returned to Farlington Marshes on March 5th and remained until April 5th. One seen flying east with Brents at two points along the north Kent coast on April 6th was probably this bird. It returned to Hampshire for its third winter on the early date of September 27th when it was seen from both Farlington Marshes and West Hayling by Tim Doran and Martin Gillingham respectively. It was found by Chris Mayhead in Fareham Creek the next day (the first record from Portsmouth Harbour), was seen by Andy Johnson on Pilsey Sands on the Sussex side of Chichester Harbour on October 3rd and subsequently was found by Olly Frampton grazing near Sturt Pond, Milford on Sea on October 10th where it remained until November 24th. It was found by the Tuesday Group at Needs Ore on November 30th and remained in the Beaulieu Estuary area until December 10th. It was at Shalfleet and Newtown, IOW, from December 17th to 20th but had crossed the Solent again by January 5th when Dave Unsworth relocated it at Park Farm. It remained in the Beaulieu Estuary area until March 6th apart from a brief visit to the fields at Lower Pennington Lane where Paul Winter found it on January 13th. It was still present the next morning although it was back at Needs Ore later that day. It was seen in Poole Harbour, Dorset from March 24th to 30th and then in Christchurch Harbour from March 31st to April 3rd; later that day it was seen passing Selsey Bill presumably having passed through the Solent.

GB status: 7, 1776–1949; 83, 1950–2019. About two per year + returnees; stable. (Breeds Taimyr Peninsula, Siberia. Migrates SW to winter coastal regions of W Black Sea in Romania & N Bulgaria. Small numbers regularly winter Netherlands, Hungary, Greece & Turkey.)

Ross's Goose *Anser rossii*

BBRC

A very rare vagrant or escape. Two records in 2001 and 2020.

The first for the county was an unringed first-winter found by Bob Chapman when it was grazing among a large group of Canada and Brent Geese at Farlington Marshes on October 29th 2001. It was found independently by Jason Crook in the same area the next day and he alerted other observers. On 31st, Jason and Kevin Crisp watched it in flight for long periods before it settled in the fields. It was still present at dusk but was not seen again. The identification was accepted by BBRC but, as with all records of this species, it was not considered to relate to a wild bird (Fraser *et al.* 2007). It is possible that this was the same bird located on November 3rd at Wells, Norfolk which associated with wintering Pink-footed Geese and remained in the area until February 2002.

An unringed adult was found by Steve Clemons grazing with Canada Geese at Lower Test Marshes on May 3rd 2020. It was subsequently seen in flight over Fishlake Meadows on May 5th and grounded there on three further dates up to May 20th. Presumably the same individual flew in to Ibsley Water from the south with a small flock of Greylag Geese on June 20th; it was back at Fishlake Meadows on July 26th.

BOURC has recently completed a review of records of Ross's Goose. The outcome is that the species has been admitted to category A of the British List with the acceptance of a bird first seen in Lancashire in December 1970 (BOURC 2022). The two Hampshire records will be assessed against the criteria recently published by BBRC for the acceptance of rare wildfowl (French 2021).

GB status: One accepted record. (Breeds in Arctic Canada, winters in S USA.)

Taiga Bean Goose *Anser fabalis*

HOSRP

A very rare winter visitor. Five records since 1985.

This species was until recently considered a race of Bean Goose along with the Tundra Bean Goose *A. serrirostris.* Both species have occurred in Hampshire since the first modern record of five Bean Goose at Langstone Harbour in March 1960 although not all sightings were ascribed to a race. Analysis of records of Bean Geese that were racially identified has shown that Tundra Beans are much more frequent and have occurred in groups of 20 or more. The five confirmed records of this species have all involved single individuals.

The first for the county was one seen by Barry Duffin, Bob Chapman and Richard Levett at Titchfield Haven in severe wintry weather on January 18th and 20th 1985 (but not the 19th). It accompanied a flock of Canada Geese but did not associate with up to seven Tundra Beans which were present from January 17th to 22nd.

The next was one found by Dave Sargeant, Dave Unsworth, Edwin Welland and the author with 530 White-fronted Geese and one Pink-footed Goose at Blashford Water Meadows on January 26th 1985. It was still present on February 2nd. The next was a first-winter found by the author with 299 Greylag Geese at Stratfield Saye Park in severe wintry weather on February 8th 1986. It was still there, along with two first-winter White-fronts, on 16th and all three birds remained until at least March 2nd.

The next was one found by Keith Wills and Graham Huxley at Wellington Country Park on November 30th 1986 which was still present the next day. Presumably the same bird was found by Bob Brunton at Tundry Pond on December 24th and seen again there by Mick Scott and the author with a single White-fronted Goose on January 5th 1987. The fifth for Hampshire was one found by John Harrington with a Pink-footed Goose at Sleaford Reservoir on February 7th 1988. It was subsequently seen by several local observers until April 2nd. A further record of one at Eversley Gravel Pits from December 14th to 21st 1996 is now known to have stayed on the Berkshire side of the River Blackwater.

GB status: rare and declining winter visitor to Norfolk and SW Scotland. (Breeds in taiga zone discontinuously from Fennoscandia to Siberia; winters NW and SE Europe, C Asia, China, Korea and Japan.)

Pink-footed Goose *Anser brachyrhynchus*

HOSRP

A rare winter visitor and passage migrant. Two records in 1940 and 1946 and 79 of 163 birds since 1952.

The first for the county were three seen by George Pierce at Winchester Water Meadows in severe wintry weather on January 16th 1940. His account of the occurrence, taken from one of a series of articles entitled *Walking in Hampshire* (Pierce 1951), is reproduced below.

> *One picture lives in my mind of College meads in the bitter freezing winter of 1940, when all the birds were unusually tame and easy to approach. The river valley was full of them, desperately engrossed in a search for food. A black north-east wind stung the cheeks and, as we took the path across the meads by the Lord Grey Enclosure, it began to snow. As we looked across the irrigated pasture, not fifty yards from us we saw three wild geese standing head to wind. They were pink-foot geese and they seemed exhausted and dispirited as they stared bleakly into the driving snow. While we watched them, a water rail walked among the brambles*

> *within five yards of us, tame with hunger, at the edge of a shrunken stream. It was a terrible time for the birds, but a wonderful opportunity for watchers to get to close quarters with timid and rare species.*

These were followed by 24 seen by Tony Norris flying north over Titchfield Haven on January 19th 1946.

The first in the modern era was one seen by Graham Bundy at Dibden Bay on October 11th 1952, followed by seven with White-fronted Geese at Ibsley on March 15th 1953. Subsequently, there have been 77 records in 44 winters involving 155 birds.

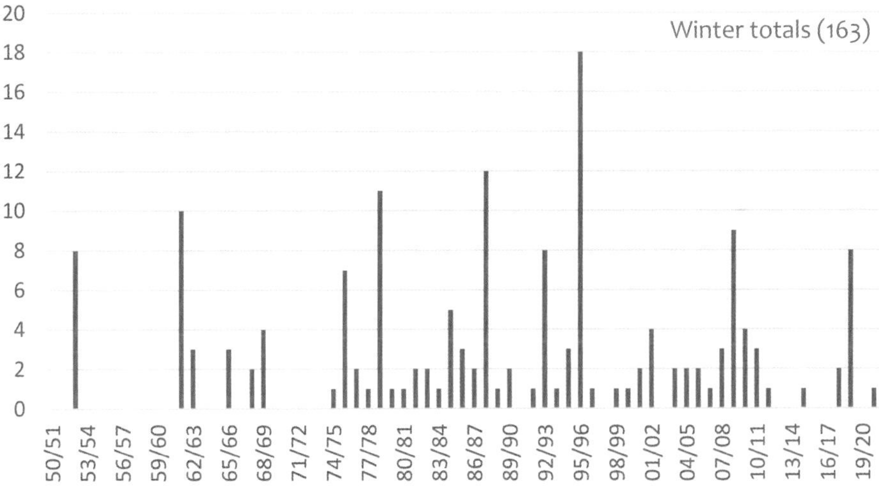

Most records were of one or two individuals, often accompanying other geese such as White-fronts, Brents, Greylags and Canadas. Many occurred during spells of severe wintry weather and some stayed for several weeks. Flocks in excess of three were of eight flying south over Hurst Beach on December 30th 1961, four at Tournerbury, Chichester Harbour on December 22nd 1968, seven at Dibden Bay on February 1st 1976, eight at Ibsley on February 23rd 1979, eight at Titchfield Haven from December 19th 1987 to March 10th 1988, seven flying SSE over Fleet Pond on February 13th 1993, 17 at Needs Ore from February 10th to March 6th 1996, four at Avon Causeway from November 23rd to 28th 2008, four flying south-west over Sway on February 19th 2009 and four at Oxey Marsh, Lymington on January 21st 2010.

Early autumn arrivals may be of suspect origin. The only August record of one flying east over Milford on Sea on 4th in 2005 was possibly that seen nearby in May that year. There are five September records from 2nd including singles with Greylags at Ibsley Water on 26th and 27th in 2010 and with Canadas at Pennington Marshes on 25th in 2020. These last two, six

Pink-footed Goose, Stratfield Saye, March 3rd 2019. *John Clark*

records of singles in October and many of those in November probably involved overshoots from the burgeoning Norfolk wintering population.

There is a clear peak in December and a larger one in February which reflect cold weather arrivals. Most wintering birds have left by March although there are eight records of singles staying into April and a further five of singles newly arrived in that month. Records for May and June span the period up to June 19th and arguably mostly refer to birds of suspect origin. However, three at Sleaford Reservoir on May 12th and 13th 1986 and one which flew east off Hurst Beach on May 12th 2005 and presumably the same with Canadas at Pennington Marshes on May 15th, when it was seen to be unringed and described as wary, may well have been wild. The monthly pattern is shown below.

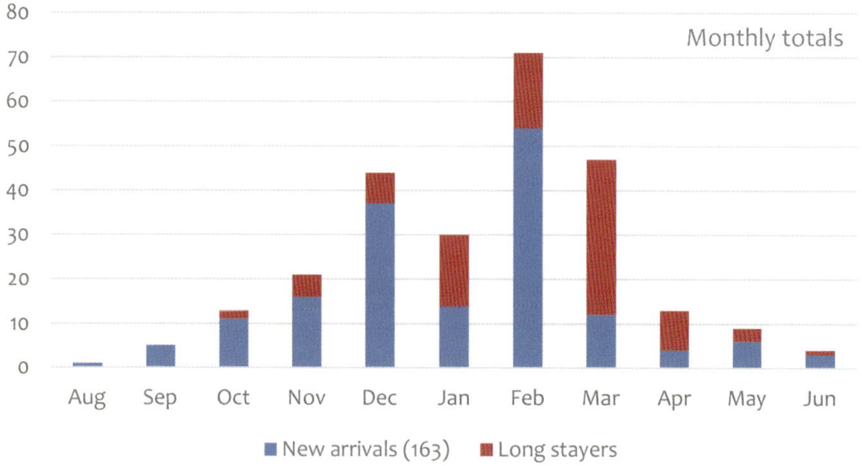

Areas with the most records were the Avon Valley (14 records, 31 birds), the north-east (15, 31), the Test Valley (11, 12), Needs Ore (7, 24), the Lymington–Hurst area (10, 24), Titchfield Haven (8, 16) and Langstone Harbour (6, 6).

GB status: Common and increasing winter visitor to North Norfolk, N England and Scotland. (Breeds E Greenland, Iceland and Svalbard; winters Britain, Netherlands and Denmark.)

Tundra Bean Goose *Anser serrirostris*

HOSRP

A rare winter visitor. 41 records of 198 birds since 1960.

Until recently the Tundra and Taiga forms were treated as a single species and so not all records were assigned to a race. All records of definite Tundra and non-identified birds are included here.

The first for the county were five found by Peter Le Brocq on Farlington Marshes on March 8th 1960 which were still present the next day. Peter's notes indicate that they were Tundra rather than Taiga Beans:

> *Three had some yellow near the tips of their bills and the other two had completely dark bills. All had brilliant chrome yellow/light orange legs. All breasts were devoid of any barring. In flight the leading edge of the wing was only very slightly paler than the rest of the wing.*

Since then, there have been 40 records in 27 winters involving 193 birds.

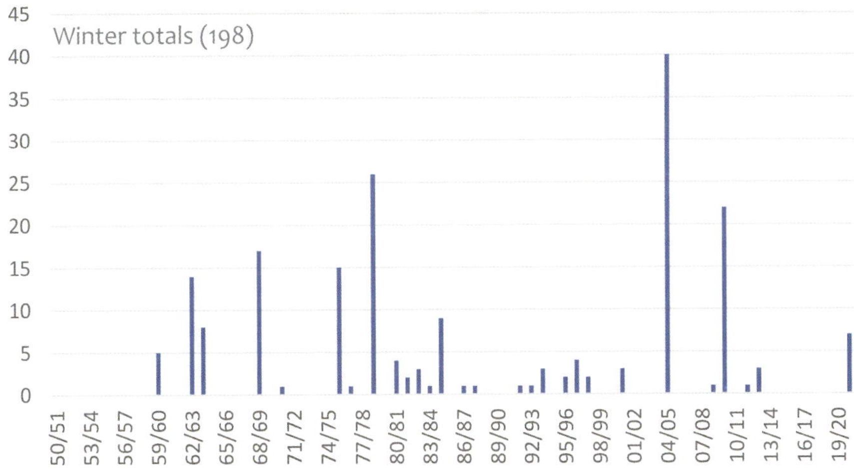

In contrast to Pink-footed Goose, occurrences often involved sizeable discrete flocks although singles accompanied White-fronts, Greylags and Canadas. Most occurred during spells of severe wintry weather. Flocks in excess of five were as follows. During the prolonged cold spell in early 1963, up to ten were grazing on Portsmouth Airfield between January 20th and February 3rd. Notes in David Billett's annual Langstone Harbour workbook suggest that at least two were of the *taiga* form. On February 9th there were nine feeding in fields at West Hayling and four, considered to be new arrivals, on Farlington Marshes. Seven were recorded from February 19th to March 3rd and the final record was of 12 flighting from Hayling into the Harbour on March 17th. A skein of eight which flew into Chichester Harbour on February 1st 1964 was presumably the same group which wintered in West Sussex

Tundra Bean Geese, Keyhaven, February 13th 2010. Keith Maycock

between December 25th 1963 and March 21st 1964. Seventeen flew SSE over Black Point on March 2nd 1969, 15 were at Titchfield Haven from February 1st to 8th 1976 with six remaining until February 22nd, 21 were at Lymore Lane and later Lower Pennington Lane from January 27th to February 3rd 1979, six moved between the East Meadows at Titchfield Haven and farmland at Brownwich between January 17th and 22nd 1985 with an extra bird present on 19th and 20th, 15 were at Needs Ore from December 12th 2004 to March 1st 2005, 23 were at Northney, Hayling Island from December 21st to 24th 2004, 13 flew up the Test Valley over Testwood Lakes on December 6th 2009 and nine were at Keyhaven from February 7th to 24th 2010.

Arrivals have not been recorded before late November with singles at Needs Ore on 27th in 2008 and at Calshot on 30th in 2011 (not 11th as in the *HBR*). There is a clear peak in February which is related to cold weather. Most wintering birds have left by mid-March but laggards include two at Needs Ore until April 2nd 2005 and one still there on April 12th 1998. Singles at Winnall Moors on May 8th 1977, Heath Pond, Petersfield on May 20th 1984 and with Canada Geese at Elvetham Park on April 11th 1992 and nearby at Yateley on May 5th 1992 were presumably escapes. The monthly pattern is shown opposite.

Areas with the most records were the Avon Valley (9 records, 14 birds), the Lymington–Hurst area (7, 42), Titchfield Haven (5, 29), Langstone Harbour (5, 23), Needs Ore (4, 20) and Chichester Harbour (3, 48).

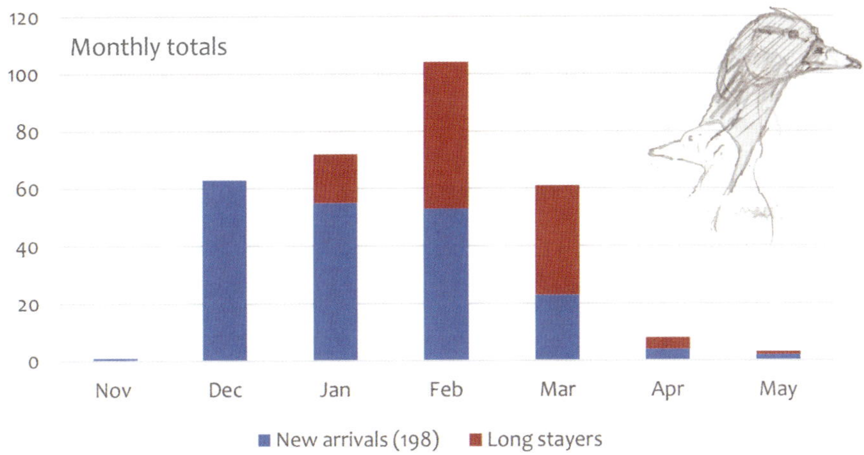

K & M considered the Bean Goose to be 'an occasional winter visitor'. They cited only one undated record from within the present-day county boundary of 'two shot by Mr. W. H. Turle out of a flock on Bransbury Common some twelve years ago'. This would have been in the early 1890s.

GB status: Very scarce winter visitor with increased numbers in cold winters. (Breeds in tundra zone across Russia. Winters in NW Europe, SE Europe, C Asia, China, Korea and Japan.)

Greenland White-fronted Goose
Anser albifrons flavirostris

HOSRP

A rare winter visitor. Eight records involving 35 or 36 individuals since 2000.

Greenland White-fronted Goose, Testwood Lakes, March 18th 2012. Paul Winter

The first record for the county of this race of White-fronted Goose was of an adult with Brent Geese found by Mark Edgeller, Ken Arber and Andy Rhodes at Keyhaven on November 19th 2000 which remained in the area until March 8th 2001. It was probably the bird with Brents at Lepe between November 8th and 15th 2000 which was not specifically identified as this race. In the same winter another first-winter was found by Jason Crook and Bob Chapman at Farlington Marshes on December 17th 2000. It was seen again on January 1st and 9th 2001. The third was an adult found by Jason Crook among a flock of Brent and Canada Geese at Farlington Marshes on November 6th 2007 but it was not seen again. The first multiple record involved a group of five seen by Ian Watts and Andy Collins flying past Weston Shore and up the Itchen Valley early in the morning on February 17th 2008.

The next was a first-winter found by Jason Crook feeding with a Greylag Goose, six Barnacle Geese and 2,100 Dark-bellied Brent Geese on Farlington Marshes on November 29th 2010. There was an influx of various wild geese at that time. It was relocated by Jason with a Greylag and 57 Canada Geese at Apuldram Manor Farm, West Sussex adjacent to Chichester Harbour on January 11th 2011. This may be the first-summer which appeared at Testwood Lakes in late June 2011 and remained there until September 14th (possibly being flightless during much of that period). Subsequently, it was seen regularly in the area between Testwood Lakes and Redbridge until June 30th 2012. It was unringed and so presumably wild.

Tim Doran saw a flock of ten White-fronted Geese, of which six had neck collars, arrive at Farlington Marshes at 0800 hrs on October 22nd 2014. They left north at 0900 hrs but returned at 1020 hrs before finally leaving to the west at 1050 hrs. Subsequent investigation revealed that these were Greenland race birds and that the six had wintered at Wexford Slobs, Ireland during the previous winter. Subsequently, one reappeared at Wexford on October 29th 2014 and another on April 3rd 2015. See Doran (2015) for full details. Finally, Steve Keen saw a flock of 16 arrive from the west over Pennington Marshes at 0825 hrs on October 26th 2018. They circled for five minutes and then carried on to the east. There was a significant influx to southern England at this time including a flock of 21 at Langton Herring, Dorset later the same day which may have included the Pennington birds.

A record of five at Harbridge on November 24th 1974 found in the HOS archives during research for *BoH* was not published in the *HBR* at the time and in the absence of details cannot be considered the first for the county.

GB status: Rare and declining winter visitor to Islay, SW Scotland and W Wales. (Breeds W Greenland, winters Ireland (principally Wexford Slobs), SW Scotland and W Wales.)

Tundra Swan *Cygnus columbianus columbianus*

BBRC

A very rare vagrant. One record in 1986–87.

The only record for Hampshire of this race of Bewick's Swan is of an adult found by the author with 138 Bewick's Swans at Ibsley on December 31st 1986. It remained in the area until January 25th 1987 when it was seen by Martin Reid and others; Martin subsequently submitted a detailed description to BBRC, extracts from which are reproduced opposite.

During the morning we visited the fields by Ibsley Bridge. While checking and counting the Bewick's Swans, a bird at the back of the group lifted its head and immediately caught my eye as it seemed to have a completely black bill. On checking with a telescope, I could see that there was a small amount of very pale yellow in the form of an elongated teardrop running in a narrow strip from below the eye, next to the feathering, about halfway down the base of the bill.

Tundra Swan (left) with Bewick's Swans, Ibsley, January 1987. Louis Rumis

This was probably the same individual that was in Somerset between January 5th and February 12th 1986 and returned there in the 1987–88, 1988–89 and 1989–90 winters. It was the first for Britain.

GB status: 2, 1986–90 and 1998. (N America.)

Whooper Swan *Cygnus cygnus*

HOSRP

A rare winter visitor and passage migrant; influxes sometimes occur in severe winter weather. Many old records and around 466 birds in the modern era.

The earliest mention of Whooper Swans in Hampshire is by Boyes (1774) who said the species had occurred at Alresford Pond sometime in winter prior to that year. Longcroft (1856) mentioned flocks of a hundred at Emsworth in 1800. K & M described the Whooper Swan as 'a winter visitor to the coast .... sometimes numerous in severe weather'. Col. Peter Hawker recorded 90–100 in the Keyhaven area between 1823 and 1845 including at least 40 on January 25th 1838. He regarded the Whooper Swan as the punt-gunner's ultimate prize and secured nine in 1829–30 and 19 in 1837–38, both of which were severe winters (Payne-Gallwey 1893). The Malmesbury collection contains one shot on the lower Avon from a herd of 30 in February 1838 (K & M). This is now held by the Hampshire Cultural Trust at Chilcomb House, Winchester. There are few records for the first half of the 20th century. In 1934, one summered at Needs Ore and was last recorded on September 5th (*Proc. BNSS* 27: 31) and in 1935, one was photographed there on June 9th (*SEBR*). Possibly the 1934 dates were in error for 1935. In the 1936–37 winter, six were on the Beaulieu River between the village and Needs Ore from November 12th to January 10th at least. A pair returned on October 29th 1937 and stayed until April 17th 1938. A pair with two first-winters were present between November 7th and 28th 1938 but only the pair remained between March 12th and 19th 1939. The only report during the 1939–40 winter was of a pair last seen there on April 11th 1940. During a cold spell in mid-February 1940 six were on a pond in Cranbury Park, Otterbourne for ten days; they left the pond each morning, returning in the evening (Vesey-Fitzgerald 1946). There are also vague records of four on flood water in the Itchen Valley for a day in 1941 and of three taking up winter quarters at Eyeworth Pond in late autumn 1944. In the severe cold spell during early 1947 four were seen from a train on the River Itchen at Shawford on March 7th and six including two first-winters were on a pond at Kingston on March 8th.

In the modern era there have been records in 51 out of 71 winters as shown opposite. The largest influx was in February 1956 when 98 were recorded, including 32 flying east at Hill Head on 4th, 18 flying west at Langstone Harbour on 18th, 20 flying south down the Itchen Valley at Swaythling on 19th and up to 17 at Fullerton in the Test Valley. The severe wintry weather in early 1963 produced another notable influx with 64 recorded including a flock of 27 on the sea off Hurst Beach on January 19th, eight in the Eling area from February 3rd to 11th and up to eight at Mottisfont from February 26th to March 10th. Totals exceeded ten in only 11 other winters. Most of the blank winters were in the 1950s, when the recording level was low, and in the 21st century, when there were sightings in only 13 winters.

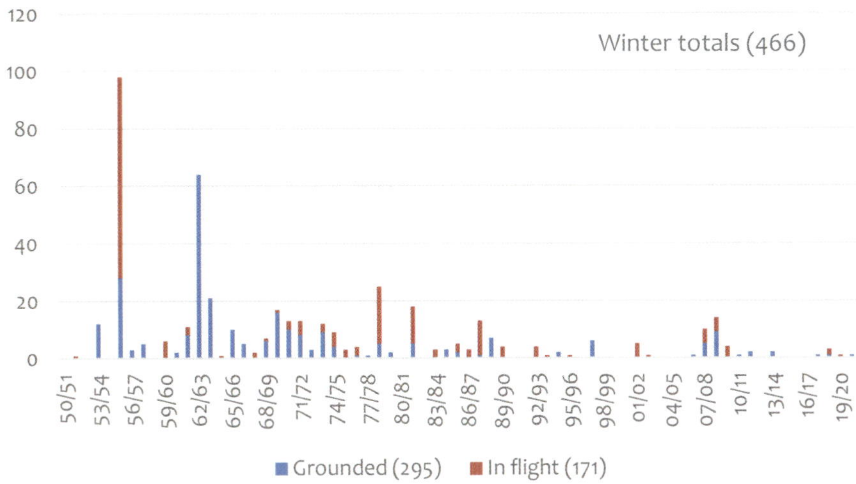

There was a series of records from the Avon Valley around Ibsley in three winters in the early 1960s with maxima of six (one first-winter) on January 6th 1962, eight on December 28th 1963 and January 19th 1964 and seven on January 29th 1966. This was at the time when the wintering flock of Bewick's Swans was becoming established in the area but most of the records submitted involved both species so clearly they were genuine. In the six winters between 1968–69 and 1973–74 up to six were resident at Titchfield Haven (earliest and latest dates November 27th 1971 and March 18th 1970).

The only records of more than five elsewhere involved five adults and four first-winters at Whitten Pond, Burley on February 19th 1954, 13 south over Titchfield Haven which landed on the sea on January 11th 1964, 14 adults north-west over Fleet Pond in severe wintry weather on December 31st 1978, eight adults west past Southsea and into Portsmouth Harbour on April 10th 1988 (following easterly winds and snow), two adults and four first-winters in Langstone Harbour on November 21st 1988 and six first-winters flying north-west over Lower Test Marshes and later on Little Testwood Lake on October 22nd 1997.

The largest groups this century were of five adults flying south over South Hayling on December 15th 2001, five flying south-west over Stubbington on January 26th 2008, two adults and three first-winters flying south down the Test Valley over Trigpoint Hill and Fishlake Meadows before landing at Testwood Lakes on March 26th 2008 where they remained until the next morning, and five adults flying north-east over Bedhampton slipway, Langstone Harbour on November 1st 2008.

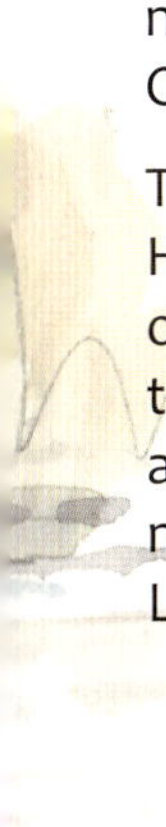

Whooper Swans, Titchfield Haven, winter 1973–74.
Painted from a teenage memory.

Analysis of arrivals dates produces the data shown below; the returning Titchfield Haven birds have been shown as new arrivals each year.

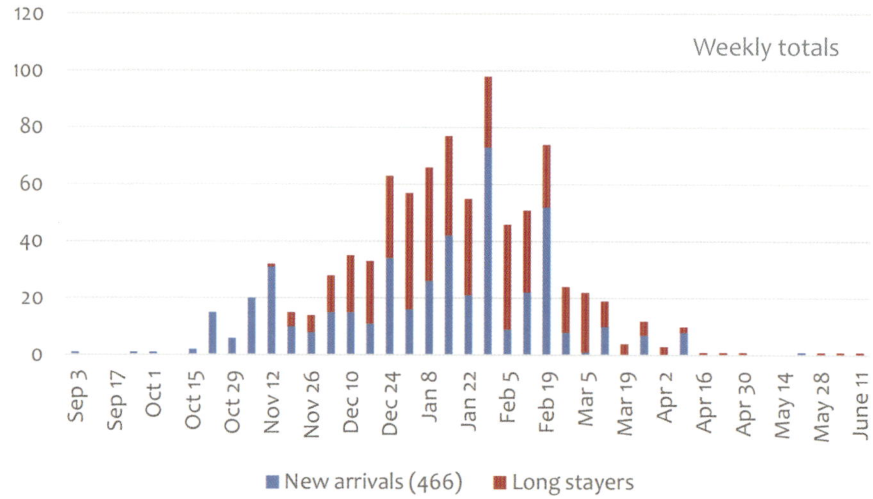

The earliest autumn records were of adults flying north-east down the Wey Valley at Bentley on September 7th 1978, at Fleet Pond on September 26th and 27th 2020 (coinciding with an early arrival in central England of birds originating from Iceland), at Bury Marsh on October 7th 2006 and at Testwood Lakes on October 17th 2010. As the chart shows, there is an initial peak in arrivals in mid-November which has been particularly noticeable in the 21st century – of 45 recorded 30 occurred in the period until mid-December. Most of these were flying over or stayed only briefly; only two singles stayed more than a day, at Titchfield Haven from November 9th to 15th 1961 and Needs Ore from November 13th to 19th 1995, while a pair at Mottisfont on November 14th 1971 remained there until January 2nd 1972 and was relocated at Timsbury on February 1st. As indicated above, sightings later in the winter usually coincide with severe wintry weather and so the prevalence of mild winters in recent years clearly accounts for the reduction in sightings at this time.

Longer stays have been recorded in the Avon Valley on several occasions with birds sometimes associating loosely with wintering Bewick's Swans.

Whooper Swans, Testwood Lakes, March 26th 2008. Richard Ford

These include a first-winter in the Ibsley/Fordingbridge area from November 26th 1984 to January 27th 1985, two adults which moved between Avon and Woodgreen from January 11th to 27th 1985, an adult which moved between Sopley and Ibsley from December 30th 1988 to January 15th 1989 and two adults in 2012, one at Ibsley from January 8th to March 29th and another at Breamore from February 11th to April 8th. Elsewhere, five adults were at Winnall Moors on March 11th and 12th 1963 one of which stayed until April 9th, and single adults were in the Sowley/Lymington area from December 6th 1970 until early May 1971, the Bishopstoke/Allington Lane Gravel Pits area from December 23rd 1981 until February 14th 1982 and at Romsey from March 27th to April 5th 2009 which had reportedly been present since November 2008.

The record of eight off Southsea on April 10th 1988 is the latest ever apart from the long-staying Sowley bird and the adult at Avon Causeway on June 2nd 2019 which was seen daily until June 8th and again on June 17th, 22nd and 23rd.

GB status: Common winter visitor to much of Scotland, N and E England; scarce further south. (Breeds Iceland, N Europe and N Asia to E Siberia; winters to south.)

Ruddy Shelduck *Tadorna ferruginea*

A rare passage migrant or escape. Three old records and around 125 individuals since 1952.

There are two records for 1892, a year in which there was a large invasion into Britain. K & M recorded a female on a lake at Burley on March 28th that was present for a 'fortnight' from that date and reported in *The Times*, and a female shot near Lymington in August. The Burley record pre-dated the invasion but the Lymington bird, which fell within the timeframe of the invasion (Ogilvie 1892) is retained in category B of the Hampshire List (Combridge & Clark 2010). The only record for the first half of the 20th century involved one seen by Col. W. A. Payn on the River Test at Chilbolton in October 1945.

The first record for the modern era was one seen by David Billett and Colin Tubbs in Langstone Harbour on April 20th 1952. The next was one at Kingfisher

Ruddy Shelduck, Sopley, Lower Avon Valley, December 6th 2021. Alan Hayden

Lake, Blashford on June 2nd 1971 and presumably the same individual nearby at Kingston on August 3rd.

Since then, a further 123 have been recorded.

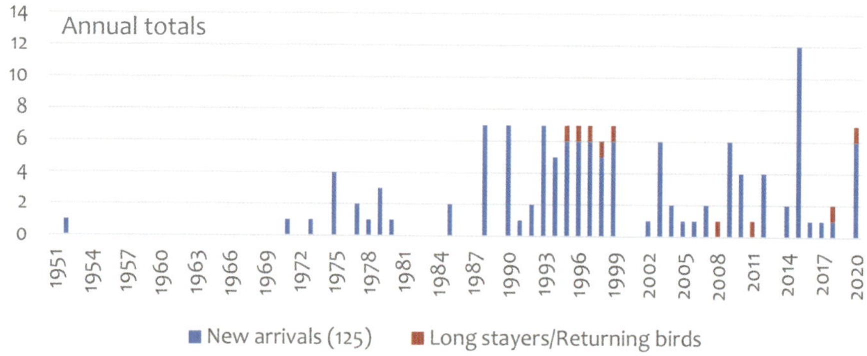

As shown opposite, analysis of arrival dates shows records in all months with a clear peak between August and November. All records of three or more were during these months. Four were at Langstone Mill Pond from September 15th to 21st 1975, four were at Mallard Sands, Langstone Harbour on October 16th, November 19th and 25th 1979 and a flock of five flew west at Pennington Marshes on August 29th 1988. A party of four which arrived at Farlington Marshes from the east on August 3rd 1996 were presumably those seen at Dungeness, Kent the previous day and at Sidlesham Ferry, West Sussex earlier on 3rd. They left high NNW at 1825 hrs. A group of three was at Farlington Marshes on August 19th 1998 which had been seen flying over Thorney Island, West Sussex earlier that day. In the following year four flew low north over the Langstone Harbour islands at 1232 hrs on October 12th, circled over the north shore of the harbour and left very high NNE. Four flew west past Hill Head at 1340 hrs on October 6th 2003. Possibly these were joined by an extra bird as five flew west along Chesil Beach, Dorset the next day and later arrived at the Exe Estuary, Devon. Three first-winters landed briefly on the sea off Hurst Beach at 0923 hrs on September 26th 2009 before continuing west. They were seen at Radipole Lake, Dorset at 1300 hrs that day and remained until October 15th before moving to nearby Lodmoor where they were last seen on 19th. Three were associating with Canada Geese at Hook-with-Warsash on December 27th 2010 before leaving west. A flock of 11 flew north-east over Locks Heath on September 2nd 2015. Four females and a male were at Ripley Farm Reservoir on November 27th 2020; these grazed in fields in the Lower Avon Valley on both sides of the Hampshire/ Dorset boundary and were present into January 2021. They did not associate with a female seen at Ripley Farm Reservoir between November 22nd and December 2nd and at Ibsley Water on November 26th 2020.

Ruddy Shelduck, Sopley, Lower Avon Valley, December 6th 2021. Alan Hayden

Records in other months were all of one or two birds, often associating with geese, at both inland and coastal locations. Sightings of two off Hurst Beach on April 3rd 1977, two at Farlington Marshes on May 22nd 1990 and a female at Sturt Pond, Milford on Sea on April 11th 2009 were possibly the best candidates for migrants returning to the continent.

BOURC have not admitted this species to either category A or C of the British List; it remains in category B on the basis of the 1892 invasion which is considered to involve wild birds (Harrop 2002). The 9th Edition of the British List (*BOU* 2017) states that all post-1950 records are regarded as of British captive origin. Although some seen in Hampshire no doubt originate from self-sustaining breeding populations on the continent, where for example they are included in category C in the Netherlands, they have not yet established such a breeding population in Britain and thus have not been added to category C. However, in 2021 BOURC asked for all records since 1994 to be submitted in order to review its status.

GB status: cat. B (no records since 1892), under review; regular in small numbers in E and S England presumed to originate from naturalised population in Netherlands, Germany and Switzerland. (Breeds Canary Islands, NW Africa and Ethiopia, SE Europe and C Asia; winters NE Africa, S and SE Asia and China.)

Baikal Teal *Sibirionetta formosa*

BBRC

A very rare vagrant. One record in 2021.

The only record for Hampshire is of a male discovered by Olly Frampton among a flock of 433 Wigeon at Ripley Farm Reservoir at 1010 hrs on March 14th 2021. It was not present when the site was visited by the author at first light when 404 Wigeon were counted. Due to Coronavirus lockdown restrictions, it was regrettably decided that news should not be released although the bird was shown to some local birders and others involved in the Avon Valley WeBS count. News was inadvertently released via eBird but the bird was not present the next day. It did reappear on March 16th and was seen by several birders.

Extracts from Olly's submission to BBRC are reproduced below.

My mid-morning visit to Ripley Farm Reservoir was for the purpose of undertaking the monthly WeBS count. Scanning with my scope along the far bank of the pit, I'd got about halfway through the Wigeon flock before noticing a grey-sided duck roosting a few feet from the water's edge, with its head tucked between its wings. I was viewing the bird side-on and immediately noticeable was the vertical white stripe separating the breast from the flanks; a feature of male Green-winged Teal, which initially sprung to mind. From what was visible of its head, a somewhat triangular-like yellow cheek bordered by a thin black line was just visible, with black surrounding the eye and green ear-coverts leading to the back of the head with a thin white horizontal stripe running from above the eye to the back of the head and a brown/blackish crown. Its eye was black, with black feather surround forming a teardrop pattern, with the base of the teardrop leading to the base of the bill, which was fairly small and solid black. Even though I have never seen Baikal Teal before, I'd seen photos of the bird that toured several counties in 2019 and had seen them in field guides, so immediately knew I was looking at a male.

The bird only made a few short flights whilst I was watching; the upperwing was barely visible long enough to note any distinguishing features, but the underwing had three distinct markings. A thick dark leading edge (but not as far along the wing as the primaries), a white 'mid-patch' extending widely from the axillaries, narrowing as it neared a tip towards the median coverts at least. Also visible were grey secondaries with what appeared to be a narrow white trailing edge.

The bird spent much of its time asleep and three long scapular feathers hung down towards the flanks which were seemingly tri-coloured: black, white and tan. Some further tan-like coloured feathers were visible between these obvious three scapulars and the bird's head. With the head turned away from me, the thin white lines running across the top of the head and from the neck towards the back of the head joined at a point on the back of the head forming a white cross. The head shape was unique, almost appearing to form a peak, in a similar profile to the head of a Ring-necked Duck. It was slightly smaller than the Wigeon with which it was associating and occasionally harassing.

Due to the presence of escapes and hybrids, BBRC recently clarified the criteria for the acceptance of rare ducks and geese (French 2021). This bird met all the criteria and has been accepted into category A.

Baikal Teal and Wigeon,
Ripley Farm Resevoir, March 14th 2021.
Olly Frampton

There is an old record of a male shot at Winchester. It was reported by A. H. Baring in *The Field* for February 9th 1928. He was responding to a notice placed on January 28th by Mr. Hugh Whistler who had shot one at Battle, East Sussex. Baring stated that he had shot one on the Itchen about 1915 but had not troubled to record it since so many had been imported alive during the years preceding the war that it must have been an escape. The specimen has not been traced and so the record was considered not proven in a review of all British records (Harrop & McGowan 2009).

GB status: 1, 1906; 6, 2001–2014. (Breeds E Siberia from Yenisey River E to Anadyr & Kamchatka, N to 70°N. Most winter South Korea & lower Yangtze River, China, with small numbers regular E & S Japan.)

Blue-winged Teal *Spatula discors*

BBRC

A very rare vagrant. Five records of six birds since 1972.

The first for the county was a male found by visiting Berkshire birders Charlie White, Nick Mutch and Jeff Hunt at Pennington Marshes at 0630 hrs on May 30th 1972. Later that day it was seen by Ed and Wendy Wiseman and Bob Dunn and remained in the area until June 28th. Extracts from Ed's notes to BBRC are reproduced below.

A fully-winged male was present in the Keyhaven–Pennington Marshes area from May 30th until June 28th 1972 and probably from May 28th, when a Garganey-sized duck was seen with 15 Mallard on Pennington Marshes. On May 30th it was first noted by observers from Reading, then later by us. It was normally seen on brackish pools on Pennington Marshes, both early in the mornings and again late in the day, when it flighted in from the east. In the early part of its stay, it could not be found during the day but later it was often seen with Mallard on Keyhaven saltings. Prior to its arrival the weather had been stormy with strong south-westerly and westerly winds for the previous five days. Its appearance coincided with the arrivals of a Red-necked Phalarope on May 27th and a White-winged Black Tern on May 29th.

Description: approx. size of Garganey, i.e. quite a bit larger than Common Teal. Head dark greyish with very conspicuous white crescent in front of eye. The bill was long and blackish. The undertail-coverts were black with a distinctive white patch in front, and very noticeable on water. In flight very striking, with large blue patches on forewing, bordered on inside with white and with distinctive green speculum. The underwing 'linings' were white – very noticeable. The neck was long and the head was often held very high. On the water it was rather 'high at the stern' in a manner recalling a Shoveler.

It was also seen by John Taverner and Derek Lister at Blackwater, Needs Ore for a few minutes from 0800 hrs on June 4th. This indicates where the bird was going when it disappeared from Pennington Marshes.

The next was a male found by David Billett among sleeping Mallard on The Lake at Farlington Marshes at 1830 hrs on May 28th 1978. It was not seen again. The third record for Hampshire was of a pair found by Tim Doran at Fleet Pond at 1000 hrs on May 4th 1980. They were subsequently seen by many observers and were still present at dusk. This was followed by a female found by Paul Bowman at Dibden Bay on May 18th 1992. It was found feeding with a pair of Teal on flooded pools and was still present on May 20th and 21st. The most recent record was of an eclipse or juvenile male found by Graham Barrett at Titchfield Haven on August 28th 2009 and refound by Tim Doran at IBM Lake, Cosham on August 31st.

Opposite: Blue-winged Teal, male, IBM Lake, Cosham, August 31st 2009. Darren Evans

An immature at Frensham Great Pond from October 29th to November 26th 1989 was accepted by BBRC but in view of its tameness regarded as an escape. The bird was in the Hampshire part of the pond but this has now been transferred to Surrey.

GB status: 10, 1858–1949; 283, 1950–2020. About five per year; stable or decreasing. (Breeds from S Alaska, across much of temperate Canada to SC USA. Migratory, wintering S USA, Mexico, Caribbean & N South America.)

American Wigeon *Mareca americana*

BBRC to 2001; HOSRP 2002 to date

A very rare vagrant. Nine records since 1963.

The first for the county was a male found by Bill Truckle and others at Titchfield Haven on December 22nd 1963. Extracts from his description to BBRC are reproduced below.

> *It was seen on the water at about 120 yards range and watched through 10x50 binoculars and a telescope. The predominant feature was a broad greyish-green stripe from the eye backwards and downwards. The forehead and crown were pale buff, almost creamy, but not yellow as in male Wigeon. The cheeks had a mottled, greyish appearance. The bill was grey. The mantle was brownish and the flanks were chestnut, tinged almost wine-coloured on the breast. After ten minutes the ducks were disturbed and the American Wigeon took off with the surrounding Wigeon and flew out into the Solent. It was not seen again.*

David Billett, in a letter dated February 1st 1964, informed Denzil Harber, the secretary of BBRC, that 'a whole brood of American Wigeon flew away from a collection at Portsmouth about 18 months ago as they were due to be ringed'. BBRC put a caveat on the record because of this information but the bird may well have been wild.

The next was a male found by Terry Martin, Ron Cooke and Rod Ship with a pair of Eurasian Wigeon at Dibden Bay on April 17th 1977 (not 14th as in *BoH*); it was still present the next day. In the following year, an adult male was found by Bob Chapman and also seen by Barry Duffin at Titchfield Haven on October 27th. It was seen displaying to Eurasian Wigeon on the North Scrape but the next day was foggy and all the Wigeon were flushed out on to the sea by a Marsh Harrier. The American Wigeon was not seen again.

American Wigeon, eclipse male, Titchfield Haven, May 1987

The first long-stayer was an adult male found by Reg Thorpe and John Walters at Bisterne on January 5th 1984 while they were watching a Red-breasted Goose. It remained on the floods there until 14th at least and was refound by the author at Ivy Lake, Blashford on February 12th and was still there on 18th and 19th. The fifth for the county was a somewhat unseasonal male found by Barry Duffin, Richard Levett and Mark Palmer at Titchfield Haven on May 7th 1987 which was last seen on July 19th. It was assumed by BBRC to be the bird at Christchurch Harbour and Brownsea Island, Dorset from May 2nd to 5th. The next was a male found by Mike Rafter with four Eurasian Wigeon just above the old A35 bridge at Lower Test Marshes on February 9th 1991. It was seen by several observers the next day but then became elusive as it accompanied a flock of several hundred Wigeon in the Eling/Lower Test Marshes area. It was last seen on April 15th (not April 11th as in *BoH*). A male was found by Steve Keen at Tanners Lane on March 13th 1994 but not seen again. It was presumably the bird at Christchurch Harbour, Dorset from March 8th to 12th. Another long-staying male was found in the Lower Test area from October 22nd 1995 when it was seen by Jess Pain, Mike Adams, Mark Edgeller and Andy Rhodes and others until the year's end. It was refound by Simon S. King and Jess Pain among a large flock of Wigeon at nearby Bury Marsh on January 22nd 1996 and was seen again by Tony Blunden at Lower Test Marshes on March 25th. A male was found by Martin Tribe and David Ireland at Farlington Marshes on December 6th 1998 and remained in the area until March 9th 1999. What was presumably the same bird was seen by Jason Crook at Farlington Marshes on April 14th and 15th although it had not been seen in the interim period despite the good coverage of the area.

There have not been any accepted records this century.

A first-winter male was at Titchfield Haven from November 21st to 23rd 2006 and again on February 18th 2007 but HOSRP felt that the hybrid possibility had not been eliminated.

A male found by Tim Doran at Eversley Gravel Pits on January 31st 1981 was in a wildfowl collection at nearby Eversley Cross the next day. The owners had no idea where the bird had come from but said it had been around for several weeks. It was seen again at Eversley Gravel Pits on February 2nd and March 15th and at Yateley Gravel Pits on February 15th. Although it was unringed and fully-winged it was felt likely to be an escape and so was not submitted to BBRC. Other non-submitted records were of a female seen by Tony Blunden at Titchfield Haven on July 17th 1993, a male reported by the bird information services at Lower Test Marshes on March 27th 2001 and a female logged on the WeBS count at the Beaulieu Estuary on January 13th 2008 which was seen again off Inchmery Quay the next day.

GB status: 15, 1907–1949; 638, 1950–2019. No clear trend, mean 18 per year in 2010–19. (Breeds C Alaska E across temperate Canada to Quebec & Nova Scotia, & S into NW & Midwest USA. Winters S Alaska, coastal British Columbia, Canada, and S throughout southern USA & C America to Colombia, Greater Antilles & West Indies.)

Green-winged Teal *Anas carolinensis*

BBRC to 1990; HOSRP 1991 to date

A rare winter visitor. 30 records involving a minimum of 20 individuals since 1956.

The first for the county was a male found by E. G. Richards at Titchfield Haven on March 18th 1956 and subsequently seen by several observers. The account of its occurrence is reproduced below (*Brit. Birds* 50: 199–120).

> *During the morning of 18th March 1956, a male Green-winged Teal was seen at Titchfield Haven by several observers including John Everett, Roger Brown, Alan Searle and the writers. It was in company with a small number of Teal and was distinguished from the males by the white vertical mark at the sides of the breast and the absence of a white horizontal line above the flanks. The buff face-markings were very obscure, being reduced to a short buff line just below the eye (E.G.R.). It also had a much more buff-coloured breast than the Teal, making it easy to pick out when head on (R.H.D.). E. G. RICHARDS, C. SUFFERN and R. H. DENNIS*

Totals in each winter are shown below. Assuming birds returning in consecutive winters are the same gives a total of 20 individuals.

The second for Hampshire was found by Anthony Cheke, a student at Winchester College, at Avington Lake on January 30th 1962 where it was subsequently seen by John Taverner and John Cloyne and stayed until March 25th. The next was not until the cold winter of 1978–79 when Peter and Alan Potts found one at Farlington Marshes on December 30th which stayed until January 14th.

The first of six during the 1980s was picked out by Mike and Pete Combridge in a large flock of Eurasian Teal at Ivy Lake, Blashford on January 12th 1980 but was not recorded again. Two were seen in 1982, one found by Terry Martin at Dibden Bay on February 6th and relocated there by Steve Smith and Keith Godfrey on March 13th and another found by the author at Stratfield Saye Park on March 14th and still present on 17th. The author found another at Dogmersfield Lake on March 24th 1985 which he showed to Andy James and Dave Unsworth and then was seen by many local observers up to 30th. The next was a one-day bird found by Mike Baker at Emer Bog on February 28th 1987.

A series of records between 1989 and 1993 is assumed to refer to the same individual. It was initially found by Barry and Tor Duffin at Titchfield Haven on May 5th 1989 and seen there again by Ian Watts on May 27th and Barry the next day. It returned in the autumn of that year and was seen by Trevor Codlin and others at Titchfield Haven between October 29th and November 13th, found by Richard White at Farlington Marshes on December 28th where it stayed until January 3rd and then refound by Dan Powell at Titchfield Haven on January 20th and seen on four more dates up to March 4th. In the following winter it was refound by Peter Marston at Titchfield Haven on January 6th 1991 and observed there by Barry Duffin from February 13th to 16th 1991. Subsequently, it was reported by Barry Duffin at Titchfield Haven between February 16th and March 16th 1992, and by Paul Bowman and others at Needs Ore between March 27th and April 18th 1993.

The next three records were all from Titchfield Haven, involving birds seen by Kevin Ilsley and Barry Duffin on April 19th and 20th 1997, Barry Duffin between November 8th and 19th 1998 and Peter and Sue Morrison between December 18th 1998 and January 2nd 1999, and Dan Houghton, Simon Hartill and Peter Milinets-Raby from February 25th to 28th 2001. Although there were no sightings in the intervening winters, it does seem possible that these sightings involved the same individual first seen in 1989.

One was found by Mick Scott with a small group of Eurasian Teal at Fox Lane Gravel Pit, Eversley on December 3rd 2001. It had ventured across the River Blackwater from Moor Green Lakes, Berkshire where it had been present between November 17th 2001 and January 27th 2002. The fourth in the north-east of the county was found by the author on a partially-frozen lake in

Bramshill Plantation on the morning of February 20th 2005 but it had moved on later in the day.

The next was found by Steve Copsey at Hook-with-Warsash on April 22nd 2008 where it stayed until April 28th. What is assumed to be the same individual was found by Jason Crook at Farlington Marshes on October 13th 2008 and last seen on November 8th. It was refound by Richard Ford at Budds Farm Sewage Treatment Works on January 9th 2010 and remained at Budds Farm and nearby reaches of Langstone Harbour until February 19th of that year. There was a two-year gap until another was located by Jason Crook at Farlington Marshes where it was present from February 23rd to March 11th and again from March 30th to April 9th 2012. It was seen to be paired with a female Eurasian Teal on March 8th and 9th (Jason Crook *in litt.*). See Riddington (2019) for discussion of the separation of female Green-winged Teal from Eurasian Teal. Another was found by Mark Rolfe at Titchfield Haven on April 7th 2012 where it remained until 14th, thus confirming it was a different bird to the Farlington individual.

The next was found by Colin Allen at Ivy Lake, Blashford on January 26th 2013 where it stayed until February 15th. It is assumed to have relocated to Fishtail Lagoon, Pennington Marshes, where Marc Moody discovered it on February 26th and it was last noted on April 12th. It returned in the following winter when David Taylor found it on floods just below Avon Causeway on January 1st 2014. It stayed in this area until March 6th and was presumably the bird found by Martin Clay at Ibsley Water on April 6th which stayed until 9th.

Another was found by Pete Combridge during a WeBS count at Bury Marsh on January 23rd 2016 and the most recent record was of one found by Marc Moody at the Upper Balancing Pond, Keyhaven on March 28th 2018 which was seen by many observers until April 14th.

Green-winged Teal, Keyhaven, March 9th 2013.
Gordon Small

Green-winged Teal with Teal, Keyhaven, March 31st 2018. Lee Fuller

Analysis of arrival dates shows records spanning the period between November 8th and May 5th although returning birds have been recorded as early as October 8th and lingering birds as late as May 28th.

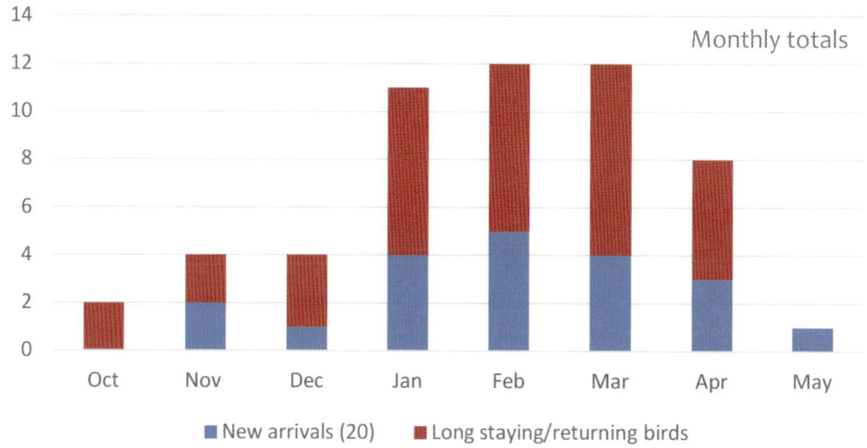

Males at Keyhaven Lagoon on January 11th 2001 and February 22nd and 23rd 2009 were considered to be hybrids with Eurasian Teal.

Records of males at Blackwater, Needs Ore on November 14th 2008 and Old Basing on October 6th 2014 were not documented.

The first mention of this species in Hampshire is in *The Field* for January 17th 1880, where Mr. Arthur Fellowes of Rotherfield, Sussex, having recently read a note in *The Field* about one being obtained in south Devon, records that he possesses one of these birds in excellent condition which was shot more than forty years ago by his father at Hurstbourne Park, Hants. The white crescent on the wing is very apparent and well defined. Given the vague date and lack of an extant specimen this record is considered not proven (Naylor 2021).

GB status: 6, 1879–1949; 1,305, 1950–2019. Stable after increase, mean 40 per year in 2010–19. (Widespread breeder throughout North America from Alaska to Newfoundland, S to N USA. Winters British Columbia, Canada, throughout USA & Mexico to southern C America & West Indies.)

Red-crested Pochard *Netta rufina*

A rare winter visitor and passage migrant, or escape. Around 181 1individuals since 1961.

The first for the county was a pair found by Norman Pratt at Stratfield Saye Park on October 28th 1961. This record has not previously been published and has been extracted from his dairies. They may have been escapes although none was reported there during the monthly National Wildfowl Counts carried out during that period.

The next was a male seen by W. Earp at Dogmersfield from April 5th to 13th 1965. Subsequently, there were records in every winter save four from 1968/69 onwards involving 178 individuals.

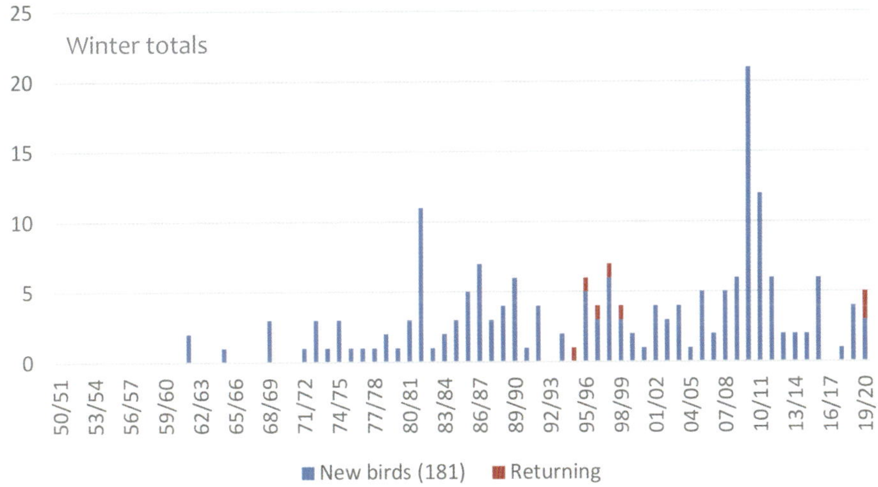

Known escapes and most long-staying individuals, some of which were ringed, have been omitted, as has a flock resident at Paultons Park between 1988 and 1995 which peaked at 50 on January 9th 1989 and July 26th 1991. The highest numbers have been recorded during severe spells of weather, particularly in January 1982 and January 2010, which suggests that birds have been displaced from further afield – possibly the Cotswold Water Park, the Royal Parks in London or naturalised populations in the Netherlands.

As shown on the next page, analysis of arrival dates shows a clear mid-winter peak (coinciding with severe weather influxes as mentioned above). However, new birds have occurred in every month; in late summer and autumn most are seen on one date or remain for a few days only but later in the winter longer stays are more frequent. In spring, further new arrivals, often present only for a single day, suggest passage. In contrast with some other wildfowl there is little evidence of birds returning in successive winters,

Red-crested Pochard, males, Bramshill Park Lake, January 8th 2019. John Clark

the only examples being of a male which wintered at Titchfield Haven from 1993/94 to 1997/98 and two males at Bramshill Park from January 2nd to April 3rd 2019 which returned in the following two winters.

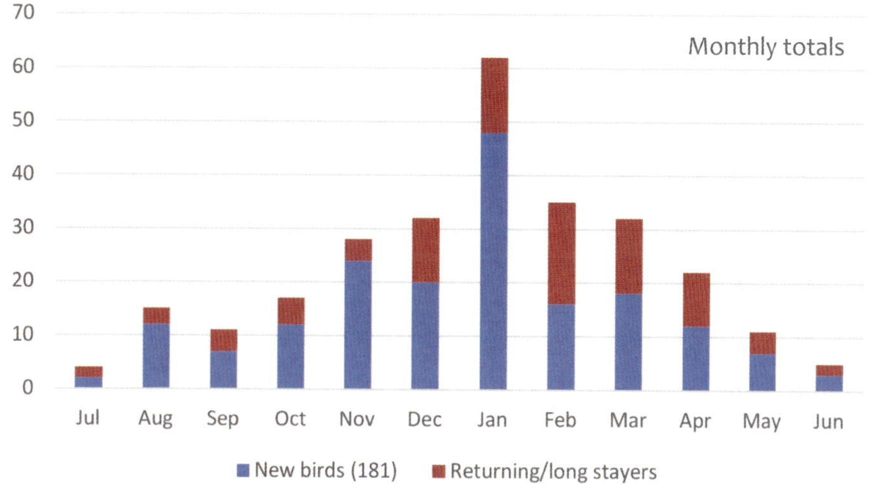

Records have come from over 40 different sites including a variety of small waters. Birds at these sites have often been suspected as being escapes and yet frequently they are present for one day only before they move on. Most have been recorded at Blashford Lakes, where there were 17 singles between 1978 and 2009 and a further 24 subsequently including six or seven (two males) in severe weather between January 10th and 18th 2010 with three until January 31st and one or two until February 13th, three (two males) on January 1st 2011, five (four males) on November 6th 2011 and three (one male) on November 11th 2019. Other totals include at least 20 in the Langstone Harbour area including three (one male) at Sinah Gravel Pit on October 12th 1974 and four (two males) in the harbour on January 7th

2010, 19 at Titchfield Haven including eight (six males) on the sea on January 13th 1982 and three (one male) on the sea on January 9th 2010, 11 at Heath Pond, Petersfield including three (two males) on February 25th 2016 and 37 at various sites in the north-east of the county. The only other records of more than two involved three at Alresford Pond on May 9th 1996, three (two males) at IBM Lake, Cosham on December 23rd 2010 and four (one male) at Avington Lake on March 19th 2011.

GB status: at least 50 breeding pairs mostly in S England; c. 1000 in winter 2019/20. (Breeds patchily in C and S Europe east into C Asia.)

Ferruginous Duck *Aythya nyroca*

BBRC up to 1968, 1999–2005, 2017 to date; HOSRP 1969–1998, 2006–16

A rare vagrant. 31 records involving a minimum of 13 individuals since 1972.

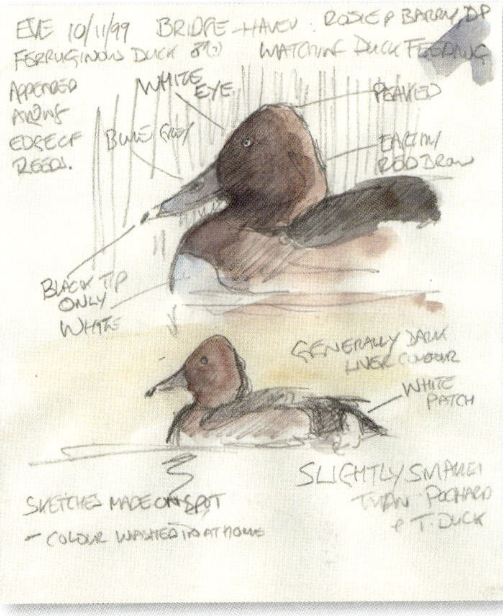

Ferruginous Duck,
River Meon, Titchfield Haven,
November 10th 1999.
"Who's writing this one up then?"

The first for the county was a male found by Philip Fawkes at Fawley Reservoir on November 25th 1972. Extracts from Philip's description are reproduced below.

> *Viewed through 10 x 50 binoculars and 15-60 x 60 telescope. It was associating with a large flock of Tufted Ducks and Pochards. Size and build as Tufted. The head was chestnut almost as rich as male Pochard. A very distinctive high raised crown. Eye white. Bill light grey but without black terminal band of Tufted. Back dark brown, flanks chestnut, not quite as rich or as dark on head. Distinctive white undertail-coverts. On preening white wing-bar shown with distinction; white underparts also noted.*

It was subsequently seen there by Philip, Stuart Sporne and John M. Jones on December 8th, 9th and 17th but was absent on other dates during that period. Presumably the same individual was found by David Thelwell at Kimbridge Gravel Pit in the Test Valley on December 16th. This was omitted in error from *BoH*.

Subsequently, a minimum of 12 new individuals has been recorded.

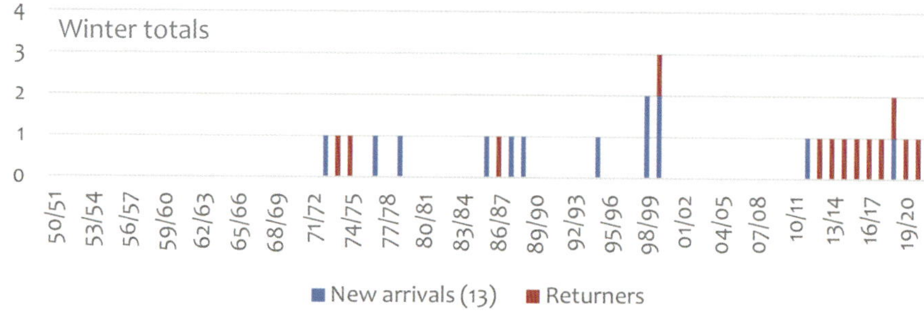

What is assumed to be the Fawley bird returning was found by Derek Lister at Allington Gravel Pits on October 1st 1973 where it remained until at least January 1st 1974. It was also seen by Norman Pratt and Jess Pain. It reappeared in the following winter when it was seen at Allington Gravel Pits by Mike Terry on November 19th and 28th and by Ron King on December 23rd and at Fawley Reservoir by Philip Fawkes and John M. Jones on December 14th. The next was a male found by John Cloyne on a new fishing lake at Winnall Moors on June 3rd 1977 which soon flew up the Itchen Valley and was never seen again. John was unable to determine if it was ringed although it was published as 'no doubt an escape' in the 1977 *HBR*. Given that it was seen once only and that summer records of the species are not unknown this seems a somewhat harsh assessment! Another single date bird, very well described by Paul Bowman, was a male on the River Test at Nursling on March 5th 1979 following a very cold winter.

The fourth for the county was a female found by Mike Taylor at Eversley Gravel Pits on December 28th 1985. It proved elusive and was seen briefly the next day and then relocated at nearby Yateley Gravel Pits by Mick Scott and the author on December 31st. It had returned to Eversley by January 26th 1986 where it remained until February 8th and was last seen at Yateley on March 16th. Later that year, what is assumed to be the same individual was found by Mick Scott and Keith Wills at Dogmersfield Lake on September 16th. It was then seen at Frimley Gravel Pits, Surrey on 18th, back at Dogmersfield on 19th and at Fleet Pond on 21st. There was then a long gap until it was seen again by Will Legge and Mick Scott at Fleet Pond on January 9th 1987 before relocating to Frimley Gravel Pits where it was seen intermittently until February 5th. The next was a male found by Clive Bealey at Mockbeggar Lake on December 13th 1987 and seen there independently by George Green on December 16th. It was relocated by Alan Hayden on the River Avon at Sopley on December 28th where it remained until February 6th 1988. Later that year a first-winter male was found by Mick Scott at Fleet Pond on October 3rd; this was omitted in error from *BoH*.

The next was another first-winter male found by Mick Scott at Fleet Pond on December 3rd 1994. It was still there the next day but was relocated at Dogmersfield Lake on December 10th where it stayed until 18th with visits back to Fleet Pond on 17th and 18th.

Ferruginous Duck with Gadwall, Kingfisher Lake, December 17th 2015, Bedhampton, December 20th 2018 and with Wigeon and Pochard, Kingfisher Lake, December 13th 2014.
Martin Bennett, Andy Johnson and Alan Lewis

There was a series of records between 1998 and 2000 which are believed to have involved four individuals. A first-winter male was found by Marc Moody at Timsbury Gravel Pit on December 8th 1998 which was seen there by Dan Philpott and others between December 21st and January 11th and by David Thelwell nearby on the River Test on February 14th. An adult male was found by Tim Lawman on the pools at Budds Farm Sewage Treatment Works on April 23rd 1999 and seen there the next day by Jason Crook. Later that year a first-winter male was found by Barry Duffin and Dan Powell at Titchfield Haven on November 10th which stayed until 12th. A pair, assumed to include the returning male from 1998, was found by Dan Philpott at Timsbury Gravel Pit on November 15th. The next day only the female was present which remained until December 26th. What is presumed to be the Timsbury male was found by Dave Unsworth at Lower Itchen Fishery between Bishopstoke and Mansbridge on January 2nd 2000; it was seen there on four further occasions until March 12th although it left when the water froze. Presumably it moved to nearby Allington Gravel Pits where it was located by Mark Edgeller on March 19th and stayed until April 1st.

There was a long wait until the next record, of a male found by David Thelwell at Fishlake Meadows on December 4th 2011. This is assumed to be the same bird which was discovered by Bob Chapman at Spinnaker Lake, Blashford on

December 30th. It was still there the next day but then moved to the adjacent Ivy Lake where it was seen intermittently until February 27th. This bird returned to Blashford Lakes in each of the following ten winters up to 2021–22, usually arriving in late September or October (earliest date September 24th 2017) and departing in March (latest date March 20th 2018). Much to the chagrin of birders it has mostly favoured Kingfisher Lake, a private water which latterly has been viewable only from an adjacent footpath by ascending a step ladder to see over the fence surrounding the site!

The first new individual since 2011 was a first-winter male found by Jason Crook at Bedhampton Mill Pond on December 16th 2018 which stayed until January 9th 2019.

Analysis of the records shows returning birds arriving from September 16th but new individuals, apart from one on October 3rd, mostly between November 10th and December 28th with nine recorded. The latest wintering bird was on April 1st; other apparent new arrivals in early March, late April and early June were presumably migrants.

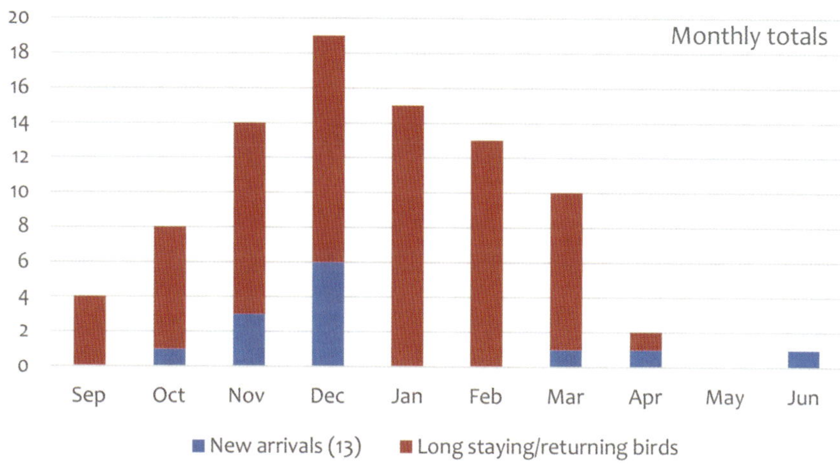

There are two earlier records from Frensham Great Pond the second of which was included in *BoH*. A male was seen there by William Shaw on October 8th 1920 and another was there from December 29th 1960 to January 25th 1961 which was certainly seen in Hampshire water by Tom Brice and Peter Le Brocq on January 11th. Subsequently, the whole of Frensham Great Pond was transferred to Surrey so the records are now excluded from the Hampshire avifauna.

Records of a female at Keyhaven on October 2nd 1972, a pair at Somerley Lake on January 14th 1973 and a male at Keyhaven on January 20th 1974 were included in the *HBR*s but excluded from *BoH* as the available notes did not rule out hybrids. The record of a male seen by Rob Macklin at Sinah Gravel Pit on February 20th 1986 was not accepted by HOSRP as it was felt that the hybrid possibility had not been eliminated. Unfortunately, he only saw the bird asleep and was not able to see the bill. Another record of a male at Charlton Gravel Pit, Andover on January 1st 2001 was not accepted as no notes were received.

A first-winter male hybrid Ferruginous Duck type was at Farlington Marshes from November 23rd to March 10th 2000. It returned to the nearby Budds Farm Sewage Treatment Works in the next ten winters (earliest date September 8th 2009, latest date March 21st 2004) but also ventured to Farlington Marshes, IBM Lake and Sinah Gravel Pit. A female hybrid also occurred at Budds Farm in three winters between 2006/07 and 2009/10. Further hybrid types continued to be reported until 2015 although they were not thought to involve the original bird.

K & M mentioned three undated occurrences between 1875 and 1893. One involved a report from George Corbin of one obtained, on the authority of Mr. J. Mills, at Bisterne about 1875. Enquiries by the author to discover more about this are still ongoing. Another shot at Ovington before 1891 was reidentified as a Scaup. None of these is acceptable as the first for the county.

GB status: 172, 1829–1949; 697, 1950–2020. About nine per year; stable or decreasing. (Breeds Poland & Hungary E through Ukraine to Caspian Sea, but distribution patchy. Elsewhere breeds S Spain, Kazakhstan, W Mongolia & Tibetan Plateau. Migratory, most winter E Mediterranean, Black & Caspian Seas, NE Africa & Indian subcontinent.)

Ring-necked Duck *Aythya collaris*

BBRC up to 1993; HOSRP 1994 to date

A rare vagrant. 39 records involving a minimum of 11 individuals since 1979.

The first for the county was a first-winter male found by the author at Ivy Lake, Blashford on December 15th 1979. It was also seen by Peter Le Brocq and Richard Leach. Extracts from the author's description are reproduced below.

The bird was located with 44 Pochards and one Tufted Duck about 75 yards from the southern shore of Ivy Lake. It was immediately picked out by its unusual bill pattern and head shape. Its size was similar to the single Tufted Duck and smaller than the Pochards. The bill was bluish-grey apart from a dark tip which joined a distinct white sub-terminal band and a thin, delicate white area around its base. The forehead was steeply angled with an almost flat crown which was angled abruptly (almost at 90 degrees) with the back of the head. The eye was yellowish/orange. The other notable feature was the flanks. They were whitish grey apart from the anterior spur which appeared whiter than the remainder of the flanks, but not as bright white as in the previous male I had seen, so I assume the bird was in first-winter plumage. In flight it was seen to have a greyish wing-bar, similar in extent to that of Pochard and unlike the white wing-bar of Tufted Duck.

It remained in the Blashford Lakes area until February 17th 1980 and was also seen on the River Avon at Avon on January 1st and Sopley on January 10th. It was refound by Jess Pain, Mick Scott and the author at North Poulner Lake, Blashford on December 29th that year and remained there until January 11th 1981.

Subsequently, a minimum of 10 new individuals has been recorded.

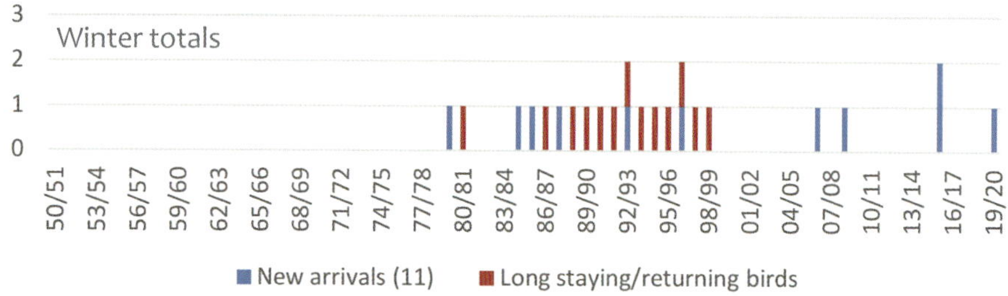

The second for the county was also seen at Blashford Lakes, a male found by Simon Woolley on Spinnaker Lake on January 31st 1985. The next was a first-winter male located by the author at Fleet Pond on November 20th 1985 which had first been found by Keith Wills at Frimley Gravel Pits, Surrey three days earlier. It remained at Fleet until November 23rd but then relocated to the Blackwater Valley gravel pits in Surrey being seen at Frimley, Ash Vale and Badshot Lea until April 16th 1986. Further sightings at Fleet Pond on May 17th and Dogmersfield Lake on May 19th, both omitted in error from *BoH*, are assumed to relate to the same bird. It returned to the Blackwater Valley at Frimley on September 9th 1986 but was found by Paul Troake at Dogmersfield Lake on September 14th. However, it promptly returned to Frimley and wintered there and at Ash Vale Gravel Pits until April 6th 1987. Subsequently, it was found by Mick Scott and Keith Wills at Fleet Pond on April 18th and by Will Legge at Dogmersfield Lake the next day, where it remained until April 25th.

The next was a first-winter male found by the author at Mockbeggar Lake on November 22nd 1987 which was still there the next day. It is presumed to have relocated to Timsbury Gravel Pit where it was found by Phil Lord on December 11th. It wintered there and was last recorded on March 20th 1988. This individual returned to Timsbury Gravel Pit for each of the next 11 winters except 1995/96 and occasionally ventured to nearby floods at Fishlake Meadows especially in 1993/94. Arrival dates were between November 2nd and December 22nd except in 1992 when it was first recorded on October 14th. Departure dates showed more variation. In 1990, 1995 and 1999 it was last seen on dates between January 17th and 28th and in other years final dates fell between March 8th and April 5th. However, in 1989 it reappeared on April 25th and 26th and in 1993 was present from May 17th to June 16th. Other sightings assumed to refer to this bird were at Alresford Pond on

January 27th 1990, Winchester Sewage Farm from March 5th to 7th 1990, Allington Gravel Pits from March 9th to 28th 1991, Linbrook Lake, Blashford from November 16th to 27th 1991 (before arrival at Timsbury on November 28th), the River Test at Broadlands on January 16th 1993, Mockbeggar Lake, Blashford on March 27th 1993, Lower Test Marshes on April 15th and 16th 1995, Titchfield Haven from March 8th to 28th 1996 (the only sightings in that winter) and April 15th 1997, Winchester Sewage Farm from April 16th to 20th 1997 and Anton Lakes, Andover from April 11th to May 9th 1999, the final occasion it was recorded.

Ring-necked Duck with Red-crested Pochard, Titchfield Haven, March 1996

The fifth for the county was an adult male found by Gilbert Rowland at Fleet Pond on May 18th 1993 which was still present the next day. This was followed by a first-winter female which Pete Combridge and Phil Toye saw 'drop out of the sky' on to a tidal creek at Keyhaven at 1050 hrs on November 8th 1996. As the tide dropped it flew on to Keyhaven Lagoon at 1324 hrs but could not be refound two hours later.

There was a long gap until the next, another first-winter female, was found by the author at Yateley Gravel Pits on November 18th 2006 where it remained until December 3rd. It was also seen at nearby Moor Green Lakes, Berkshire on November 19th. A mobile first-winter male was found by Mike Wearing at Passfield Pond on December 10th 2008 where it stayed until December 21st. It was then at Frith End Sand Pit on December 29th and 30th, back at Passfield Pond from January 1st to 14th, at Passfield Park Farm Pond on January 21st, back at Frith End from March 3rd to 27th and finally at Frensham Great Pond, Surrey from March 29th to April 3rd 2009.

The first multiple occurrence was of two first-winter females found by Kevin Haggar and David Piper at Rooksbury Mill, Andover on November 20th 2015 where they stayed until January 20th 2016. The most recent record was of an adult male found by Bob Chapman at Ibsley Water on April 16th 2020 which was still present the next day.

Ring-necked Ducks: male, Frithend Sand Pits, March 25th 2009 and females, Rooksbury Mill, Andover, November 27th 2015. Steve Copsey (top two) and John Hilton

As shown opposite, analysis of the records shows that arrivals of new individuals peaked in November with six, then two in December and one each in January, April and May. The species has been recorded in every month except July and August with earliest and latest dates of September 14th and June 16th respectively.

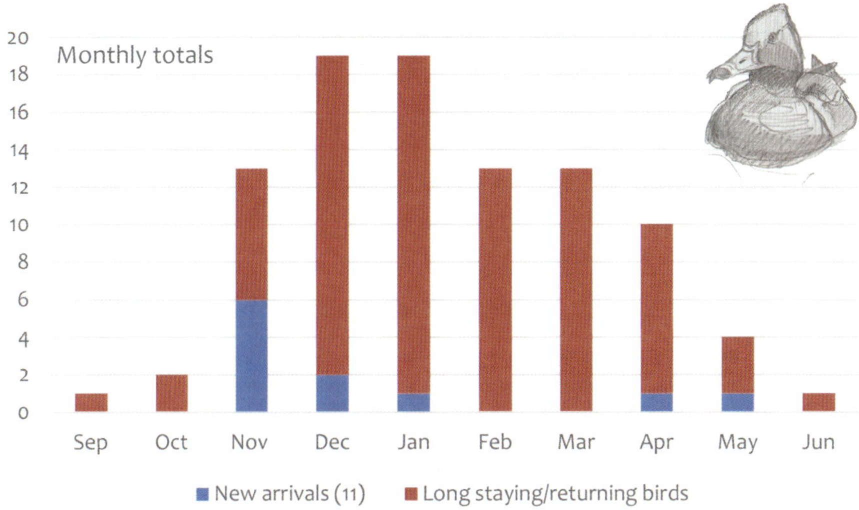

GB status: 902, 1959–2019. Stable after increase, mean 24 per year in 2010–19. (Breeds locally E Alaska, throughout cool temperate Canada from S Yukon to Maritime Provinces, and N USA from Washington to Maine. Winters in ice-free regions of southernmost Canada, S throughout southern USA to S Mexico, Costa Rica & Greater Antilles.)

Lesser Scaup *Aythya affinis*

BBRC up to 2014; HOSRP 2015–19; BBRC 2020

A very rare vagrant. Three records since 2017 probably involving only one individual.

The first for the county was an adult male found by Alan Lewis on Spinnaker Lake, Blashford on October 28th 2017. Extracts from Alan's account of his find (Lewis 2018) are reproduced below.

Of all the gaps in the Hampshire List, one of the most glaring omissions at the start of 2017 was Lesser Scaup. Records have come from most other parts of Great Britain and even inland counties such as Hertfordshire have attracted this species. Since moving to the county in the late 1990s, I've always checked any flocks of Aythya ducks for rare species but with little to show for the effort. When a male Lesser Scaup turned up at Longham Lakes, Dorset on October 23rd, I speculated whether it might move further east. It was seen up to the evening of Thursday October 26th but there was no sign of it there the next day and so I made plans to search for it at Blashford Lakes on Saturday 28th.

Arriving at dawn at Ibsley Water, a low mist was settled across the lake and was slow to clear, but eventually I was able to satisfy myself that the Lesser Scaup was not present. I set off south along the footpath, checking both Rockford and Ivy Lakes without success and then set up my scope on the footpath along the western edge of Spinnaker Lake. Again there was a low mist across most of the lake and the low sun from the east made viewing difficult, but a flock of about 30-40 Tufted

Ducks was stretched in a line along the distant eastern shore. As I was scanning the lake with binoculars, a lone Aythya flew through my view and I followed it as it pitched down at the southern end of the lake. In flight it seemed to show a grey back and a short white wing-bar, but the light was so poor I was unsure if these features were real. As I fixed my scope on the lone duck, I was delighted to confirm that this was a male Lesser Scaup and noted the key features including the grey mantle, the all-grey bill with a small black nail and the peak-headed profile so different to the rounded head of the (Greater) Scaup. Having secured a few acceptable images, I released the news and set about taking more images, waiting for the first birders to arrive.

Lesser Scaup, male, in flight and with Tufted Ducks, Blashford Lakes, October 28th 2017.
Alan Lewis and Lee Fuller

This was presumably the individual which appeared at Chew Valley Lake, Avon on November 12th and subsequently wintered there. It was at Ham Wall, Somerset on February 2nd 2018 before it was found by Dave Taylor at Ibsley Water the next day. It was seen by several observers and remained on February 4th until at least noon. It was again at Ibsley Water on January 20th 2019; Dan Houghton found a bird at long range at 0950 hrs which he considered to be of this species and later in the day it was seen by several observers who agreed with his identification.

GB status: 215, 1987–2020. Declining after increase, mean 6 per year in 2010–19. (Breeds from C Alaska through Canada to Hudson Bay & S to Washington & South Dakota. Isolated populations E of Great Lakes. Winters along both coastlines of USA, in E from New Jersey to Mexico, West Indies, C America to N Colombia.)

Surf Scoter *Melanitta perspicillata*

BBRC up to 1990; HOSRP 1991 to date

A very rare vagrant. Seven records of six birds since 1976.

The first for the county was a male that flew east past Hurst Beach on May 15th 1976. It was first seen by Mike Combridge, who alerted Pete Combridge, Ed Wiseman and Dave Wooldridge. Dave was elected to submit a description to BBRC, extracts from which are reproduced below.

> *At 0710 hrs a flock of 12 scoters flew east in good light consisting of 4 males, 7 females and 1 male Surf Scoter. The Surf Scoter was entirely black except for white on the head. Clear white patch on forehead, easily seen as bird approached and at side view. As the flock was going away to the left from us two observers through scopes noted white patch on nape (clearly separated from forehead patch). Size was not greatly but noticeably larger and bulkier than male Common Scoters in the flock. The Surf Scoter occasionally isolated itself from the rest of the flock. Eventually the flock settled on the sea off Hurst Castle and was not seen again.*

Surf Scoters: female, Langstone Harbour, December 18th 2006 and male, Stokes Bay, February 25th 2015. Jason Crook and Andy Johnson

The next was a first-winter seen by Mike and Pete Combridge and Phil Toye in flight with a small group of Common Scoters off Hurst Beach at 1050 hrs on November 17th 1996. Wiseman (2017) drew attention to this record which had not previously been published; it was subsequently accepted by HOSRP following the receipt of notes from Pete Combridge. A male was seen by Tim Parminter moving east with two Common Scoters off Hurst Beach at 0834 hrs on April 23rd 1999. The same birds were seen by Richard Levett and Sue Morrison as they landed briefly on the sea off Hill Head before leaving east at 0908 hrs.

Mike Wearing discovered a first-winter female in Broom Channel, Langstone Harbour on December 18th 2006 where it was present for several hours, allowing others to see it, but it was gone the next day. Marc Moody found a female or first-winter with Eiders off Hurst Beach on November 19th 2011 which left west and was not relocated. Dave Stevenson discovered a male with two Common Scoters in Stokes Bay on February 25th 2015 where it remained until March 4th. This was the first to stay for longer than a day and enabled most Hampshire birders to catch up with it.

Surf Scoter, male and Common Seal, Stokes Bay, February 26th 2015

GB status: 49, 1837–1949; 673, 1950–2019. Stable, mean 16 per year in 2010–19. (Breeds coastal Alaska from Kotzebue Sound to Alaska Peninsula, & Canada from Mackenzie River delta S to S Yukon, NE British Columbia, C Alberta, N Saskatchewan, N Manitoba, & E to C Quebec & Labrador. Winters on coast from S Alaska S to Baja California, Mexico, and Canadian Maritime Provinces S to Texas Gulf coast, USA.)

Smew *Mergellus albellus*

HOSRP

A rare and declining winter visitor. At least eight old records and around 660 birds in the modern era.

The first for the county appears to be a female shot by Col. Peter Hawker at Keyhaven on January 15th 1830. In his diary Col. Hawker stated, 'Made a long flying shot with a musket at a lough-diver or female smew, a bird I have long wanted for my collection. The first I ever saw, or heard of, on our coast.' K & M recorded four specimens in the Hart collection, which may all have been obtained in present-day Dorset, and one killed at Breamore in November 1855. Three adult males on display in the Science School at Winchester College were obtained at Totton in 1895.

Smew, male

Records for the first half of the 20th century are very sparse: one at Stockbridge on February 1st 1939, in the severe winter of 1947 two redheads at Keyhaven on February 5th, a female shot at Fordingbridge on February 14th and an undated record from Ibsley, a male at Keyhaven on March 6th 1949 and another male there on December 29th 1950. Subsequently, there have been records in 56 winters involving at least 660 birds.

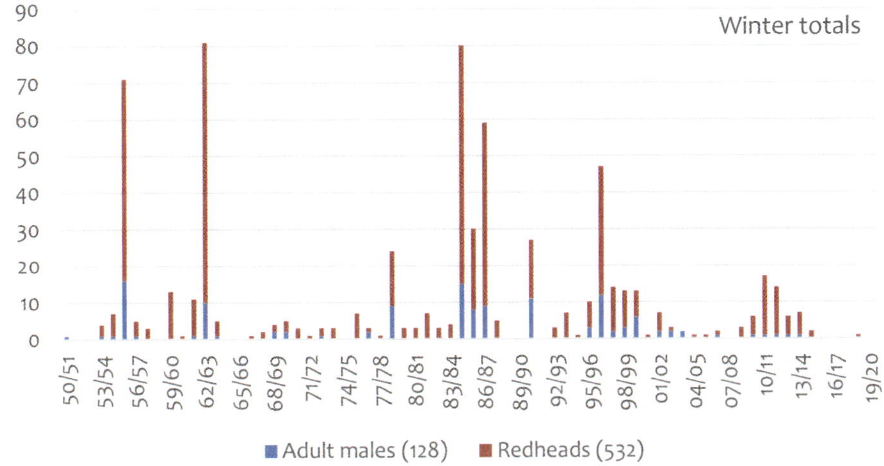

The highest numbers of Smew occur during severe winters. At the onset of harsh conditions, a sudden influx may occur, with ones, twos and small parties occurring on unfrozen lakes and gravel pits, rivers and in sheltered estuaries and harbours. Especially at the start of an influx, many records refer to birds staying for only a few hours or flying over. However, once settled, they may remain in the same general area for several weeks until a thaw takes place. The return of mild weather usually results in their disappearance, equally as rapidly as their arrival.

The true numbers present during the severe spells in early 1956 and 1963 were almost certainly greater than shown in the chart, as coverage, especially of inland localities, was considerably less thorough than today. The largest gathering seen in the county was in February 1956, when numbers at Titchfield Haven rose to a peak of 34, including nine males, recorded by Roy Dennis on 24th. In 1963, the highest counts were of eight flying east off Hurst Beach on January 6th, up to ten at Titchfield Haven from January 14th to March 16th, up to ten at Eling from January 19th to March 2nd and up to seven at Needs Ore and Sowley Pond from January 19th to March 9th.

In the remainder of the 20th century there were sizeable influxes in six winters with severe spells of weather. Numbers have been much reduced during the mostly mild winters of the 21st century with totals exceeding ten in only two winters and none recorded in five. Maxima at the most frequently used sites in eight winters with the highest numbers since 1978–79 are summarised below.

	7879	84/85	85/86	86/87	90/91	96/97	10/11	11/12
Langstone Harbour/Sinah GP		1	2	1	1	2	1	
IBM Lake/Paulsgrove Reclamation		5	2	5	6	1		
Titchfield Haven	2	6	2	3		1	2	
Eling/Broadlands Lake	5	6	6	4	8	3	1	7
Sowley Pond/Needs Ore	2	3	3	5		3		
Lymington/Hurst		5	1	6		1	2	
Blashford Lakes	3	10	8	6	5	5	3	3
Allington GP/Bishopstoke /Itchen Valley CP	5	6	1		7	7		
Eversley/Yateley GPs	1	1	1	2	7	6		1

The highest individual counts in recent years were of seven (one male) on the River Itchen at Itchen Valley Country Park on January 11th 1997, seven (two first-winter males) at Normandy Lagoon on December 1st 1998, seven redheads on the River Avon at Sopley on December 19th 1999, six (one male) at Spinnaker Lake, Blashford from January 4th to 6th 2002 and seven redheads at Testwood Lakes on February 5th 2012. The only record in the last five winters was of a redhead off West Hayling in Langstone Harbour on January 29th 2019.

Comparison of the cumulative monthly totals for two periods (see opposite) shows greater numbers and a later arrival in 1950/51–1991/92 which is the result of cold weather influxes in January and February. In 1992/93–2019/20 there were more in December and January than in February which probably reflects the earlier incidence of cold spells during that period.

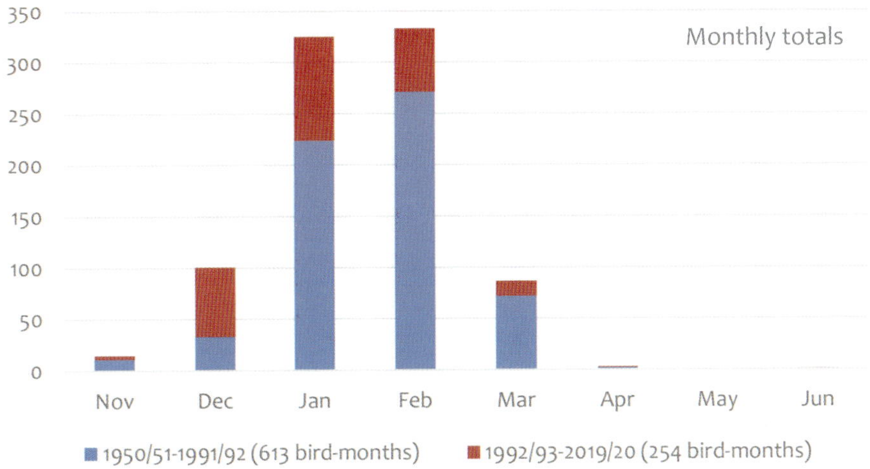

The earliest recorded were two at Paulsgrove Reclamation on November 8th 1982. One or two were present at this site or nearby IBM Lake in November in five consecutive years from 1981 to 1985 and subsequently wintered; it seems likely that the same individuals were involved each year. Otherwise, the earliest were at Langstone Harbour on November 17th 1990 and November 19th 1999. Departure is usually complete by mid-March with the latest at IBM Lake on March 21st 1981 apart from redheads at Alresford Pond on April 10th 1963 and flying east with Common Scoters off Hurst Beach on April 8th 1979 and a first-winter male which stayed at Wellington Country Park until April 1st 2013.

Smew, female, The Vyne, February 12th 2012. Barry Stalker

There has been one summer record: Nigel Jones saw a female fly in off the sea and up the Beaulieu River on June 13th 2009. He later photographed it on Blackwater, Needs Ore where it was in view for an hour but not seen again.

GB status: scarce and declining winter visitor. (Breeds taiga zone, Scandinavia to E Russia, Winters Europe and China/Japan.)

Black Grouse *Lyrurus tetrix*

Formerly resident but died out by the 1930s.

Gilbert White knew of Black Grouse in Woolmer Forest. In one of his early letters to Thomas Pennant, written prior to 1767, he wrote:

> *'But there was a nobler species of game in this forest, now extinct, which I have heard old people say abounded so much before shooting flying birds became so common, and that was the heath-cock or black-grouse. When I was a little boy, I recollect one coming now and then to my father's table. The last pack remembered was killed about 35 years ago, and within these ten years one solitary greyhen was sprung by some beagles in beating for a hare.'*

Gilpin, writing in 1834, described how the Black Grouse was formerly found in great abundance in the New Forest, but had become much scarcer. This decline, which was traced in some detail by Tubbs (1968), continued through the 19th century. Considerable numbers were killed by sporting licensees and keepers, and this certainly hastened the decrease of the species, perhaps significantly. By 1905, it had 'not quite died out', but the surviving population included stock introduced by Lascelles, the Deputy Surveyor at the time. The species hung on until the 1930s; the last record appears to be of a greyhen seen at Fernycroft on October 26th 1936 (J. H. Branford *per* Colin Tubbs).

Elsewhere in the county, it was reintroduced to Woolmer Forest after White's time and also occurred in the Aldershot and Yateley areas but had disappeared by 1899 (K & M). A greyhen obtained at Crawley in 1882 formed part of the Winchester College collection and is displayed in the Science School there, along with a pair obtained in the New Forest in 1880.

GB status: Resident uplands of N Wales, N England and Scotland, formerly more widespread. (Northern Eurasia.)

White-throated Needletail *Hirundapus caudacutus*

BBRC

A very rare vagrant. One record in 1879.

The only record for Hampshire is of one shot near Ringwood on July 26th or 27th 1879. The specimen was retained by George Corbin and bequeathed to the Natural History Museum on his death in 1914. It is held at Tring with accession number NHMUK 1914.3.17.1 (Dr. Alex Bond, Senior Curator in charge of birds, *in litt.*).

The record was fully endorsed in a review of records of category B species undertaken by HOSRP (Combridge & Clark 2010). An extract from the panel's report is reproduced opposite.

The first published mention of this record appears to be in the Proceedings of the Zoological Society for 1880 (No 1, January), which stated: 'Prof. Newton... drew the attention of the meeting to a specimen of... [Needle-tailed Swift], which had been intrusted [sic] to him... by Mr. G. B. Corbin, of Ringwood, near which place it had been shot on the 26th or 27th of July last [1879]. About the middle of that month Mr. Corbin saw one evening two strange birds flying over the river Avon in company with Swifts, and in the course of the following week had better opportunities of observing at least one of them. A few days after, Mr. Corbin held in his hand the bird now exhibited, which had been shot in the meantime, and was, he had no doubt, one of those he had previously watched.'

This version of events was repeated in essence by K & M, who added that 'The specimen is still in Mr. Corbin's possession.' Corbin's own account, published in The Zoologist in March 1880, differs in certain respects from that given in the Zoological Society's Proceedings. Corbin reported that 'The unusual abundance of Swifts... during the summer of 1879 has been the subject of remark in these pages, and in this neighbourhood they were extremely numerous... on 27th July I saw two strange birds flying about... in company with Swifts... they appeared larger than the rest... and their flight... was somewhat different.' Corbin looked unsuccessfully for them the next day, but 'In the course of the week' saw 'one so near that the white throat, the white under the tail, as well as the pale patch on the back were plainly visible.' A 'few days later' he was brought the corpse of a Needle-tailed Swift, 'killed about 26th or 27th July... about two miles from where I first observed it... the man who shot it did not see another like it'.

Corbin's account implies that he did not identify the 'two strange birds' on 27th July, and it is unclear as to when he identified, or how often he observed, the lone Needle-tail during 'the following week'. It is also curious that by the shooter's uncertain reckoning the specimen may have been obtained a day before Corbin's first observation. It is a pity that the editor of The Zoologist did not query these points and clarify the issue for posterity.

These anomalies may seem suspicious to some and so it is worth reviewing what little we know of Corbin. Research by Alan Cox has revealed that George Bentley Corbin (1841–1914) was a life-long resident of Ringwood and a cabinet maker and upholsterer by trade, while K & M wrote that he was a long-time contributor to The Zoologist who had 'given us much valuable information'. Perusal of K & M does not reveal Corbin to have been connected with an unusual number of rare birds, and thus fraud on his part seems unlikely, especially as he retained the specimen and therefore did not gain financially. If Corbin was intent on perpetrating fraud, he would surely have been at pains to place the earliest possible date of the Needle-tail's shooting after, not before, his initial sighting. It is perhaps more likely that the circumstances were not correctly reported in the Zoological Society's Proceedings, and that Corbin failed to express himself clearly in The Zoologist but simply (and honestly) added the shooter's uncertain dating without proper explanation, never considering the record might be queried in future. Perhaps due to the confused nature of Corbin's account, Hartert et al. (1912) accepted just the specimen, noting that it was 'said to have been seen with another'; this decision was followed by Witherby et al. (1938–41).

The panel (which was not aware that the specimen still existed) considered that it would have been correctly identified, as it was examined by Professor Alfred Newton (then Britain's leading ornithologist) and publicly exhibited; that there is no evidence of sharp practice on Corbin's part; and that if the unnamed shooter was committing fraud his timing was remarkably fortunate, given Corbin's sightings and the influx of Swifts. The balance of opinion within the panel was to follow the decision taken by Hartert et al. (1912), though a minority of members voted for the acceptance of two birds.

The Hampshire bird is the second ever for Britain, following the first obtained near Colchester, Essex on July 8th 1846.

GB status: 2, 1846-1949; 9, 1950–2017. Very rare. (Nominate race breeds W Siberia from region of Tomsk, E through S Siberia to Sakhalin, Kuril Islands, Japan and N China; winters Australia S to Tasmania. Race H. c. nudipes breeds Himalayas to N Yunnan, SW China, to Assam, NE India, perhaps also N Burma.)

Alpine Swift *Tachymarptis melba*

BBRC to 2005; HOSRP 2006 to date

A rare vagrant. Three old records and 13 since 1964.

The first for the county were two seen by James Salter flying over a pond a mile to the east of Basingstoke on April 28th 1886 (*The Field*, May 15th 1886). The description lacks plumage details but they were evidently seen at close range and the record was accepted by K & M. In the early 20th century, one was seen by R. S. Mitford and E. A. Waterhouse at Brockenhurst on May 28th 1911. The description given in *The Zoologist* (1911: 278) is brief but they also saw the bird at close range. The next was seen by Peter Scott and Terry Jones at Leckford on July 18th 1946. Peter's note in *British Birds* (40: 21) is reproduced below.

> *At about 6.30 p.m. on July 18th, 1946, Mr. Terry Jones and I were looking at the ornamental waterfowl on the water meadows of the Test on Mr. Spedan Lewis's estate at Leckford, near Stockbridge, when we observed amongst a number of Common Swifts which were hawking up and down the river, a substantially larger bird with a white belly. During the next half hour this bird frequently passed within 10 yards of us in excellent light, so that its identification as an Alpine Swift (Apus m. melba) was certain. The most noticeable features of the bird were its large size (about one third as large again as the common Swifts), its stronger and more purposeful flight with much slower wing beats, and its brilliant white belly. When the bird banked towards us, the tawny colour of the back, much paler than that of the Common Swift, was observed. The white chin was not at all prominent and was only seen when the bird was at its very closest. For the half hour that it was under observation the bird was apparently entirely silent. It was feeding in the same way as the 50 or 60 Common Swifts in the same area. The Handbook gives 'about 40' records for the Alpine Swift in England.*

One claimed by Richard Meinertzhagen at Mottisfont on September 9th and 10th 1894 is not considered acceptable (Naylor 2021).

The first modern record was of one seen by Robert Cheke, a student at Winchester College, flying over Winnall Moors on April 30th 1964. Extracts from Robert's description are reproduced below.

> *I spent the afternoon of Thursday 30 April 1964 birdwatching next to and beyond Winnall Lake where, using 8 x 30 binoculars, I had seen a Grasshopper Warbler and two Yellow Wagtails. I was at the far end of the area with public access and began to walk back to the entrance when my attention was drawn to a group of Common Swifts about 50-60 metres up in the sky, among which I was astonished to see an Alpine Swift, which I immediately recognised given my experience with the species the year before. The bird was clearly larger than the Common Swifts by a factor of about 10-15%, with very much longer wings and darker underwings and an extensive clear white belly. No other details were discernible given the speed of the bird's movements and the short time that I could see it. The bird flew from west to east towards me and directly over me, before turning back to the west and disappearing due north and I did not see it again.*

The bird was seen again nearby over the river below Easton Down by Air Marshal Sir Robert Saundby on May 7th who in his letter to HFC describing the bird also mentioned that his wife Lady Saundby saw it on May 4th. Robert's description was submitted to and accepted by BBRC in 2020.

Subsequently, a further 12 have been recorded.

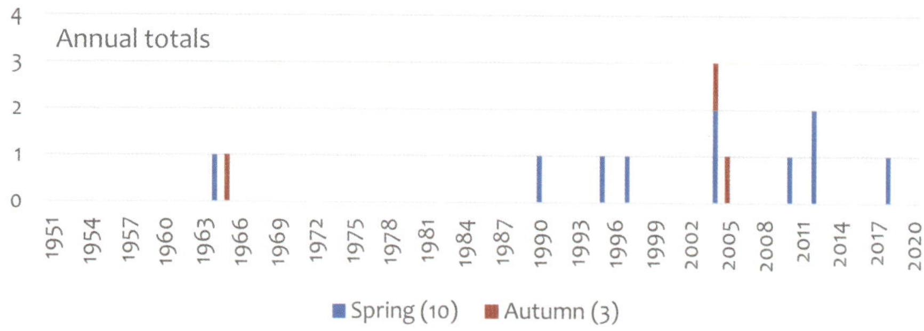

The second was seen by Barry Duffin at Gilkicker Point on October 4th 1965. Extracts from Barry's description are reproduced below.

> *At 0915 GMT I observed a large swift descending from a great height towards the Solent from a northerly direction. For the first minute or so the bird appeared to be a fast-flying raptor-like bird, but as it descended I could see that it was of a typical swift shape and that the underparts were white. The flight was strong and direct with rapid wingbeats, and on two occasions it gave a brief glide. The white belly was very conspicuous and so too was the brown breast band. The throat was less conspicuous and appeared a creamy buff colour, suffusing into the brown of the breast. It appeared considerably larger than the Common Swift and seemed to dwarf a Swallow. The bird was viewed through 10x50 binoculars for approximately three and a half minutes.*

Another in March 1990 was part of an unusually early influx of at least six into southern England. Colin Savage found one trapped in an office at Titchfield on March 26th which was ringed and released later the same day at nearby Titchfield Haven. The next was seen by Dave Treacher with Common Swifts over Peel Common Sewage Treatment Works at 0900 hrs on June 4th 1995 and relocated by Barry Duffin along the eastern edge of Titchfield Haven at 1425 hrs later that day. Dave was fortunate to find another feeding on insects above rotting vegetation at low tide at Hook-with-Warsash at 1810 hrs on May 26th 1997. Unfortunately, it departed north-east before others could see it.

The next were not until 2004 when three were recorded. The first was reported over the M3 adjacent to Winnall Moors on the evening of March 17th and then relocated nearby at Abbots Worthy by Graham and Debbie Barrett at 1000 hrs the next morning. It was apparently seen later that day 6 km away at Sparsholt and then back at Abbots Worthy daily until March 22nd enabling many observers to connect with it. Its presence coincided with the HOS Annual General Meeting on March 20th and news of its reappearance precipitated a rapid exodus from the meeting! Later that spring one was seen by Pete Durnell moving south-east over Normandy Lagoon at 1830 hrs on

May 3rd. The third was found by Kevin Crisp over Farlington Marshes at 1545 hrs on October 12th. It was also seen by Jason Crook but was lost to the north in poor visibility a few minutes later. A year later one was seen by Graham Osborne at very close range over Brownwich Cliffs at 0925 hrs on October 30th. What was presumably the Brownwich bird was seen by nine-year-old Fergus Scutt as it flew east over Sandy Point shortly afterwards but his notes and sketch were unfortunately not sufficiently detailed to submit the record.

Five years elapsed until the next was found by Alan Bines at Fleet Little Pond on March 31st 2010 while he was dining at the adjacent *Heron on the Lake* pub. He saw it intermittently between 1245 and 1445 hrs as it fed low over the water with Swallows during heavy showers. He put the news out but it was not seen again that day. Fortunately, it was relocated over Fleet Pond on April 3rd when it showed well from mid to late afternoon and was seen by many observers. The next was one watched for five minutes by Peter Milinets-Raby as it soared high over his garden in Havant with two Hobbies at 1330 hrs on March 29th 2012. He had returned home for a five-minute coffee break! Later that year Alan Dawson saw one in flight over the reedy ditch at Needs Ore at 1145 hrs on June 30th. Brian Cozens managed to get his binoculars on to it and watched it fly off east over Needs Ore Point. Finally, Greg Pullin watched one for a few seconds as it flew rapidly north-east over Compton Memorial Playing Fields near Winchester at 1435 hrs on May 1st 2018.

Alpine Swifts: Fleet Pond, April 3rd 2010 and Abbots Worthy, March 20th 2004. Ian Wiiliamson and Adrian Webb

As shown on the next page, analysis of arrival dates shows ten in spring between March 17th and June 30th and three in autumn between October 4th and 30th.

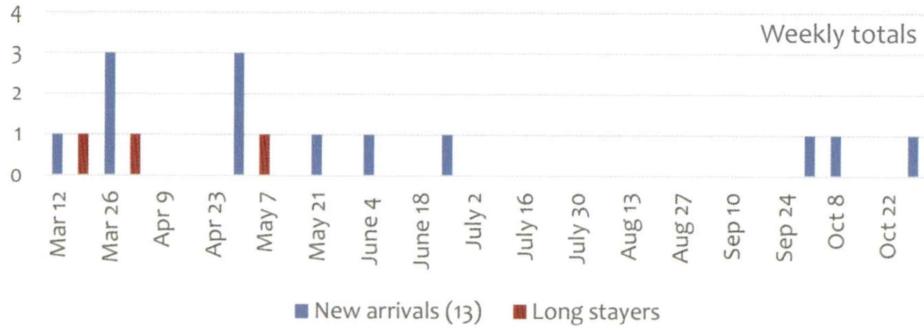

Several other records are probably acceptable. One seen with Common Swifts by Joan Bennett and Marion Campbell over Toyd Down on June 30th 1988 was considered not proven by BBRC. Their notes are very brief but they were not inexperienced observers and this was probably a genuine record. During the influx in March 1990 singles over New Milton on 19th and nearby Bransgore on 20th were not documented and Mike Wearing saw one over Lymington on 21st which was harshly considered not proven by BBRC.

GB status: 81, 1830–1949; 695, 1950–2019. Stable, mean 16 per year in 2010–19. (Breeds discontinuously from NW Africa, throughout S Europe, N to C France & Switzerland, to Ukraine. To E, breeds locally through Turkey & Caucasus to Iran, Afghanistan & N Pakistan. Winter range unknown but assumed to lie within Afrotropics or W India.)

Pallid Swift *Apus pallidus*

BBRC

A very rare vagrant. Four records since 1983.

The first for the county was seen by Andy Hanby flying over Farlington Marshes with over a hundred Common Swifts between 1755 and 1815 hrs on May 20th 1983. He left the site to phone other observers but it had gone, along with most of the swifts, on his return. This was only the second for Britain. Extracts from Andy's description are reproduced below.

> *Whilst looking through the 100–200 swifts gathered over the lake at Farlington, I noticed a strikingly paler swift which looked exactly similar to birds I had seen in Israel one month before. I was able to watch the bird from above and below using my binoculars and Kowa scope mounted on a rifle grip from the sea wall/dyke with the sun shining on it from behind. It came as close as 20 feet at times.*
>
> *Overall it was a sandy brown colour approaching Sand Martin in tone though perhaps a shade buffer (no Sand Martins present). There was a marked dark brow through the eye and a very large silky throat patch. The wings were pale but offset by darker primaries and a darker body. The wing coverts, particularly the upper, appeared mottled, due it seemed to pale edgings. The abdomen appeared to have some scalloping. The forked tail was paler than the body with apparently blunter tips. The jizz was different to Common Swift with long glides over the lake.*

Due to panic, unfortunately my description is not fuller and I was unable to pay full attention to body structure and to whether the white throat patch continued above the eye. I am 100% convinced this is a Pallid Swift, in plumage and colour the same as the Stodmarsh (Kent) bird in May 1978 although possessing a more obvious throat patch. It is this similarity that to me rules out pekinensis Common Swift (though I have no experience of that race).

The second record was of two, or possibly three, seen by Jason Crook, Kevin Crisp and Bob Chapman over Bedhampton from 1630 to 1730 hrs on October 12th 2004. Jason and Kevin were attempting to relocate the Alpine Swift they had seen earlier over Farlington Marshes when they saw a small swift drifting east to the north of the reserve. They then met Bob by chance and agreed to drive to Budds Mound. Once there, they got good views of two of a group of three swifts which confirmed that they were Pallids. Eventually the swifts moved purposefully away as a group of three high to the east/south-east. They scanned around further and quickly located another much darker swift which was obviously 'just' a Common Swift, clearly lacking any of the contrasting pale plumage features of the previous birds. It was joined by a second Common Swift after which both were soon lost to the north-west.

Pallid Swift, Posbrook Floods, Titchfield Haven, October 22nd 2020. Mark Francis

Most recently, Mark Francis saw and photographed one feeding with Swallows over Posbrook Floods between 1655 and 1725 hrs on October 22nd 2020 which was also seen by four other observers. It was last seen flying away to the south-east and was lost behind trees. The next day Philip Fawkes and Nigel Jones were on the sea wall at Pennington Marshes when Philip picked up a distant swift over Oxey Marsh at 1150 hrs. Nigel located it in his telescope and was confident it was a Pallid. They ran for five minutes to Oxey Marsh and were able to watch it for a further ten minutes at distances down to 80 metres before it disappeared ahead of a light shower which came in from the west. Despite searching with other observers for the next hour, there was no further sign. BBRC considered that both records referred to the same individual.

GB status: 132, 1978–2020. About four per year; stable. (Breeds throughout Mediterranean basin from Iberia to Greece but absent from many regions. Outside Europe, breeds locally from Mauritania & Canary Islands across NW Africa & Middle East to Arabian Peninsula & coastal S Iran. Most winter N African tropics, but some remain S Europe.)

Little Swift *Apus affinis*

BBRC

A very rare vagrant. One record in 2000.

The only record for Hampshire is of one found by Jonathan Stokes feeding with large numbers of Swifts over Bishop's Dyke, New Forest on June 4th 2000. Extracts from Jonathan's account of the events that day from Clark (2002) are reproduced below.

> *Paul Stanbury and I were leading a Naturetrek tour at Beaulieu Road. The group started to disperse and I walked down to a boggy area. About 30 feet overhead there were many swifts and I immediately picked up one that was 'odd'. It had a square tail and the wings looked slightly longer than the body. My immediate reaction was Chimney Swift, a species I recently had seen in Canada. I shouted to Paul to look at the bird. His immediate reaction was, 'That's no Chimney Swift, look at the rump!'. At that moment, the bird turned and a clear white rump was visible. 'It's a Little Swift'. The bird circled overhead for about three minutes and then disappeared. We ran over to the group of 50 birders watching a Honey-buzzard. There then followed an agonising five minutes when there was no sign of the bird at all. People obviously thought we were 'stringing', and it began to feel very uncomfortable. Then I picked it out flying low across the heath towards the group from exactly the opposite direction to where everyone was looking. 'Little Swift', I shouted, and to a man they swung on to the bird. This time the view was excellent, and from the side, clearly showing the white rump. It passed over the group at about 30 feet and was gone. Despite suggestions that the bird was seen again that afternoon, I'm not sure it was refound, and so the 50 or so of us were probably the only people to see it.*

Little Swift, Shatterford Bottom, Bishop's Dyke, June 4th 2000. Stood out among the hundreds of swifts and hirundines, hawking insects over cotton grass.

Ian Watts and Peter Hooper saw one feeding over Southampton University campus between 1700 and 1710 hrs on February 6th 2004 before apparently roosting on the Chemistry Building. It could not be relocated the following morning. A short extract from Ian's notes is reproduced below.

Size of House Martin – flight quicker and more bat-like. No notch in tail which was short and square but more fan-like when spread. Broad white rump. Body short, wings with long primaries, overall dusky brown. Small head with small bill. Saw touch of white above bill but didn't note chin patch. Kept flying up to building to snatch insects off warm surface. Presumably roosted there but not found in morning. This bird arrived on the same southerly winds which brought a sprinkling of House Martins to the south.

Despite the detailed notes of two observers the record was harshly considered not proven by BBRC. It occurred during unprecedentedly warm weather and coincided with an influx of at least 35 House Martins into southern England between February 1st and 10th including up to six together at Bembridge, IOW throughout the period.

Claims of another seen briefly over the scrapes at Titchfield Haven on September 17th and 18th 2015 were considered not proven by BBRC. On the first day Mark Barrett found the bird at 1410 hrs and reported the following:

A chunky all dark swift with a striking white rump and a short blunt tail and blunt wings in actively fluttering flight flying at a height of 10 feet and briefly seen when it flew across the front of the Spurgin Hide. It flew in a direct line. The white rump was large and obvious and the bird was twice the size of a House Martin. The view was too brief to notice whether the white rump wrapped around the tail and I was not able to see the throat. The bird was seen from 50 metres in bright sunshine and there were no other hirundines in the vicinity. The bird was in view for no more than a couple of seconds and reported to staff at the information centre.

The next day David Wallace and Brian Goddard arrived at the Meon Shore Hide at 0930 hrs. David's notes were as follows:

We concentrated on scanning the few hirundines flying over the South Scrape. About 15 minutes later about 40 or so House Martins and a few Swallows started to feed low over the scrape in front of us. Brian then drew my attention to a bird in with the House Martins which looked different from the others. I locked onto it instantly and we both came to the rapid conclusion that we were looking at a small Swift. It was associating with the House Martins and flying very low back and forth across the scrape about 40 yards away. Because the hide is raised, we were viewing the bird from a slightly elevated position and therefore looking down on it. It was perhaps slightly shorter than a House Martin but appeared long-winged with a broad white rump patch. It was also surprisingly easy to follow (but not to photograph!) as its wings were scythe-shaped like other Swifts and it did not glide on straight wings like the accompanying House Martins. Its upperparts looked dark-brown rather than black and I didn't notice any forking in the tail. I did not hear it call. After about 30 seconds it was unfortunately lost to view and was not refound. One other observer was present, but he was trying unsuccessfully to obtain a photo of the Swift and only viewed the bird with the naked eye. However, Brian and I are happy that the bird was a Little Swift.

Given the brief views obtained and the size discrepancy noted on the first day it is perhaps not surprising that BBRC considered it not proven.

GB status: 26, 1981–2020. Rare, about one every 1–2 years. (In WP breeds locally in S Spain & Morocco. Widespread throughout Middle East from Israel to SE Iran & N along Euphrates River to SE Turkey. Largely resident, but some Middle East populations migratory. Elsewhere, resident or dispersive throughout sub-Saharan Africa & Indian subcontinent to Sri Lanka.)

Great Bustard *Otis tarda*

Formerly resident, last recorded in 1910. Recent sightings are from the Salisbury Plain reintroduction scheme.

Rosemary Powell

The earliest record for Hampshire is described by Chafin (1818) who records how in November 1751, while living at Wallop, he managed one day to kill five dotterels (*Lapwings?*) out of a flock of twelve when out shooting on the downs near Winterslow Hut, Wiltshire. The report of his gun scared a flock of no less than 25 bustards, which flew over the hill called Southern Hill. Having returned to his inn, loaded his gun with swanshot and put a new flint in it, he mounted his horse and went in pursuit of his noble quarry. He went on, 'When I had come to the brow of the hill over which I had seen them pass, I rode very cautiously, looking carefully before me, when on a sudden I espied them nearly within shot, and they espied me also and rose for flight. The motion of the wings frightened my horse, he started back, threw me down, and ran way.' Chafin adds that when he got up, he had a shot at the birds.

Waters and Waters (2005), reviewing the former status of the Great Bustard in Britain, opined that it was uncertain that Chafin's birds had crossed into Hampshire. However, James Ferguson-Lees, writing in *Birds of Wiltshire* (WOS 2007), stated that Southern Hill was probably Suddern Hill, just west of Middle Wallop and well inside Hampshire.

Gilbert White tells of a day spent on the downs between Andover and Winchester on November 17th 1782. While being entertained to lunch at a remote farm, he was told that a flock of 18 bustards had been seen there in about 1770, but since then only two. White did not include it in the list of birds he saw at Selborne up to his death in 1793. His brother, the Rev. Henry White, in his diary for October 10th 1780, recorded 18 on a trip to Stonehenge and Wilton but did not mention on which side of the Hampshire/Wiltshire border he saw the birds (Dewar 1899).

K & M mentioned one recorded in the *Hampshire Chronicle* of March 9th 1795 which was 'lately shot at Froyle' and only two somewhat vague records for the first half of the 19th century. It seems likely that breeding had ceased well before that time. J. E. Harting, writing in *The Zoologist* for March 1891, details the occurrence of seven females procured in Britain between December 9th 1890 and February 5th 1891 including one shot by Mr. J. Bourne near Romsey on January 2nd 1891. Mr. Bourne reported his in *The Field* for January 10th 1891.

The only record for the 20th century was one reported by Philip Munn in 1910 (*Brit. Birds* 4: 190) as follows:

> *A Great Bustard was shot on January 12th 1910 in a turnip-field on Jamaica Farm, St. Mary Bourne, Hants. The bird is an adult female and in good condition; the state of its plumage does not show any signs of its having been in captivity, nor was it at all tame, and I cannot hear that any have been kept in captivity in the neighbourhood. The weather in England at the time it was obtained was very unsettled and changeable, with heavy gales from the S.W., W. and N.W. and snow, hail and rain; and possibly severe weather may have driven it from its usual haunts.*

In 2004 the Great Bustard Group started its reintroduction project on Salisbury Plain. There have been several records of birds from this project in Hampshire: a female at Harbridge from January 7th to 16th 2009, a male at Sparsholt on May 13th 2009, a female at Harbridge on December 30th 2010 (red tag 28), a first-winter male at Barton Stacey from November 29th 2012 to March 18th 2013, two at Hannington Farm on April 5th and 6th 2018 and singles at Holmsley on June 13th 2018, Vidle Van Farm, Keyhaven from November 16th to December 30th 2019 and intermittently between February 12th and March 21st 2020 and Toyd Down between January 4th and February 22nd 2020.

GB status: 64, 1820–1949; 21, 1950–1987. (C and S Eurasia.)

Little Bustard *Tetrax tetrax*

BBRC

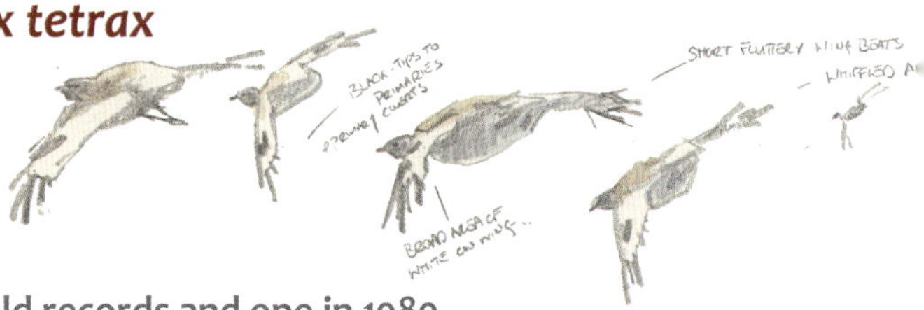

A very rare vagrant. Six old records and one in 1989.

The first for the county was a female shot by Lord Palmerston at Broadlands in September 1810. It was in a turnip field; it rose and pitched again, flying heavily, and had a strong and most remarkably unpleasant smell when killed. K & M reported that it was in the Earl of Malmesbury's collection at Heron Court. The next was a female shot on January 4th 1873 in a turnip field on Mr. Twitchin's farm near Whitchurch and afterwards presented to the Natural History Museum. This was confirmed by Henry Reeks in *The Zoologist* for April 1873. Another was shot on January 9th 1879 near Portsmouth Harbour between Fareham and Gosport by Mr. M. Dexter, who reported the occurrence in *The Field* for March 25th 1879. K & M recorded that in 1897 a Mr. Metcalf, of Roche Court, Salisbury, flushed a bird, which the Rev. A. Morres thinks must certainly have been a Little Bustard, out of a turnip field about two miles south of Grately Station; it passed so close that there was no difficulty in seeing its colour and markings distinctly. No date was given, but this is clearly the same bird as reported between Roche Court and Over Wallop on September 27th in *The Field* for October 30th 1897 and on September 29th in the *Salisbury & Wiltshire Journal* for October 16th 1897 (WOS 2007).

In the first half of the 20th century, one was reported by K. J. Acton Davis as shot during a Partridge drive at The Manor Farm, Stockbridge on December 14th 1935 which was donated to the Natural History Museum (*Brit. Birds* 29: 252) and another was found by Maj. M. Portal in a turnip field at Baring's Farm, Avington in January 1944 (*HBR*). It was later seen at Avington Park and may have been the bird subsequently reported at Bighton and Chawton Park. The only modern record was of an immature male found by Alan Hayden at Coward's Marsh, Dorset at 1200 hrs on December 30th 1988. It remained in the area until at least January 5th 1989 and was seen in fields at North Ripley, on the Hampshire side of the county boundary, on January 1st.

Little Bustard, Sopley, a hangover-free dawn twitch on New Year's Day, January 1st 1989

Extracts from Alan's account of the events are reproduced below.

On Wednesday 30th December at 12 noon I arrived at Coward's Marsh, an area overlooking the River Avon north of Christchurch, that I study on a regular basis. The weather was very dull with light rain. The wind was south-west 2–3 and the visibility moderate. Whilst wiping the mist off the lenses of my bins, I was looking over the marsh with the naked eye. I saw what I first thought was a Shelduck coming in to settle on some pools. This Shelduck impression was given by the large amount of white on the wing, and its mode of flight suggesting that its wings were 'set' prior to landing. I had by this time focused my bins on the bird. I was mystified for a few seconds because it seemed to be a cross between a duck and a game bird in shape, having an extended head and neck, rounded 'drooping' wings and short, squarish tail. Plumage-wise, it had an ochre-coloured head, neck, body, tail and inner forewing, a dazzlingly conspicuous broad white wing flash over the secondaries and primaries, and jet-black wing tips. It suddenly dawned on me that I was looking at a Little Bustard, either in female or first-winter male plumage, a bird that I was familiar with in southern France a few years previously.

The bird flew several times in front of me for a few minutes, and eventually settled into cover on the opposite side of the river. Despite an extensive search for the rest of the day by three of us, in atrocious conditions, it was not relocated. It was rediscovered the following morning and later that day it flew to the east and settled on large arable fields near the village of Burton. On January 1st, it was seen on the ground and flying between a belt of similar fields extending approximately five miles northwards into Hampshire. It was seen by 1,000–2,000 twitchers across a wide area (5 miles square) over the next few days and was last seen on the afternoon of January 5th, close to the point of discovery.

News of the sighting including video of the bird in flight was broadcast on BBC *South Today* and can be found here www.youtube.com/watch?v=tv_yffOco2E

GB status: 171, 1751–1949, 29, 1950–2019. Very rare, last in 2019. (Locally numerous on steppe grasslands of Iberian Peninsula with smaller populations S & C France, Italy, Sardinia & Ukraine. French populations migratory, wintering S France & Iberia, but Iberian breeders largely resident or disperse locally. To E, breeds widely S Russia to E Kazakhstan & winters SW Caspian Sea.)

Great Spotted Cuckoo *Clamator glandarius*

BBRC

A very rare vagrant. One record in 2000.

The only record for Hampshire is of a first-summer found by two unknown observers at Keyhaven Marshes at 1130 hrs on April 2nd 2000 (Clark 2002). They did not know what it was but on showing a sketch of it to Glyn Horacek-Davis he suspected its identity. This was confirmed by reference to a field guide. Glyn and his brother Jonathan rediscovered the bird at 1230 hrs and it was subsequently seen by Russell Wynn, Tim Parminter and others before disappearing at 1340 hrs, much to the frustration of the rapidly assembling crowd. There were no further reports until a claim of it in Lower Pennington Lane at dusk on April 8th. Keith Betton and John Eyre were on site at first light the next day and quickly relocated the bird. It performed well for hundreds of twitchers during the rest of the day and remained in the area until April 17th. It was the only Great Spotted Cuckoo recorded in Britain in 2000. It seems likely that it arrived in a strong S/SSW airflow from western Iberia to southern England on the night of April 1st and 2nd. This was a brief weather window in between periods of northerly winds.

Great Spotted Cuckoo, Lower Pennington Lane, April 9th 2000. Photos opposite Keith Stone

GB status: 3, 1870–1949; 51, 1950–2019. Rare, about one per year; stable. (Breeds Spain, Portugal, S France & E to Greece. W Asian population uncommon, breeding discontinuously from C Turkey, Cyprus, Israel & Jordan to N Iraq & SW Iran. European breeders winter sub-Saharan Africa but range uncertain owing to presence of African populations.)

Yellow-billed Cuckoo *Coccyzus americanus*

BBRC

A very rare vagrant. One old record and three in 1973–85.

The first for the county was shot near Avon Castle on October 30th 1901. It was recorded by George Corbin of Ringwood in *The Zoologist* of December 1901. Extracts from his notes are reproduced below.

> *On October 26th, a gentleman told me he had, whilst standing on one of the bridges watching some pike-fishers, seen a strange bird settle upon and apparently scrutinize a bush at no very great distance from him. His description was that the bird was about the size of a thrush but seemed to have more the colour of a nightingale on its sides, and a very long tail. On October 30th, a specimen of the above-named cuckoo was shot about half a mile from the spot. I saw it soon after it was killed, and I may note the following particulars: 'Except where shot through the neck, the plumage appeared to be perfect, with no sign of abrasion of either wings or tail, as are seen in an 'escape', and the body was fat and well-conditioned, weighing just over 2½ oz; it measured 11¾ in. from beak to tail. Upper mandible and base and sides of the upper yellow, the rest of the beak black. Eyes dark brown; eyelids bright yellow. Back and two middle tail feathers dark mouse-colour, with a tinge of reddish, especially on the tail; four outer feathers on either side blackish with graduated dirty white markings, the longest being merely and indistinctly tipped, and the shortest and outermost white its entire length, at least on the outer web. The reddish tawny mark on the wing was large and conspicuous even when the wings were closed, but this may be a sexual characteristic, as on dissection it proved to be a male. The under parts from beak to tail were of a uniform pale grey, with a slight tinge of brown on the breast and sides. The legs were bluish lead-colour, with a silvery bloom on them, which latter soon faded; the claws were black, and it seemed to me the scales on the legs were remarkably large, as only five in number occupied the bare space. I should have mentioned, perhaps, that the beak was longer and more decurved than in the common species, and the inside of the beak, which is well-known to be bright orange yellow in C. canorus, was conspicuously spotted with black, especially on the lower part of the palate. The tongue also had black markings on it. The bird had been feeding freely upon the grubs of the sawfly, as the distended gizzard proved.*

The record was included by K & M but no information about the subsequent whereabouts of the specimen was given.

© Portsmouth Museums

The first modern record involved one discovered at Naish Holiday Village, Chewton Bunny by R. Jones on October 20th 1973 and found dead there by Martin Banks two days later. Subsequently, the skin was sent to Cumberland House Museum, Portsmouth by Mike Stewart. Although it was included in the 1973 *HBR*, the locality was cited as Highcliffe which passed into Dorset as part of county boundary changes in 1974; therefore, it was not included in Clark (1981) or the subsequent *BoH*. Ed Wiseman *(in litt.)* recently highlighted the omission hence its inclusion in the present volume.

Three years later one was found by Dave Bishop at Pennington Marshes on November 6th 1976. Extracts from Dave's description make gripping reading, even over 40 years after the event.

> *As I was about to go home at around 2.30 pm, I took one last look at Pennington Lagoon from the end of the track to the Jetty. As I stepped into a gap in the gorse bushes, I chanced to flush a bird I could not identify at first. It flew over the sea wall and I was surprised to see it had large rusty patches on its wings. It looked something like a large, long-tailed thrush. As it flew back across the sea wall again it looked more cuckoo-like and I started to wonder if it might be a Yellow-billed Cuckoo, a bird with which I was not familiar. It landed on the sea wall about 150 yards away. I hastened to the spot but before I reached it the bird was flushed by a passer-by and it flew into an isolated bush just behind the Butts Lagoon. I sat on the sea wall and after a few moments the bird showed itself again and I was able to identify it as a Yellow-billed Cuckoo. I then watched the bird for about an hour as it caught insects. The bush was approximately 70 yards away but I was able to see it well in good light through my Hertel Televari telescope. It acted rather like a shrike at times sitting high on the top of a gorse bush while at other times it sat motionless in the centre of the bush, frequently out of view.*

After some time, Dave found Phil Keane and his wife, who were able to get good views of the bird, but unfortunately no other birders were in the area before it disappeared just before dusk. It was not relocated the next day despite the efforts of many visiting birders. Dave commented in 2020:

> *This remains my most exciting UK birding experience, partly because I was young and inexperienced. I can still remember quite a lot after all these years – especially the first few seconds when I knew I had something amazing but not sure what. What is incredible now is that no birders (or anyone else) came by except the Keanes in what seemed like an eternity as I was desperate for someone else to see it. I've always been disappointed I couldn't contact Mike Stewart and Pete Combridge who I parted company with shortly before I found it – these days a mobile phone would have meant they and lots of other people could have seen it. The Keanes were carrying a Peterson guide which had a great picture of a Yellow-billed but only a two-line description in the appendix for Black-billed. How things have changed! The twitch the next day was one of the biggest at the time as I understand there had never been a twitchable Yellow-billed – probably over 50 birders including some from Norfolk!*

The final record was of one found freshly dead in the garden of Mr. C. Lloyd in Woolston late on October 17th 1985. It had not been there in the morning. The bird was subsequently identified and photographed by Dave Unsworth before it was sent to Cumberland House Museum. Unfortunately, the skin could not be located at the museum in 2021 (Christine Taylor *in litt.*).

GB status: 15, 1832–1949; 55, 1950–2020. Rare, about one every 1–2 years. (Breeds across S Canada from British Columbia to New Brunswick, & through USA S to C Mexico & Greater Antilles. Winters throughout South America to Argentina.)

Pallas's Sandgrouse *Syrrhaptes paradoxus*

BBRC

A rare vagrant not recorded since 1908. Twelve records of at least 105 birds.

There were six well-documented invasions into Britain in 1859, 1863, 1872, 1876, 1888 and 1908 but birds were recorded in Hampshire only in 1863, 1888 and 1908.

The first for the county were two, one of which was shot, near Aldershot on May 26th 1863. The account from *The Zoologist* is reproduced below. Subsequently, a further flock of nine was seen in the same area one of which was shot.

> *Pallas's Sandgrouse in Hampshire. – I have recently had the pleasure of examining, in the flesh, and dissecting a beautiful example of Pallas's Sandgrouse. On the 26th of May my brother was told by the major of his regiment that he had disturbed a brace of sand grouse that morning while riding on the heath, at no great distance from the camp at Aldershot. My brother at once went to the spot where the birds had been seen; they were still there but rose wild and out of shot. They were marked down, and this time my brother had a long shot and killed one of them; the other escaped. My brother at once despatched to me the bird he had killed. When I handled it, I was struck by the fitness of its specific name of 'paradoxus,' for to look upon this little sand grouse is something of a puzzle. The beak and the horseshoe markings on the breast remind one of a partridge; but the head and neck remind one of a dove ; the wings are like a swift's, or perhaps more closely resemble those of a shearwater, but have a peculiarity of their own in the pinnated first primaries; the pinnated tail feathers put one in mind of the mallard pintail ; while the feet, with the exception of wanting a hind spur, resemble those of a grouse. The plumage (dissection proved the bird a female) is of a uniform isabelline colour, prettily barred with umber-brown; on either side of the head, and extending a little beyond the ear-coverts, is a patch of beautiful orange-yellow: the horseshoe marking on the breast is of a very dark chestnut, almost black. The bird was extremely plump, and so fat that it was not easy to skin it. The crop was full to distention of the seeds of a species of grass which grows commonly on heaths; these, a few small pebbles, and one or two tops of grass-blades, were all I could discover. There were no traces of insect remains. I was much struck by the great depth of the keel of the sternum*

in proportion to the other bones of the bird; I did not measure it but am confident that it must exceed an inch or even an inch and an eighth in depth. The shape of the wings, their very powerful muscles, and the depth of the sternum, would indicate that the bird is possessed of great powers of flight. My brother noticed the great rapidity with which the birds he saw flew. In the bird I dissected I found eggs larger than peas in the ovary. There seems no reason why this pretty species should not be naturalized in this country. From several notices which have recently appeared in the 'The Times', it is evident that a considerable flight of this little grouse has been driven by some cause or other to visit England. The birds appear to have no difficulty in finding an abundance of congenial food on our heaths. Although smaller in appearance, I should say they would weigh nearly as much as a partridge, and their plumpness would render them a delicious bird for the table. I noticed that the flesh on the breast, like that of some of our most favourably known game birds, is of two colours, dark and white, the whitest flesh being closest to the bone. I take it that this is a very good sign of the bird's fitness to form a delicacy for the table. I may observe that there is an excellent figure of the bird, from the pencil of Mr. Wolf, in the second volume of the 'The Ibis.' – Murray A. Mathews; Weston-super-Mare, May 29, 1863.

PS. Since writing about the specimen of Pallas's Sandgrouse which was shot at Aldershot, I have received another, also a hen bird, and also shot by my brother near the camp. On the occasion when this last bird was killed a covey of nine were seen. The birds were very wild, and only allowed one long shot to be obtained at them. Among the feathers of the first bird my brother sent me, and clinging to the skin, I noticed a very singular tick, of large size, the body of a leaden blue colour, with lighter veins; the legs and sucking apparatus of a bright lake colour. Another singular thing about Pallas's Sandgrouse I omitted to mention in my first communication is the under surface of the feet. This is covered with a rough serrated skin, which may well be described by comparing it to the surface of a large rasp: it is an admirable provision for enabling the bird to run with ease upon the hot desert sands. Mr. Bree, in his 'Birds of Europe,' vol. iii., does not mention this peculiarity of the under skin of the foot as pertaining to the genus Pterocles. I should be glad to know if it is confined to Pallas's bird alone of the different species of sandgrouse. My brother writes to me that the birds he saw ran with great rapidity upon the ground, and that they skulked very low, seeming at a distance to be more like rats than birds. – Id.; May 30, 1863.

There were many records during the irruption of 1888 which are recorded in *The Zoologist* for that year. A flock of 20 at Itchen Stoke on May 15th, of which one was shot by Lord Ashburton's gamekeeper, were apparently the first to be recognised in England. An earlier record of at least one shot at Itchen Stoke on May 5th may be erroneous. A flock of 30–40 flew north-east over Broxhead Common on the evening of May 20th. Two were seen in a taxidermist's shop in Portsmouth on May 31st which had been shot out of a flock of about 12 a few days previously on Sinah Common, Hayling.

Single hens were picked up near Winchester prior to May 31st and in the New Forest on June 2nd both having flown into telegraph wires. A flock of 17 flew in a westerly direction over the River Avon near Ringwood on June 8th. Subsequently, a male was shot near Stubbington on December 14th. K & M also mentioned that three specimens were obtained at Hayling Island as late as December 1888 and January 1889 (but this may be an error for the May record above) and that Edward Hart informed them that some were still to be found in the New Forest until May 1889.

The most recent invasion in 1908 produced three records in Hampshire of five near East Liss in mid-April (*The Field,* June 20th 1908), two flying north-east over Havant on July 8th (*The Field,* July 18th 1908) and five at Burley in early August (*Brit. Birds* 2: 98).

GB status: many, 1859–1909; 7, 1950–90. (C Asia east to Manchuria and China).

Corncrake *Crex crex*

HOSRP

Formerly a moderately common summer visitor which last nested in 1959. Now a rare passage migrant with 70 recorded since 1951.

The species was well known to Gilbert White in the late 18th century but it was clearly scarce around his home village of Selborne. He wrote in his journal in October 1789, 'A man brought me a land-rail, a bird so rare in this district that we seldom see more than one or two in a season, and those only in autumn.' Harting, in his edition of *The Natural History of Selborne,* remarked that the scarcity of it at Selborne in White's time is not a little remarkable. He describes it 'as by no means scarce there at the present day' and records shooting three brace in one day in September 1863 within ten miles of the village. K & M described the Corncrake as a common summer visitor to all parts of the county, occasionally remaining for the winter. George Corbin, contributing to *The Zoologist* in May 1891, recorded three specimens procured near Ringwood between December 9th and 24th 1890.

Corncrakes were evidently in decline in Edwardian times. Ethel Williams, an all-round naturalist who lived in Winchester and submitted a regular article 'Out of Doors' to the *Hampshire Chronicle,* wrote on December 25th 1909 that the species had vanished around Winchester. It was certainly local by the 1930s although observer coverage was limited at this time. Between 1934 and 1938 there were several reports from Ashmansworth, Highclere and Ecchinswell. At the latter site, an adult and three young were destroyed by a mowing machine in a hay field on June 14th 1938 (Summers-Smith 1950). In the New Forest, first arrivals were noted at Fritham on April 30th 1935 and at both Brockenhurst and Godshill on April 19th 1936, while at Brockenhurst the latest was recorded on September 18th 1936. Other areas where the species was reported during the decade include the Anton Valley, Leckford, Petersfield, West Tytherley and West Dean, where a pair and six young were recorded in 1938 and said still to be present on October 22nd. One at Medstead on August 28th 1936 was the 'first there for many years' (*SEBRs*).

In the late 1940s, it was still present at Rockbourne and Damerham in summer and continued to be recorded almost annually in that area up to 1960. Breeding was confirmed in 1957 and 1959 (Cohen 1963) although most records were for August–October and may have involved migrants. Since 1950, the only other confirmed cases of breeding have been at Fawley in 1951, Park Gate in 1957, when Roy Dennis found a nest with three eggs, and Brownwich in 1959, when Sean White found a nest with ten eggs. One was seen and heard at Petersfield on July 12th 1951, a male was heard at Hill Head from July 2nd to 6th 1954, the first there in 34 years, and a male held territory in overgrown water meadows near St. Cross from late April until July 12th 1957.

Apart from records suggestive of breeding, a total of 70 was recorded between 1951 and 2020 with only 15 since 1981.

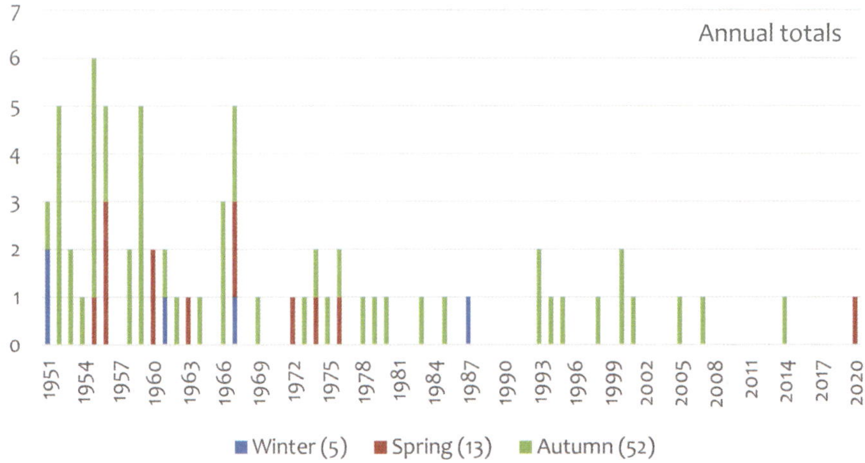

In addition to the Damerham records already mentioned, which account for 15 birds, eight were seen at Farlington Marshes (none since 1980) and the rest were widely spread around the county, with 24 at coastal sites and 23 inland. Unfortunately, eight of those since 1978 were found dead, having been run over or flown into overhead cables.

Analysis of arrival dates shows spring passage between April 16th and May 31st although only one has been recorded at that season since 1976. Singles heard at St. Leonards on July 27th 1959 and Danebury on July 11th 1976 may have been wandering unmated males but otherwise autumn passage extended from August 4th to November 25th. Winter records were of singles at Bedhampton on February 11th 1951, Overton in December 1951, Brownwich on March 25th 1961, Middle Wallop on March 16th 1967 and Needs Ore in December 1987, the latter being found freshly dead.

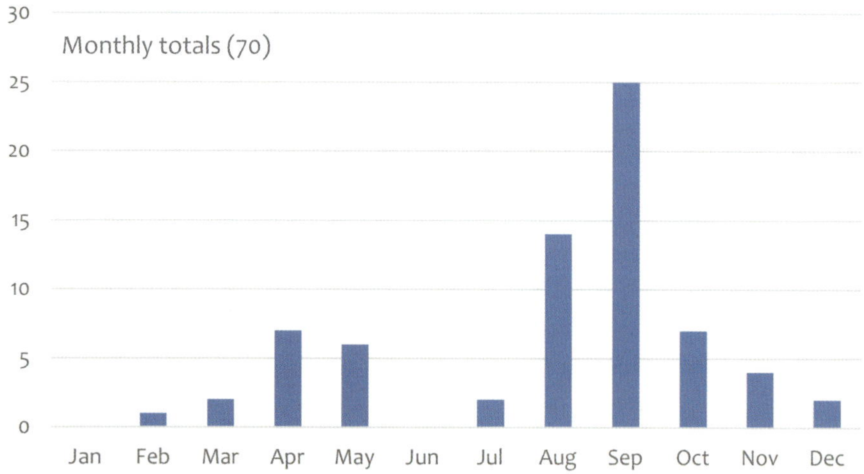

Corncrake, Titchfield Haven, October 8th 2000. Trevor Codlin

The only records this century have been of one seen by Ted Raynor at the Woodland Trust, Burkham on September 9th 2000, a juvenile trapped by Duncan Bell and Trevor Codlin at Titchfield Haven on October 8th of that year, one seen by Melvyn Brickwood at Brownwich on October 13th 2001, one seen by Tim Lawman as it crossed a path at The Kench, Langstone Harbour on September 17th 2005 which was not relocated despite a thorough search, one disturbed by Brian Plank who was driving a tractor during clover cropping at the Cholderton Estate on September 11th 2007, one flushed by Mick Hay at Itchen Valley Country Park on September 22nd 2014 and one heard by Peter Milinets-Raby calling at close range in a meadow at Warblington for several minutes from 0930 hrs on April 24th 2020. This was the first in spring since 1976 but it was not heard again. This is clearly now a difficult species to add to one's Hampshire list!

GB status: rare summer visitor, mostly to NW Scotland where c. 1000 pairs in recent years, and passage migrant. (Breeds UK through C Europe to C Siberia; winters S and E Africa.)

Spotted Crake *Porzana porzana*
HOSRP

A very scarce passage migrant and winter visitor; has possibly bred. At least 24 old records and a minimum of 169 birds in the modern era.

The first for the county was shot by Gilbert White in the sedge of Bean's Pond, near Oakhanger, on November 4th 1774. He sent it to London to be stuffed and preserved. White also records a Spotted Water-hen shot in the Forest (presumably Woolmer Forest) on August 26th 1777. K & M described the species as 'a very local resident', although their only reports of confirmed breeding were from Hart, who collected mainly in the Christchurch area, outside the present-day county boundary. K & M also included at least 23 obtained at various places in Hampshire between 1811 and 1897; 12 of these were in September–November, two in April or May and eight undated. Two specimens obtained at Newton Stacey in 1889 and one at Kings Worthy in 1890 formed part of the Winchester College collection and are on display in the Science School there.

The only dated records for the first half of the 20th century were of one shot at Stockbridge on October 4th 1904 (K & M) and one seen by Tony Norris on the edge of a small pond by a golf course at Gosport on September 3rd 1948.

Since 1951 Spotted Crakes have been noted in every year except nine.

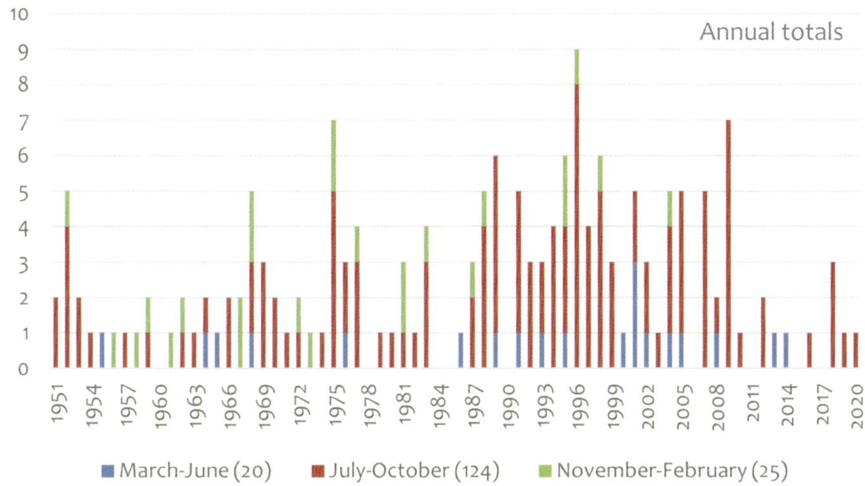

Analysis has produced a total of 169 although this is a minimum since records of birds from a site over a period of weeks are assumed to be the same unless it is clear that more than one is involved. Numbers peaked in the 1990s and 2000s but have declined in the 2010s with only ten recorded since 2012.

Most have occurred in autumn between July and October with records in 50 years accounting for 124 individuals. The earliest was an adult in moult at Farlington Marshes on July 15th 1999 which stayed until at least August 14th. Other early birds were at Hook-with-Warsash on July 20th 1992 and Farlington Marshes on July 30th 2010 but most arrivals have been in August and the first half of September as shown on the next page. Farlington Marshes has been by far the most popular location with 61 recorded in 39 years but with none since 2012 until one there on October 16th 2020. At least five were present in 2007 with four on September 1st and 2nd. A further 50 have occurred at 14 coastal sites including 16 at Titchfield Haven, 12 in the Keyhaven area, five at Hook-with-Warsash and five at Lower Test Marshes. Inland, 13 have been recorded at eight sites including six at Fleet Pond. The only multiple reports away from Farlington Marshes were of two at Fleet Pond on August 15th 1975 (with one from 8th to 21st), an adult and a juvenile at Lisle Court, Lymington from August 27th to September 2nd 1989, two at Lower Test Marshes on September 12th and 13th 2004 (with one from 10th to 15th) and two at Butts Lagoon, Pennington on October 14th 2009.

Spotted Crakes: Titchfield Haven. adult, September 12th 1993 (left) and juvenile, August 25th 1996

Some birds may have remained to winter; between November and February there have been records in 18 years involving 25 birds. These comprised nine at Farlington Marshes, six at Titchfield Haven (including two on December 25th 1975) and seven at four other coastal locations. Most were seen on one or two dates only but one was at Normandy Farm, Lymington from November 12th to December 3rd 1967. Inland, there were singles at Alresford Pond on November 26th 1977, Fleet Pond from January 25th to February 2nd 1972 and Funtley on February 29th 1996. One at Lower Test Marshes on November 11th 2004 was the only such record in the 21st century. Several records (mostly of birds heard only) from the late 1950s and early 1960s have been excluded.

There have been 20 spring records. Singles were at Farlington Marshes on March 20th 1955, March 29th 1964 (found dead), March 27th 1965, March 24th to 31st 1968 and April 5th 1986. Other records of presumed migrants, mostly of birds heard singing at dusk or during the night, came from Lower Test Marshes on April 21st 1976, Fleet Pond on April 12th 1989, Itchen Valley Country Park from April 12th to 16th 1995, Bickerley Common on April 24th 2001 and Highbridge from May 7th to 11th 2001. At Titchfield Haven, there were records in 1991, 1993, 2000 and 2001 on dates between March 29th and May 14th and at Fishlake Meadows in 2002, 2004, 2005, 2008 and 2014 on dates between April 27th and May 19th. One heard singing during a nocturnal survey near Hill Top in the New Forest on June 10th 2013 is the only record for that month.

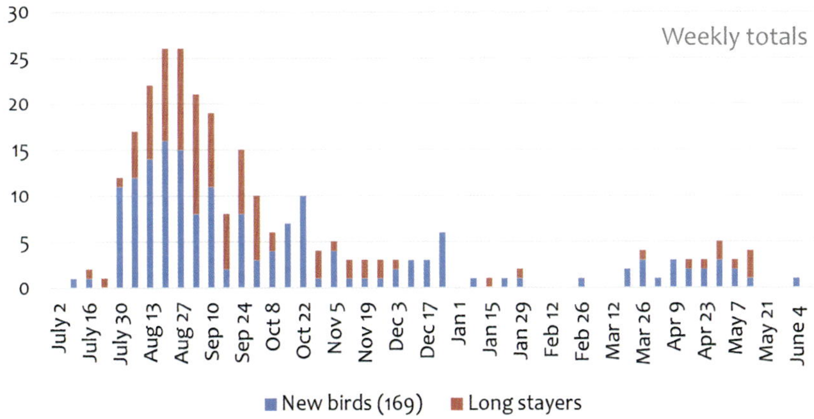

Duffin (1991) stated that Spotted Crakes had occurred in the breeding season at Titchfield Haven on two occasions, and young were seen once only. Barry Duffin has since confirmed (*in litt.*) that he considers unreliable the record of young being seen. However, in 1993, two juveniles, one of which was trapped, were present in August in the area where song was heard between April 18th and May 14th. This is the best evidence that breeding has occurred in Hampshire in the modern era.

An earlier record of breeding in the Avon Valley has arisen from the transcript of the BBC Radio 4 programmes *Sounds Natural* broadcast on May 24th 1971 when Kenneth Allsop was interviewed by Derek Jones. Extracts are reproduced overleaf.

Jones: Did you ever come across a nest of a particularly rare species?

Allsop: Yes. Not through skill, just through blundering upon it. But I remember in, I think it must have been about the mid-thirties, and I was on a school holiday down in Hampshire with a school friend and we were on one our great egg forays, coming back with boxes stuffed with them. Fearful really. But we were down beside the River Avon looking for Redshanks' nests and there were also Lapwings nesting on the higher meadows. And we were tramping through the rather boggy, soggy meadows, with cattle in them, and parting the rushes and dead rank grass and we came upon this nest which looked a bit like a Moorhen's. It was quite a deep cradle and had, I think, eight or ten eggs, a large number of eggs and they were a bit 'moorhenish' – about that size and speckled and blotched with liver and on a sort of buff background. But we knew they weren't Moorhen's. They were the wrong size and the colouring was a bit different, and anyway it was not the place you'd find a Moorhen's. So we retreated and circled back half an hour later in the hope of finding the bird on it. Once again the eggs were hot, they'd obviously been sat upon until a moment ago – but the bird was not to be seen. And in a rather remarkably self-disciplined way we didn't take any. We thought, no, we'll leave it. And we spent three days doing this and we went back on successive days each time trying to surprise the bird. And we never did and eventually took an egg each and we did limit ourselves to one each on this occasion, and I remember having blown it with my little brass blowpipe and drill which I bought through a mail order firm in some magazine. And I took the egg to the British Museum to the curator of the eggs, of the bird section there, and it was the Spotted Crake's, which is rather exceptional. I've never since in my life seen a Spotted Crake or found its nest.

Another record of five seen together by D. Whiting at Redbridge in severe weather on January 5th 1963 was published in the *HBR* but perhaps harshly not accepted for *BoH*. Whiting's notes are reproduced below.

Observed through binoculars, seemed to be about the size of a Ringed Plover or perhaps a little larger. They had a jerky tail action like a Moorhen showing buff undertail-coverts, clearly barred flanks, bill yellowish, breast grey spotted, legs greenish. They were all feeding and seemed to be a flock not individuals. The weather had been hard for some time and the birds had all congregated on a piece of marsh washed by a high tide. Visibility was good and I am in no doubt at all that they were Spotted Crakes. I checked my opinion on the spot with the Field Guide to the Birds of Britain and Europe and double checked at home with British Birds by Kirkman and Jordan.

Spotted Crake, Farlington Marshes, September 14th 2007. Richard Ford

GB status: 1,662, 1986–2019. Declining, mean 26 per year in 2010–19. (Breeds locally in Europe from S Spain & Atlantic regions of coastal France to N Gulf of Bothnia, Finland, E in C Russia to c. 100°E. Some winter Mediterranean basin, others in sub-Saharan Africa & Indian subcontinent.)

Baillon's Crake *Zapornia pusilla*

BBRC

A very rare vagrant. One record in 1990.

The only record for Hampshire is of an adult caught by Elly Wiseman's cat Arthur at Normandy Farm on March 17th 1990. Wendy Wiseman takes up the story.

> *I was looking out of the window of Normandy Farmhouse into the garden and saw Arthur and a small bird which I didn't recognise, squaring up to him. I called Edward over, who, on looking out of the window, uttered some expression of disbelief, and rushed outside to rescue the bird which he could see was some sort of crake. It was apparently unharmed but kept allowing it to recover from its ordeal. It was shown to Barry Duffin and Capt. Leslie and Mrs. Beryl Masters who were able to photograph the bird before it was released a short while later in the day.*

Baillon's Crake, Normandy Farm, March 17th 1990. Barry Duffin

GB status: 62, 1819–1949; 30, 1950–2020. Rare, about one every 1–2 years, bred in 2012. (Locally common breeder from Spain E through Mediterranean Basin to Ukraine, occasionally N to Netherlands. To E breeds warm temperate Asia E to Japan. European breeders migrate throughout S Europe to winter sub-Saharan Africa. Elsewhere, winters Indian subcontinent & SE Asia to Singapore. Other races breed Australasia & southern Africa.)

Little Crake *Zapornia parva*

BBRC

A very rare vagrant. One record in 1959.

The only record for Hampshire is of a probable male found by Michael Bryant at Farlington Marshes on September 14th 1959. Two days later improved views were obtained by several observers which confirmed its identity. It remained until September 20th.

Michael Bryant's account of the events, as submitted to BBRC, is reproduced below.

> *September 14th. I saw a very small crake, smaller than a Spotted Crake, for a very few minutes, flying and swimming. All I could see was that the head was buff. Later E. J. Wiseman and I saw a very small dark crake flying low and diving into cover. It appeared only half the size of a Moorhen.*
>
> *September 15th. I saw a very small dark crake but could not see details properly. No sign of buff undertail-coverts, however, so far as I could see.*
>
> *September 16th. N. Money, EJW, John Simons and I saw a male Little Crake at Farlington Marshes. Description: Very small crake smaller than Spotted Crake with short bill. Upperparts dark brown, underparts dark blue grey – the same colour as breast of water rail but darker, no sign of bars on flanks, undertail-coverts dark. September 18th. JS saw the same or a very similar bird and confirmed absence of bars on flanks and confirmed that legs were green.*
>
> *September 20th. Alan Walker and Michael Burnop saw the same or a very similar bird again.*

A Spotted Crake was also present at the same time.

K & M mentioned that John Kelsall was informed in 1889 by Mr. J. G. Millais that he had a specimen obtained at West Liss, Petersfield. John Millais was a respected ornithologist and bird artist but in the absence of a specimen and even an approximate date, this record cannot be accepted as the first for Hampshire. Saunders, in a list of British records, includes one 'near Fordingbridge in Hampshire' but without reference or date (Yarrell, 1871-85). It was not included by K & M. However, a record of two shot, pre-1880, at Alderholt near Fordingbridge, included by Mansel-Pleydell (1888) and subsequent Dorset authors, may have been the source of Saunder's reference.

GB status: 59, 1791–1949; 44, 1950–2020. Rare, about one every two years. (Breeds across temperate steppe region from Austria E through Ukraine & European Russia to W Siberia, C Kazakhstan & NW China. Small numbers occasionally breed to N & W, reaching Netherlands, Finland & Spain. Most winter NE & E Africa, although some W to Senegal.)

Crane *Grus grus*

BBRC to 1987; HOSRP 1988 to date

A rare but increasing passage migrant. 47 records of at least 148 individuals since 1959.

The first for the county involved one at Needs Ore which was reported to Edwin Cohen on November 5th 1959. It was seen by Dave Wooldridge on 21st and was still present on 28th when it was evidently in poor health. It was found dead the next day. A government post-mortem concluded that the bird had died due to injuries to its neck and left leg.

Annual totals since then are shown below.

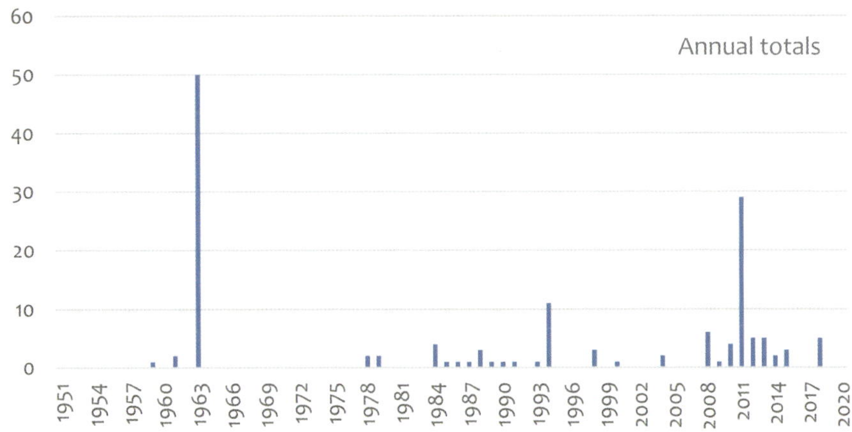

The second record involved two watched for 30 minutes by Gerald Westerhoff as they circled over woodland at Chandlers Ford on June 2nd 1961 before soaring off to the north-east. In late autumn 1963 there was a large influx into southern England involving some 500 birds (Harber 1964). These were probably en route from Fennoscandina to wintering grounds in Iberia and were drifted westwards by east winds. In Hampshire the first were on October 30th when 25 flew south-east over Otterbourne and two east over Farlington Marshes. The following day 25 were seen over Bishop's Dyke, New Forest and c. 50 off Milford on Sea. On November 1st, flocks of 25 and 19 were seen over Bishop's Dyke and the next day 17 moved east over Farlington Marshes. These were also seen over Sidlesham, West Sussex. It is impossible to calculate the total seen in Hampshire but it could be anywhere between 50 and 163.

The next were not until 1978, when Tim Lawman found an adult and a juvenile feeding in a field at Tanners Lane, Stubbington on December 3rd. They flew off at 1020 hrs and ten minutes later Dennis Bill watched them as they flew north over Fareham Creek. Later that winter Bob Chapman, Bob Gomes and Barry Duffin saw two adults arrive from the north over Upper

Titchfield Haven at dusk on February 27th. These were presumably the same two birds seen flying east over Winspit, Dorset two days earlier. They were not seen the next day.

A run of sightings between 1984 and 1994 was possibly connected with the re-establishment of a small breeding population in East Anglia. However, there then followed a dearth of records despite the gradual expansion of the British population. It was not until 2008 onwards that records became somewhat regular once again. The Great Crane Project in 2010–14 involved the release of 93 young cranes on the Somerset Levels. As of December 2019, 52 of these were alive and they had raised 27 young of which 23 survived (Antliff-Clark & Bridge 2019). All are colour-ringed but the rings are not obvious on flying birds. As a result, since 2015 HOSRP has accepted records only of birds where there is strong evidence that they originated elsewhere.

Cranes, Tanners Lane, September 26th 2018. Alan Lewis

Most records were of one to three but there have been four sightings of higher numbers. E. and M. Snudden saw a noisy flock of 11 which circled low over Vidle Van Farm, Keyhaven at about 0800 hrs on October 17th 1994. They did not land and headed high south-east towards the Isle of Wight and were later seen over Hengistbury Head and the Fleet, Dorset. Pete Wickham and Mike Mockler saw a group of four circling over The Bench, Lyndhurst before flying off south-west on December 27th 2008. A flock of 24 was seen circling over Wickham and Regent's Park, Southampton between 1420 and 1435 hrs on November 17th 2011 by three independent observers. These were no doubt the group of 23 adults and one juvenile, all unringed, which had been in North Cornwall from November 12th and were tracked eastwards over Devon early on 17th. They were not seen again after appearing in Hampshire and presumably left the country southwards. Most recently, five were seen circling over Finkley near Andover at 1623 hrs on September 25th 2018 and then at 1900 hrs flying east off Pennington Marshes. The next morning they were located nearby in the saltmarsh at Tanners Lane. After a false start they left south-east at 1040 hrs and were later seen leaving south from St. Catherine's Point, IOW at 1125 hrs. They were first recored on the Ythan Estuary, Aberdeenshire on September 21st where they were still present at 1030 hrs the next day. They were subsequently tracked south over Saltholme RSPB reserve, Cleveland at 1137 hrs on 24th and over Knap Hill, Wiltshire around midday on 25th.

Analysis of arrival dates shows that in spring there were 21 records totalling 33 birds, all of one to three, between April 6th (one on a small pond at Michelmersh in 2013) and June 16th (one in flight over Bishop's Dyke in 1990). Later in the year, excluding the 1963 influx, there were 17 records totalling 62 birds. The earliest was a juvenile seen moving west over Eastney, Hill Head, Hook-with-Warsash, Normandy and Hurst Beach between 0830 and 1010 hrs on September 16th 2011. It was seen again later that day drifting back eastwards towards Lymington and roosted in the saltmarsh off Pylewell. There are just two late winter records of one flying west over Farlington Marshes on January 6th 2012 and the two at Titchfield Haven on February 27th 1979 mentioned earlier.

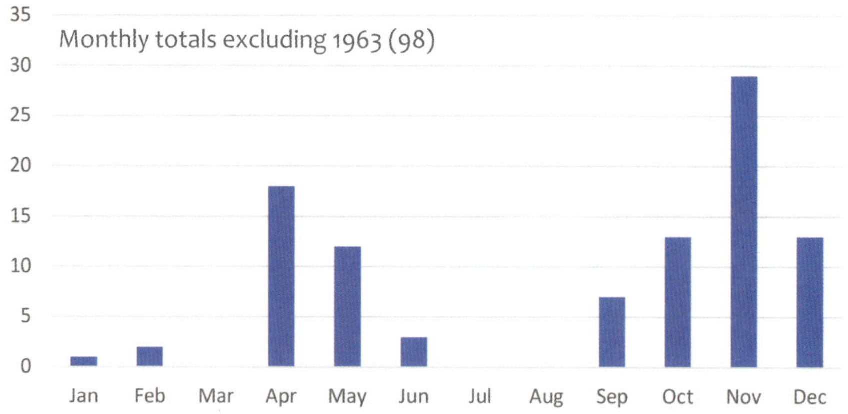

A record of one seen by Trevor Codlin at Upper Hamble Country Park on April 26th 1984 was harshly considered not proven by BBRC. When Trevor pointed out that it was presumably the same bird seen soaring over nearby Titchfield Haven by Richard Levett the next day, BBRC replied that no records had been submitted for Titchwell (Norfolk)!

One seen flying north over Stockbridge Marsh by Ralph Cook and Ted and Penny Raynor and others during a HFC Ornithological Section field meeting on September 11th 1976 was not submitted to BBRC.

There is an old undated reference to the species occurring at Alresford before 1774 (Boyes 1774).

GB status: 115, 1767–1949; 1,269, 1950–88; recolonised from 1981, 22 pairs by 2013, re-introduction subsequently. (Breeds northern Europe across most of Asia, winters SW Europe, NW and E Africa, China.)

Black-winged Stilt *Himantopus himantopus*

BBRC to 2016; HOSRP 2017 to date

A rare vagrant. Two records of five birds in 1923 and 1945 and 19 of 26 birds since 1962.

The first for the county was one seen by Cecil Paddon at Needs Ore from November 23rd to 26th 1923 (*Brit. Birds* 17: 212). One to four were at nearby Warren Flats in mid-May 1945 (*HBR*). There was an influx into England in spring 1945 with at least 25 at eight sites culminating in two pairs breeding successfully at Nottingham Sewage Farm. The group in Hampshire may have been part of the flock of ten at Bovisand, Devon on May 10th.

Black-winged Stilts, Pennington Marshes, May 4th 1987

The first modern record was of one found by Mrs. J. Edom at Badnam Creek on the west side of the Hamble Estuary on October 3rd 1962. Extracts from her description are given below.

> *It was a large wader, considerably larger and taller than Redshank with which it was feeding. Very white neck and body. Top of head pearly grey. Jet black wings which in flight were also black on undersides. Bill black. Legs very long and pink extending far beyond tail in flight.*

Since then, there have been a further 18 records of 25 birds.

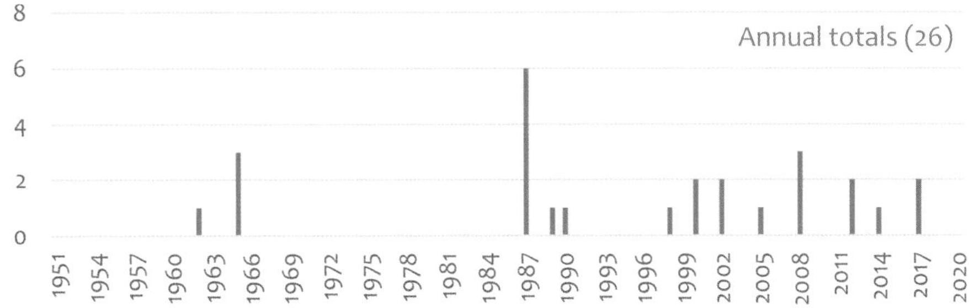

The next was a group of three, described as an adult male, adult female and immature male, found by Mrs. E. Miller and Norman Pullen on a freshwater pool at Dibden Bay on April 10th 1965 and still present on April 13th.

There was an influx into southern England in spring 1987 involving around 30 individuals. Dick Barrett went to Stanpit Marsh, Dorset on May 3rd and heard about two present earlier in the morning which had left east. Dick fortuitously checked the Lower Pennington Lane floods on his way home and found them. They remained in this area until May 7th. A male, found by an unknown birder, was at Farlington Marshes from May 10th to 20th and an adult was watched by John Cloyne at Winchester Sewage Farm from 1100 to 1300 hrs on May 17th. Further records in that year involved a first-summer seen by David Billett and Rob Clay at Farlington Marshes on June 16th and a long-staying male there, first found by Rob Clay, from July 11th to 25th which flew off south on the final date.

Another adult was found by Chris Cockburn at Farlington Marshes on November 8th 1989 and this was followed by a male there from April 30th to May 2nd 1990. This was seen by many observers but not submitted to BBRC until 2019 when notes from Graham Stephenson were accepted.

There was an eight-year gap before the next, a first-summer, was found by R. Weaver at Normandy Lagoon on March 30th 1998. It moved to Oxey Barn fields the next day where it remained until April 5th. A pair was found by Mike Baker at Normandy Lagoon on April 29th 2000 which later moved to floods at Pennington Marshes. They were then seen at Farlington Marshes during the late morning of April 30th before departing east and arriving at Sidlesham Ferry, West Sussex in the afternoon.

Black-winged Stilt, Beaulieu Mill Pond, June 30th 2008. Steve Lavington

A pair was located by Peter Milinets-Raby and Ken Martin on the South Scrape at Titchfield Haven at 0600 hrs on April 17th 2002 where they stayed all day and departed overnight. The next was a male found by Tony Blakeley at Fishtail Lagoon, Pennington Marshes on June 21st 2005 which was not

Black-winged Stilt, female, one of two, South Scrape, Titchfield Haven, April 17th 2002

present the next day and was presumably the bird at Dungeness, Kent on 22nd and 23rd.

The first inland record was of one found by Tim Doran at Ibsley Water at 1015 hrs on May 10th 2008 which remained in the area for the rest of the day but departed overnight. Later that year a pair was discovered by Jeff Ross at Beaulieu Mill Pond at 0730 hrs on June 30th where they remained until 2040 hrs before leaving south-east. These were almost certainly the pair which bred at Ashton's Flash, Cheshire and were last seen in their nesting area the previous afternoon – their young having been predated.

A female was found by Steve Piggott at Shoveler Pools, Pennington Marshes on May 27th 2012. It was still present on 29th and was joined by a male from June 2nd to 9th. The birds were observed copulating and nest scraping but poor weather and the constant attention of corvids led to them abandoning their breeding attempt and leaving the area. The next was one found by Barry Duffin at Titchfield Haven on May 6th 2014 which stayed all day. Another was found by Michael Finlason at Titchfield Haven on March 30th 2017 which then moved to Posbrook Floods where it remained until April 2nd. Another was found at Fishtail Lagoon by Yvonne Holloway on May 9th 2017 which was still there the next day.

The weekly distribution of records shown opposite confirms that most have been seen in spring between March 30th and July 25th with just two in autumn on October 3rd and November 8th.

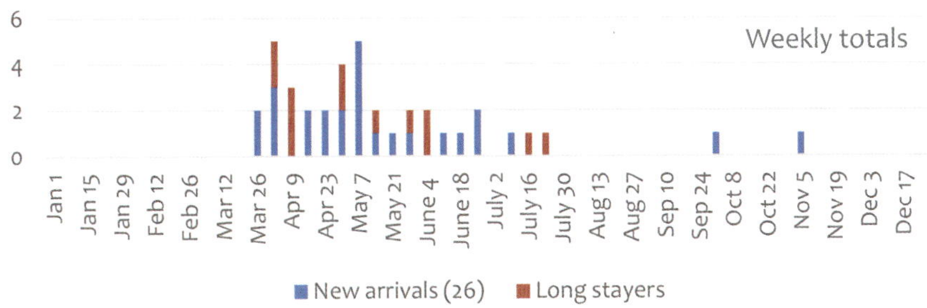

K & M included two undated 19th century records from near Havant prior to 1845 and at Gosport 'many years ago' but in the absence of even approximate dates and extant specimens neither of these can be considered as the first for the county. Also excluded is the famous sighting reported by Gilbert White of a flock of six, of which five were procured, on the county boundary at Frensham Great Pond on April 28th 1779, as there is no evidence that they were seen in Hampshire.

GB status: 125, 1684–1949; 523, 1951–2019. Large increase, mean 27 per year in 2010–19. (Breeds along Atlantic coast of France & locally throughout Mediterranean basin to Black Sea. To E, breeds from S Siberia & C Asia to NW China & S to Hong Kong. Most European birds winter sub-Saharan Africa &, increasingly, in SW Iberia. Asian breeders winter S & SE Asia & S China.)

Sociable Plover *Vanellus gregarius*

BBRC

A very rare vagrant. One record in 1986.

The only record for Hampshire is of a juvenile found by Barry Duffin on the scrapes at Titchfield Haven on October 17th 1986. Extracts from his submission to BBRC are reproduced below.

At approximately 1445 hrs a Sociable Plover was discovered feeding on an expanse of freshly exposed mud in the southernmost of two freshwater scrapes located some 150 m inside the seawall. My first impressions on seeing this wader were that it was a winter plumage Golden Plover with a creamy white supercilium. My attention was next drawn to the black primaries on the closed wing. The bird was watched for about three minutes before it flew some 200 m away and joined Lapwing feeding on a grassy area in the northernmost of the two scrapes. On flying the wings showed a striking contrast between the black primaries and white secondaries. The bird was watched 30 minutes later feeding with Lapwing from another observation hide at a range of around 60 m. When feeding close to Lapwing it was noticeably longer-legged and appeared plumper. It was frequently aggressive towards Lapwing when feeding close to them, whilst in flight it would itself be harassed by the Lapwing. Following the finding of the bird it put in only two more brief appearances in the reserve on the mornings of October 18th and 22nd. Most of the time up to 24th was spent on farmland at Brownwich and Chilling to the west.

Sociable Plover, North Scrape, Titchfield Haven, October 17th 1986. photo Barry Duffin

Subsequently, it was refound by Richard Fairbank on October 25th on fields at HMS Daedalus, the naval air station to the east of Titchfield Haven. It remained in this area, usually in the company of Lapwings, until November 8th.

GB status: 3, 1860–1949; 41, 1950–2020. (Breeds from Volga & Ural Rivers E across steppes of SE Russia & W/C Asia to E Kazakhstan; now rare and declining throughout much of range. Most migrate to winter in NE Africa, smaller numbers to Pakistan & NW India.)

Pacific Golden Plover *Pluvialis fulva*

BBRC

A very rare vagrant. One record in 1992.

The only record for Hampshire is of an adult in summer plumage found by Dave Unsworth at Needs Ore on June 27th 1992 but not conclusively identified until the next day. Extracts from his submission to BBRC are reproduced below.

Pacific Golden Plover, adult, Needs Ore, June 28th 1992

The upperparts had a rather dark golden spangled appearance and in general were not as bright as I had seen on individuals previously and this led me initially to try and convince myself that I was only watching a particularly stunning Grey Plover. However, my nagging doubts were put to rest when I decided to flush the bird to check the axillaries and underwing colour. To my pleasant surprise the bird did not have black axillaries and the underwing colour was greyish rather than white so I was quite certain that I was dealing with a 'Lesser Golden Plover' (possibly American rather than Pacific). Added to this, the bird appeared different structurally although with no Grey Plover nearby for comparison I chose not to rely too heavily on this in case it was just 'wishful thinking' on my part.
I took a plumage description on the evening of June 27th when I found the bird but it was not until I returned home that I realised that I needed to check the structure of the tertials/primaries/tail tip to be sure of the identification. This, coupled with other factors such as the sensitivity of the site, led to me telling only a very small number of people of the bird's presence that evening. The following day I was able to clinch the identification and was able to phone the news out to a few local birders who fortunately arrived in time to see it that afternoon. The warden could still not be contacted for arrangements to be made for the news to be released more widely but as the bird departed high to the west at 1845 hrs on June 28th this was no longer a problem.

Underparts: black extended from just above the bill to the eye and then curved around the ear-coverts and down the side of the neck, along the flanks to the undertail-coverts. The vent was whitish and the undertail-coverts showed quite broad black barring. The white line down the neck and flank did bulge slightly more on to the side of the neck and was present all the way down the flanks.

A second record of a moulting adult at Hook-with-Warsash on August 21st 1992 was submitted to BBRC although the observer was not 100% certain and so the record was deemed not proven.

GB status: 2, 1870–1949; 105, 1950–2020. About three per year; stable. (Breeds across Siberian tundra from Yamal Peninsula E to Chukotskiy Peninsula, including New Siberian islands, & W Alaska. Winters Indian subcontinent, S China & S Japan, S through SE Asia to Australia, New Zealand & islands in C Pacific, small numbers regular Kenya & Persian Gulf.)

American Golden Plover *Pluvialis dominica*

BBRC to 2005; HOSRP 2006 to date

A very rare vagrant. Eight records since 1994.

The first for the county was a juvenile found with Golden Plovers by Mike and Pete Combridge in a field at Lower Pennington Lane at 1145 hrs on November 20th 1994. Extracts from Pete's description to BBRC are reproduced below.

> *The morning of November 20th was grey, windy and showery but was interesting for birds, including a small westerly passage of Golden Plovers. We were checking the area inland of Pennington Marshes when we discovered a tight flock of Golden Plover resting in a grass field. From a footpath, screened by a hedge, we were able to view the flock side-on through binoculars and telescopes at 100 yards range. We soon noticed an odd-looking plover and, although we've seen several grey-looking Goldens in Britain and the Netherlands, this one did look different. After watching it for some time and seeing it wing-stretch to reveal its underwings we were sure that it was an American Golden Plover. Just after 1200 hrs the flock flew off westwards and the bird was not seen again. It was slimmer and slighter than Golden Plover with proportionately longer legs and a long primary projection (tertials falling well short of the tail tip), the primary tips projecting noticeably beyond the tail, giving an attenuated look to the rear end. Its plumage was duller and greyer than Golden Plover without obvious yellow tones. It had a black cap with an obvious white supercilium running from above eye to rear of ear-coverts; area at base of bill whitish; 'face' greyish with dusky mark in front of eye; hindneck grey; mantle and upper scapulars very dark, remainder of upperparts greyer, 'spangled' with white; primaries black; chin whitish; fore neck and sides of neck, and breast, greyish with blurred striations; flanks greyish with dull barring; belly greyish-white, vent and undertail-coverts whitish. Bill, legs and eye black. In flight, underwings and axillaries dusky grey; wing-bars duller and less obvious than Golden Plover.*

A *Pluvialis* plover in summer plumage was on The Deeps at Farlington Marshes on May 13th 1998 but not identified. Two mornings later, Pete Gammage found it asleep but eventually it raised its wings briefly to reveal dusky axillaries; he thus identified it as a Lesser Golden type. Pete had to leave but returned later that afternoon having researched the differences between the American and Pacific species. The bird was actively feeding on

The Deeps and proved straightforward to identify as an American Golden Plover. By the time other observers had arrived the plover had disappeared and it was not seen again. The record was unfortunately not submitted to BBRC. Five days later an adult in summer plumage adult was found by Tim Parminter at Pennington Marshes on the afternoon of May 20th 1998; his identification was confirmed by Simon Ingram that evening and the news released. The bird remained in the area until June 2nd and was seen by hundreds of observers. This is assumed to be the same individual that originally occurred at Farlington Marshes.

American Golden Plover, Pennington Marshes, May 24th 1998

The third was a juvenile Lesser Golden type found by Reg Russell at Needs Ore on October 30th 1999 which is assumed to be same individual found by Tim Parminter with two Golden Plovers at Normandy Lagoon the next day and confirmed as an American Golden Plover. It remained in the area until November 14th.

The fourth was another juvenile, again found by Tim Parminter, with Grey Plovers along the shore at Pennington Marshes on October 4th 2006 which stayed until October 13th.

The second in spring was in 2007. Jason Crook found one at The Deeps, Farlington Marshes at 0928 hrs on May 11th; he and others continued to watch it until 1230 hrs when it was spooked by a passing Peregrine and flew off high to the east. The same bird was refound by Pete Gammage on The Lake at Farlington Marshes at high tide on May 20th. It remained faithful to the lake at high tide until May 24th and on the final day was also seen feeding on the fields. Photographs taken by Jason on May 11th and during the second period confirmed that the same individual was involved.

In 2019, Clay Jones found an adult in summer plumage at Keyhaven Lagoon at 0845 hrs on August 13th. He was studying a Little Stint and noticed what he presumed to be a European Golden Plover out of the corner of his eye. Clay took some record shots with his bridge camera before it flew off towards the Isle of Wight at 0856 hrs. He showed the photos to Steve Keen who suggested its identity which was confirmed by Nigel Jones and Alan Lewis. It was never seen again.

American Golden Plovers: adult, Pennington Marshes, May 24th 1998 and juvenile, Pennington Marshes, October 9th 2006. Keith Stone and Linda Miller

The next was an adult or second-winter which Jason Crook found roosting on The Lake at Farlington Marshes at high tide on September 21st 2020. Originally found asleep and only half-visible, it soon woke up, flew around calling and re-settled with the roosting Grey Plovers. Just over an hour later it flew into Langstone Harbour. Jason subsequently relocated it at low tide in the Chalk Dock Lake and North Binness Island area on September 25th where it was seen by many observers until October 24th. It was last recorded a week later on 31st when it flew low east, calling loudly, over Farlington Marshes at 0905 hrs. It had re-visited The Lake at Farlington Marshes during very high tides on October 19th and 20th. The most recent record was of an adult in summer plumage found by Jason Crook off the eastern side of Farlington Marshes on June 23rd 2021. It remained in the area until June 27th. This may have been the bird of the previous year returning.

GB status: 2, 1883–1949; 631, 1950–2019. Increasing, mean 25 per year in 2010–19. (Breeds on coastal tundra from extreme NE Siberia, E across N Alaska & Canada to Baffin Island. Migrates over W Atlantic to wintering grounds in S South America.)

Semipalmated Plover *Charadrius semipalmatus*

BBRC

A very rare vagrant. One record in 2013.

The only record for Hampshire is of a first-winter found by Andy Johnson in the high tide roost at Black Point on October 17th 2013 which stayed until October 29th. Extracts from Andy's account of his discovery and the subsequent events published in the 2013 *HBR* (Johnson 2014) are reproduced below.

Around 0900 hours on Thursday October 17th, whilst doing the rounds of my patch at Sandy Point, I decided to see what was gathering at the Black Point high tide wader roost on the spit to the west of Hayling Island Sailing Club. There were over two hours to go to high tide but from the bottom of the causeway I could see a fisherman at the end of the spit had already flushed everything, and a couple of dog walkers were finishing off the job (both frequently a problem here). Nevertheless, I went up to Sparkes Marina to see if there was anything left. When I got there, I could see there were about 20 Ringed Plovers flying around the spit. Then I heard a clear, unequivocal 'chewit!'. As soon as I heard the call, I thought 'This is it!!'.

I looked to where the noise came from to see the last two remaining 'ringed' plovers take flight – one clearly a lot smaller than the other! They flew round and round for what seemed an eternity, but eventually landed on the near edge of the mud, around 60 m away. I swung my scope onto them and got onto the smaller bird to specifically look for a white wedge above the gape... it was there! Clear and obvious!! Elation turned to edginess, however, as they quickly took flight once more, and joined three more flying round and round again. Eventually all five settled at the far end! My only choice was to walk all the way down Wittering Road and back up the causeway to Hayling Island Sailing Club. This takes around 15 minutes, during which I was thinking 'If they've flown off by the time I get there, then that's that! If they're still there, I'll nail it!!'

I got to the spot, and they were still there... got myself into position about 45 m away, scope up, start from the left 'Nope... nope... no... yes!!!'. The bird stood out like a sore thumb – tiny, compared to the accompanying Ringed Plovers, with an equally tiny bill, and the obvious wedge of white feathering above the gape. I found the yellow eye ring rather difficult to discern, except when the bird glanced upwards at the sky (as the waders here often do, on the lookout for raptors). I even managed to convince myself I could see the necessary webbing between the toes as the bird took a few steps, but before long it hunkered down into a hollow in the sand.

I rattled off a load of digiscoped photos, hoping they would be sharp enough to show all the relevant features. With restless birds still coming and going – and the tide now very high – the Semipalmated Plover took off at 1024 hrs and flew off high east with six Ringed Plovers, its diminutive size again obvious as they departed.

I stayed for a further two hours, during which time birds were continuously coming and going, and by 1230 hrs, around 50 Ringed Plovers had returned as the tide began to recede. However, THE bird hadn't returned, and with views becoming increasingly distant I was confident that the show was over, at least for today (at low tide Ringed Plovers are scattered far and wide throughout Chichester Harbour, most at considerable distance). With the pressure off, the most important issue now was to make sure I hadn't completely lost my marbles – to check the photos, and to get several second opinions, and then to put the news out. Nevertheless, the first thing I did (as I was already there) was to call in to Hayling Island Sailing Club, who own the land, to tell them what I'd found, and ask whether they had any events on at the weekend! Fortunately, they didn't and were happy for the overflow car park to be used for the impending twitch.

Confident that I had seen all the relevant features, I sent photos to a few contacts; the responses came back quickly and favourably, and the news was put out immediately. The next day, there were around 50 people watching the gathering high tide roost. Sadly, another very high tide forced the birds that gathered to go elsewhere, and it seemed the Semipalmated Plover had not joined them. The crowd dispersed. Remarkably, however, the Semipalmated Plover apparently flew in just after high tide and spent the next couple of hours in the same spot, giving the best showing of its stay.

Semipalmated Plover with Ringed Plover,
Black Point, Hayling Island,
October 17th and 18th 2013.
Andy Johnson

On the Saturday, the Semipalmated Plover thankfully appeared amongst 76 Ringed Plovers and 300 Sanderlings on the rising tide for the assembled 120 birders but the waders were flushed by an approaching (very unpopular) wind-surfer and the whole flock headed purposely off. Fortunately, I knew exactly where the flock would head and hot-footed it to groyne 24 on the seafront. Sure enough, the waders were present, and the Semipalmated Plover was practically the first bird I looked at. I phoned back to the crowd at Black Point and 20 minutes later there was a stampede on the beach as the crowd descended. However, there was no need to panic, as once again the birds remained in place for the next couple of hours to see out the high tide period.

Subsequently, the bird appeared in most high tide gatherings until Tuesday 29th, which proved to be its last day.

Semipalmated Plover, Eastoke Beach, Hayling Island, October 19th 2013

It was also located on the rising and falling tide on Pilsey Sands, off Thorney Island, West Sussex from October 18th to 21st.

This was the fourth for Britain and followed previous records in the Isles of Scilly in 1978, Devon in 1997–98 and the Outer Hebrides in 2012.

GB status: 4, 1978–2013. (Breeds Alaska & across N Canada to Baffin Island & Newfoundland. Winters from Atlantic & Pacific shores of southern USA to S South America.)

Killdeer *Charadrius vociferus*

BBRC

A very rare vagrant. One record in 1980.

The only record for Hampshire is of a first-winter heard calling by Mike Stewart and Bob Dunn before it landed at Iley Point, Keyhaven at 1600 hrs on September 28th 1980. It was subsequently seen by Mike and Pete Combridge and a dozen or so observers that evening and by many more the following day. It was last observed on a shingle spit off Pennington Marshes. Extracts from Pete Combridge's notes submitted to BBRC at the request of the finders are shown below.

The initial impression was of a 'stretched' Ringed Plover, due to its strikingly long tail and legs, an effect heightened when it stretched its neck when alert. Its most obvious plumage feature was the two black breast bands; the upper band narrow continuing around top of mantle; the lower band thicker and crescentic in shape. The rump and tail were coppery red with a thick black subterminal band and broad white tips. In flight, which was powerful, the rump was very noticeable as was the striking white wing-bar which was broad and long. The bird fed by picking at the mud in typical plover fashion but due to its long legs had to pivot the body quite noticeably to do so. It quite often used 'foot stirring' as it fed in a similar fashion to Ringed Plovers.

Killdeer, Iley Point, Keyhaven, September 29th 1980. Tony Croucher

GB status: 4, 1859–1949; 55, 1950–2020. Rare, about one per year, stable or decreasing. (Breeds S Alaska, S Canada & throughout USA to Mexico. Northern breeders migratory, wintering S USA & Mexico to Columbia. Other races resident Caribbean & South America.)

Kentish Plover *Charadrius alexandrinus*

HOSRP to 2019; BBRC 2020

A rare and declining passage migrant which has occurred once in winter. Two in 1948 and 1949 and 49 records of 59 birds in the modern era.

The first for the county was found by Tony Norris at Titchfield Haven on August 30th 1948. His description is reproduced below.

> *Breast, belly, flanks, collar, top of head, 'undertail', eyestripe, white; back, top of head, wings, breast patches, light brown; bill and legs, black. Very active. Thin white wing-bar. Range 20 yards. Telescope used. Light good. Size 6 inches. In company with five young Ringed Plover.*

Another was seen there on June 23rd 1949.

Kentish Plover, male, Hill Head, May 21st 1999

Subsequently, there have been sightings in 29 years with a total of 59 recorded.

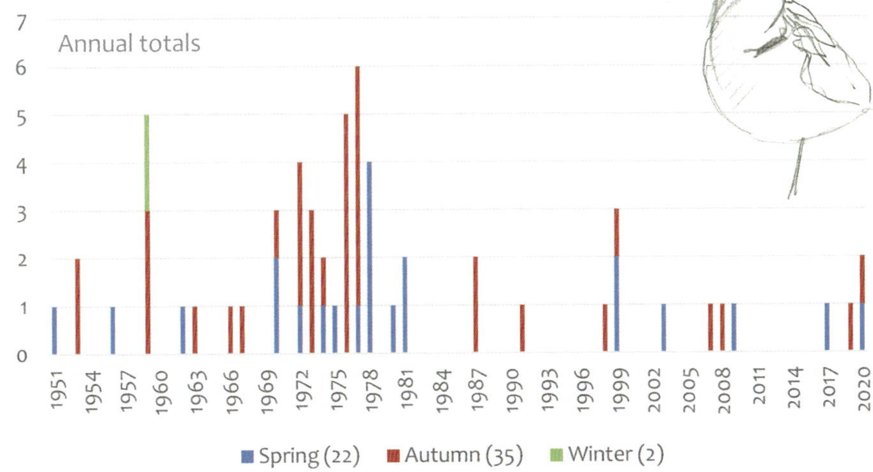

The peak during the 1970s can be attributed to the attractiveness of the reclamation pools at Dibden Bay at the time. Since 1982 it has become much scarcer with only 14 records in 11 years involving 15 birds. This reflects the increased rarity of the species in Britain.

Reports came from 11 localities, Keyhaven/Pennington Marshes (14 birds), Dibden Bay (14), Langstone Harbour/Farlington Marshes (9), Fawley Reclamation (5), Hill Head/Titchfield Haven (5), Black Point/Eastoke (5), Needs Ore (3), Hook-with-Warsash (2, including one also seen at Calshot), Weston Shore (1) and Ibsley Water (1). All were coastal apart from a male at Ibsley Water on June 8th 2003. The record in *BoH* of a male at Eversley Gravel Pits on April 11th 1986 is excluded as it was only seen in Berkshire.

Kentish Plover, juvenile, Calshot, October 25th 1987

Analysis of arrival dates shows spring records between April 5th (female, Pennington Marshes, 1978) and June 11th (female, Weston Shore, 2020) and autumn records between July 3rd (female, Titchfield Haven, 1999) and October 18th (juvenile, Calshot, 1987, until 28th; also seen at Hook-with-Warsash on 26th).

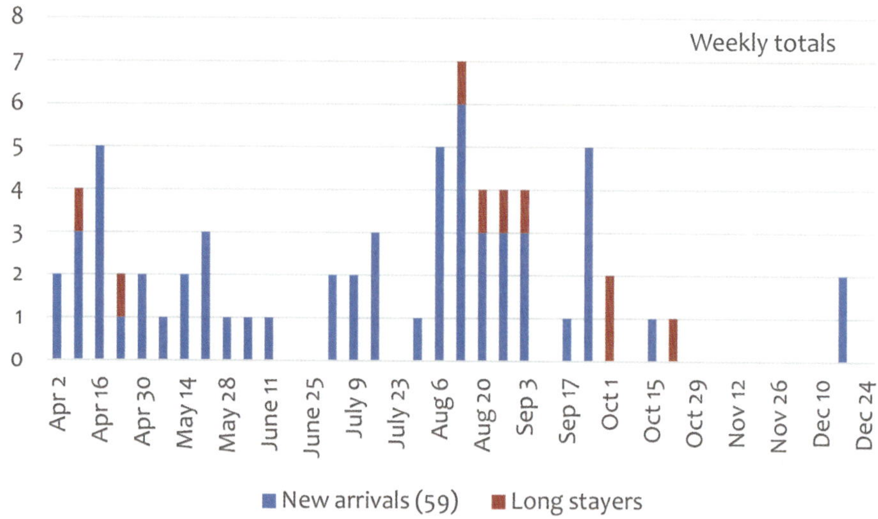

In spring most sightings were of singles on one date only, the exceptions being of a female at Farlington Marshes from April 16th to 19th 1970, a male at Pennington Marshes on April 21st and 29th 1974, a female there on April 5th and 9th 1978, two on the beach at Hill Head on May 21st 1999 of which one, a female, relocated to the scrapes at Titchfield Haven where it remained until May 24th, and a male at Pennington Marshes on April 20th and 21st 2017.

In autumn four were at Fawley Reclamation on September 25th 1976, of which two were trapped overnight on 28th/29th, with two staying until October 3rd, and a female and three juveniles were at Dibden Bay on August 17th 1977, one remaining until 27th. All other records were of singles on one date only except for two at Needs Ore on August 28th 1953, two at Langstone Harbour on July 15th 1959, the Calshot bird mentioned above and a first-winter in the high tide roost at Eastoke, Hayling Island on September 2nd and 7th 2019.

Kentish Plovers: male, Pennington Marshes, April 21st 2017 and female, Black Point, Hayling Island, May 4th 2009. Lee Fuller and Andy Johnson

There is one winter record of two seen by Tony Norris at Black Point, Chichester Harbour on December 23rd 1959.

Interestingly, almost as many have occurred in the Sussex part of Chichester Harbour as in the whole of Hampshire. Since 1949 a total of 60 has been recorded there comprising 36 in spring, 21 in autumn and three in winter. Forty-four of these were between 1988 and 2011 but only two were seen subsequently in 2018 and 2020. The most recent Hampshire record involved a juvenile seen by Andy Johnson in the high tide roost at Black Point on September 25th 2020 which was doubtless the same individual present at Pilsea Island, Chichester Harbour between September 19th and October 7th.

GB status: 807, 1986–2020. Declining, mean 12 per year in 2010–19. (Breeds Atlantic coastal regions of Iberian Peninsula N to Netherlands, Germany & Denmark, Mediterranean basin, & across inland C Asia to Japan and coastal China. Winters coastal regions of Mediterranean basin, Arabian Peninsula & S & SE Asia N to S Japan.)

Lesser Sand Plover *Charadrius mongolus mongolus*

BBRC

A very rare vagrant. One record in 2003.

The only record for Hampshire is of a male of the race *mongolus* in summer plumage found by Lawrie Chappell at Keyhaven Marshes at 0930 hrs on July 22nd 2003. Amusing extracts from his account of its discovery and identification (Chappell 2004) are reproduced below.

Stanpit Marsh, my usual haunt, had not been producing the goods of late and so I decided that a wander around Keyhaven Marshes would perhaps be a better bet. At least there would be a different selection of dogs to be harassed by, even if the birds were no better. So I headed east over the county border. However, it wasn't long before I began to regret my decision. There was little to see and I was becoming very disillusioned. I was wishing I had stayed at home – at least the grass would have been cut! I was now a mile or so from the car, fed up, hungry and on the point of giving up the ghost. The 'proverbial' one last look had never worked for me but, in a desperate attempt to see something, I gave it a go. I quickly scanned the muddy shore beyond the marsh and was astonished to find myself eyeball to eyeball with a sand plover. I wish I could say that I picked it out from the crowd by its call, leg length, footprints or droppings size, but I didn't. It was just there, on the mud, bright orange and looking at me. It was a stunning summer-plumaged bird – the orange breast and black and white head pattern made identification as a sand plover straightforward – but which one?

Lesser Sand Plover, male, Keyhaven Marshes, July 22nd 2003

Lawrence released the news and birders started to arrive on the scene. Initially the balance of opinion favoured Greater Sand Plover but once photographs had been taken and shared on the internet its identification as a male Lesser Sand Plover of the nominate race *mongolus* was agreed. It was seen by hundreds of birders on the first day and continued to show very well to numerous visitors until its departure some time on July 26th.

It was the fourth for Britain and, at the time, the first record of the race *mongolus*. Subsequently, an earlier record of one at Donmouth, Aberdeenshire in August 1991 was confirmed as the same race. The IOC has recently proposed splitting Lesser Sand Plover into Siberian Sand Plover *C. mongolus* and Tibetan Sand Plover *C. atrifrons*.

Lesser Sand Plover, male, Keyhaven Marshes, July 24th 2003.
Richard Ford and Jim Lawrence

GB status: 6, 1991–2013. Very rare. (Nominate mongolus group breeds mountains & tundra of Arctic E Siberia, Kamchatka & Komandorskiye (Commander) Islands. Distinct atrifrons group breeds from C Pamir & Tien Shan Mountain ranges S to N Kashmir & E to the Tibetan Plateau. Winters along tropical coasts of S & E Africa, Persian Gulf, Indian subcontinent, S China, SE Asia & Australia.)

Dotterel *Charadrius morinellus*

HOSRP

A rare passage migrant. Two old records of eight birds and 38 of 117 birds since 1955.

The first dated record for the county is of two shot at Ovington by Captain Phelps on September 30th 1893 which were added to the Winchester College collection (K & M, Fearon 1900). Unfortunately. the specimens have not survived. In the first half of the 20th century the only record was of six on a barren hill-top in the New Forest near Gorley on May 6th 1934 (*Brit. Birds* 28:

175). It is given as May 13th in the *Proceedings of the Bournemouth Natural Science Society* which is probably where the sighting was first put on record.

The first record for the modern era involved three found by David Billett at Farlington Marshes on September 26th 1955. His notes stated:

> *They were in winter plumage; small, dark plovers between Ringed and Golden Plover in size, slim and long-winged in appearance. The breast was a ruddy buff, the belly grey shading to white on the undertail-coverts; rump and tail dark and the outer tail feathers white, rather like a Turtle Dove; buffish eyestripe; shade of wing inconspicuous.*

There was a long gap until the next, one seen by Ed Wiseman at Keyhaven on August 27th 1969 and still present two days later when Edwin Cohen observed it. Between 1971 and 2020 there have been a further 36 records in 24 years involving a total of 113 birds.

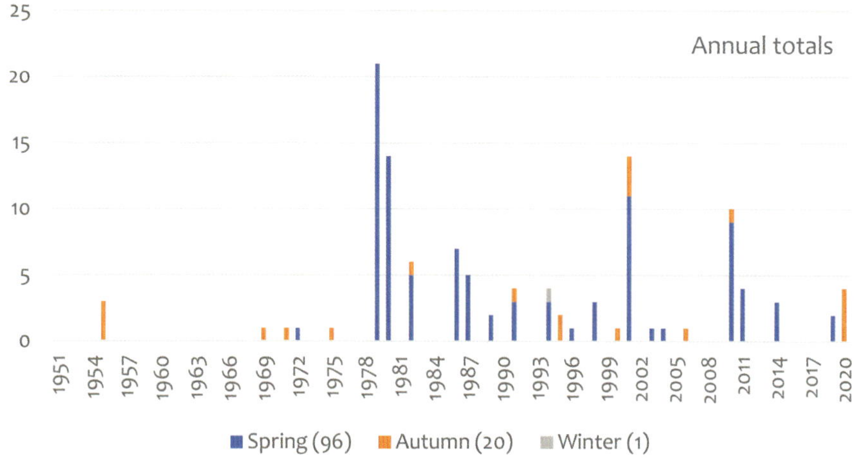

Most occurred in spring with 22 records totalling 96 birds spanning the period between April 6th and May 23rd. The peak period for passage was between May 1st and 9th with 12 records of 72 birds. There has been a tendency to an earlier arrival in recent years; the first ever in April was in 1996 and since then there have been five records of 17 birds in that month with the earliest ever of two at Cheesefoot Head from 6th to 8th in 2019. Most records have come from pea or bean fields on the chalk or burnt or short-cut heather moors in the New Forest. Favoured locations have been the Greywell–Odiham–Crondall area (4 records totalling 27 birds including 14 at Greywell on May 3rd 1980), Cheesefoot Head/Gander Down (5 records totalling 25 birds including 17 on May 6th 1979) and the New Forest (5 records totalling 19 birds including 11 at Ridley Plain on May 9th 2001). The only inland records away from these sites were of seven at Danebury on May 5th 1986, nine at Preston Candover on April 27th 2010 which were relocated at nearby Bradley Hill the next day and singles at Old Winchester Hill on May 5th 1991 and Over Wallop on April 21st 2014. There were four records from coastal locations involving three at Farlington Marshes on May 5th 1979, two at Dibden Bay on May 17th 1991 and singles at Hook-with-Warsash on May 19th 1979 and at Sandy Point on April 15th 1996.

Dotterels: female, Cheesefoot Head, May 1st 2011 and male, April 6th 2019. Keith Betton and Simon M. King

In autumn there were 15 records involving 20 birds spanning the period between August 17th and October 14th. Apart from the 1955 record at Farlington Marshes and the most recent of four moving south over Trigpoint Hill on August 24th 2020, all were of singles and most were present on one date only. Four occurred in the Keyhaven area, two at Needs Ore and Nether Wallop (the latter both associating with Stone-curlews), and one each at Fleet Pond, Hillside (Odiham), Long Down, Old Winchester Hill and Taddiford Gap. The longest stay was by one at Needs Ore from August 23rd to 26th 1991.

There is one record of a wintering bird. A juvenile was found by Ed Wiseman, Mike Baker and Adrian Clark among a flock of Golden Plovers at New Lane, Milford on Sea on December 2nd 1994 which remained in the area until at least January 29th 1995. Possibly the adult seen at nearby Keyhaven on October 14th 1995 was the same individual returning.

Dotterel, male and four females, Snatchangers Farm, Odiham. May 4th 1987

The weekly breakdown of records is shown below.

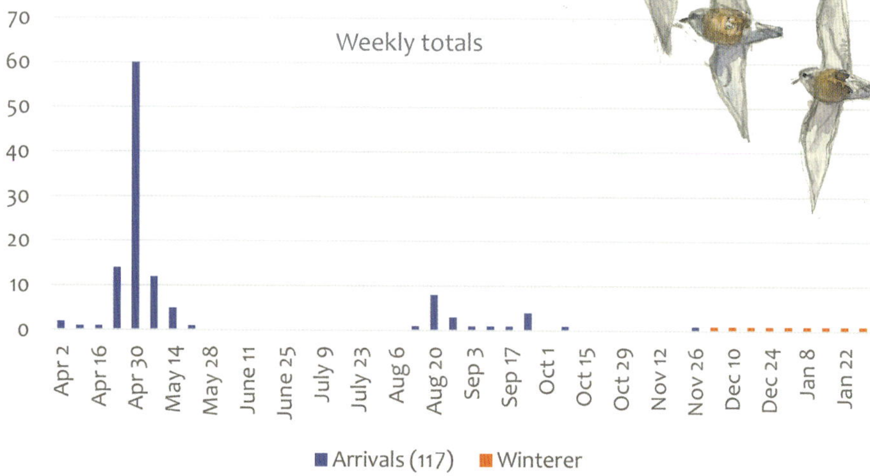

GB status: Very scarce and declining summer visitor to Scotland and N England; passage migrant elsewhere. (Breeds from W Europe to E Siberia; winters Mediterranean Basin to Middle East.)

European Black-tailed Godwit *Limosa limosa limosa*

HOSRP

Status uncertain; probably a very scarce passage migrant. At least nine between 1989 and 2005 and seven since 2017.

There was little awareness of this race of Black-tailed Godwit until the late 1990s. Since then, interest has increased, especially given the sharp decline of its breeding population in the Netherlands which fell from 47,000 pairs in 2007 to 33,000 in 2015 (Keller *et al.* 2020).

A series of records from Farlington Marshes between 1998 and 2003 has shown that the first juvenile *limosa* appear from late June onwards whereas juvenile *islandica* rarely appear before early August, the earliest known being on July 31st 2003. Juvenile *limosa* were recorded by Bob Chapman on July 8th 1998 and July 16th 2000, Bob, Jason Crook and Pete Potts on August 1st 2001 and Jason Crook on August 7th and 8th 2002 with two on August 11th of that year, and one or two on ten dates between June 28th and July 21st 2003. An adult there on September 24th, 26th and 27th 2005 was found and photographed by Jason Crook.

This suggests that juveniles recorded at Farlington Marshes by David Billett on July 8th 1989 and Jason Crook on July 13th, 15th and 17th 1995 and the Itchen Estuary by Ian Watts on July 25th 2011 were *limosa*.

The next record was of a male feeding at low tide with 98 *islandica* off North Hayling on April 11th 2015 which was found by Pete Potts; the flock flew off west towards Farlington Marshes. Another or the same male was found by Pete and also seen by Ruth Croger at Farlington Marshes on April 21st 2015. Pete's notes say, "very long bill (bright yellow), flat forehead, very long legs especially tibia, plumage paler than summer plumage *islandica*, stands taller than *islandica* when alongside, longer, thinner necked".

European Black-tailed Godwit, juvenile, Hook-with-Warsash, August 27th 2017. Andy Collins

There were no further records until two juveniles were identified by Ashley Saunders, a visiting birder, at Titchfield Haven on August 9th 2017. They were still present the next day and one remained until 12th. Another was found by Andy Collins at Hook Links with a small group of *islandica* on August 27th and what was presumably the same individual was found independently by Dan Houghton among a flock of 60 *islandica* at nearby Bunny Meadows on the same day. In the following year Ashley Saunders again found a juvenile among a flock of 20 *islandica* at Titchfield Haven on July 13th. Later that month, a colour-ringed juvenile was at Titchfield Haven on 28th. In 2020, two including the same colour-ringed individual, were seen by Ivor McPherson at Titchfield Haven on July 6th, the colour-ringed bird was seen again by Chris Button on 23rd and two days later Amy Robjohns located an unringed juvenile there. The colour-ringed bird is a male from the Nene Washes named Morgan. In May 2018, his parents were unable to nest on the washes due to flooding and so they bred on nearby farmland. His egg, caked in mud, was carefully cleaned and incubated at WWT Welney. He hatched on May 19th, was released on June 19th and was last seen on July 18th. He returned to his natal area in summer 2019 and 2020 and was also sighted at Cara Larga, Portugal on February 4th 2020 and at Pagham Harbour, West Sussex on July 1st 2020 (Jennifer Smart *in litt.*).

In 1981 a territorial pair was at Farlington Marshes in May and June. The male displayed over the main open pasture area in May and the female was missing for a period, suggesting incubation, but there was no further evidence of breeding. The following year a displaying male was present for some time but no female was seen. In 1990 a pair was seen displaying on May 31st but there were no further records. It is assumed these records refer to nominate *limosa* rather than *islandica*.

GB status: Very rare breeder, mostly in Ouse and Nene Washes. Passage migrant, under-recorded. (Breeds W and C Europe to Russia; winters in SW Europe, West Africa, Middle East and Indian subcontinent.)

Broad-billed Sandpiper *Calidris falcinellus*

BBRC

A very rare vagrant. Five records since 1957.

The first for the county was one found by Cliff Henty and David Billett at Farlington Marshes on October 5th 1957 where it remained until the next day. Their note to *British Birds* and its editorial response (51: 194–195) are reproduced below.

> *On 5th October 1957§, by the lagoon at Farlington Marshes, Hampshire, we found a small wader which we later identified as an adult Broad-billed Sandpiper in transitional plumage. The following day, 6th October, the bird was seen by Messrs. A. Allen, R. H. Dennis, G. Kinsey, B. W. Renyard, G. H. Rees, A. Searle and S. White. The following detailed description was made.*
>
> *Head: crown densely streaked dark brown or blackish; a conspicuous creamy stripe over the eye and, in addition, an inconspicuous, short, pale stripe passed backwards from the base of the bill, above the supercilium and separated from it by dark feathering (this secondary pair of crown stripes was difficult to see from the side, but clear enough from a head-on view); back of head and nape lighter and slightly greyer than mantle or crown; ear-coverts dark. Back and wings: boldly patterned, similar to Jack Snipe or Pectoral Sandpiper, both of which species were present that weekend; centre of back very dark, with a pair of conspicuous cream stripes from shoulders to upperparts of rump; another lower and broader 'V' was formed by pale edgings to scapulars; feathers of folded wing edged buff-brown, secondaries with bright chestnut (in sunlight this was a predominant feature of the bird); carpal joint showed as a dark patch; centre of rump and tail blackish, lateral tail-coverts white, outer tail feathers pale greyish. Underparts: chin and upper breast pale buff (chin paler, but not pure white); streaking on sides of breast, which just met in a narrow band across the centre; belly and vent silvery white. Bill: relatively long and heavy, stout at base and decurved at tip rather like that of a Ruff; colour of bill dark brown, base paler. Legs: colour varied with light but was generally agreed to be olive.*

It was in general squat and short-necked, appearing about the size of a Little Stint but relatively longer-bodied (possibly an illusion due to the low carriage). It showed a preference for reed and spartina stubble, where it moved with a crouching gait and was difficult to find or flush, having a strong tendency to walk away and hide out of sight. In flight the bird appeared very dark-backed and occasionally uttered a quiet 'chree' or 'chree-chree'. D. F. Billett and C. J. Henty

§ erroneously given as October 12th 1957 in British Birds *and corrected in 51: 526. [We showed this record to Mr. P. W. P. Browne, in view of his particular experience of Broad-billed Sandpipers out of breeding plumage, and he commented as follows: 'I presume the bird was thought to be an adult because the chin was not white. However, this is not a feature which could be relied upon (from the point of view of field-observation) and I should be surprised if an adult still had a bright chestnut patch on the secondaries by 12th October; a juvenile would be less likely to have the chestnut edges worn off'.–EDS.]*

A further quarter of a century was to pass until the next, an adult, was found by Dick Barrett and confirmed by Dave Unsworth at Paulsgrove Reclamation on May 22nd 1982. It remained until May 25th enabling many to observe it at close range. Another was found by K. Evans and two visitors from north-east England at Farlington Marshes on June 4th 1988. It was confirmed by David Billett and remained until the next day. An adult in partial summer plumage was found by John Eyre and confirmed by Dave Gumn and the author at Butts Lagoon, Pennington on May 2nd 1994 while scouting for a 24-hour bird race. It stayed until May 7th. Finally, one was seen and photographed by Russell Wynn at Pennington Marshes at dusk on May 2nd 2006 but was not present the next morning.

Broad-billed Sandpiper, Pennington Marshes, May 2nd 2006. Russell Wynn

GB status: 15, 1836–1949; 262, 1950–2020. About 4–5 per year; decreasing/stable. (Nominate falcinellus breeds N Norway, Sweden & Finland, & Arctic Russia where distribution uncertain. Migrates through E Mediterranean, Black & Caspian Seas to winter Persian Gulf & W India to Sri Lanka, with small numbers coastal E Africa. Race sibirica breeds from Taimyr Peninsula E to Kolyma River delta, & winters from Bay of Bengal through coastal SE Asia to Australia.)

Stilt Sandpiper *Calidris himantopus*

BBRC

A very rare vagrant. Three records in 2002, 2016 and 2017.

The first for the county was an adult moulting out of summer plumage found by Russell Wynn on Butts Lagoon, Pennington on July 21st 2002. It remained in the area until August 3rd and was seen by hundreds of observers. Russell told the story of his discovery in the 2002 *HBR* (Wynn 2003b), extracts from which are reproduced below.

At about 0750 hrs on Sunday July 21st 2002, I noticed a medium-sized greyish wader feeding on the far side of Butts Lagoon. At first glance I thought it was a Ruff but its grey tone and overall jizz made me suspicious so I walked round to the bank adjacent to the bird for a better look. An initial look through the scope revealed an elegant-looking wader with a long, decurved black bill, long greenish legs and lots of lovely dark barring extending all the way from the breast to the undertail-coverts. Although I had never seen one before I knew straight away that I was dealing with an adult Stilt Sandpiper.

Stilt Sandpiper, adult, Pennington Marshes, July 22nd 2002

I proceeded to spend the next two hours in a dream-like state writing a detailed set of notes and taking a series of digiscoped images. The bird was obviously in largely summer plumage, but the presence of some new feathers on the scapulars and wing coverts indicated that it was beginning its post-breeding moult. At 1000 hrs I saw another birder walking along the seawall with a mobile 'phone on his belt, so I rushed over and asked him if I could make a call. I contacted the reserve warden and a few local birders who arrived within minutes in various states of disrepair. Once we were all happy that the identification was watertight and the site was ready for a big weekend twitch, we 'phoned the news out at 1100 hrs. I estimated that several hundred people visited the site during the day; happily, everyone saw the bird and there were no problems at all apart from the bird sleeping for long periods in the afternoon! It was not shy, happily feeding within 30–40 m of the assembled crowds and continued to show well until it was last seen on August 3rd.

The same area also hosted the second when Richard Coomber discovered an adult in summer plumage on Fishtail Lagoon, Pennington Marshes on May 22nd 2016 which remained until the following day.

Stilt Sandpipers: Pennington Marshes, July 21st 2002 and with Black-tailed Godwit, May 23rd 2010. Paul Gale and Mike Trew

The first for Langstone Harbour was a juvenile found by Tim Doran at Farlington Marshes on November 7th 2017. This is assumed to be the same bird present at various sites in Dorset between September 11th and October 7th and again from November 24th until March 1st 2018.

GB status: 40, 1954–2020. Rare, about one every 1–2 years. (Breeds North America from NE Alaska to Hudson Bay, Canada. Migrates through interior & E USA to winter C South America from E Bolivia & S Brazil to NE Argentina. Occasionally winters N to Mexico, Caribbean & S USA.)

Temminck's Stint *Calidris temminckii*

HOSRP

A very scarce passage migrant. Three records of seven birds in 1935–1948 and around 142 since 1951.

The first for the county were four watched by Ken Smith at Keyhaven Marshes on May 12th 1935. He saw another there on May 25th of that year. An extract from his note in *British Birds* (29: 324) follows.

On May 12th, 1935, K.D.S. flushed a party of four Temminck's Stints from a muddy, disused salt-pan on Keyhaven Marshes. They were extremely tame, and for at least an hour he was able to watch them at a range of only a few yards. When flushed,they would tower high into the air before flying in a definite direction. Although they flew together, their flight was not well-ordered, as it is with most small waders, and on descending they would often separate and shoot headlong to earth, to rejoin again when this was reached. Their call could only be described as a succession of rippling notes, quite unlike that of the Little Stint, and it was usually uttered as they rose from the ground. They seemed to prefer marshy pools to the tide-line. When at rest their resemblance to minute Common Sandpipers was most striking, particularly with regard to the under parts. In flight the broad white sides of the tail contrasted well with the dark centre, and the habit, on landing, of expanding the tail, showed to advantage the pure white outer feathers. Their legs were tinged with brownish-yellow. On May 25th, Ken again watched a Temminck's Stint in the same place and was able to confirm his former notes.

Two were watched by Tony Norris at Titchfield Haven on and off all day on October 30th 1948. Although the notes included in the 1948 *HBR* include the pertinent field points, it is a very late date.

Between 1951 and 2020 there were sightings in 53 years with a total of 142 recorded, 79 in spring and 63 in autumn. The species has become much rarer in the last decade, with only five reported since 2011.

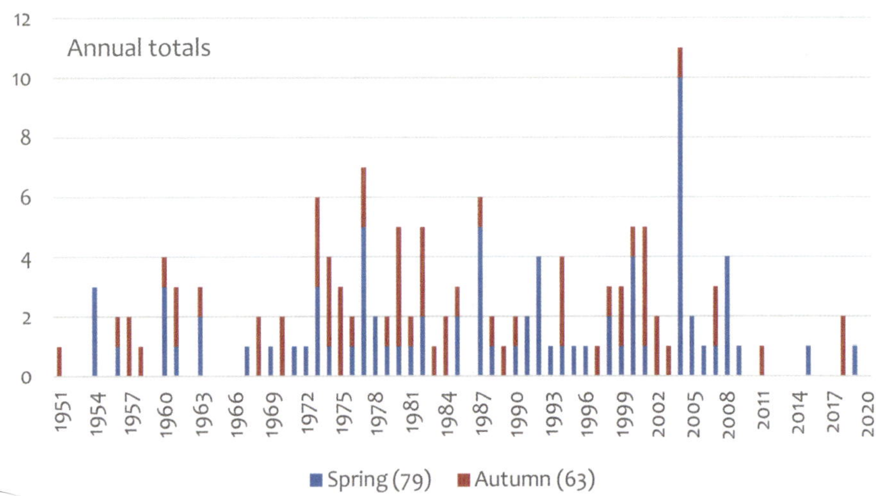

Spring records covered the period May 1st (1960, Farlington Marshes) to June 6th (1971, Farlington Marshes) with the majority in mid-May. Records were for Farlington Marshes (35 birds), Keyhaven/Pennington Marshes (18), Titchfield Haven (9), Ibsley Water (7), Dibden Bay (3), Needs Ore (2) and Fleet Pond, Lower Test Marshes, Paulsgrove Reclamation, Hook-with-Warsash and West Hayling (1 each). All sightings were of one or two birds apart from those of four at Farlington Marshes on May 15th 2004, three there from May 16th to 18th 1954 and on May 20th and 21st 1992, and three at Titchfield Haven from May 4th to 6th 2000. Most were present for one or two days but some remained for up to five. Two were at Farlington Marshes from May 17th to 22nd 1977 and one was still present on May 24th and the most recent record was of one at Keyhaven from May 17th to 25th 2019, although it was not seen from 19th to 22nd.

Temmink's Stint, Needs Ore, May 22nd 2007. Nigel Jones

Autumn arrival dates extended from July 8th (1980, Farlington Marshes) to October 1st (1957, Farlington Marshes) apart from an adult at Titchfield Haven on June 20th 2000 and one there from October 15th to 25th 1999 which was also seen at Hook-with-Warsash on 16th and 25th. One first noted at Dibden Bay on September 26th 1973 was still there on November 3rd. Records were for Farlington Marshes (25 birds including one also seen at West Hayling), Keyhaven/Pennington Marshes (13), Dibden Bay and Titchfield Haven (7 each), Lower Test Marshes and Milton Reclamation (2 each) and Ashlett, Eling, Ibsley North Lake, Lepe, Needs Ore, Hook-with-Warsash and Winchester Sewage Farm (1 each). All records were of one or two birds. Stays of up to ten days were not infrequent and, in addition to the long-staying Dibden Bay bird, one was at Farlington Marshes from July 20th to August 11th 1974 and another was at Keyhaven from September 24th to October 9th 2018.

The weekly breakdown of records is shown on the next page.

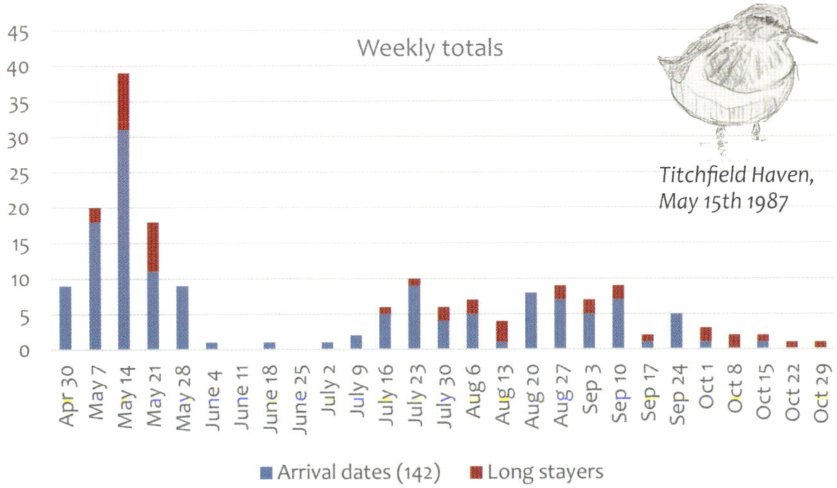

GB status: 5.027, 1968–2019. No clear trend, mean 96 per year in 2010–19. (Breeds mountains of S Scandinavia N through Fennoscandia to Barents Sea, N Gulf of Bothnia, & cool temperate Russia E to Chukotskiy Peninsula, E Siberia. Winters S of breeding range from Mediterranean basin, N Afrotropics, Middle East, S & SE Asia N to Japan.)

Long-toed Stint *Calidris subminuta*

BBRC

A very rare vagrant. One record in 2011.

The only record for Hampshire is of an adult found by Jason Crook on the Scrape at Farlington Marshes at dusk on September 8th 2011. The record was finally accepted by BBRC in 2022, along with that of one at Weirwood Reservoir, West Sussex from September 15th to 21st 2011, which was adjudged to refer to the same individual (Paul French, BBRC Chair, *in litt.*). Jason commented as follows:

It was a very dull, cool and overcast evening when I found a pale-legged stint, in worn adult plumage, while looking from the hide at the scrape. In fading light, I knew that I was up against the clock to identify and document it. I decided to mostly focus on obtaining digiscoped videos of it but my time was also used, away from the scrape, consulting a couple of field guides at the reserve's building and negotiating access for other birders to see it. The general jizz of the bird, in particular its crouching posture on bent legs as it crept around feeding, did not seem to fit the typical field guide descriptions and illustrations of Long-toed Stint, a species that I had never seen before. The details of the bird's structure had not been discernible until it was closer to the hide but by this time it was already dusk. It was questionable whether the last videos would prove to be any more useful than the first ones when the bird was more distant. In near-darkness, and with the bird cowering from the presence of a sparrowhawk, I also managed to get a couple of still images. Next day, despite extensive searching on the reserve, it was not seen again.

Long-toed Stint, The Scrape, Farlington Marshes, September 8th 2011. Jason Crook

For several years the videos and images resided in an ever-increasing file of stint identification papers, photographs and videos which I had sourced from books, journals and the internet. More time elapsed before I looked at the videos again and realised that the weight of evidence pointed towards Long-toed Stint. For example, I now considered the feeding posture and behaviour that had earlier detracted from its identification were not inconsistent with the species. I was in a quandary. Had I got enough evidence to prove, beyond doubt, that the Farlington stint had indeed been a Long-toed? In early 2018 I had the opportunity to see Long-toed Stints in Thailand and this further cemented my thoughts that the Farlington bird had been of this species. But I was still unsure if I had enough to proceed further. There were key identification points, such as the tibia and toe lengths, just about visible in the videos, in support of Long-toed Stint, but would these be enough for acceptance of such a rare vagrant to the UK?

More recently I discovered that the BBRC, through a separate avenue of working, had also been pursuing a path towards the correct identification of the Farlington stint. And then, in 2021, the committee received new documentation of a Long-toed Stint in Sussex and used this opportunity to finally re-assess the video evidence of the Farlington bird. They have concluded they were the same individual. Measurements of bill, toe and leg lengths and ratios, coupled with increased confidence in some features of Long-toed Stint identification, proved critical in their decision.

GB status: 5 records, 1970-2021. (Breeds boreal zone of Siberia, winters mostly SE Asia and Australia.)

Baird's Sandpiper *Calidris bairdii*

BBRC

A very rare vagrant. Eight records since 1986.

The first for the county was a juvenile found by David Christie at Hook-with-Warsash on October 7th 1986. Extracts from his submission to BBRC are reproduced below.

> *Size that of a smallish Dunlin; a slender looking calidrid with noticeably long wings extending well past tail and creating very long, slim appearance. Head rather small and rounded. Bill fine and rather straight, with fine tip; length less than diameter of head; black. Legs blackish, shorter than Dunlin'. Upperparts with prominent feather edgings; clear but narrow pectoral band. Flew a few metres, once. Rump and tail centre obviously very dark with narrow white side to rump. Called once, a short, low, trilled 'kreeet'.*

The next was a juvenile found by Paul Bowman at Pennington Marshes on September 7th 1988 which was seen by many others during its stay until September 18th. This was almost certainly the same individual which was on West High Down, IOW from September 3rd to 5th. A year later another juvenile was found by Terry Jennings and confirmed by Dave Unsworth at Farlington Marshes on September 20th 1989; it was last seen on 22nd. The fourth for the county was yet another juvenile found by Philip Hack at Blackwater, Needs Ore on September 13th 1994. It remained until 24th enabling many observers to see the bird during its stay.

The pattern of autumn juveniles was broken in 2000 when Jason Crook found an adult in summer plumage roosting with Dunlin and Ringed Plovers at Hayling Oyster Beds, Langstone Harbour at 1415 hrs on May 8th. Five other birders managed to arrive in time to see it before it was flushed by a male Peregrine at 1645 hrs and flew off south into the harbour. It was not seen again.

Another juvenile was found by Graham Barrett and Martin Cooper at Fishtail Lagoon, Pennington Marshes on September 29th 2005. Its identity was confirmed by Marc Moody and the bird remained in the area until October 16th.

Two further records also involved autumn juveniles in the same area. The first was found by Jim Waterman on Shoveler Pools on September 14th 2011 where it stayed until 17th and a year later one was videoed by Andy Olden in the same place on September 14th 2012 but was not seen by any other observers.

GB status: 1, 1911; 294, 1950–2020. About seven per year; stable. (Breeds extreme NE Siberia on Chukotskiy Peninsula & Wrangel Island, E across N Alaska & Arctic Canada to N Baffin Island & NW Greenland. Migrates through North American interior to winter South American Andes, from S Ecuador to Tierra del Fuego.)

Baird's Sandpiper, juvenile, Pennington Marshes, September 14th 2011. Simon Johnson

Least Sandpiper *Calidris minutilla*

BBRC

A very rare vagrant. One record in 1977.

The only record for Hampshire is of an adult in summer plumage with two Temminck's Stints found by David Billett at Farlington Marshes on May 22nd 1977. It was also seen by Rosemary Billett, Dick Barrett and Barry Collins and filmed by Tony Juniper. Extracts from David's submission to BBRC are reproduced below.

Circumstances of the Observation. While watching two Temminck's Stints at some distance at the edge of the lagoon at Farlington Marshes, an odd stint occasionally came into view. At this range, c. 150 yards, it briefly caused confusion with the Temminck's if seen head on. It was assumed, at this stage, to be a Little Stint in rather dull summer plumage. Later in the day, under perfect light conditions, all three were watched at close range down to 15 yards and filmed. At this range four characteristics which ruled out Little Stint were obvious.

The upper parts were darker and duller though still richly patterned. The breast was closely streaked brown, terminating fairly abruptly but with a suggestion of some faint marks towards the flanks. The bill was fine and distinctly down tilted slightly at the tip. The legs and feet were a striking warm ochre colour also described as a light brownish orange with the feet distinctly brighter.

Description.
Size. As Little Stint as less stocky with finer head shape.
Upperparts. Dull brown with some warmer areas. Darker on head – crown, coverts picked out with warm buff tips to give a typical summer stint pattern. A faint broken V discernible when viewed from the rear. Pale supercilium. Ear-coverts rather warmer than rest of head.

Underparts. Very white; upper breast strongly streaked brown.
Tail. In flight as Little Stint.
Legs and feet. Warm ochre with brighter feet.
Bill. Black. Fine and slightly downward tilted at tip.
Call. Only heard once. Somewhat like a Temminck's Stint but higher pitched and weaker.

Behaviour. Typically stint-like on the ground and in flight. When alert head and neck drawn up in Pectoral Sandpiper like stance. Fed in company with the Temminck's and a northern race Dunlin. This bird was not present on the following day and only one Temminck's remained.

Unfortunately, the film could not be located in the BBRC archive. Paul French (BBRC Chair) recently informed the author that of the ten voting members, eight accepted the record without comment, one abstained as he couldn't view the film, and one said: 'yes, Long-toed is a big problem with Least records but this bird fits Least so nicely that it is acceptable as such. I would think Long-toed to be blacker on the mantle than this bird'.

A moulting adult stint found by Marc Moody at Pennington Marshes on September 9th 2005 was subsequently identified as a Least Sandpiper and almost certainly that which had been at South Milton Ley and Thurlestone Marsh, Devon from August 28th to September 7th of that year. Marc was initially unsure about its identification but became certain when he compared sketches made at the time with photographs of the Devon bird which appeared in the media. However, BBRC harshly considered the record not proven. Marc has recently confirmed that he remains 100% happy with the identification.

GB status: 4, 1853–1949; 37, 1950–2017. Rare, about one every 1-2 years. (Breeds C & S Alaska, E across N Canada to Labrador & Newfoundland. Winters S USA, C America, the Caribbean & South America, S to Brazil & N Chile.)

White-rumped Sandpiper *Calidris fuscicollis*

BBRC to 2005; HOSRP 2006 to date

A rare vagrant. 20 records of 19 birds since 1963.

The first for the county was found by Canning Suffern and three members of Winchester College NHS, Robert Cheke, Jeremy Davidson and Thomas Sparrow, at Titchfield Haven on October 17th 1963.

Canning tells an interesting story in his submission to BBRC from which extracts are reproduced below.

This bird was resting and feeding with a flock of nine Redshanks on mud in shallows of the River Meon until we flushed them down river. When we came off the marshes the three Winchester boys went off to Farlington Marshes, but I went the other way just in time to see a helicopter flush these Redshanks again. With them still was the sandpiper, but they did not go far and soon returned, but without the sandpiper, which was not seen again.

I took care not to identify this bird to the other observers; and, without saying what I thought, I asked them for their opinion on the bird's bill. Without hesitation they all agreed that it was short and straight. I asked them for three separate descriptions, to be written without collusion. They came today, and I enclose them herewith.

The three descriptions all confirmed that the bird had a short, straight bill and showed a white rump in flight.

Subsequently, a further 18 have been recorded.

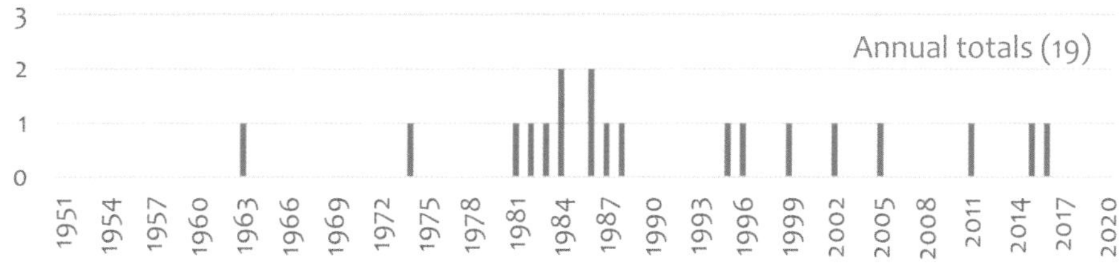

The next was found by Bob Dunn and Ed Wiseman at Pennington Marshes on August 10th 1974. It could not be relocated later that day or on subsequent visits. The third was an adult in winter plumage found by David Christie at Hook-with-Warsash on October 12th 1981. This was followed by a juvenile found by Reg Russell at Eling Great Marsh on October 25th 1982 and still present two days later, when it was described as lame. David Christie had the good fortune to find another adult at Hook-with-Warsash on October 27th 1983. Two were found in 1984, the fourth consecutive year with records. An adult was found by Graham Stephenson and seen by Will Legge and the author on The Lake at Farlington Marshes on August 27th 1984 and was still present the next day. David Christie completed his hat-trick with a juvenile at Hook-with-Warsash on October 12th 1984. The next was a moulting adult found by Roger Wilkinson on The Lake at Farlington Marshes on September 3rd 1986. He watched it with Ian Calderwood and Kevin Ilsley for an hour up to 1230 hrs when it flew off into Langstone Harbour as the tide dropped. It was not seen again. Later that autumn Reg Russell found another juvenile among a flock of Ringed Plovers at Eling Great Marsh on September 29th. Farlington Marshes' third bird of the 1980s was an adult found by Richard White on The Lake at high tide on August 10th 1987. It was confirmed by Terry Jennings and John Smith and seen by many observers in the period up to August 17th. The final bird of the decade was a juvenile found by Dan Powell at Titchfield Haven on October 2nd 1988 which was seen by many others up to 5th.

White-rumped Sandpipers: adult, North Scrape, Titchfield Haven, October 2nd 1988 and juvenile, South Scrape, August 8th 1995

There was a seven-year gap until the next, an adult found by Barry Duffin at Titchfield Haven on August 7th 1995. It was subsequently seen by many observers and last seen on 18th. In the following year Bob Chapman found an adult which had almost completed its moult to post-breeding plumage at Farlington Marshes on August 16th where it stayed until September 7th. The following day at 1615 hrs Paul Bowman found a similarly-plumaged adult at Blackwater, Needs Ore which was seen the next day by Dave Unsworth and others. BBRC considered it only possible that it was the Farlington bird but here it is treated as the same individual. The next was an adult in summer plumage found by Ron King at Titchfield Haven on July 28th 1999 which was seen by many observers until August 3rd. Ron believed it was the same bird he saw at Ferrybridge, Dorset on July 26th.

An elusive bird was found by Paul Winter and seen by Ted Rolfe at Testwood Lakes at 0945 hrs on September 15th 2002. Paul managed to get further views later that day and obtained some photographs on 16th which helped clinch the identification. An adult in summer plumage was found by Jason Crook in the high tide roost at Farlington Marshes at 1810 hrs on July 27th 2005. It was subsequently seen by Pete Gammage and Tim Lawman before flying off over Langstone Harbour with two Dunlin at 1923 hrs.

The next was a remarkable record of one picked up by Tim Parminter and Marc Moody as it flew east with a small group of Dunlin past Milford on Sea on September 24th 2011. They searched for it without success at the nearby Keyhaven/Pennington Marshes but did enjoy ample compensation when they located a Semipalmated Sandpiper! An adult was found by Pete Gammage on the stream at Farlington Marshes on October 25th 2015 which later moved to The Deeps and then flew out into Langstone Harbour at low tide. Finally, a rather mobile bird was first identified by Jim Waterman at Pennington Marshes on September 19th 2016. It was seen at various sites between Normandy Lagoon and Keyhaven until September 24th.

All occurred between July 27th and October 27th.

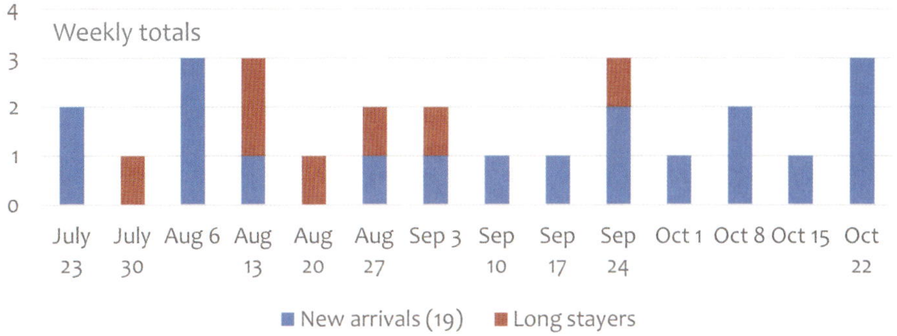

Of the 19 birds identified, ten were aged as adults and four as juveniles. All the adults were found on dates between July 27th and September 3rd apart from those on October 12th and October 27th; the juveniles were found between September 29th and October 25th.

White-rumped Sandpiper, adult, Normandy Lagoon, September 19th 2016. David Ryves

GB status: 16, 1837–1949; 679, 1950–2019. Increasing, mean 19 per year in 2010–19. (Breeds N Alaska & Arctic Canada, from Mackenzie River E to S Baffin Island. Overflies W Atlantic to winter S South America.)

Buff-breasted Sandpiper *Calidris subruficollis*

BBRC to 1982; HOSRP 1983 to date

A rare vagrant. 13 records since 1975.

The first for the county was a somewhat flighty individual which was seen by Bob Dunn, Ed Wiseman and Mike and Pete Combridge at Pennington Marshes on August 31st 1975. It was also seen by David and Keith Godfrey. Extracts from Ed's description are given below.

Our first impression was of a long-winged dark looking wader recalling, superficially, a Pectoral Sandpiper. It continued to fly around until after about half an hour we were able to get views of it feeding on the mudflats on the outside of the seawall. It eventually flew off eastwards. In size the bird was a little larger than a Ringed Plover and in flight recalled a Golden Plover-like bird. On the ground the underparts were an unrelieved warm buff colour, and the upperparts very scaly looking and boldly marked blackish brown. The bill was short and blackish and the legs long and noticeably yellowish. As it stopped feeding and raised its head a long-looking neck was revealed. With a small head it then resembled a Reeve. In flight there were no pale markings on the upperwings but the underwing was whitish with dark edgings. The outer tail feathers were greyish with dark central feathers. Movement on the ground was Ringed Plover-like with little runs then tipping-up to feed. Eventually we were able to view it down to 35–40 metres.

The second was found by Andy Williamson at Farlington Marshes at 1930 hrs on June 11th 1976. It was watched at close range where it was persistently chased by a pair of Ringed Plovers. After 20 minutes it disappeared and was not seen again.

Later that year Stuart Sporne trapped an adult during an overnight wader ringing session at Fawley Reclamation at 0200 hrs on October 8th. The next day Philip Fawkes, Norman Pratt and John M. Jones searched for the bird but found an unringed individual that was still present on October 10th. The next was watched by Mike and Pete Combridge at Pennington Marshes from 1645 to 1750 hrs on September 4th 1977. Another was found by Nigel Rogers at Keyhaven on September 12th 1980 and seen the next day by Ed Wiseman. It remained in the area until 22nd. The fourth for that area was found by Len

Mummery and Ed Wiseman at Pennington Marshes on September 25th 1982 and stayed until 29th.

The eighth for the county was found by John Walters at Sinah Common on September 17th 1983 and seen by many observers before departing the next day. There was a gap of eight years before the next, found by Ed Wiseman at Pennington Marshes on October 6th 1991. It was taken by a Sparrowhawk the next day.

Another eight-year gap ensued before the next was found by Jason Crook at Farlington Marshes on October 3rd 2001. It was initially seen at 1310 hrs as it flew in and settled with roosting Dunlin before it became mobile and elusive. After two more views of it settled on the reserve in different locations Jason last saw it heading southwards across the harbour towards Hayling Island at 1710 hrs. It was relocated by Tim Lawman on Sinah Common at 1800 hrs. It remained there until October 5th when it departed high to the north-east at 0815 hrs.

Three subsequent sightings were somewhat brief encounters. One was observed flying east by Steve Keen over Barton on Sea Golf Course at 0845 hrs on September 26th 2008 and back west ten minutes later. An adult was found by Martin Cooper and seen by Marc Moody at Pennington Marshes on August 20th 2013. They watched it from 1330 to 1425 hrs but it was not seen again. One was seen briefly at Titchfield Haven by Dan Houghton on September 15th 2016. It was spooked with other waders but flew around allowing the observer long enough to note the key features before it departed to the west.

Buff-breasted Sandpiper, Pennington Marshes, August 20th 2013. Barry Lintott (far left) and Marc Moody

GB status: 14, 1829–1949; 1273, 1950–2019. Stabilising after increase, mean 38 per year in 2010–19. (Breeds Chukotskiy Peninsula & Wrangel Island, E Siberia, & Arctic North America from N Alaska to Yukon, Northwest Territories & Queen Elizabeth Islands, Canada. Most migrate S through Midwest USA, with a few following Atlantic seaboard, to winter on grasslands of Uruguay, Paraguay & Argentina.)

Pectoral Sandpiper *Calidris melanotos*

BBRC to 1962; HOSRP to date

A very scarce passage migrant. One in 1947, then 96 records of 100 birds since 1952.

Arguably the first for the county was one (of two present) which James Ferguson-Lees saw fly from Thorney Island, West Sussex across Emsworth Channel towards Hayling Island on October 12th 1947. However, it was not seen to land (Cohen 1963). A full description is given in *British Birds* (41: 186–187).

The first grounded bird was found by David Billett and Graham Rees at Farlington Marshes on September 7th 1952 where it remained until 11th. David's description, taken from the Langstone Harbour yearbook, is reproduced below.

> *Upperparts – richly coloured and patterned like a Jack Snipe. In flight showed an inconspicuous light wing-bar. Tail whitish with dark centre. Underparts – neck and breast a warmish buff (sandy) closely streaked with brownish/grey – this terminating suddenly to form a 'bib' similar in extent to a Common Sandpiper. Throat white. Bill – Blackish and straight. Legs – not clearly seen but appeared to be colour of dead reed.*
>
> *In flight appeared slightly larger than Common Sandpiper. Short flights similar to this species but when disturbed flew rapidly with a Snipe-like action. On each occasion, when flushed, uttered a rapidly repeated, almost twittered note – rendered by observers as 'skcheuc – skcheuc' or 'kchic – kchic'. When suspicious adopted a very Ruff-like stance. It was in fact quite shy. Fed by rapidly picking small objects from surface of ooze or water.*

Between 1957 and 2020 there were records in 48 years involving a minimum of 100 birds. Annual totals exceeded three in seven years with a maximum of six in 1974.

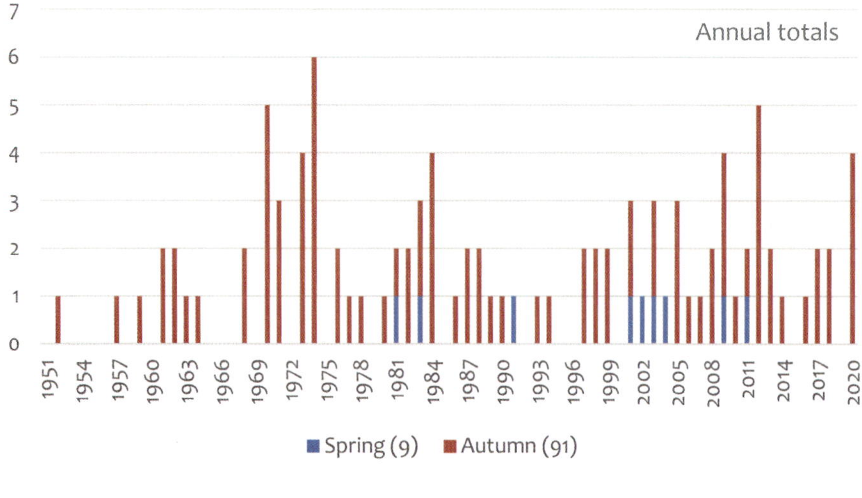

Pectoral Sandpiper, juvenile, West Hayling Shore, September 26th 2009.
Nigel Jones and Steve Bassett

Nine occurred in spring, at Pennington Marshes on May 16th 1981 and May 29th 1983, Hook-with-Warsash on May 21st 1991 and then Titchfield Haven on May 22nd and 23rd, Farlington Marshes from June 7th to 10th 2001, on May 16th 2002 and from June 3rd to 5th 2003, and Pennington Marshes on May 3rd 2004, from May 9th to 11th 2009 and on May 7th 2011. Possibly the records at Farlington Marshes in three consecutive years referred to the same individual although they have been treated separately in the analysis.

The remainder were in autumn with arrival dates between July 2nd (1977, Farlington Marshes) and October 30th (1976, Keyhaven Marsh). The pattern reflects the national picture with small numbers of adults in July and August, a peak in September when juveniles occur and very few after mid-October. The main localities were Keyhaven/Pennington Marshes (37 including up to five between September 6th and 16th 1970 and two on September 1st and 2nd 1982), Farlington Marshes (21), Dibden Bay (7, including two on September 20th 1968), Needs Ore (6), Titchfield Haven (5) and Paulsgrove Reclamation (4, including two from September 1st to 8th 1984 and one first seen at Farlington Marshes). Other records, all of singles, were from Eling Great Marsh (October 6th–10th 1961), Calshot Reclamation (August 24th–26th 1974), Hook-with-Warsash (August 17th 1974 and September 23rd–October 2nd 1986), Mockbeggar Lake (August 30th–September 20th 1981 and September 14th–21st 2003), Winchester Sewage Farm (September 10th–27th 2001), The Vyne Water Meadows (September 23rd–27th 2006 and September 27th to October 14th 2018), West Hayling Shore (September 25th–27th 2009), Black Hill, New Forest (September 29th and 30th 2016) and Sandy Point (August 24th and 25th 2020).

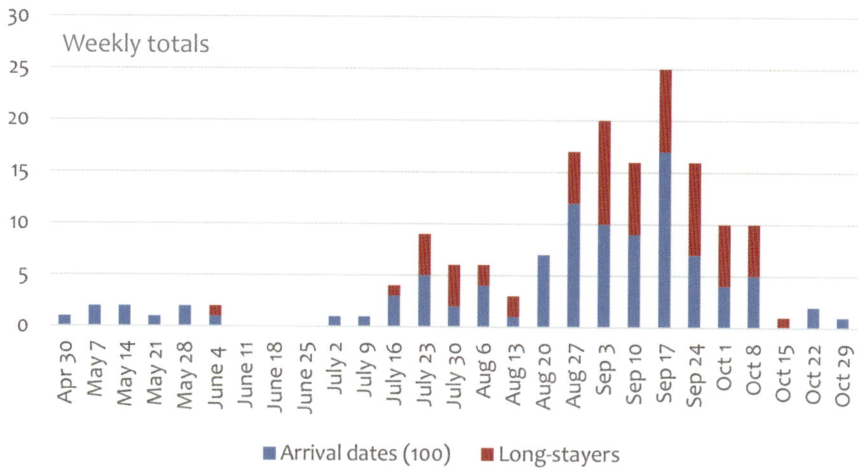

One which was at Farlington Marshes from October 25th to 27th 1969 was almost certainly a mis-identified Sharp-tailed Sandpiper. It has now been excluded from the records of this species. See Appendix II.

GB status: 74, 1830–1949; 4.345, 1950–2019. Stabilising after increase, mean 126 per year in 2010–19. (Breeds Arctic Siberia from Taimyr Peninsula E to Chukotskiy Peninsula, & North America from Alaska E to Southampton Island & S shore of Hudson Bay, Canada. Migrates through North America to winter South America from Peru & N Argentina to Patagonia.)

Semipalmated Sandpiper *Calidris pusilla*

BBRC

A very rare vagrant. Seven records since 1994.

The first for the county was an adult in summer plumage found by Paul Stancliffe at Keyhaven Lagoon late in the morning on July 17th 1994. He was unsure of its identity at first and other birders dismissed it as a Little Stint. However, Paul suspected it was something rarer and, having gone home to check the literature and make a few phone calls, he returned to Keyhaven where he met up with David Smith and John Hall. With views down to 25 m they were able to clinch the identification and news was released. It remained through the next day and was seen by many observers. Extracts from Paul's account of its discovery and identification from the 1994 *HBR* (Stancliffe 1995) are reproduced below.

I decided to head for Keyhaven Lagoon. Getting comfortable on the sea wall I started scanning the waders and found a juvenile Wood Sandpiper at close range. Felling pleased with myself I resumed scanning and located two more Wood Sandpipers and an 'odd' Little Stint. Although fairly distant, it showed a strong white supercilium, an obvious broad, black scapular line, long tertials and a distinct 'pec' band. It obviously wasn't a Little Stint, but what was it? It then flew closer, showing a narrow white wing-bar and white sides to a dark tail. In flight it looked quite grey. Could it possibly be a Temminck's Stint? This would explain the long tertials, pectoral band and possibly the black scapular line, but not the white supercilium, face pattern and black legs. I was sure I had one of the rarer stints or 'peeps' and set about taking notes. I called two other birders over and explained what I'd seen. After about 15 minutes they both agreed it was a Little Stint and left. At this point, I wondered if I'd made a big mistake and worked through the bird again. It definitely wasn't a Little Stint; in fact, I suspected that it might be a Semipalmated Sandpiper. I was joined by three other birders who wanted to see the Little Stint! They admitted they didn't have any experience of the rarer species and didn't know what to look for. It became clear I had to get help, so I left to make a few phone calls and collect Shorebirds from home. I arranged to meet Dave Smith at Keyhaven later. Returning to the spot, I was the only birder there but was relieved to find the bird still present although right at the back of the lagoon. With the help of Shorebirds, I was convinced it was a Semipalmated Sandpiper in breeding plumage but it was impossible to see some features I needed to clinch its identity. Dave arrived with John Hall and we decided to get closer to the bird. With views down to 25 metres the bill looked distinctly blob-ended and webbing could be seen between the toes, more obviously as the bird walked away. The dark centred scapulars were edged pale grey with no rufous tones at all. It was without doubt an adult Semipalmated Sandpiper in breeding plumage, and news of the bird's presence was subsequently put out.

The next was an adult trapped at Farlington Marshes at 0440 hrs on September 3rd 1995. Duncan Bell extracted it from the mist net and handed it to Trevor Codlin to process as a 'Little Stint'. Trevor noticed that it had palmations between its toes and thus his suspicions that it was the rarer species were confirmed. Processing was completed by Duncan, Trevor and Guy Thompson and it was released on the main lake. It remained there until 0820 hrs, enabling several birders to connect with it, before flying off, not to be seen again in Hampshire. However, it was subsequently controlled ten days later in Majorca, Spain and again in spring 1997 in southern Spain, on both occasions being identified as a Little Stint. The Spanish ringers have refused to amend the identification despite the existence of photographs showing the palmations.

Semipalmated Sandpiper, juvenile with Little Stints, Pennington Marshes, September 4th 2006

The first for the 21st century was a juvenile discovered by Marc Moody at the Jetty Lagoon, Pennington Marshes at 0920 hrs on September 4th 2006. His identification was confirmed by Tim Parminter. The bird remained in the area until 19th although it relocated to nearby Normandy Lagoon from 9th to 11th and on 13th. Another juvenile was found by Alan Lewis, Marc Moody and Tim Parminter at Fishtail Lagoon, Pennington Marshes on September 24th 2011. It was seen by many observers and was last seen on October 2nd. The third juvenile in the area was found by Mark Hastings and identified by Lee Fuller at Keyhaven Lagoon on September 6th 2013. It was still present the next day. The sixth for the county was a juvenile at Titchfield Haven on September

10th and 11th 2016. It was initially identified as a Little Stint but reidentified from photos by Alan Lewis. A juvenile stint found by Tom Jordan and seen by Dan Hoare at Winchester Sewage Farm at 1630 hrs the next day was thought to be the same bird. It was flushed by a Hobby at 1745 hrs. Unfortunately, BBRC considered the photographs obtained not sharp enough to confirm the identification.

The seventh for the county and fifth for Keyhaven/Pennington Marshes was a first-winter found by Chris Rose and Ian Smith at Oxey Marsh on November 16th 2019. Initially there was some discussion that it might be a Western Sandpiper due to its rather long bill but it was soon confirmed as this species. It was seen by many observers up to November 25th. Further claims of it between January 10th and 18th 2020 were not substantiated.

Semipalmated Sandpipers: juveniles, Keyhaven Lagoon, September 7th 2013 (top left), Shoveler Pools, Pennington Marshes, September 27th 2011 (left) and South Scrape, Titchfield Haven, September 11th 2016 (top right). John Hilton, Martin Bennett and Barry Stalker

GB status: 166, 1964–2020. About 4–5 per year; stable or increasing. (Breeds on tundra of W Alaska, E across Arctic Canada to S Baffin Island & coastal Labrador. Has bred extreme NE Siberia. Migrates across Great Plains & E seaboard of USA to winter C America & coasts of tropical South America to Brazil & Peru.)

Western Sandpiper *Calidris mauri*

BBRC

A very rare vagrant. One record in 2021.

The only record for Hampshire is of an adult in winter plumage found by Martin Cooper at Normandy Lagoon on October 4th 2021. Initially, there was uncertainty about its identity, Semipalmated Sandpiper also being possible, but on the second day Simon Buckell, who has considerable experience of the species in Alaska, was unequivocable in its identification as Western.

Martin described the circumstances of the bird's discovery as follows:

I arrived at Normandy Lagoon about 0900 hrs to check the waders coming in at high tide. I noticed some Dunlin among which there was a Little Stint, a well-marked juvenile with obvious white braces. I moved closer and settled down with the telescope and then thought I saw a second stint. This was quite dark backed, unlike the other bird, and it had a very long bill – too long for any Little Stint. It had black legs, black bill and was very white on the belly unlike most of the Dunlin which still retained some dark feathers. My thoughts turned to Semipalmated or Western Sandpiper. In 2019 there had been a Semipalmated on nearby Oxey with an unusually long bill. I showed the bird to Martin King and Kevin Kearns and also contacted reserve warden Pete Durnell to come to see the bird. We discussed the possible identity of the bird but couldn't reach a firm conclusion. Pete released the news via Going Birding stating that it was a 'SemiP' but that, because of its long bill, Western had not been ruled out. Several birders arrived that afternoon but the bird had flown out to feed on the intertidal mud with the Dunlin. I returned at 0800 hrs the next morning and joined Simon Buckell, who was watching a handful of Dunlin and a juvenile Little Stint. I had to leave for work soon after.

Simon takes up the story:

Around 0840 hrs several small mixed flocks of Ringed Plover and Dunlin began to fly in to roost on the lagoon. I scanned through a group which landed on the nearest island and immediately could see the bird in question from yesterday, a pale stint/peep. It was walking along the edge of the island and stopped to feed directly head on; through the scope I could see the long, fine-tipped bill pointing towards the ground. As the bird turned side on I noted the attenuated jizz, egg-shaped body and tall, long structure which hit me immediately and made me realise it was a Western. I was able to compare it with the juvenile Little Stint and could see that it was larger, taller, longer-legged and more attenuated. The bill length looked good and the plumage tied in nicely for an adult Western moulting to winter plumage.
I advised several local birders that I believed it was a Western Sandpiper.

More birders arrived during the morning although the strong winds and distance of the bird made observation difficult. Photos and videos were shared on social media and much debate ensued but by the third day, when viewing conditions were much improved, all were agreed that it was a Western Sandpiper, confirming Simon's initial call the previous day. The bird continued to feed and roost on the lagoon at high tide and was also seen by Simon feeding at low tide in Oxey Creek on October 11th. It was last seen on October 14th.

Western Sandpiper, adult, Normandy Lagoon, October 11th 2021. Jeremy McClement

GB status: 10, 1969–2016. (Breeds NE Siberia, W & N Alaska. Winters California to Peru and N Carolina to Venezuela)

Long-billed Dowitcher *Limnodromus scolopaceus*

BBRC

A rare vagrant. Eleven records, characterised by long stays, since 1975. In addition, two records of unidentified dowitchers.

The first for the county was found by George Johnson and Malcolm Theobald at Keyhaven Marshes on the morning of September 20th 1975. It was relocated the following morning by Bob Dunn and subsequently seen by many observers during its stay until 29th. A detailed description was submitted by Ed Wiseman, extracts from which are reproduced below.

> *On the ground the bird was Snipe-like and about the size of nearby Redshank. The bill was straight and very long, at least one and a half times the length of the head, blackish in colour and flattened at the tip. The upperparts were brownish and richly marked with blackish brown centres to the feathers. There was distinctive greyish feathering on the scapulars. The underparts were a warm buff-white colour, darker on the breast and flanks. The flanks were barred, particularly towards the tail. The tail was barred, blackish, on the uppersides – the alternate blackish and white bars appeared to be equal in width. The undertail was also barred, and there was barring on the underside of the wings. The bird was grey about the face and head and showed a distinctive white eyestripe.*

The legs were fairly short and yellowish olive in colour. In flight it was heavy looking and dumpy. A long white 'V' showed up the back and there was a prominent white rear edge to the secondaries. Voice: during its stay the only one basic call was heard by RD and EJW – a single high pitched 'keek', sometimes uttered two or three times in quick succession.

A dowitcher was seen in flight by Mike Terry at Keyhaven Marshes on April 5th 1978. Subsequently, it was confirmed by Mike Combridge and Ed Wiseman as an adult Long-billed moulting into summer plumage. It remained in the area until 24th. Another adult in summer plumage was found by Pete Potts on the scrape at Farlington Marshes on May 5th 1979 but it flew off before other observers could see it. The same thing happened the next morning but eventually good views were obtained by Pete, Wilf Nelson and John Marchant on the evening of 6th. It was eventually seen by many observers and was last reported on the evening of 12th flying off east over Langstone Harbour.

Long-billed Dowitcher, adult summer plumage, displaying and singing, Farlington Marshes, May 11th 1979

The next was a juvenile found by S. West at Pennington Marshes on October 3rd 1981 and seen the same day by Mike and Pete Combridge and Ian Watts. It over-wintered in the area and was last seen on May 11th 1982. Later that year on November 4th, Barry Duffin saw and heard an unusual wader fly over while working on the Titchfield Haven reserve. Subsequently, he and Richard Levett obtained excellent views of the bird on the South Scrape and heard it call and were able to confirm it as a juvenile Long-billed Dowitcher. Unfortunately, it was not seen again. The next was found by Trevor Codlin at Curbridge on the Upper Hamble Estuary on March 10th 1986. It was subsequently seen by many observers during its stay until April 16th, by which time it was moulting into summer plumage. Two years later one moulting into summer plumage was found in the same location by an unknown observer on April 4th. It was confirmed by Dave Unsworth who submitted a description to BBRC; it remained in the area until April 20th. It may well have been the 1986 bird returning but is counted as a different individual. The fifth for the 1980s was an adult still in summer plumage found by Adrian Clark, Bill Rutherford and Ed Wiseman at Keyhaven Marshes on September 5th 1989. It remained in the area until September 12th. The record was not submitted to BBRC at the time but thanks to a recent submission by Marc Moody it has now officially been accepted.

The first for the 21st century, after a twenty-year gap, was a first-winter found by Rob Hume on Ibsley North Lake on October 11th 2009; it stayed until dusk and was seen by many observers but left overnight. Another first-winter was found by Lee Fuller and Andrew Davidson at Blackwater, Needs Ore on March 10th 2012 but like the previous bird it stayed only for the day.

Long-billed Dowitcher, adult, Pennington Marshes, August 2013.
Marc Moody (above) and Gordon Small

Fortunately, the next eventually proved to be far more obliging: an adult found by Martin Cooper and confirmed by Marc Moody at Pennington Marshes on July 31st 2013. It was seen again from August 2nd to 23rd, on September 9th, 27th and 28th and from October 6th to November 19th. What is assumed to be the same bird reappeared on January 20th and 21st 2014 and then intermittently between February 1st and April 14th. The most recent occurrence was of another first-winter found by Marc Moody at Keyhaven Lagoon on September 23rd 2015. It was seen daily until October 31st, then on December 4th and 8th, and then again regularly between January 12th and April 28th 2016.

Two earlier records were accepted as Dowitcher *species* by BBRC. The first was found by John Taverner and seen by Norman Pullen at Blackwater, Needs Ore on October 5th 1963. It was also seen by John on October 12th, 19th and November 2nd. On the basis of calls heard he thought it may have been a Short-billed. The second was seen in flight with two Knot and a Bar-tailed Godwit by Barry Duffin and Bob Dunn at Pennington Marshes on September 13th 1970 and later by Peter Mays and David Peart. It was still present the next day when it was seen by Peter Mays and Chris Fox. In the description submitted by Barry he considered it to be a Long-billed.

Long-billed Dowitcher, first-winter, Keyhaven Lagoon, September 24th and 25th 2015 and with Lapwing, February 24th 2016. Alan Lewis, David Ryves and Gareth Rees.

GB status: 5, 1801–1949; 276, 1950–2020. About five per year; stable. (Breeds Arctic Siberia W to Lena River delta, & coastal tundra of W & N Alaska, E to Mackenzie River, Canada. Migrates through USA to winter coastal S USA to N/C America.)

Great Snipe *Gallinago media*

BBRC

A very rare vagrant. Four records in 1901–1934.

The first for the county was a specimen recorded by George Corbin as killed near Ringwood on October 12th 1901 which is described in *The Zoologist* (1901: 428). He wrote:

> *On Oct. 12th a specimen of the Great Snipe was shot in this neighbourhood, but, having been killed with 'duck-shot', it was very much mutilated, part of the tail having been carried away, and the dark barred underparts of the body were very torn. It weighed exactly 8 oz. and under the broken skin appeared to be a mass of oily fat. The whole plumage was much darker than the Common Snipe, the lengths of the beaks being about equal; but in the larger species the legs were of a greener hue, and much more robust, and from the toes to the first joint measured three-eighths of an inch more than in the commoner bird. The outer tail feathers were not wholly white but had indications of dark bars across their whole width, an indication of immaturity.*

was a juvenile moulting to first-winter plumage found by Tim Parminter and Martin Cooper at Butts Lagoon, Pennington at 1900 hrs on September 9th 1998. Identification was not confirmed until the next morning when news was released and it was seen by many observers over the next two days. This bird was also reported by Helen and Paul Theobald as present on August 27th 1998 but BBRC rather harshly considered their description not proven.

Wilson's Phalarope, juvenile, Titchfield Haven, September 14th 1989

The first for the 21st century was a juvenile moulting to first-winter plumage found by Hugh Venables on The Lake at Farlington Marshes on August 21st 2005. It was subsequently seen by many observers during its stay until August 25th. This was presumed to be the bird present at Bowling Green Marsh, Devon from August 18th to 20th. The most recent was an extremely confiding first-winter found by Nigel Jones and Philip Fawkes at Fishtail Lagoon, Pennington Marshes on October 9th 2020 which was subsequently seen by hundreds of observers during its stay until October 20th. It was often in the company of a Grey Phalarope.

Wilson Phalaropes: juvenile, Farlington Marshes, August 28th 2005 and first-winter, Fishtail Lagoon, October 12th 2020. Paul Gale and Mike Creighton

GB status: 247, 1954–2020. About 3-4 per year, decreasing. (Breeds interior W Canada S to California & throughout Midwest states of USA; also S Ontario. Most migrate through interior USA & winter South America from Peru S to Argentina & Chile.)

Red-necked Phalarope *Phalaropus lobatus*

HOSRP

A rare passage migrant, recorded once in winter. One record of two birds in 1890 and 52 records of 54 birds since 1950.

The first for the county were two, one of which was shot, on the River Avon near Ringwood on October 27th 1890. Although this seems a late date for the species, the description from George Corbin in *The Zoologist* for May 1891, where he compares the specimen with one of Grey Phalarope, is convincing.

Red-necked Phalrope, juvenile, Great Salterns Lake, October 4th 1983

On October 27th last a friend of mine was in the meadows, not far from the river, when he observed a bird which he supposed, at first sight, was a wagtail running about on some weeds which were floating down the stream. On drawing closer he saw there were two birds, one of which slipped off the weeds, drawing its head over its back like a Teal when it swims. He shot it, and I had the chance of seeing the elegant little bird when it was still warm; the other fortunately escaped.

At first sight any lover of birds would see it was a phalarope, regardless of its characteristic feet; but on comparison of a specimen of the grey species which I chanced to possess, its specific differences were at once apparent, although the colours of both, being in winter garb, are somewhat similar.

The smaller size and more slender build of the red-necked one were noteworthy, and a closer look at the beaks of the two species seems almost to support the generic distinction which has been proposed; for whilst that of the grey is broader and compressed towards the point, that of the red-necked is gradually tapering through its entire length from base to point.

The little bird in question weighed just one ounce, and on dissection it proved to be a male.

The next was not until 1950 when Canning Suffern and J. C. Draper saw one at Titchfield Haven on October 14th.

Subsequently, there have been 50 records in 32 years totalling 52 birds as shown opposite.

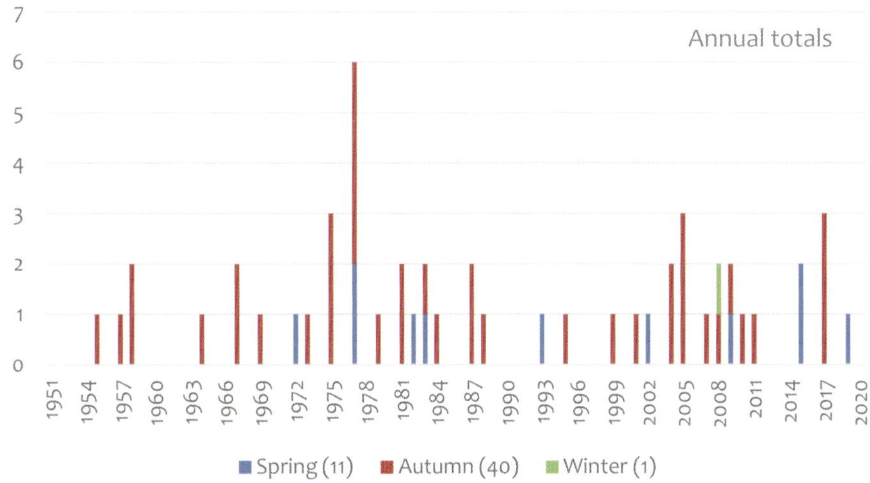

Eleven have been in spring on dates between May 27th and June 20th. Records came from Keyhaven/Pennington Marshes (May 27th–31st 1972, May 29th–June 1st 1977 and May 31st and June 1st 2019), Needs Ore (June 19th 1977), Titchfield Haven (June 9th 1982 and June 20th 2009), Farlington Marshes (June 1st–6th 1983, June 13th 1993 and June 7th 2002) and Ibsley Water (2 on June 2nd 2015).

Red-necked Phalaropes: juvenile, Keyhaven, August 26th 2009 and adult female, Keyhaven, May 31st 2019. Marcus Ward and Lee Fuller

Forty-one were in autumn. One was at Pennington Marshes from July 10th to 18th 1975 and the remainder were on dates between August 7th and November 11th. Multiple records came from Keyhaven/Pennington Marshes (15 including long-stayers from September 23rd to 28th 1975 and October 7th to 17th 2001 and a colour-ringed juvenile on September 19th 2012 which was

omitted from the *HBR*), Dibden Bay (9 including two on October 6th 1958 and possibly four individuals between August 7th and October 2nd 1977), Farlington Marshes (6), Hurst to Milford on Sea (3) and Needs Ore (October 4th 2005 and November 2nd 2017). Singles were at Stokes Bay from October 10th to 12th 1957, Overton Lagoons on September 20th 1967, Titchfield Haven from September 27th to October 4th 1983, Great Salterns Lake from September 25th to October 11th 1984, Hilsea Moat on October 23rd 1987, Hook-with-Warsash from September 30th to October 4th 1999, Ober Heath, New Forest on October 5th 2004 and Cherque Gravel Pit, Gosport on October 7th and 8th 2005. The juvenile at Hilsea Moat relocated to Farlington Marshes where it was present from October 25th to 28th 1987. Another juvenile which left south-east from Keyhaven Lagoon on August 22nd 2004 is assumed to be the same bird which appeared later that day at Farlington Marshes.

There was a single winter record of one with two Grey Phalaropes seen off Hurst Beach by Nigel Jones, Mike Duffy and Colin Allen during gales on January 21st 2008. There are no other UK records between December 7th and April 25th.

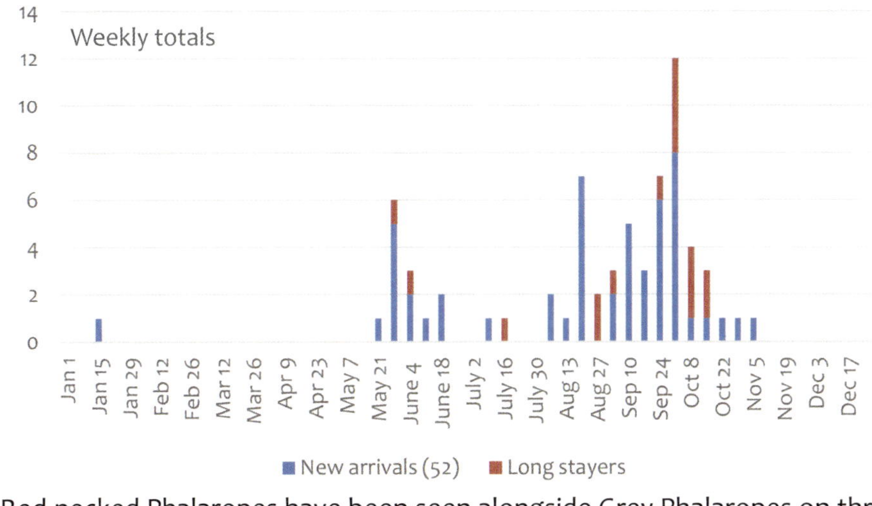

Red-necked Phalaropes have been seen alongside Grey Phalaropes on three other occasions: at Great Salterns Lake, Portsmouth between October 4th and 11th 1984, in a ditch at Hook-with-Warsash between September 30th and October 4th 1999 and at Pennington Marshes between October 3rd and 15th 2001.

Red-necked Phalarope, first-winter, Ober Heath, New Forest, October 5th 2004. Mike Read

GB status: 1,182, 1986–2019. No clear trend, mean 41 per year in 2010–19. (Circumpolar low Arctic breeding range: N Scotland, Iceland, Fennoscandia, Svalbard, N Russia, Alaska, N Canada and Greenland. Migrates overland to winter in cold current upwellings in tropical seas, particularly off W Pacific coast of South America, also Arabian Sea, Sundas & Moluccas, Indonesia, E to Papua New Guinea.)

Grey Phalarope *Phalaropus fulicarius*

HOSRP

A very scarce autumn and early winter visitor, usually occurring after gales. Rare in December–February and recorded once in April. Many old records and around 361 birds in the modern era.

The first to be recorded in the county may well have been one which dropped into a farmyard puddle and remained for a week or two at Mr. Thresher's at Bentley in the autumn of 1825 (Bucknill 1900). Col. Peter Hawker shot one near Keyhaven on September 26th 1839 and two on successive days at Longparish on October 15th and 16th 1846 (Payne-Gallwey 1893). There was a large invasion in September 1866 following gales which is summarised by Gurney (1867). At least 38 were obtained including a number at inland localities and many more were seen. Gurney described how one gentleman gave him notice of three shot on the 10th of September, very near Lymington: there were twenty in one flock; but when he obtained one sufficiently clean for stuffing, he desisted from shooting any more. There were further large invasions after autumn gales in 1886 and 1891. In the latter year, Coles saw many in the Milford on Sea area between October 12th and 24th including six on Sturt Pond on 12th and 39 there on 18th (Kelsall & Coles 1913). Seven were reported on a pond at Milford on Sea (presumably Sturt Pond) on September 20th 1896 (*The Field,* October 3rd 1896).

Grey Phalarope, Great Salterns Lake, October 4th 1983

K & M also mentioned further inland occurrences including those of two obtained at Kings Worthy and St. Cross in 1883, which became part of the Winchester College collection and are on display in the Science School there, and one shot at Avington Park on September 15th 1896.

There are several records for the first half of the 20th century. Most were from the Keyhaven area, where there were singles on September 24th and October 11th 1933, three on October 16th 1933 (described as being on a pool by the Solent), four to six between September 21st and October 4th 1935, one from September 22nd to 27th 1946 and another on September 24th 1950. Elsewhere, singles were at Damerham Watercress Beds on September 17th 1935, on a pond at Hiltingbury for several days about October 6th 1935, on the River Avon near Fordingbridge on November 7th 1936, at Ellingham Gravel Pit in late November 1943, at Hill Head on January 18th 1948 and at Romsey on October 30th 1949.

Between 1951 and 2020 there were records in 59 years involving at least 361 birds.

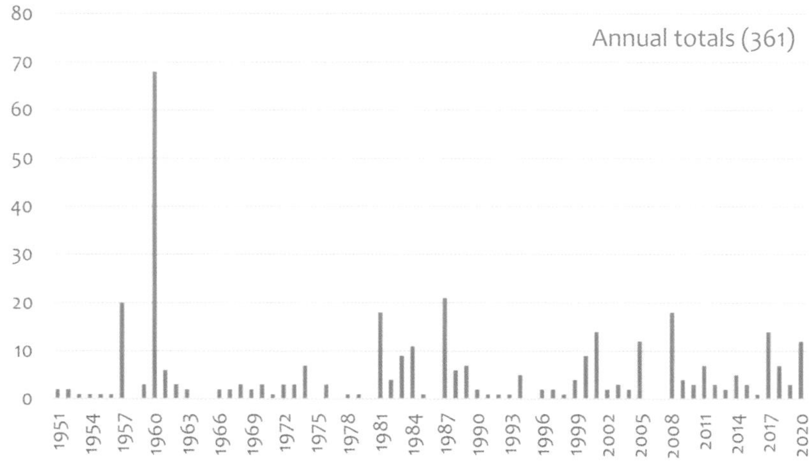

Most years produced a few birds but double-figure totals occurred in ten, usually following severe gales, including 20 in 1957, 68 in 1960, 18 in 1981, 21 in 1987 and 18 in 2008.

Coastal records came from the Lymington/Hurst area (172, including 18 on October 9th 1960 and six on September 27th 1981 and October 19th 1987), Langstone Harbour/Farlington Marshes (52, including four on September 24th 1983 and five on October 18th 1987), Hill Head/Brownwich (30, including three on October 18th 1987), Gilkicker Point/Browndown (23, including six on October 3rd 1957 and seven flying west on October 8th 1960), Needs Ore (12), Eastney/Southsea (10), Hook-with-Warsash (9), Sandy Point/Black Point (7), Portsmouth Harbour and Lepe (6 each), Dibden Bay (5) and Badminston Gravel Pit, Calshot, Eling, Langstone Mill Pond, Riverside Park (Southampton), Sowley Pond and Weston Shore (1 each). Inland, singles occurred at Ibsley Water (5), Fleet Pond (3), Alresford Pond and Ripley Farm Reservoir (2 each) and Bisterne, Burley Golf Course, Exton, Holly Hatch, Janesmoor Pond, Kingston Common, Southampton Common, Winchester, Winchester Sewage Farm and Woolmer Pond (1 each).

As shown on the next page, analysis of arrival dates shows that until 1997 there were no autumn records before September 1st but since then there have been six August records including a juvenile at Pennington Marshes from 20th to 25th in 2008 and an exceptionally early bird off Hurst Beach on August 4th 2013. Influxes, nearly always during or following gales, may occur at any time between early September and early November with few seen after the middle of that month. Small numbers have occurred in mid-winter, notably during a gale on January 21st 2008 when two were seen with a Red-necked Phalarope off Hurst Beach. Later records were of one at Calshot on

January 30th 1972 and presumably the same bird on February 20th 1972 and one off Hill Head on February 23rd 1996. There were no spring records until 2020 when Douglas Marcuse found a female moulting into summer plumage at Fishtail Lagoon, Pennington Marshes on April 16th.

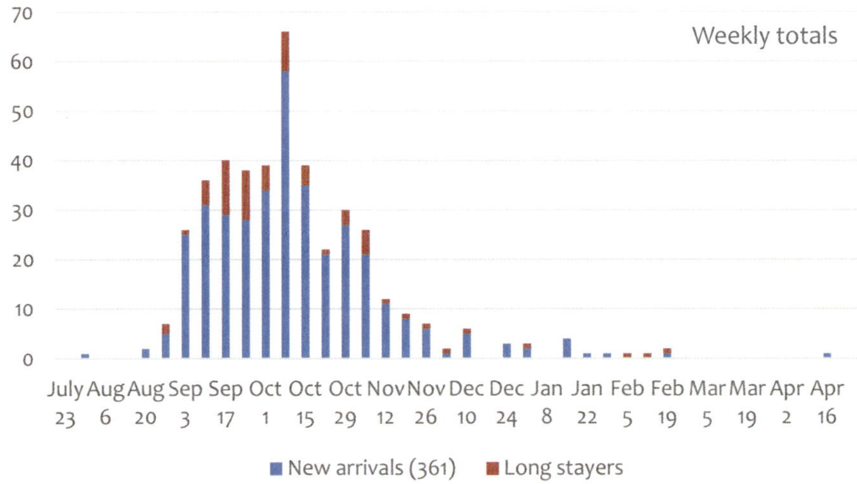

First-winter Grey Phalaropes with Black-headed Gulls, Hill Head, after the storm, October 18th 1987

Grey Phalarope, Burley Golf Course, New Forest, November 14th 2005.
Mike Read

Grey Phalarope, first-winters: Fishtail Lagoon, September 22nd 2018 (top left) and October 10th 2020 (left), and Oxey Marsh, November 6th 2017 (above). Lee Fuller, Mike Trew and Gareth Rees

GB status: 11,066, 1986–2019. No clear trend, mean 339 per year in 2010–19. (Circumpolar high Arctic breeding range on coastal tundra, from Iceland and Svalbard E across Arctic Russia and islands, to N coastal Alaska and parts of N Canada and W Greenland. Mostly pelagic in winter at edge of Gulf Stream from North Carolina to Florida, USA, & cold current upwellings off W coast of Africa, & Pacific coast of South America.)

Spotted Sandpiper *Actitis macularius*

BBRC

A very rare vagrant. Two records in 1986 and 2009.

The first for the county was found by Dan Powell on the South Scrape at Titchfield Haven at 1230 hrs on October 4th 1986. Soon after, Dan was joined by Barry Duffin and John Walters. The bird resembled a Common Sandpiper but the observers noticed the shortness of the bird's tail, the distinct barring on the wing coverts and the bright yellow legs and thus identified it as a Spotted Sandpiper. The news was broadcast and the first twitchers arrived later that afternoon. Over the next eight days the bird frequently fed close to a Common Sandpiper and Barry was able to make detailed comparisons which were included in his description to BBRC. It was initially thought to be a juvenile, but when trapped on November 25th examination of the moult condition of the wings caused Barry and Pete Potts to age the bird as an adult. It was last seen on January 12th 1987 with ice encrusting its legs and presumably succumbed to the severe wintry weather at the time.

Spotted Sandpiper, adult,
South Scrape, Titchfield Haven, October 1986.
Trevor Codlin (top) and Barry Duffin.

Spotted Sandpiper, first-winter, River Test,
December 6th 2009. Philip Underwood

A putative Spotted Sandpiper was found by Mike Rafter by the River Test at Lower Brook House on November 20th 2009. Other observers viewed the bird in the next few days; some dismissed it as a Common Sandpiper but David Thelwell thought it was a Spotted Sandpiper. Nigel Jones and Simon Ingram visited the location on November 25th and were able to get close views and take photographs which confirmed it as a first-winter Spotted Sandpiper. Thanks to the owners of the house who allowed access to their riverside garden, it was subsequently enjoyed at close range by many observers. It remained in the area until at least December 17th.

GB status: 1, 1924; 231, 1956–2020. About five per year; increasing. (Breeds North America from W Alaska to Newfoundland & S to California, Texas & North Carolina. Some winter coastal USA to S of breeding range but most winter C America, Caribbean & N South America, S to N Argentina & Chile.)

Lesser Yellowlegs *Tringa flavipes*

BBRC to 2018; HOSRP 2019 to date

A rare vagrant. 16 records of 15 birds since 1953.

The first for the county was one seen by Bob Sharland at Keyhaven Marshes on September 29th (not 28th as in *BoH* and other recent publications) and October 2nd 1953. Extracts from his description are reproduced below.

On 29th September 1953 I walked along the seawall by the salt marshes at Keyhaven, Hants. At about 5 pm I saw a flock of Green Plover [Lapwing] on a lagoon inside the sea wall. I approached to within 50 yds and saw a wader with yellowy legs which looked like a slim Redshank. When I tried to get nearer, I disturbed the flock and the slim wader flew upwards showing grey streaks below as far as the breast, the rest of the underparts being white. It then circled to fly out to sea and I clearly saw a white tail, uniform back and wings with no wing-bar and a very slight spotted appearance. The bill appeared to be shorter and slimmer than a Redshank's.

I next visited the area on October 2nd and at about 10 a.m. about 300 yds further east I disturbed a similar bird from a small pond inside the seawall. The build of the birds was suggestive of a Marsh Sandpiper (I am familiar with this bird in Nigeria) but the plumage was as previously mentioned. As it got up it gave three distinct 'teuks', louder than a Knot's call but not as ringing as a Redshank's or Greenshank's. The bird flew rising out to sea.

The bird was viewed through 10 x 40 binoculars in sunlight. On the 2nd there was a slight haze. This would appear to be a Lesser Yellowshank and Mr. Edwin Cohen states that there has been no previous record of this species in Hants.

Subsequently, a further 14 have been recorded.

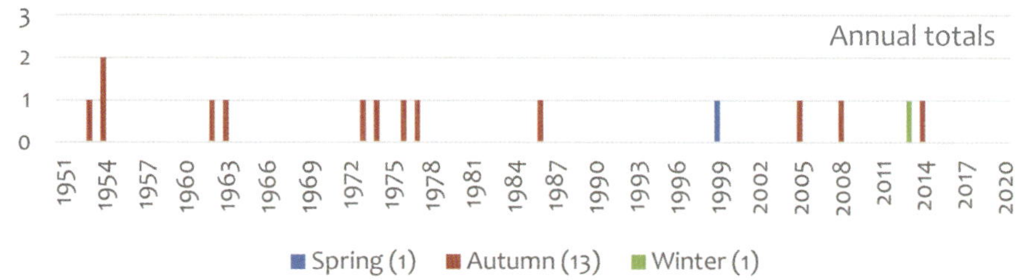

The next was watched by Gerald Kinsey at Keyhaven Marshes for a few minutes on August 29th 1954 before it flew off towards the Isle of Wight. Gerald's description stated that it was a little larger than a Redshank and gave the impression of a small Greenshank and called frequently with a shrill 'tew … tew … tew'. The editors of *British Birds* queried the size of the bird and in his reply Kinsey said that it was unfortunate that no other wader was present and that he had to estimate the size … but it was certainly smaller than a Greenshank. Later that year one was found at Farlington Marshes on the evening of September 20th but 'not properly sorted out'. It was identified beyond question by Cliff Henty early on 21st and was subsequently seen by many people until the evening of 28th, when it flew off high to the east.

All three of the above records were included in an article in *British Birds* reviewing occurrences in the British Isles in 1953–54, when there were eight and six records respectively, more than had ever been recorded previously (*Brit. Birds* 48: 363–366).

Despite the glut of records in the early 1950s the next was not until 1962 when David Billett, George Clay, Michael Bryant and Ed Wiseman found one at Farlington Marshes on the evening of July 19th. According to the *HBR* it was last seen on 21st but Anthony Cheke has recently informed the author that he and Aidan Sudbury saw it on July 24th.

Another was found by Bryan Renyard at Farlington Marshes at 1500 hrs on June 29th 1963 and subsequently seen by Michael Bryant before it flew off into Langstone Harbour at 1615 hrs. However, it turned out to be a long-staying bird and was subsequently seen by many local and visiting observers until September 1st.

The sixth for Hampshire was seen by Bob Dunn and Ed Wiseman at Pennington Marshes for a few minutes on September 18th 1973 but was not seen again. This was followed almost a year later by one seen at low tide (apparently one of the lowest tides of the century) at East Winner Bank, Hayling Island by T. F. Twist on September 19th 1974. The next was another one-day bird, seen briefly by Mike Stewart and Mike and Pete Combridge at Keyhaven at 1600 hrs on October 9th 1976 and then watched closely from 1740 to 1815 hrs when the observers, by now including Ed Wiseman, left it feeding in a saltmarsh creek. It may well have been the bird at Stanpit Marsh, Dorset from October 16th to 31st. A juvenile was found by Philip Fawkes and Graham Armstrong at Dibden Bay on August 29th 1977 and seen by many observers up to September 9th 1977.

A presumed Lesser Yellowlegs was seen briefly by Richard White at Farlington Marshes on August 15th 1986; it was also seen by Terry Jennings on 17th who showed it to Dave Unsworth and Geoff Farwell but they did not see the wing-pattern to rule out juvenile Redshank. It was found and identified by Pete Potts on 19th who released the news. It was subsequently seen by many observers until 27th and trapped by Pete Potts and Duncan Bell at 0200 hrs on 25th when it proved to be an adult.

The next was the first in spring. It was found at Pennington Marshes late on May 5th 1999 by Nigel Jones, Richard Coomber and Mike Witherick while they were engaged in a 24-hour bird race. They released the news and it was refound the next morning by Paul Winter and subsequently seen by many birders until 9th. The next occurred at Titchfield Haven. Doug Robertson had

Lesser Yellowlegs, Stansore Marshes, Lepe, January 22nd 2014. Martin Bennett

a brief view of a possible Lesser Yellowlegs there on October 22nd 2005. The next day Richard Levett and David Parsons had arranged to meet at the reserve at 1030 hrs. As the Meon Shore hide was busy, they went to the Pumfrett Hide where the first bird they noticed was a Lesser Yellowlegs in the north-west corner of the South Scrape. No-one else was looking at it; apparently the birders in both hides had identified it as a Wood Sandpiper! News was released and a steady stream of birders arrived but all the waders on the scrape were flushed by a Sparrowhawk at 1350 hrs. The yellowlegs was refound by Keith Maycock on the South Scrape at 1607 hrs but it flew off high to the west six minutes later. There were no further sightings until 25th when Bob Marchant saw it on the nearby Posbrook Floods where it remained until 28th. It was also seen briefly at Titchfield Haven again on 26th. The next was an adult found by Dave Unsworth during a WeBS count at Park Shore, Needs Ore on September 21st 2008. He tried to contact one of the other counters but his phone was switched off, so Dave had to leave the site to find them. On their return the bird was nowhere to be seen.

Just over five years later Philip Fawkes found a first-winter on Stansore Pools, Lepe on November 10th 2013. It overwintered and remained faithful to the pools apart from occasional forays to the nearby Dark Water. It was last seen on April 15th 2014. Later that year another first-winter was found by John R. Jones, Marcus Ward and Steve Piggott at Normandy Lagoon on September 5th where it stayed until 9th. No doubt the same bird was relocated by Andrew Madgwick at Titchfield Haven on September 11th where it remained until 29th (not 28th as in the *HBR*).

Lesser Yellowlegs, North Scrape, Titchfield Haven, September 27th 2014. John Dodds

Another found by Mick Scott at Fox Lane Gravel Pit, Eversley on the evening of June 13th 2001 and subsequently seen by Graham Stephenson, was harshly considered not proven by BBRC. The author received a mobile call about this bird but unfortunately was locked inside the Eelmoor Marsh NR to do a Nightjar survey and unable to leave until a pre-arranged time. Needless to say, it was nowhere to be found the next morning.

Lesser Yellowlegs, North Scrape, Titchfield Haven, September 2014

GB status: 16, 1854–1949; 401, 1950–2019. No clear trend, mean 12 per year in 2010–19. (Breeds throughout much of subarctic Alaska & Canada, E to James Bay. Migrates through USA, where some overwinter, but majority winter from Caribbean & C America to Chile & Argentina.)

Marsh Sandpiper *Tringa stagnatilis*

BBRC

A very rare vagrant. Five records since 1976.

The first for the county was found by M. W. Ball on The Lake at Farlington Marshes at 0500 hrs on June 27th 1976, while photographing birds from a hide. Extracts from his description to BBRC are reproduced below.

> *My first impression was of a small yellow-legged Greenshank with a thin bill, more streaked and spotted generally. The appearance of a Greenshank about 15 minutes later confirmed my thoughts about the bird. It spent much of its time feeding. Occasionally it would stand on one leg and half stretch a wing. It frequently uttered a 'tchup' type call, this being accompanied by a bobbing of the body. It was seen to fly once when it flew directly away from me and I noticed a white rump and vee extending up the back.*

The observer left the hide at 0800 hrs and accidentally flushed the bird. He reported his observation to David Billett, the warden, who was initially unable to relocate it. Fortunately, he found it in the same area at 1900 hrs and was able to take a detailed description. He saw it again from 1600 to 1630 hrs the next day.

Conceivably the same bird returned in 1977, when R. A. Jones saw an unusual wader fly into The Lake at Farlington Marshes at 1900 hrs on July 5th. He showed the bird to Malcolm Arnott and John Smith who identified it as Marsh Sandpiper; shortly afterwards Andy Williamson arrived and concurred with the identification. It fed with a Redshank and a Spotted Redshank but at 1940 hrs flew off south-west into Langstone Harbour. It was then seen intermittently at Farlington by many observers until July 11th, when it flew in with a Ruff at 1040 hrs but flew off strongly ten minutes later.

The next was an adult found by Humphrey Girdlestone and identified by John M. Jones at Pennington Marshes at 1430 hrs on July 28th 1983. It was subsequently seen by many observers and remained in the area until August 4th although apparently was missing on 1st and 2nd.

The fourth was watched for 45 minutes by Darren Fry at Farlington Marshes on the evening of July 28th 2006. Despite phoning the news out to another observer, no one was able to reach the site before it disappeared. Presumably, the same individual was present at Sidlesham Ferry, West Sussex from July 29th to August 6th.

The most recent occurrence was of one found by Steve Keen at Fishtail Lagoon, Pennington early on June 10th 2018 which was seen by many observers during the day but departed overnight.

Marsh Sandpiper, Fishtail Lagoon, June 10th 2018. top and bottom left Martin Bennett, middle top and right Ian Williamson

GB status: 5, 1937–1949; 144, 1950–2019. About three per year; stable or decreasing. (Breeds Ukraine & W Russia, & occasionally Finland & Baltic countries. To E, breeds in forest-steppe region of Siberia to Mongolia & NE China. Winters throughout sub-Saharan Africa, especially E Africa, & Indian subcontinent E to S China & SE Asia; also Australia.)

Greater Yellowlegs *Tringa melanoleuca*

BBRC

A very rare vagrant. Two records in 2007 and 2015.

The first for the county was found by Jason Crook and Kevin Crisp on The Deeps, Farlington Marshes at 1510 hrs on September 26th 2007. Extracts from Jason's detailed account of its occurrence (Crook 2008) are reproduced below.

While searching for a Richard's Pipit I'd seen earlier in the day, we found ourselves looking over The Deeps. A Spotted Redshank was feeding on the far side of the pool but as I started scanning the edge of the mud for more waders, a movement to the left and much closer caught the corner of my eye. I saw a medium-sized wader with dark wings, square white rump, barred tail and long trailing yellow legs fly off over a bank and land out of sight. I was fully aware of what I had just seen, and suggested to Kevin that even with such a brief view, the features noted, combined with the size of the bird, could only add up to a Greater Yellowlegs. He agreed. The question was: how do we see it properly? It soon flew off strongly, calling as it went, over the western seawall of the reserve and into Langstone Harbour. The tide was falling rapidly and there was already a lot of mud exposed. We needed to relocate the bird, so rushed to the other side of the reserve while calling several local birders to get the news out. It was finally relocated at 1850 hrs on The Lake at Farlington Marshes where it was watched until dusk feeding alongside two Spotted Redshanks. We again released the news but only Tim Lawman was able to make it back in daylight. A crowd of observers was in position to observe the lake prior to dawn the next morning but unfortunately the views of the bird were into the strong early morning light and disappointingly it flew off west calling with Spotted Redshanks at 0652 hrs, never to be seen again.

The second for Hampshire was an adult discovered by Mike Howard and Dan Houghton and independently by Mark Rolfe at Posbrook Floods, Titchfield Haven on the morning of January 11th 2015. It was seen by many observers during the day but departed overnight. What was presumably the same bird was found by Howard Menhenet on the Eleven Acre Mere at Titchfield Haven on April 11th and remained on the reserve until 21st. It was then rediscovered by Mark Rolfe on Posbrook Floods on May 9th and commuted between there and Titchfield Haven until it was last seen on July 24th. Fuller details of its discovery and subsequent return are described by Rolfe (2016). What was no doubt the same bird then appeared on the Medina Estuary, IOW where it remained from September 23rd to March 11th 2016.

Greater Yellowlegs, South Scrape, Titchfield Haven.
photos Mike Trim, May 20th, Dan Powell, July 21st and David Aitken, June 14th 2015

GB status: 6, 1906–1949; 29, 1950–2020. Rare, about one every 1–2 years. (Breeds from S Alaska across subarctic Canada E to Labrador & Newfoundland. Migrates throughout USA to winter coastal S USA, C America, Caribbean & South America.)

Cream-coloured Courser *Cursorius cursor*

BBRC

A very rare vagrant. One record in 1845.

The only record for Hampshire is of one shot by a shepherd in the employ of Mr. W. Tice at Sopley in 1845 (K & M). The specimen became part of the Hart collection and is now held by the Hampshire Cultural Trust at Chilcomb House, Winchester.

GB status: 28, 1785-1949; 8, 1950–2012. (Breeds Cape Verde & Canary Islands E across deserts of N Africa, S to Chad & Sudan, & through Middle East to Turkmenistan, Iran, Afghanistan & Pakistan. Western populations resident, or dispersive to deserts S of breeding range, while eastern breeders winter from S Iran to deserts of NW India.)

© Hampshire Cultural Trust

Collared Pratincole *Glareola pratincola*

BBRC

A very rare vagrant. One record in 1857 and three since 1974.

The first for the county was one shot by Lt. H. Henn at Barton on Sea in May 1857. The specimen passed into the Hart collection and is now held by the Hampshire Cultural Trust at Chilcomb House, Winchester.

The first modern record was of one found by Graham Stephenson in flight over Fleet Pond at 0945 hrs on May 28th 1974. It was also seen by Tim Henderson and Richard Millington. Extracts from Graham's description are given opposite.

© Hampshire Cultural Trust

The bird first appeared flying in low from the north. It briefly circled overhead at close range before appearing to fly off south. It called several times during this first period of observation. It then returned and began hawking the abundant flying insects low over the south-east corner of the pond. After ten minutes it left low to the south, calling. It was very tern-like in structure with long, slender pointed wings, a very pronounced forked tail and short and slightly de-curved bill. The forehead, crown, nape and mantle were a uniform rich sandy brown. Creamy chin/throat patch bordered by a thin black line – beginning as the malar stripe and extending in a deep 'U' across the lower breast. Breast lighter sandy brown becoming paler towards belly. Belly and flanks white. Upper wings sandy brown with white trailing edge to secondaries and darker primaries. Underwing sandy brown with chestnut red axillaries and wing linings – seen very well when bird was overhead. Rump and uppertail-coverts white, tail black with white outers. Call notes harsh and tern-like – 'kitti, kitti, kerrick, rick, rick'.

The next was reported as found by Mike Read at Mockbeggar Lake on May 6th 1987. News was released to local observers including Martin Reid who saw it later in the day and submitted a description to BBRC. Mike (*in litt.*) has recently confirmed that he first saw the bird in the water meadows above Ibsley Bridge on May 5th, when it dropped into the pond there and subsequently start aerial feeding. It was still present when he left but could not be found later in the day.

Collared Pratincole, Farlington Marshes, May 1st 2005. Jason Crook

The only 21st century record was of one initially spotted by Bob Chapman some distance away at Farlington Marshes on May 1st 2005. It was then jointly identified by Bob and Jason Crook through telescopes as it headed towards The Deeps where much closer views were obtained. After 15 minutes it was flushed by a Peregrine, at which time Pete Gammage also saw it, and was then lost to view low down over South Binness Island, never to be seen again.

The Zoologist for 1872 contains a note written by Col. H. W. Feilden, according to K & M a well-known naturalist, extracts from which are reproduced below.

> *A brother officer of mine. Lt. Telfer, informs me that in October 1864 he was strolling along the beach at Stokes Bay, Gosport when he saw a bird fluttering along the shore, evidently maimed or very weary, as it was unable to fly but a few yards at a time. Not having a gun with him, he pursued it, throwing shingle at the bird, finally knocking it over, when it proved to be a fine example of Glareola pratincola. Lt. Telfer preserved it, but the skin is no longer in his possession. Having shortly after gone abroad to join his regiment, he lost sight of the specimen and can now find no trace of it. Lt. Telfer could have made no mistake in the species, for, besides being a fair ornithologist, he had just returned in 1864 from Corfu, where the pratincole is abundant, and where he had procured and skinned many of the birds.*

Given the vagueness of the date, hearsay nature of the report and lack of a specimen this record could at best be considered as Pratincole *species.*

The 1944 *HBR* included the report of a pair in the week of May 15th at Somborne near Stockbridge. They were seen by a rabbit catcher who noticed their swallow-like tail feathers and sharp call note in flight, and at once recognised them from an illustrated book. Even by recording standards at that time this record appears unacceptable.

GB status: 20, 1805–1949; 79, 1950–2020. Rare, about one per year; stable. (Breeds locally throughout Mediterranean basin from N Africa & S Iberia to Black Sea, most in S Spain, Portugal & Greece. To E, breeds SW Asia to Pakistan & Kazakhstan but distribution highly fragmented. Winters sub-Saharan Africa. Another race resident Africa.)

Collared/Black-winged Pratincole
Glareola pratincola/nordmanni

BBRC

A very rare vagrant. One record in 1983.

A pratincole watched by Nigel Crook for three to four minutes off the sewage outfall at Pennington Marshes on August 6th 1983 was accepted by BBRC as Pratincole *species*. However, Nigel's description, extracts from which are reproduced below, suggests it was probably a Black-winged.

> *I was walking along the sea wall when I saw a dark-coloured bird flying amongst a flock of about 20 Black-headed Gulls. I raised my binoculars and almost instantly realised it was a pratincole. I watched it through my scope at about 200m. The pratincole was hawking over the sea among the gulls which were circling low over the water near the sewage outfall. I watched it flying around in circles amongst the gulls for 2½ minutes before it then flew more purposefully in a south-westerly direction. I lost it when it disappeared behind a mud bank. Just in case it had landed on the mud out of sight, I waited until the tide had come in during the evening,*

hoping that it would be pushed up by the rising water. Although there was no shortage of eyes looking for it, the Farlington and Southampton crowd having joined me, it did not reappear.

Size, structure and flight action – It was noticeably smaller than the gulls and was a long, slim bird. The long body ended in a long, deeply forked tail. The fork was not often evident due to the tail being closed most of the time, giving the rear end a long, pointed appearance. The wings were long, narrow and pointed and for most of the time were sharply angled. The bill was small and pointed. Its wing beats were deep and slow while flying away but shallower when it was amongst the gulls. Its flight resembled that of terns.

Description of plumage – The top of the head, back and upperwings were a dark brown colour. The uppersides to the wings looked uniform in colour. I could not detect any difference in tone of colour between the primaries and the rest of the wing, despite looking for it. I could also not detect a pale trailing edge to the secondaries. The underwings were very dark and looked uniform in colour. I could not detect any chestnut colouration; this could have been due to the distance. The breast was pale brown with the chin and throat a paler shade brown. The pale brown of the breast merged into the white belly. The white of the underparts extended onto the undertail-coverts and up onto the rump. The tail feathers were a dark brown colour. Because of the distance I could not see a red base to the bill and could only just make out a thin black line along the edge of its throat. Due to there being no contrast of colour on the uppersides to the wings, I think there is a possibility that it could have been a Black-winged Pratincole.

Sabine's Gull *Xema sabini*

BBRC to 1962; HOSRP to date

A rare visitor, usually storm-driven. One record in 1981 and at least 168 birds since 1954 including 120 or more in October 1987.

The first for the county was an adult, shot among some Grey Phalaropes, by R. E. Coles, at Sturt Pond, Milford on Sea on October 21st 1891 (*The Field*, November 7th 1891). Coles provided further details in *The Field* for November 28th 1891 as follows:

> *It has been preserved by Mr. Hendy of Southampton. Length, nearly 13 in.; bill, black tipped with yellow; crown of head and neck, on the sides and front white, dusky-black markings on nape, behind the eye, also extending faintly to sides of neck in form of a collar; back, bluish-grey; breast, white; the wings, when closed, extend rather more than an inch beyond the tail; wing coverts, bluish-grey; the first five primaries black tipped and edged inside with white; tail, forked, pure white; legs, toes and claws, black. The tips of several of the primaries are slightly damaged, and the feathers on the nape are ragged in appearance.*

He sent the specimen to Howard Saunders, who wrote in the *The Field* for December 19th 1891:

It is a thoroughly adult bird; in fact, among a somewhat large series in my collection, there is only one specimen which shows a trifle less black on the outer web of the sixth primary. The hood, characteristic of summer plumage, has now given place to irregular dusky black markings on the nape, while, as Mr. Coles observes, the primaries are abraded, especially their tips ... Reverting to the occurrence of adults in the British islands, I may mention that only two examples are known to me... Immature birds are of not infrequent – though irregular – occurrence; but Mr. Coles may be congratulated on the acquisition of a third adult British specimen.

The first modern record is arguably not a Hampshire one. An adult was seen by Mike Adams in Hurst Narrows on November 28th 1954 but he was watching from Warden Point, IOW. Extracts from his description published in *British Birds* (48: 83-84) are reproduced below.

I watched on several occasions between 0930 hours and 1200 hours what I identified as a Sabine's Gull working the rough sea of Hurst Narrows. The bird was readily distinguished from other gulls by its buoyant, tern-like flight and striking wing pattern, in which the whole of the leading edge of the wing appeared dark, in strong contrast to the white secondaries. The primaries only were definitely black. No terminal tail-bar was present, but I could discern, at times, a dark nape. The Sabine's Gull dropped momentarily to the surface on several occasions but did not settle whilst under observation. Present for comparison were two adult Little Gulls and both adult and immature Kittiwakes in addition to the commoner species of gulls.

Annual totals, excluding the influx of 120 in October 1987, are shown below.

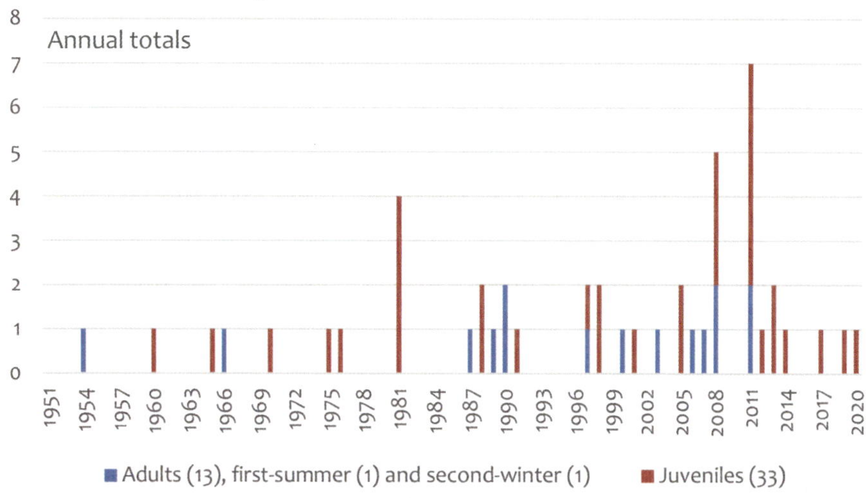

The second was a juvenile seen by B. E. Cooper and M. Nolan at Eastney sewage outfall on December 4th 1960. Nolan's description to *British Birds* (54:187) was as follows:

A small, dark gull was watched for about three minutes as it flew from the direction of the harbour entrance, passed within 30 yds of us, through feeding gulls, and out to sea in a south-west direction. It fed frequently from the surface of the water as do Little Gulls and several times seemed to land on the water for a time without closing its wings. Its flight generally more buoyant than B.H. gulls. Wings less rounded than Little Gull's but more so than B.H. Gulls. Primaries black, contrasting slightly with sooty grey/brown mantle and coverts and very strongly with triangular patch on secondaries and tertials. This pattern visible through underwing. Nape, crown, chin and breast washed with paler grey/brown and there was a dark terminal band to its conspicuously forked tail. Forehead and rest of plumage appeared white. Bill 2/3 length of B. H. Gull and slenderer. Legs possibly pale grey/green.

The next two records were not published in *HBRs*. A juvenile was seen by Barry Duffin and Martin Terry flying east off Gilkicker Point on September 6th 1965 and an adult was seen by Phil Agland off Hurst Castle on December 31st 1966. Good descriptions of these birds are in the HOS archives.

Three occurred in the 1970s. A juvenile was found by Barry Duffin at Sturt Pond on September 12th 1970; it was seen soon after by Bob Dunn and Ed Wiseman and remained in the area between Pennington Marshes and Sturt Pond until October 11th. Another juvenile was seen there by Ed Wiseman on September 28th 1975. Inland, Spike Millington found a juvenile at Fleet Pond on October 30th 1976 which was also seen by Tim Doran.

Sabine's Gull, juvenile, Sturt Pond, September 1970. Geoffrey Fisher

In late September 1981 probably four were in the Keyhaven area. Bob Dunn found a juvenile at Sturt Pond on 25th, the next day two juveniles were seen separately by Owen Haisell and others at nearby Pennington Marshes and on 27th Mike and Pete Combridge saw a juvenile at Hurst Castle.

An amazing wreck of this species in 1987 involving a minimum of 120 birds (about 95 adults and 25 juveniles), and probably in excess of 140, occurred in the wake of the famous hurricane which hit south-central England during the night of October 15th/16th. The first were an adult and juvenile discovered by Phil Kerslake among a flock of Black-headed Gulls on floods at Queen's Parade, Aldershot. They were seen by most local observers including the author who had to absent himself from his sister's wedding reception to see them! They flew off south-west at dusk. Also on this date, Bryan Renyard found two adults off the Naval Base in Portsmouth Harbour and Graham Stephenson and Mick Scott saw two adults fly south-east over Eversley Gravel Pits at dusk.

The next day, October 17th, 42 were recorded at coastal sites, mostly feeding offshore. They comprised six at Sandy Point and Black Point, one off Eastney, seven off Hill Head, eight off Netley, one at Eling, 12 off Lepe and seven which moved west off Pennington Marshes. All were adults apart from four juveniles. Inland, two adults were seen separately at Fleet Pond and a juvenile was reported nearby on the Queen's Parade floods.

Sabine's Gulls, adults and juveniles, Titchfield Haven, October 17th 1987

On October 18th, 112 were recorded; as on the previous day, few of these appeared to be moving through the county. They comprised 19 from Black Point to Eastney and along the Hayling shore of Langstone Harbour, four at Farlington Marshes, two at Hill Head/Titchfield Haven including one which fed over farmland at nearby Chilling all day, one at Hook-with-Warsash, three off Netley, two at Calshot, 30 at Lepe and the Beaulieu Estuary mouth including five west and two east and 50 at least between Pennington Marshes and Hurst of which fewer than ten moved off west. Only 14 juveniles were reported including ten at Pennington. One juvenile was seen inland on a flooded playing field at Mayfield School, Farnborough.

The next day coastal records totalled 66. In the Hurst area 36 were present including 33 which moved off west in the morning; at the same time at least 19 were milling about off Lepe although subsequently only one was seen there. Later in the day eight moved west off Calshot (possibly including three seen off Hill Head), two were still in Southampton Water and four were around Langstone Harbour entrance.

Numbers declined rapidly thereafter. On 20th four adults were between Black Point and Eastney, two adults were off Netley and single adults were at Eling and off Lepe. The next day only one was at Eastney, three adults and a juvenile were off Lepe and in Southampton Water a juvenile was at Eling, an adult was off Mayflower Park (probably the Eling bird from 20th) and two adults were off Netley. On 22nd two adults were off Hill Head and at least one remained at Eastney. The final sighting was on 23rd when an oiled adult was at Eastney; this was not the bird seen on 22nd.

An adult seen by Mike and Pete Combridge at Pennington Marshes on November 14th was well after birds from the wreck had dispersed.

In the following year, a juvenile was seen flying east off Hill Head by Rob Clay and Richard Levett and later by Pete Durnell off Lee-on-the-Solent on September 2nd and another was seen by Mike and Pete Combridge off Hurst Castle on October 8th. The last for the decade was an adult watched by David Glue for 20 minutes on flooded fields at Keyhaven on December 20th 1989.

Six occurred in the 1990s. In 1990 Mike and Pete Combridge saw an adult at Hurst on August 19th and Reg Russell recorded another at Hill Head on October 6th. In the following year Pete Durnell saw a juvenile in Hayling Bay on November 12th. The next was a juvenile seen by Steve Mansfield at close range off Southsea on September 2nd 1997 and later that autumn Tim Parminter watched an adult fly west off The Butts, Pennington Marshes on November 12th. In 1998 J. A. C. Patrick and John Vickerman saw a juvenile off Pennington Marshes from 1300 to 1330 hrs on August 22nd and Mike Rafter saw another fly east through Hurst Narrows in a strong south-westerly wind at 1200 hrs on November 3rd.

The 2000s produced twelve. Pete Gammage and others watched an adult at Eastoke, Hayling Island for three hours on the afternoon of November 6th 2000. Peter Milinets-Raby saw a juvenile fly east at Stokes Bay between 0858 and 0910 hrs on October 8th 2001. The first ever January record was on New Year's Day in 2003 when Steve Keen watched an adult in winter plumage move east over Milford on Sea and Sturt Pond between 0827 and 0830 hrs. The next was a juvenile seen at close range by Marc Moody flying east past Milford on Sea on November 7th 2005. Four days later Andy Johnson found another juvenile at Sandy Point which stayed in the area from 0850 to 1020 hrs;

Sabine's Gulls:
above second-winter,
Sopley, August 19th 2011
and juveniles, Sturt Pond,
September 21st 2011.
Alan Hayden (top)
and Simon Johnson

it was refound further west in Hayling Bay where it was present until dusk. It was seen briefly the next morning. George Spraggs found the corpse of a juvenile at Sandy Point on November 18th which was presumably the same bird. The next was an adult seen by Richard Coomber and Nigel Jones flying west off Milford on Sea during gales at 1320 hrs on December 5th 2006.

The only spring record to date involved a first-summer which Simon Ingram and Marc Moody located flying west past Hurst Beach at 0740 hrs on May 6th 2007. It remained in view for seven minutes as it continued past Milford on Sea and was lost to view off Taddiford Gap. It occurred as the wind had swung round to WSW earlier that day following a period of sustained northerlies. Five occurred in 2008, the most in any year until then apart from 1987. The first were two adults found by Marc Moody and Tim Parminter as they moved west past The Butts, Pennington Marshes at 1255 hrs on September 7th following gales. These also produced a juvenile found by Andy Collins off Weston Point on 7th which was seen again the next morning. It moved upstream and was relocated by Simon Hartill off Riverside Park at 1220 hrs. It was subsequently seen by many observers as it commuted between the two areas, favouring Weston at high tide and Riverside Park at low tide. It was last seen off Weston on the evening of September 12th. Further gales on October 7th produced two juveniles found by Marc Moody off Milford on Sea; one flew east at 1040 hrs and the second flew east at 1302 hrs and back west at 1350 hrs.

A good total of 13 was recorded in the 2010s including seven in 2011. In that year Dave Taylor found a second-winter feeding with Black-headed and Mediterranean Gulls at a pig farm at Sopley on August 18th; earlier that day it had been seen flying north from Christchurch Harbour, Dorset. The next day it was relocated by Mark Halligan at a pig farm at Hinton Admiral but then returned to Sopley where it was also seen on unseasonal floods at nearby Avon Causeway. It was last recorded on August 22nd. It returned to Christchurch Harbour to roost where it was seen on August 21st and 22nd. Following gales, Andy Johnson watched a juvenile in Chichester Harbour Entrance for ten minutes on September 6th before it headed back out to sea. Further changeable weather in that month produced a juvenile seen by Andy Johnson heading north-east over Milford on Sea at 1705 hrs on 10th, another juvenile found there by Dave Stevenson on 13th, and two juveniles found at the adjacent Sturt Pond by Colin Allen on 19th which were subsequently enjoyed by many observers during their stay until 28th. In the meantime, the Needs Ore Tuesday Group saw an adult fly west 50 yards offshore at Needs Ore at 1045 hrs on September 20th.

There were records in each of the next three years. A juvenile was seen by Nigel Jones flying east into Hurst Narrows on August 29th 2012. Two juveniles were watched by Pete Combridge and Phil Toye flying east off Keyhaven

Marshes at 0820 and 0923 hrs on October 29th 2013. Yet another juvenile was seen by Marc Moody moving east past Hurst Castle at 1121 hrs on October 6th 2014 and probably the same bird moving east again at 1330 hrs. In 2017 a juvenile was watched by Pete Combridge and Phil Toye off Milford on Sea on September 12th, in 2019 a juvenile was seen by Steve Keen moving west of Pennington Marshes on September 30th and in 2020 Mark Francis saw a juvenile off Hill Head on September 29th.

Analysis of arrival dates shows autumn and winter records between August 10th and January 1st with a peak between early September and early October. The 1987 influx in mid-October (not included in the chart below) occurred during a two-week period when no other birds have been recorded. The only other record involved the first-summer moving west off Hurst Beach on May 6th 2007.

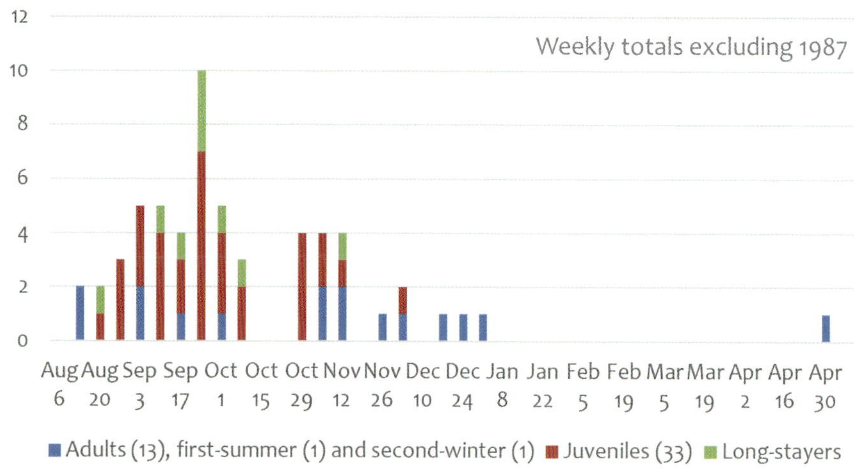

A further 12 records of 14 birds between 1992 and 2020 have been omitted. Seven were considered not proven by HOSRP, two, both of two birds, were submitted without documentation and requests for notes were ignored, and three originated from *Birdguides* and could not be traced.

GB status: 7,658, 1968–2019. Stable, mean 187 per year in 2010–19. (Circumpolar high Arctic breeding range on coastal tundra from Svalbard, E across Arctic Russia and islands, to W & N coastal Alaska, parts of N Canada & N Greenland. Winters in South Atlantic off W Africa, off SE Africa in Indian Ocean, and in Pacific Ocean from N Mexico to Peru.)

Bonaparte's Gull *Chroicocephalus philadelphia*

BBRC

A very rare vagrant. Seven records since 1980.

The first for the county was a first-summer moulting into second-winter plumage found by Bill Oddie at Farlington Marshes on July 23rd 1980 (not July 15th as in *BoH*). Extracts from his description are reproduced opposite.

At about 1600 hrs I was scanning through a flock of Black-headed Gulls with my scope hoping to find a Med. Gull when I picked up a small gull which for a moment I expected to be a Little Gull. However, I immediately noted that despite traces of a hood it had black primary tips and didn't seem quite as tiny as Little Gull. Suddenly it flexed a wing, showing a glimpse of a white leading edge. I then realised it had to be a Bonaparte's Gull – or an extremely tiny Black-headed. I watched the bird for about five minutes through my scope in perfect sunlight at 100 yards or less. The whole flock then took off. The Bonaparte's was the last to fly and I managed to hold it for a brief while before the flock flew off towards Portsmouth. I managed to take the following description:

Size and jizz: Clearly considerably smaller than BHG, but a little larger than a Little Gull. Head and chest correspondingly slimmer than BHG. Head pattern: a faded rather blotchy hood – pale grey, darkest just behind the eye, not extending far down nape. Eye: dark. No white eyering. Bill: shorter and finer than BHG. Mainly black but dark red at base. Mantle, scapulars, coverts: all pale grey, similar to BHG's colour. No obvious dusky shading on the coverts at rest. Primary tips black. Closed wings occasionally showed white flash of leading edge. When the bird flew the wing pattern was clearly obvious. Upperwings: similar to BHG with white leading edge and black primary tips and dusky inner primaries. Underwings: vivid white 'triangle' on leading edge – much more extensive than BHG – almost translucent. Primary tips black. Grey secondaries and upper part of inner primaries offset the white leading edge strikingly. Underparts, nape and tail: all white. No dark terminal band on tail or greyness on nape. Legs: seen very briefly as it took off, looked mid reddish brown. Flight-style: daintier than BHG but unable to watch at length.

Pete Potts saw a bird which he identified as a Bonaparte's Gull at Farlington Marshes on July 15th and was accepted by BBRC as the same individual seen by Bill. However, Pete subsequently felt uncertain about his identification and withdrew the record. Bill, in correspondence with Mike Rogers, BBRC secretary, following Pete's change of mind in 1990, stood by his record.

The next was a first-summer found by Jeff Lawson and subsequently identified by Roger Wilkinson at Gilkicker Point during a seawatch on March 21st 1986. It was first seen flying in from the west and landing offshore with Black-headed Gulls at 0820 hrs but although they suspected it was a Bonaparte's Gull their views were brief before it flew out to sea. At 1000 hrs Roger refound it feeding in the same area and watched it for ten minutes and confirmed its identity. It then flew off again but an hour later was again discovered standing on a nearby car park and then watched on the adjacent lakes. It was looked for the next day but not seen again.

Another first-summer was found in Langstone Harbour by Jason Crook on June 1st 2003. The bird was picked out at 2 km range from the southern tip of Farlington Marshes at 1255 hrs. Jason phoned Bob Chapman to alert him of the bird's presence. They were both able to obtain much improved views

Bonaparte's Gull, first-summer, Langstone Harbour, early evening, June 1st 2003

as the bird came within 250 m in the next hour but then disappeared. Despite extensive searching it was not seen the next day but Jason refound it on Farlington Lake at 1300 hrs on 3rd. As the tide was dropping at 1645 hrs it joined other gulls and terns in the middle of the harbour to feed before returning at 1730 hrs. It moved on to the mud flats at 1830 hrs and finally flew off west over Portsea Island at 1915 hrs. It was again on the lake between 1400 and 1745 hrs on 4th and at 1650 hrs on 5th and was last seen feeding at very close range over the adjacent harbour at 1650 hrs on 6th.

The fourth for the county was a first-summer found by Andy Collins among Black-headed Gulls off Weston Shore on May 3rd 2015. It remained in the Itchen Estuary until May 22nd allowing many Hampshire listers to catch up with this hitherto difficult species. It was seen mostly at Riverside Park where it took bread and allowed excellent views.

The next was another first-summer found by Bob Chapman at Ibsley Water at 1730 hrs on April 17th 2017. It was not seen on 18th but was refound by Bob at 1715 hrs on 19th. It was then seen daily until May 11th although it did go missing occasionally, e.g. on April 29th when Tim Doran found it at Hamer Warren on the west side of the Avon Valley.

The sixth for the county, another first-summer, was found by Keith Simpson from the Tern Hide at Ibsley Water on April 25th 2019 and confirmed by Bob Chapman and Paul Winter. It remained until May 12th and was seen by many birders during its stay. Yet another first-summer was found by Simon Woolley at Ibsley Water on April 20th 2020; it was confirmed by Bob Chapman and remained until 27th. Although it was seen by some local observers the news was not released more widely due to the Coronavirus lockdown.

An older record of a second-winter seen by Graham Rees and Pete Le Brocq in Russell's Lake, Langstone Harbour on October 6th 1956 was apparently accepted by *British Birds* but not published. Graham subsequently withdrew the record in 1971 (David Billett *in litt.*).

Bonaparte's Gull, first-summer, Itchen Estuary, May 4th–7th 2015.
top, Andy Collins and Alan Lewis and bottom, Barry Stalker

GB status: 8, 1850–1949; 275, 1950–2020. About seven per year; increasing. (Breeds widely across N North America from W & C Alaska through Canada to James Bay. Winters locally on ice-free rivers & lakes in N USA, & S along both coasts of USA to Mexico & Caribbean.)

Laughing Gull *Leucophaeus atricilla*

BBRC

A very rare vagrant. Two records in 2005 and 2009.

The first for the county was a first-winter found by John Norton and Peter Milinets-Raby at Haslar Creek, Gosport at 1300 hrs on November 5th 2005. The account of its discovery is given by Norton and Raby (2007). Extracts are reproduced overleaf.

During the previous afternoon, while driving past Stokes Bay in the course of his work as a driving instructor, PM-R briefly saw a medium-sized dark gull. He thought no more of it until that evening when he learnt of an influx of Laughing Gulls into south-west England. Discussing their Gosport area WeBS count scheduled for the next day, Peter suggested to John Norton that he may have seen a Laughing Gull. They checked out several known gull feeding areas without success before arriving at Walpole Park. Walking towards Haslar Creek, a noticeably dark gull landed on the grass about 10 m away. A quick check through binoculars revealed the most obvious features, especially those to eliminate Franklin's Gull: a long slightly drooping dark bill, dark rakish looking legs, elongated tail end and white eye crescents, and we both realised straight away that it was a first-winter Laughing Gull. It stayed on the grass for three minutes or so until it was enticed into flight by someone feeding the Mute Swans on the boating lake. It flew around for ten minutes, giving excellent views, before settling on the water. The news was released and the first local birders arrived within the next half hour.

Laughing Gull, first-winter, Walpole Park, Gosport, November 6th 2005

It remained in the area until 13th although it went missing at times. Other records of first-winters seen by Mike Rafter flying west off Hill Head at 1500 hrs on 7th, Simon Woolley and Julia Casson at Salterns Eight Acre Lake, Normandy on 20th, Trevor Carpenter on a playing field at Henry Cort School, Fareham on 22nd, John Chapman at Langstone Mill Pond on December 3rd and Trevor Carpenter back at Haslar Creek on 11th are assumed to relate to the same individual. Photographs of the Haslar and Salterns birds indicate that this was the case. A claim of a second individual at Haslar Creek on November 10th was considered not proven by BBRC. This occurrence was part of an unprecedented influx into Britain and Ireland involving in excess of 50 individuals which was the result of Hurricane Wilma tracking up the eastern seaboard of the USA to Newfoundland and the subsequent depression which moved across the Atlantic.

The next was a second-summer found by Ian Pibworth at Testwood Lakes at 1500 hrs on May 24th 2009. He was able to watch and photograph it from close range from a hide but unfortunately it flew off south-east at 1525 hrs and was not seen again.

Laughing Gulls: first-winter, Walpole Park, Gosport, November 12th 2005 and second-summer, Testwood Lakes, May 24th 2009. Richard Ford and Ian Pibworth

In 2021, Graham Stephenson picked up an unusually dark-looking gull flying past the visible migration watchpoint at Tweseldown Hill near Fleet at 0840 hrs on November 18th. He was able to put the author on to the bird and both watched it through telescopes for two minutes until it disappeared behind trees. They identified it as a second-winter Laughing Gull. The record is currently under consideration by BBRC.

GB status: 1, 1923; 206, 1966–2020. About five per year, influxes 2005 (58), 2006 (22); stable. (Breeds along E seaboard of USA from Nova Scotia, Canada, S to Florida & Gulf coast, the Caribbean, & C America to N Venezuela. Southern populations largely resident but N breeders winter within southern breeding range.)

Franklin's Gull *Leucophaeus pipixcan*

BBRC

A very rare vagrant. Two records in 1970 and 2014.

An adult moulting into summer plumage was found by John Smith and David Billett off the southern tip of Farlington Marshes at 1640 hrs on February 21st 1970. It was not refound until March 1st but then was seen regularly at Farlington Marshes and on the playing field at Springfield School, Drayton until May 16th. An account of the finding and identification of the bird was given by Billett & Grant (1971) and extracts from this are reproduced below. It was the first to be recorded in the Western Palearctic.

Franklin's Gull, adult, Farlington Marshes, March 3rd 1970. J.B. & S. Bottomley

At 16.40 on 21st February 1970 JTS and DFB were at the southern tip of Farlington Marshes when a small, dark-winged gull approached low along the shore and settled immediately in front of them about 100 yards away. They were able to watch it for about ten minutes before it circled and flew off. During this time, it made several short flights of a yard or two when harassed by Black-headed Gulls nearby. Several features suggested that it was something unusual. It was slightly smaller than the Black-headed Gulls and its mantle was rather dark grey. Its bill was short, stout and blackish; its dark red legs were also short and this accentuated a hunched appearance. The bird had an incomplete black hood and was presumably acquiring summer plumage. In flight its tail appeared pale grey down the centre, contrasting with a pure white rump. Most startling of all, however, was its wing pattern in flight: a rather small black tip, with a white terminal spot or spots, separated from the dark grey of the remainder by a broad white area across the primaries which joined a continuous white border along the trailing edge. Reference to literature confirmed it as an adult Franklin's Gull.

The Franklin's Gull was not located again until 1st March, but it was then seen regularly around Farlington Marshes until 16th May. News of its presence soon spread and Farlington quickly became an essential destination for birdwatchers from all over Britain: during the following weeks many hundreds must have gone there, including Peter Grant who was able to make very detailed sketches and notes. The bird seemed to establish a regular pattern of movement during the daytime: with other local gulls it would visit a marsh lagoon during the early morning, then move a short distance inland to a feeding area of playing fields, rough pasture and adjacent factory sites, and in the evening again visit the lagoon where it would bathe and preen before flying off in a south-westerly direction. Where it roosted was never discovered.

It was a long wait until the county's second, an adult, was found by Rob Hume hawking over trees to the west of Ibsley Water between 1200 and 1300 hrs on October 19th 2014 before flying off south. Many thought this one had got away but fortunately it was relocated by Olly Frampton in the gull roost at Ibsley Water on October 28th. It continued to roost there on most evenings until November 24th and was enjoyed by many birders.

Franklin's Gull, adult, Ibsley Water, Blashford, cracking views in the dark, November 16th 2014. photo Steve Bassett

GB status: 84, 1970–2020. About 2–3 per year; stable. (Breeds locally throughout interior provinces of temperate W Canada, E to Great Lakes & S to Midwest USA. Winters along Pacific coast of South America, from Guatemala to Chile.)

Mediterranean Gull *Ichthyaetus melanocephalus*

BBRC to 1962; HOSRP 1963–1989

Formerly a rare visitor; 18 records between 1954 and 1965. First nested in 1968 and now breeds in substantial numbers.

The first for the county was an adult in winter plumage found by Graham Rees at Eastney sewage outfall on April 27th 1954. Extracts from his description published in *British Birds* (48: 89) are reproduced below.

> *On 27th April 1954, while watching Black-headed Gulls at Eastney Point, Langstone Harbour, Hampshire, I identified an adult Mediterranean Black-headed Gull. The flight of this bird was different from the others, the wings being shorter and more rounded at the ends. The primaries were pure white, and the rest of the wing was grey (not blue-grey) having no striking white leading edge, as in* ridibundus. *There was a black mark behind the eye and the nape was grey. The bill was stouter than those of the Black-headed Gulls, and like the legs and feet, which were often dangled when the bird was hovering, was dark red. In size it appeared slightly larger than* ridibundus, *and when on the water appeared shorter in the neck.*

Between 1956 and 1965 there were 17 reports from the Langstone Harbour area with sightings in every month except May, June and September. In the next three years there were reports from Gilkicker Point, Normandy Marshes, Titchfield Haven, Hook-with-Warsash and, most significantly, the Black-headed Gull colony at Needs Ore Point.

Mediterranean Gull, adults, South Scrape, Titchfield Haven, May 2014

At the last-named site, single adults were present on July 18th 1966 and from April 23rd to 30th 1967. In 1968 an adult pair bred and raised two young, a second-summer paired with a Black-headed Gull and raised three young, and an adult bred with a Black-headed/Mediterranean Gull hybrid. In addition, one second-summer and one first-summer were seen in the gullery but did not stay. This was the first record of breeding in Britain; full details are given by Taverner (1970, 1972).

Mediterranean Gull: adult, Needs Ore, May 1968 and juvenile, South Scrape, Titchfield Haven, July 31st 2004. Geoffrey Fisher and Dan Powell

For the next thirty years the picture was of occasional breeding attempts and slowly increasing numbers outside the breeding season. During the 1990s up to five pairs attempted breeding at Needs Ore but had little success. In 1997 breeding was first recorded on the islands in Langstone Harbour with two pairs raising three young; by 2000 there were 38 pairs, this then increasing with fluctuations to 110 pairs in 2005, 834 pairs in 2017 and 1,736 pairs in 2018. Until the turn of the century the species was predominantly regarded as a winter visitor, mostly at coastal locations. Since the expansion of the breeding population the highest numbers have occurred in spring between mid-February and April when nesting birds return and migrants pass through en route to colonies around the Baltic Sea. In summer substantial numbers fly inland from Langstone Harbour to feed. Three-figure gatherings are now commonplace and as many as 812 were noted flying over Winchester Sewage Farm between 1400 and 2029 hrs on July 7th 2015 (Cloyne & Pearson 2016).

Mediterranean Gull, adults displaying, North Scrape, Titchfield Haven, April 5th 2012. John Wichall

GB status: 5, 1866–1939; 95, 1954–1963; c 2,400 breeding pairs in 2018. (Breeds NW, C and SE Europe, Black Sea and Azerbaijan; winter range extends to Mediterranean Basin and NW Africa.)

Ring-billed Gull *Larus delawarensis*

BBRC to 1987; HOSRP to date

A very scarce visitor recorded in every month except May. Around 82 records involving a minimum of 36 individuals since 1977.

The first for the county was a first-summer identified by Dick Barrett at Bedhampton Rubbish Tip on June 1st 1977. It remained in the area until at least August 1st and was seen by many birders. There was much discussion about whether two individuals were involved. Dick's original description has not been traced but in a second description he stated:

I last saw the immature bird on 24 June but while watching on 6 July I noted a Ring-billed Gull in a plumage state so different from the immature bird that it could only be a different bird. In size, shape and actions it did not vary in any way from the previous bird but from its plumage it was obviously approaching full adulthood.

A detailed description taken by Peter Grant on June 18th describes it as in active moult from first-summer to second-winter plumage. A further description provided by Dave Flumm which was taken on July 30th describes a sub-adult type plumage. BBRC concluded that only one individual was involved.

Since then, a minimum of 35 new individuals has been found.

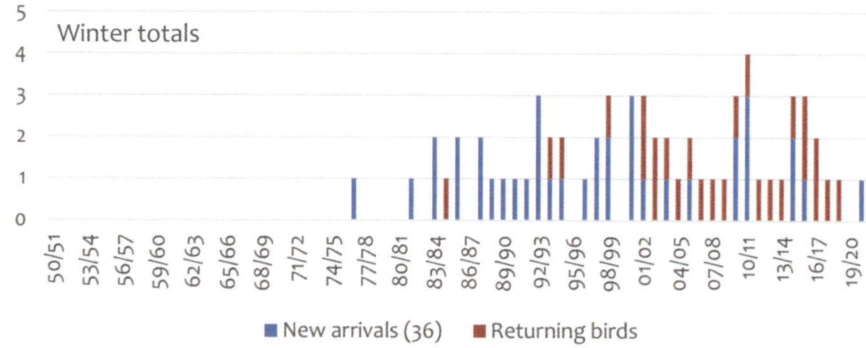

The next was an adult found by Keith Godfrey at Dibden Bay at 1530 hrs on March 17th 1982; he was able to telephone Dave Bishop and they watched it together from 1700 to 1800 hrs. What is assumed to be the same individual was found there by Paul Pearson and confirmed by Mike Terry on April 10th. The next was a first-winter found by Dave Bishop at Titchfield Haven on December 26th 1983. It was seen again there by Bob Chapman on January 8th and 20th and then relocated on floods at nearby Stubbington Green on January 31st. It was then seen intermittently at Stubbington and at Titchfield Haven and Hill Head until February 19th. An adult was found by Andy Hanby at Stubbington on February 4th 1984 and later seen by several members of the Hayling Group; it was subsequently seen at Hill Head by Bob Chapman on February 17th and Roger Wilkinson on March 16th.

It is assumed that the 1984 adult returned in the following winter. It was seen by Barry Duffin, Bob Chapman, Richard Carpenter and others at Titchfield Haven on several dates between January 18th and February 17th 1985. Details were submitted to BBRC but confusion evidently reigned as the only record published was a not proven decision on a second-winter also seen at Titchfield Haven on January 19th. However, BBRC did accept and publish a sighting of an adult by Roger Wilkinson at Paulsgrove Reclamation on April 4th which is assumed to be the returning Titchfield Haven adult. The next was an adult found by Richard White at Bedhampton Rubbish Tip on February 15th 1986. Later that year a second-summer was found by John Walters and seen by Trevor Codlin at Sinah Gravel Pit on March 28th; two days later Pete Potts saw presumably the same bird off Eastney which flew off east towards Hayling Bay. It was not relocated. The final record to be accepted by BBRC before responsibility for the species passed to HOSRP was an adult found by Pete Potts and seen by Keith Turner at The Kench, Langstone Harbour on November 29th 1987; it flew off into the harbour but later that day it was seen by several local birders in the same area. It was not seen again despite being searched for.

Ring-billed Gull, first-winter, Stubbington Green, January 31st 1984

Since 1988 HOSRP has accepted some 66 records which relate to a minimum of 29 new individuals.

In the years up to 2009 most records were from Langstone and Portsmouth Harbours with a scattering elsewhere along the coast and inland at Alresford Pond, Aldershot and Testwood Lakes. Since then, most sightings have been

Ring-billed Gull, adult with Black-headed Gull, Walpole Park, Gosport, January 8th 2008. Richard Ford

at Ibsley Water where scrutiny of the evening gull roost has revealed numerous observations of this and other rare gull species. There have been several instances of birds returning in successive winters which can be followed by tracing the same bird number in column 1 of the table below. The longest-lived of these was an adult found by Peter Milinets-Raby at Walpole Park, Gosport on November 20th 2003 which stayed until March 29th 2004 and returned in every winter until 2014/15 when it was last seen on March 26th (number 25). It is likely that this was the first-winter found by Jason Crook at nearby Tipner Lake on February 25th 2001 which he recorded again in early 2002 and 2003 (also shown as number 25). At Ibsley Water, an adult first seen in 2014/15 returned for the next four winters (number 34) and a second-winter recorded on one date in early 2015 is assumed to be one of the adults present there in the next two winters (number 35). Records for the period between March 4th and 12th 2016 indicate that three different identifiable adults were present at Ibsley. Another adult first discovered by Bob Chapman at Farlington Marshes on January 9th 1993 (number 14) was seen in the following two winters. Adults seen in 1997/98 and 1998/99 (number 19) and from 2000/01 to 2002/03 (number 23), although treated as separate individuals, were probably the same as number 14 based on small size and mantle colour (Jason Crook *in litt.*).

Analysis of arrival dates shows five in the period between June 1st and September 11th which were all first-summers or second-winters. Numbers of new birds were evenly spread over the winter period although all records including returning birds peaked in February. Several lingered into April while new arrivals in that month were a first-summer at Pennington Marshes on 8th in 1988 and a first-summer at Farlington Marshes on 24th in 2004.

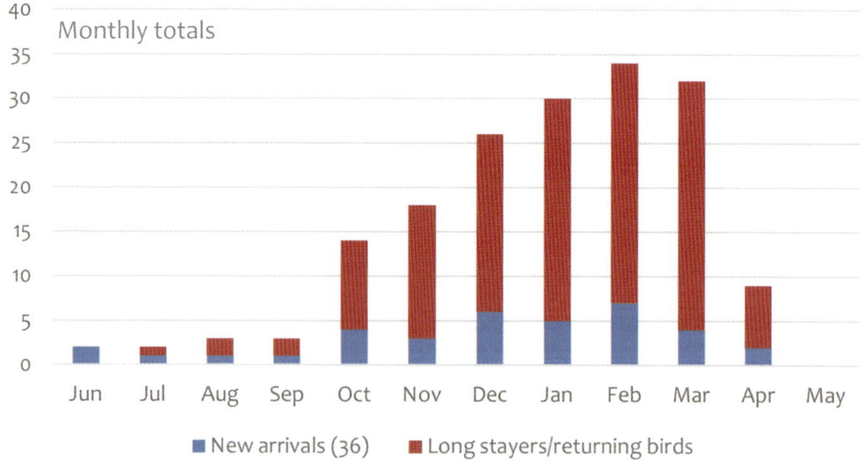

Records accepted by BBRC are tabulated opposite. The numbers in the first column identify returning birds.

	Date(s)	Age	Location	Observers
1	1977: June 1st–August 1st	1s	Bedhampton RT	Dick Barrett *et al.*
2	1982: March 17th	Ad	Dibden Bay	Dave Bishop and Keith Godfrey
2	1982: April 10th	Ad	Dibden Bay	Paul Pearson and Mike Terry
3	1983/84: December 26th–February 19th	1w	Stubbington Green/Hill Head	Dave Bishop and Bob Chapman
4	1984: February 4th February 17th March 16th	Ad	Stubbington Green Hill Head Hill Head	Hayling Group Bob Chapman Roger Wilkinson
4	1985: January 18th–February 17th	Ad	Titchfield Haven	Bob Chapman, Barry Duffin
4	1985: April 4th	Ad	Paulsgrove Reclamation	Roger Wilkinson
5	1986: February 15th	Ad	Bedhampton RT	Richard White
6	1986: March 28th March 30th	2s	Sinah GP Eastney sewage outfall	John Walters and Trevor Codlin Pete Potts
7	1987: November 29th	Ad	The Kench, Langstone Harbour	Pete Potts and Keith Turner

HOSRP took over from 1988. Accepted records are tabulated below.

	Date(s)	Age	Location	Observers
8	1988: April 9th	1s	Pennington Marsh	Mike and Pete Combridge
9	1988: December 26th	Ad	Farlington Marshes	Terry Jennings, Doug Radford, Pete Potts and Geoff Farwell
10	1989: June 7th	1s	Hayling Oyster Beds	Graham Stephenson
11	1990: December 30th	1w	Sinah GP	Robert Clay
12	1992: February 28th	1w	Titchfield Haven	Pete Potts
13	1992: October 30th	Ad	Camber Docks, Portsmouth Harbour	Christopher Wear
14	1993: January 9th–March 19th	Ad	Farlington Marshes	Bob Chapman
15	1993: February 26th, March 29th and 31st	Ad	Alresford Pond	Tony Blunden and Guy Thompson
14	1993: December 9th and 10th	Ad	Farlington Marshes	Terry Jennings, Jason Crook and Bob Chapman
16	1994: February 19th	Ad	Hill Head	Nigel Jones

14	1994: February 23rd	Ad	Farlington Marshes	Jason Crook
17	1994: October 13th	2w	Weston Shore	Clay Jones
14	1995: January 13th and 16th	Ad	Farlington Marshes	Bob Chapman, Jason Crook, Kevin Crisp and Terry Jennings
18	1996: October 2nd	Ad	Aldershot	Mick Scott
19	1998: February 22nd–March 23rd	Ad	Bedhampton Creek and Farlington Marshes	Andy Hanby *et al.*
20	1998: March 1st	3w	Farlington Marshes	Ken Arber, Mark Edgeller and Andy Rhodes
21	1998: December 19th	Ad	Sturt Pond	Mike and Pete Combridge and Phil Toye
22	1998: December 20th	Ad	Little Testwood Lake	Ian Pibworth
19	1999: January 4th–March 28th	Ad	Bedhampton Creek and Farlington Marshes	Bob Chapman
23	2000/01: December 31st–April 1st	Ad	Bedhampton Creek	Jason Crook
24	2001: January 5th February 5th February 15th–March 30th	1w	Tipner Lake Broadmarsh, Langstone Harbour Tipner Lake	Jason Crook Jason Crook and Pete Durnell Jason Crook
25	2001: February 15th–April 19th April 26th	1w	Tipner Lake South Moor, Langstone Harbour	Jason Crook Jason Crook
26	2001: July 1st–August 22nd	2w	Beachlands, Hayling Bay	Simon Wright
23	2002: January 2nd–March 22nd March 30th	Ad	Broadmarsh area Farlington Marshes	Jason Crook
25	2002: January 27th	2w	Farlington Marshes	Jason Crook
23	2002: April 9th–13th	Ad	Hayling Bay	Simon Wright and Tim Lawman
23	2002/03: December 28th–April 2nd March 4th and 10th	Ad	Broadmarsh area Farlington Marshes	Jason Crook
25	2003: March 24th	3w	Tipner Lake	Jason Crook
25	2003/04: November 20th–March 29th	Ad	Gosport	Peter Milinets-Raby
27	2004: April 24th	1s	Farlington Marshes	Jason Crook and Bob Chapman

25	2004/05: October 18th–March 27th	Ad	Gosport	
28	2005: August 27th	2w	Gilkicker Lake	Ian Calderwood
25	2005/06: October 5th–March 27th	Ad	Gosport	
25	2006/07: October 3rd–March 28th	Ad	Gosport	
25	2007/08: October 31st–March 27th	Ad	Gosport	
25	2008/09: October 10th–March 14th	Ad	Gosport	
29	2009: October 11th	2w	Ibsley Water	Rob Souter
30	2009: November 6th	1w	Ibsley Water	Rob Hume
25	2009/10: October 20th–March 13th	Ad	Gosport	
31	2010: September 11th	2w	Sopley Pig Farm	Graham Armstrong
25	2010/11: September 27th–March 16th	Ad	Gosport	
32	2010: November 7th	Ad	Ibsley Water	Lawrie Chappell
33	2011: January 22nd	1w	Ibsley Water	Rob Hume
25	2011: September 4th	Ad	Eastoke, Hayling Island	David Ball
25	2011/12: October 23rd–March 17th	Ad	Gosport	
25	2012/13: October 21st–March 24th	Ad	Gosport	
25	2013/14: November 3rd–March 17th	Ad	Gosport	
25	2014/15: October 26th–March 12th	Ad	Gosport	
34	2014/15: December 23rd–March 7th	Ad	Ibsley Water	Rob Hume
35	2015: January 4th	2w	Ibsley Water	Rob Hume
34	2015/16: November 14th–25th and December 20th–March 25th	Ad	Ibsley Water	
35	2016: March 4th–12th	Ad	Ibsley Water	Rob Hume and Dave Taylor
36	2016: March 4th–12th	Ad	Ibsley Water	Dave Taylor and Rob Hume
34	2016/17: November 18th–February 15th	Ad	Ibsley Water	

35	2016: December 1st–9th and January 2nd–February 14th	Ad	Ibsley Water	
34	2017/18: December 10th–April 2nd February 10th–March 6th	Ad Ad	Ibsley Water Blackheath Down	
34	2018/19: November 15th–December 2nd; February 3rd–March 6th	Ad	Ibsley Water	

Two records of first-summers found by Keith Godfrey at Eling Great Marsh on March 28th and April 5th 1981 and at Farlington Marshes from June 24th to 28th 1982 have not been submitted to BBRC. The 1981 record was published in the *HBR* and *BoH* as accepted by BBRC but this was erroneous. During the period up to 1987 when records were adjudged by BBRC, eight were considered not proven (see Appendix VIII). Since then, HOSRP considered a further 15 to be not proven. Failure to eliminate sub-adult Herring Gulls which often have a black sub-terminal band on their bill was a frequent pitfall for inexperienced observers. A further 23 claims during this period were not documented.

GB status: 1,978, 1973–2019. Large decline, mean 16 per year in 2010–19. (Breeds temperate North America from British Columbia E to Maritime Provinces, Canada, S to NE California, Nevada & Wyoming to Michigan, also N Maine, USA. Winters to S of breeding range south to S Mexico & Gulf of Mexico, & inland on Great Lakes.)

Glaucous Gull *Larus hyperboreus*

HOSRP

A very scarce visitor, usually in winter but recorded in every month. One in 1948 and around 88 individuals since 1953.

The first for the county was an adult found by Tony Norris at Titchfield Haven on October 16th 1948. His description from the 1948 *HBR* was as follows:

> *This bird was about the same size as the four Great Black-backed Gulls with which it was in company. Most salient points were grey wings and mantle, otherwise it was white all over. Legs pink. Bill yellow with red spot on lower mandible. Head streaked with brown. Adult. Very high-pitched voice. No black on wings.*

The first in the modern era was an adult seen by David Billett and Graham Rees in Langstone Harbour on October 3rd 1953. Their description was as follows:

> *A large gull, about size of Great Black-back. Light grey mantle, pure white primaries, short wings not protruding beyond tail when bird was at rest; heavy bill; peculiar posture when on the water.*

Since then, a minimum of 87 individuals has been recorded.

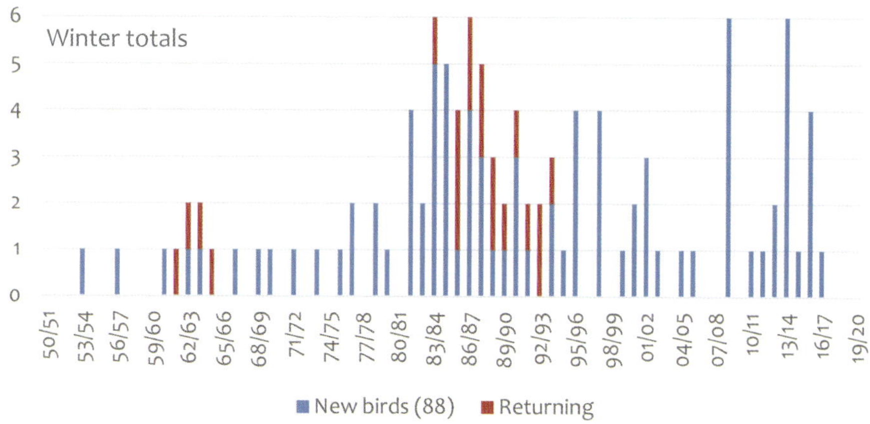

In the period up to 1980 around 15 were recorded with no more than two in any winter. Most were seen on one or two dates only but a third-summer in the Keyhaven area between April 23rd and June 18th 1961 returned in the next three winters (earliest date November 23rd 1963; latest April 15th 1962). Further sightings of an adult there on September 30th and October 1st 1964 presumably referred to this individual but it was not seen in the subsequent winter when observer coverage was limited (Ed Wiseman *in litt.*). Other records during this period were of singles on one or two dates only apart from a juvenile at Black Point on December 16th 1956 and nearby in Langstone Harbour entrance from January 1st to 13th 1957, a juvenile at Titchfield Haven from February 8th to March 13th 1977 and presumably the same bird at Bedhampton Rubbish Tip on February 9th.

The period between 1981/82 and 1997/98 saw an increase in numbers with four to six seen in five consecutive winters including five in early 1984 which were associated with an arrival of Iceland Gulls following north-westerly gales. Five individuals are adjudged to have returned in successive winters. The most long-lived of these was the fourth-winter first found in Portsmouth Harbour at Hardway, Gosport on November 14th 1984 which returned in the next eight winters. It was relocated in August or September in seven of these winters (earliest date August 8th 1992) and was last recorded in February or March (latest date March 14th 1987 except in 1991 when it stayed until April 24th). When last seen on February 14th 1993 at the age of 13 years it had a broken leg and unfortunately did not return. Another fourth-winter found at Browndown Rubbish Tip on October 27th 1984 was considered different to the Hardway individual; this returned for the next five winters and was most frequently recorded at Hill Head. It usually arrived later in the winter (earliest date November 14th 1985 but in three years not seen until the new year) and stayed until March (latest date March 29th 1987). Other long-staying birds were a first-winter at Paulsgrove Rubbish Tip from February 2nd to March 3rd 1983 which relocated to the Weston Shore area between April 4th and July 3rd (with a day out to Lepe on June 22nd), returned to Paulsgrove between August 22nd and December 28th and was last seen in Hayling Bay on January 29th 1984, a second-winter at Hill Head between March 13th and May 4th 1985 which returned there between January 13th and March 31st 1986 and finally an adult first seen at Sandy Point on December 20th 1986 which was at nearby East Winner on January 11th 1987 and then seen in Chichester Harbour on October 7th 1992 and December 2nd 1993. These records are assumed to relate to one which wintered in the Pagham Harbour/Selsey Bill area of West Sussex from 1984 to 1996.

Glaucous Gulls: adults, Hardway, Gosport, March 2nd 1991 and Southsea, January 31st 2009. John Eyre and Darren Evans

In the period from 1998/99 to 2007/08 the species returned to its former status of less than annual records with most seen on one or two dates only. The only longer stay was by a juvenile at Efford Rubbish Tip (Pennington) between December 26th 1997 and January 9th 1998 which was seen with a second-winter on the final date. In 2008/09 six new arrivals were recorded, the most ever. The first was a juvenile at Keyhaven Lagoon on November 15th and off Hurst Beach on November 27th. Four more arrived in late January following north-westerly gales. A juvenile and a second-winter were seen within five minutes of each other flying west off Milford on Sea on 25th; on the same day another second-winter was off Sandy Point and an adult was at The Kench, Langstone Harbour on 27th. A second-winter off Pennington Marshes is assumed to be one of the Milford birds but may have been different. Both the Sandy Point and Kench birds were refound at Southsea on 30th where they were seen by many observers and presented good photographic opportunities. The second-winter was last seen on February 7th and the adult on February 16th. The second-winter was refound in Langstone Harbour where it was seen until March 1st and presumably nearby at Priddy's Hard, Gosport on February 21st before moving to Emsworth Harbour on the last date. Finally, a juvenile was seen off Hurst Beach flying towards The Needles on April 10th.

Only four were recorded in the next four winters including a juvenile which stayed in the Test Estuary between Town Quay and Eling from January 29th until February 26th 2012. A further influx of six occurred in early 2014. An adult, a second- or third-winter and two juveniles were recorded in the area between Hayling Bay and Gosport between January 4th and February 2nd, a third- or fourth-winter was between Milford on Sea and Hurst Castle on January 25th and 26th and a juvenile was in the same area between February 20th and March 10th where it was regularly seen feeding on dead auks. Since then, the only records have been of a juvenile at Ibsley Water on January 1st

2015, a total of four in early 2016 comprising juveniles off Hurst and Milford on Sea between January 1st and 8th, off Southsea on 10th and 18th and at Riverside Park, Southampton on February 14th, a wide-ranging second-winter seen between Southsea and Southampton Docks between March 21st and May 2nd, and a second-winter in the Black Point/Chichester Harbour entrance area on January 25th and February 10th 2017 and presumably the same individual at Titchfield Haven and Hill Head on January 26th.

Glaucous Gull, second-winter, Beachlands, Hayling Island, January 21st 2014. Alan Lewis

This species is predominantly coastal and in the late 20th century rubbish tips proved particularly attractive. Since their closure birds have occurred in estuarine habitats but also on grasslands such as at Southsea. Approximate totals at coastal sites are as follows: Lymington/Hurst, 29; Titchfield Haven/Hill Head and nearby sites, 17; Langstone Harbour, 13; Chichester Harbour, 12; mid-Southampton Water (mostly off Weston Shore and at Newtown Rubbish Tip), 9; Southsea, 8; Portsmouth Harbour, 5; Hayling Bay and Lee-on-the-Solent/Gilkicker Point, 5 each; Eling, 3; and the Beaulieu Estuary, Calshot, Fawley, Lepe and Sinah Gravel Pit, 1 each. Inland records are rare. In addition to those at Ibsley Water and Riverside Park mentioned above, there have been three juveniles at Bramshill Rubbish Tip on December 16th and 17th 1978, between December 30th and January 7th 1980 and on January 31st and February 1st 1982, a second-winter at Little Leigh Farm near Havant on April 9th and 10th 2001 which was also seen feeding in Langstone Harbour at low tide from 9th to 12th and a juvenile in the enclosures at Marwell Zoo on February 15th 2005. Note the totals of records listed here exceed the overall total for the county as several individuals were seen at multiple locations. However, returning birds have been included once only.

As shown opposite, analysis of arrival dates shows that most new birds occur between December and April. Apart from returning and long-staying individuals, the only earlier arrivals were of a third-winter at Pennington Marshes on August 19th 1987, a second-winter flying over the Beaulieu

Estuary on October 2nd 2001, the first for the county in Langstone Harbour on October 3rd 1953 and four in November. Most have left by the end of April but there have been several long-staying birds which have remained until July or later. The only new arrivals in May to July were a first-summer which flew north over Hurst Beach on May 4th 1967, a sub-adult at Efford Rubbish Tip (Pennington) on May 10th 1976 and a second-summer at Hill Head on June 25th 1979.

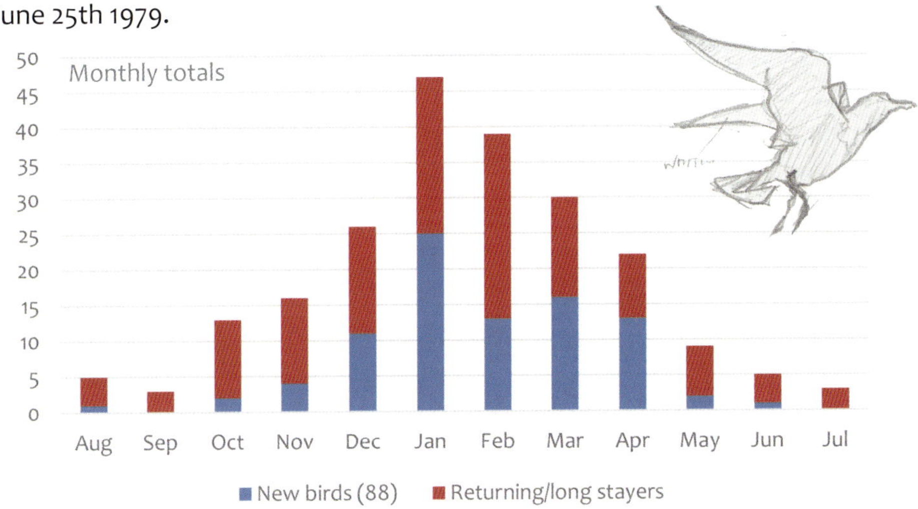

The approximate totals of each age class when the birds were first recorded were as follows: adult or fourth-winter, 14; third-winter, 5; second-winter or -summer, 16; juvenile or first-summer, 53.

K & M recorded a specimen obtained in the neighbourhood of Portsmouth which was held in the museum of Haslar Hospital, Gosport. In the absence of any details or an extant specimen this cannot be accepted as the first for Hampshire. Two were claimed by Enid Gaynor Davies near Lymington on May 24th 1941 (*Brit. Birds* 35: 39), not 1921 as in Cohen (1963) and *BoH*. She wrote:

> *I watched a pair swimming on one of the marshy inlets near Lymington. There were a number of Herring Gulls in the vicinity, and the large size, very pale grey back and wings, yellow bill and absence of black in the wings was very noticeable at a distance of not more than 30 yards. The birds appeared to be adults as I could see no dark markings on their tails or bills.*

This seems unacceptable by any standards.

Also of interest is a record of a juvenile Glaucous x Herring Gull hybrid, colloquially known as 'Viking Gull', found by Andy Collins at Weston Shore on March 15th 2013. It was never seen again.

GB status: scarce winter visitor, most frequent in north, occasional larger influxes following NW gales. (Breeds throughout Arctic, wintering south along all coasts.)

Iceland Gull *Larus glaucoides*

HOSRP

A very scarce visitor, usually in winter but recorded in every month. Three old records and around 106 individuals since 1955.

The first for the county was one obtained at Portsmouth in 1860 which K & M stated was held in the Charterhouse School Museum at Godalming, Surrey. The next was not until 1937, when one, almost adult, was at Beaulieu on January 31st and possibly the same individual was at Langstone Harbour on February 14th (*Brit. Birds* 30: 353–354). Cohen (1963) included an immature seen by Col. and Mrs. Highway by the toll-bridge to Hayling Island on February 2nd 1952. Size was that of Herring Gulls with it; no dark feathers anywhere. This was omitted by Cohen & Taverner (1972) and BoH. The first acceptable modern record was in 1955, when John Ash recorded a second-winter feeding around a threshed stack at Damerham on February 22nd.

Iceland Gull, first-winter, Hill Head, February 14th 1984. Freezing!

Since then, a minimum of 105 individuals has been recorded.

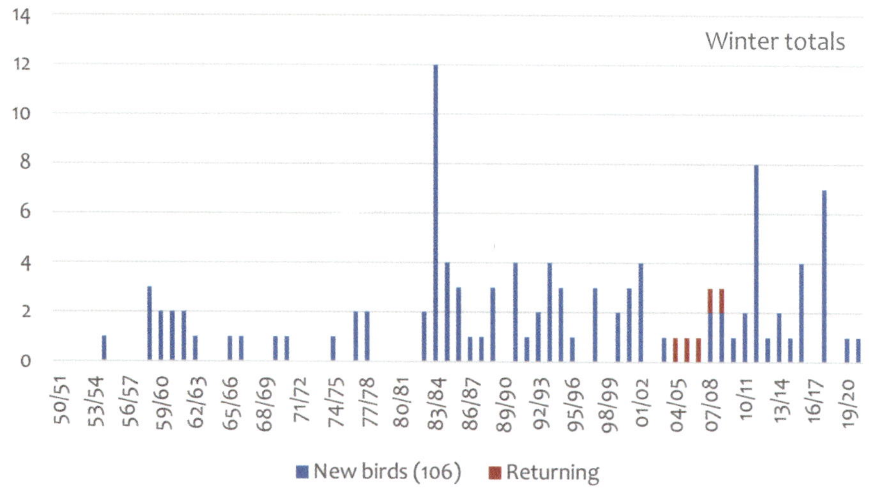

Records have increased in frequency since the 1958/59 winter. In the 34 winters covered by *BoH* up to 1991/92 there were records in 21 of them. Subsequently, there have been just five blank winters up to 2020/21. Most records were associated with influxes into Britain following north-westerly gales.

Early records were all for one date only apart from a juvenile found by Dick Barrett at Bedhampton Rubbish Tip on March 7th 1977 which was seen by other observers until March 17th.

In early 1984, at least 12 were seen. These arrived in the wake of severe north-westerly gales which drove hundreds into Britain and Ireland from their usual wintering grounds in Iceland. The first was a second-winter at Wrecclesham Rubbish Tip, Surrey on January 6th which flew off west into Hampshire; this was followed by a first-winter at Sinah Gravel Pit on January 17th and nearby at The Kench on January 22nd. Subsequently, an unprecedented concentration was in the Southampton Water area where probably three adults and seven juveniles were present between January 19th and March 27th. Most records came from rubbish tips at Browndown, Hook and Newtown and the shore at Hill Head, Calshot and Weston Shore. One juvenile continued to feed at Newtown Rubbish Tip until May 4th.

Following the 1984 influx up to four individuals were recorded in most winters but with higher totals of eight in 2011/12 and seven in 2017/18. In contrast to the earlier period, several long stays were recorded, although this probably reflects improved observer coverage. These included a juvenile at Walpole Park, Gosport between February 23rd and March 17th 1991, another at IBM Lake, Cosham on three dates between January 30th and February 26th 1994, a second-winter at various sites in Southampton Water between February 2nd and March 15th 2002 and a juvenile first seen at Eastney on January 31st 2004 which was seen at various sites in the East Solent area until May 4th of that year. This individual is assumed to have returned in the following five winters, usually arriving in late December (earliest date 23rd in 2005 and 2007) and staying until late March or later (latest date May 6th 2008). It was most frequently seen in the area between Hill Head and Portsmouth Harbour but also ranged west to Hurst Beach, east to Sandy Point and inland to Winchester Sewage Farm where it was seen on seven dates between January 3rd and February 9th 2006 and on January 22nd 2007.

In the influx in 2011/12 the first was an adult found along the Gosport shore of Portsmouth Harbour on December 31st which remained in the area until February 29th; there was also a second-winter there between January 8th and February 11th. A juvenile was in Langstone Harbour between January 18th and April 1st while further records, mostly on one date only, came from Fishlake Meadows, Chichester Harbour entrance, Ibsley Water, Bishopstoke, Titchfield Haven and Winchester Sewage Farm. A second-summer in

Portsmouth Harbour on June 4th and July 4th presumably summered and may well have been the individual first recorded there in January.

In early 2018 at least seven were seen including a long-staying juvenile in the Hurst/Keyhaven area between January 14th and February 4th with presumably a different individual on April 1st and another in the Ibsley Water gull roost between January 31st and March 31st. The latter was seen feeding with other gulls at a pig farm at Blackheath on several dates in February.

Most records have been coastal and particularly in the late 20th century many occurred in the vicinity of rubbish tips. Approximate totals from coastal sites are as follows: Lymington/Hurst, 24; Titchfield Haven/Hill Head and nearby sites, 20; Langstone Harbour including Sinah Gravel Pit, 19; mid-Southampton Water (mostly off Weston Shore and at Newtown Rubbish Tip), 17; Portsmouth Harbour, 11; Lee-on-the-Solent/Gilkicker Point and Chichester Harbour entrance, 6 each; Eastney/Southsea and Fawley/Calshot, 5 each; Eling/Redbridge, 2; and Beaulieu Estuary and Tanners Lane, 1 each. Inland records have come from Ibsley Water, 6; Winchester Sewage Farm, 3; Avon Causeway and Wrecclesham and Bramshill Rubbish Tips, 2 each; and Allington Gravel Pits, Bishopstoke, Damerham, Fishlake Meadows and Rownhams, 1 each. Note the total of records listed here exceeds the overall total for the county as several individuals were recorded at multiple locations.

As shown opposite, analysis of arrival dates shows that few are recorded before late December. There are several early autumn records of birds seen by experienced observers: adults in Langstone Harbour on August 4th 1958 and August 19th 1985, two adults which flew east off Eastney 20 minutes apart on September 18th 1960 which should perhaps be regarded as one individual, an immature at Milford on Sea on October 11th 1970, a third-winter at Iley Lane Lake on October 21st 1994 and an adult in Langstone Harbour on October 23rd 1976. Numbers of new arrivals peak in January but the actual total recorded peaks in February with a gradual decline through to early May. Five have been recorded in the first week of that month with the latest an adult at Keyhaven on May 6th 2008. In addition to the summering bird already mentioned there is one further July record of a first-summer seen by Philip Fawkes at Fawley Reclamation on July 17th 1977.

Iceland Gulls: adult, Curbridge, April 21st 2014 (far left) and juvenile, Langstone Harbour, March 12th–15th 2012.
David Ryves, John Wichall, Richard Ford and John Hilton

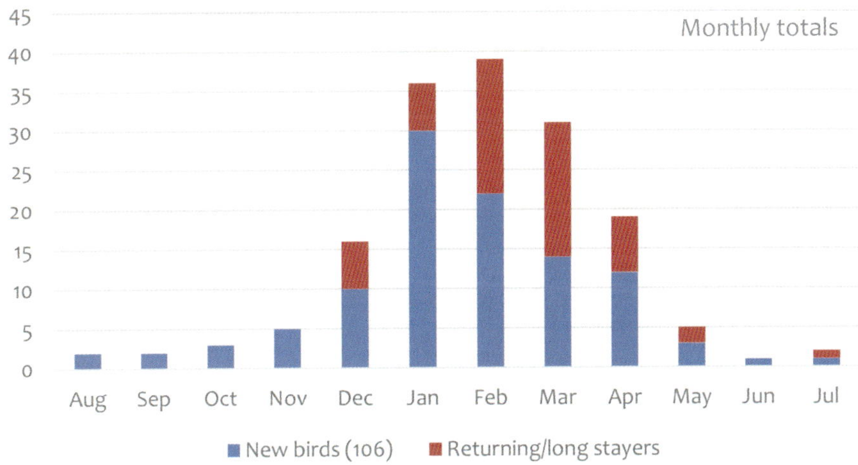

The approximate totals of each age class when the birds were first recorded were as follows: adult or fourth-winter, 26; third-winter, 5; second-winter or summer, 17; juvenile or first-summer, 58.

GB status: very scarce winter visitor, most frequent in north, occasional larger influxes following NW gales. (Breeds Greenland, wintering south to NW Europe and rarely E USA.)

Thayer's Gull *Larus glaucoides thayeri*

BBRC

A very rare vagrant. One record in 2018.

The only record for Hampshire of this race of Iceland Gull is of a juvenile found by Alan Lewis in the gull roost at Ibsley Water on January 28th 2018. It stayed in the area until March 8th. Extracts from his account of its discovery in the 2018 *HBR* (Lewis 2019) are reproduced below.

Arriving at the Tern Hide at Ibsley Water at about 3 pm, I began scanning the gulls which had already arrived. The hide soon began filling up with people keen to see the long-staying Ring-billed Gull for their year lists, and it was soon standing room only. Once people had had their fill of the Ring-billed Gull, they began to leave and by 4.15 pm I was the only one left in the hide.

At about 4.20 pm I noticed an interesting juvenile large gull on its own behind the closest Black-headed Gulls. The primaries were dark brown with narrow pale fringes, rather than the solid black of most juvenile large gulls, the body plumage was rather smooth mid-brown and relatively unstreaked and the bill was rather stocky and black. Having seen two before, I thought there was a possibility that it could be a juvenile Thayer's Gull and so took a series of images with my DSLR. After a few minutes it flew past the hide with other gulls and I managed to get a reasonable series of images in flight before it landed in a group of other large gulls. Seeing it in direct comparison with other gulls for the first time, it was clearly an 'Iceland Gull type' with rather long wings, a relatively short stocky bill and a rounded head, and I was convinced by now that it must be either a very dark Kumlien's L. g. kumlieni *or a Thayer's* L. g. thayeri. *I left the hide as the light faded and the bird was still present.*

Returning home, I forwarded a selection of the images to several friends and discussed the bird with them. The flight shots were important and showed the extensive dark outer webs on the outer five primaries contrasting with the paler inner webs (the oft-quoted 'venetian blind' effect of Thayer's) and the extensively dark tail, something I hadn't noticed in the field. We were all certain that this was indeed a juvenile Thayer's Gull and the news was released.

Despite many others checking the gull roost on subsequent evenings, it was not seen again until Friday February 2nd when I re-visited the roost and picked up the bird as it dropped in very briefly allowing six other lucky birders to see it. I also managed to fire off a few more distant flight shots as it flew off.

To the relief of many, the bird then showed very well in the roost over the weekend of February 3rd and 4th, enabling many more the chance to see it.

On February 5th, Olly Frampton located the Thayer's at the pig fields at Blackheath Down, approximately 12 km north-west of Ibsley Water, where it was then seen regularly (and on floods close to Martin village) during the day until March 8th. On most evenings, the bird returned to the Ibsley Water roost where it was last seen on March 6th.

Thayer's Gull, juvenile, Blackheath Down, February 8th and 9th 2018. Alan Lewis and Gareth Rees

Thayer's Gull is currently regarded as a race of Iceland Gull although it has previously been considered a separate species, a race of American Herring Gull *L. smithsonianus* or even a hybrid between the two. The record was the eighth for Great Britain, the first having been as recently as the 2007/08 winter.

GB status: 8, 2007-17. Very rare. (Breeds Canadian Arctic islands & NW Greenland, wintering on W coast of North America; less frequently Great Lakes and E coast.)

Caspian Gull *Larus cachinnans*

BBRC to 1999; HOSRP 2000 to date

A very scarce passage migrant and winter visitor. Around 39 records involving a minimum of 33 individuals since 2001.

Awareness of this taxon increased in the mid-1990s when it was regarded as the nominate race of Yellow-legged Gull *Larus cachinnans*. It was subsequently admitted to the British List and in 2007 *cachinnans* and the race *michahellis* were officially split as separate species Caspian and Yellow-legged Gulls.

The first for the county was a first-winter found by Mark Edgeller at Eling Great Marsh on November 24th 2001. Extracts from Mark's description are reproduced overleaf.

The bird was first noted at c.100 yards range when the long neck, 'snouty' head shape and very dark tertials were noted. It was watched through 8 x 32 bins and x 30 scope for 20 minutes swimming and picking objects from just below the water surface. The bird was not seen in flight but lifted its wings on one occasion. The long neck was held extended with the head held down making it look like a giant phalarope. Bill – long and parallel-sided, lacking the prominent gonydeal angle of Herring and Yellow-legged Gulls. Head – clean white with limited dark smudging around the eye. Back of head and upper nape also clean white with streaking on base of hind neck only. Head shape was distinctive – a long, sloping forehead with the crown peaking well behind the dark eye. Rear crown squared off. Underparts (breast and flanks only visible) – also clean white. Upperparts – tertials very dark, almost black, with narrow clean white fringes not extending round the whole feather border. Scapulars and mantle appeared greyish with dark internal markings, although exact pattern was not discernible due to the distance. Lesser, median and greater coverts formed a solid brown block on the closed wing with only narrow pale fringing. Underwing – pale (milky?) with pale bases to primaries.

Since then, at least 32 further individuals have been recorded.

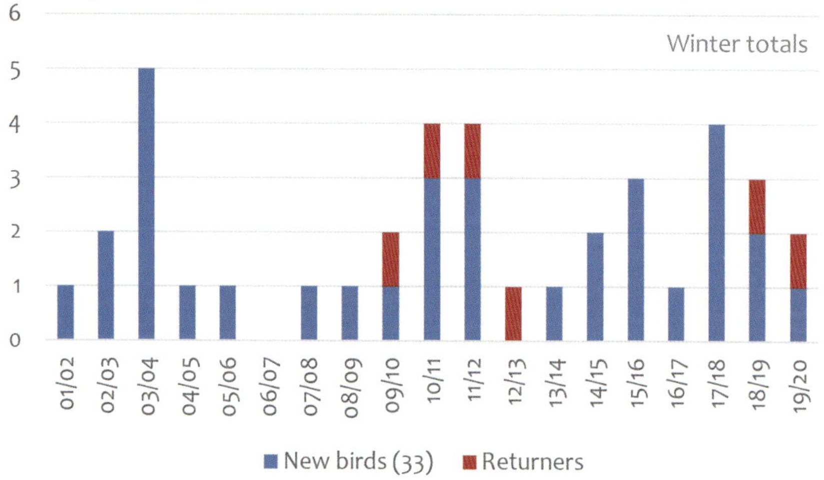

Most records have been from the gull roost at Ibsley Water and nearby lakes where at least 17 have been identified. These included at least four in early 2018 when it was discovered that many gulls were feeding around the pig farm at Blackheath Down, about 12 km north-west of Ibsley Water. Two individuals returned in consecutive winters, an adult from 2008/09 to 2012/13 (number 12 in the table below) and a second-winter from 2017/18 to 2019/20 (number 29). Other records have come from the Keyhaven area (6), the Eling/ Lower Test area (3), Black Point (2) and Avon Causeway, Eversley Gravel Pits, Fleet Pond, Langstone Harbour and Portsmouth Harbour (1 each).

Arrivals have occurred from September 25th but of seven seen in that month and October four were recorded on one date only. Most are seen from November to March with several individuals at Ibsley Water staying for several weeks or more. Photos taken have shown that single first-winters were there from November 5th 2017 to March 3rd 2018 and from October 18th 2018 to February 17th 2019. At Ibsley none has been recorded after March 6th but there are three later records during that month with the last on 30th. There is one late spring record of a first-summer at Lower Test Marshes from May 28th to June 15th 2011. Monthly totals are shown below.

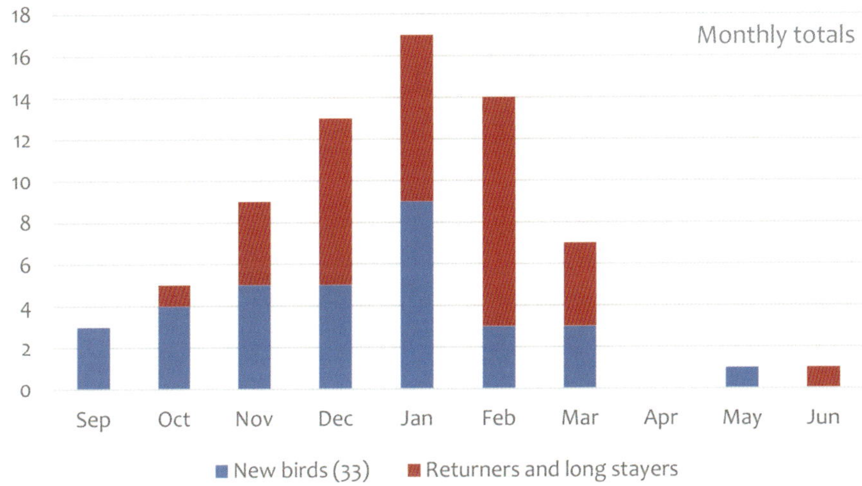

The approximate totals of each age class when the birds were first recorded were as follows: adult 8; third-winter, 2; second-winter, 3; first-summer, 1; first-winter, 19.

Most records away from Ibsley Water have been on one date only. This reflects the mobility of gulls and the difficulty of picking out Caspian Gulls among large gatherings of larids. Coastal records in September, October and March may well refer to migrants. Even at Ibsley Water, where the gull roost has been scrutinised on a regular basis for several years, this is a difficult species to find, especially when the roost often forms at ranges of up to 1 km and in poor light and/or windy conditions. The older, less distinctive age classes, particularly adults, and any short-staying individuals are likely to be under-recorded.

The first claims in Hampshire were in late 1997: a third-winter watched for 90 minutes by Pete Combridge and Simon S. King at Eling Great Marsh on November 12th 1997 and three seen by Mick Scott at Bramshill RT: different first-winters on November 26th and December 2nd and a second-winter on December 3rd. In the following year Mike and Pete Combridge and Simon S. King identified four individuals at Eling: a first-winter on eight dates between February 2nd and April 21st, an adult on four dates between August 18th and September 4th, a first-winter on August 28th and September 2nd and another first-winter on December 18th and 27th. Further records in 1999 involved at least two at Eling and Lower Test Marshes, second-winters on February 14th and March 26th and adults on four dates between July 30th and October 26th, and a first-winter at Blackbushe Airfield on August 29th.

Caspian Gulls: adult, Ibsley Water, January 2nd 2009 and first-winter, Sturt Pond, October 18th 2019. Richard Ford and Lee Fuller

By July 2000, BBRC had established a clear set of identification criteria and requested the submission of any records up to the end of 1999 supported where possible by photographs and/or video (Anon 2001). Following the first record for Britain in Essex in August 1995, BBRC subsequently accepted a total of 84 for the period up to the end of 1999 including 43 in Essex, the majority of which were seen by Martin Garner and supported by photographic or video evidence (Rogers *et al.* 2003, 2004). The four 1998 Hampshire records were submitted but were among 35 considered not proven as they fell short of the strict criteria applied and were not supported by photographs. A first-winter at Radipole Lake, Dorset from December 20th 1992 to January 3rd 1993 was retrospectively accepted as the first for Britain (Hudson *et al.* 2014).

HOSRP assumed responsibility for *cachinnans* with effect from 2000 although descriptions were not requested until 2001. All records in 2000 were from the north-east with an adult at Fleet Pond on January 17th, 31st and March 2nd and a fourth-winter there on February 12th and March 2nd, two first-winters at Blackbushe Airfield on September 17th and a first-winter at Winchfield on October 10th. In 2001 records of five in the north-east which lacked documentation were not accepted although they were published in the *HBR*.

All accepted records are tabulated below. The numbers in the first column identify returning birds.

	Date(s)	Age	Location	Observers
1	2001: November 24th	1w	Eling Great Marsh	Mark Edgeller
2	2002: September 25th, 27th and October 5th.	1w	Keyhaven	Marc Moody and Russell Wynn
3	2003: February 9th	1w	Tipner Lake, Portsmouth Harbour	Jason Crook
4	2003: September 27th	1w	Baker's Island, Langstone Harbour	Jason Crook
5	2003: December 8th	1w	Keyhaven Lagoon	Tim Parminter
6	2003/04: December 14th–January 24th	Ad	Ibsley Water	Daniel Houghton
7	2004: January 4th	Ad	Ibsley Water	Lawrie Chappell
8	2004: March 21st	1w	Keyhaven Marsh	Russell Wynn
9	2005: March 30th	1w	Black Point	Andy Johnson
10	2005: September 25th	1w	Fleet Pond	Graham Stephenson
11	2007: January 21st	Ad	Ibsley Water	Bob Chapman
12	2008/09: November 8th–January 18th	Ad	Ibsley Water area	Nigel Jones *et al.*
12	2009/10: December 12th–January 28th	Ad	Ibsley Water area	Nigel Jones *et al.*
13	2010: March 14th	1w	The Butts, Pennington	Alan Lewis, Tim Parminter, photo
14	2010: October 12th–November 14th	2w	Ibsley Water	Bob Chapman *et al.*
12	2010/11: November 7th–February 7th	Ad	Ibsley Water	Lawrie Chappell *et al.*
15	2011: January 5th–8th	Ad	Redbridge	Paul Winter
16	2011: May 28th–June 15th	1s	Lower Test Marshes	Simon S. King
17	2011: December 29th	2w	Normandy Lagoon	Marcus Ward
12	2011/12: November 25th–February 18th	Ad	Ibsley Water area	Rob Hume *et al.*
18	2012: January 29th and February 12th	Ad	Ibsley North Lake and Ibsley Water	Alan Lewis
19	2012: January 16th	1w	Fox Lane GP, Eversley	Graham Stephenson and John Clark
12	2012: December 24th	Ad	Spinnaker Lake, Blashford	John Clark
20	2013: November 15th	1w	Avon Causeway	Lawrie Chappell
21	2014: October 27th	Ad	Ibsley Water	Simon Woolley
22	2014: November 28th	1w	Black Point	Andy Johnson
23	2015: December 6th	3w	Ibsley Water	Alan Lewis, photo
24	2016: January 8th–February 28th	1w	Ibsley Water	Lee Fuller *et al.* photo

25	2016: January 30th–March 6th	1w	Ibsley Water	Alan Lewis *et al.* photo
26	2016/17: December 12th–February 7th	1w	Ibsley Water	Rob Hume *et al.*
27	2017/18: November 5th–March 3rd	1w	Ibsley Water and Blackheath	Alan Lewis *et al.* photo
28	2018: January 22nd and February 8th	Ad	Ibsley Water	Dave Taylor
29	2018: February 8th–March 3rd	2w	Ibsley Water and Blackheath	m.o., photo
30	2018: February 11th–March 3rd	3w	Ibsley Water and Blackheath	m.o., photo
31	2018/19: November 5th–February 17th	1w	Ibsley Water	Dave Taylor *et al.* photo
32	2019: January 1st–29th	1w	Ibsley Water	Alan Lewis *et al.* photo
29	2019: February 9th–17th	3w	Ibsley Water	Alan Lewis *et al.* photo
33	2019: October 17th and 18th	1w	Sturt Pond, Milford on Sea	Steve Keen *et al.* photo
29	2019: December 3rd–5th	Ad	Ibsley Water	Paul Winter, photo

Caspian Gull, third-winter, Blackheath Down, February 17th 2019. Alan Lewis

GB status: 88, 1992-1999; c. 7,000 records on Birdguides, 2000-2020, mostly in E, C and SE England. (Breeds from C Europe to C Asia; winters Persian Gulf, Red Sea and Black Sea coasts, N W Europe.)

Gull-billed Tern *Gelochelidon nilotica*

BBRC

A rare vagrant. 15 records of 16 birds since 1958.

The first for the county was observed by K. Brown in Chichester Harbour on May 18th 1958. His description, which was accepted by *British Birds*, was as follows:

> *I had perfect views of a Gull-billed Tern in Chichester Harbour at Hayling Island, Hants on May 18th 1958. I was first attracted by its call note which resembled a 'Comic' Tern's rather than a Sandwich Tern's. Views of the all black very stout short bill were obtained in flight and when it rested on spartina. In flight its tail was seen fully spread and was definitely less deeply forked than a Sandwich Tern's, with also an apparent slower wing beat or heavier flight than that species.*

The next occurrence constituted a far more convincing record. David Billett, Michael Bryant and Graham Rees saw two fly east over Farlington Marshes during a heavy passage of terns on May 1st 1965. Their description of events was as follows:

> *Between 1400 and 1800 hrs a total of 439 Black Terns and 281 Common or Arctic Terns (of which 49 were proven to be Arctic and most, if not all, were this species) moved east from Ports Creek across Farlington Marshes. At about 1530 hrs a party of 'Comics' flew past at 50 yards range, accompanied by two larger terns identified as Gull-billed Terns. The party had been watched flying towards the observers over a distance of about half a mile and MB, who has had considerable experience of Gull-bills abroad, gave his opinion that they were of this species at an early stage of the observation.*
>
> *Description: About size of Sandwich Tern, very white compared with accompanying 'Comics', and of a decidedly plump appearance, tail seeming shorter than Sandwich and less deeply forked. Wings less pointed and straighter with shallower beats. Short and stout entirely black bill was very noticeable as birds passed close by and appeared distinctly different even at a distance and gave the birds an un-tern-like appearance and probably added to stocky impression. At close range bill could be seen to be angled in the lower mandible.*

The next was seen flying past Hurst Castle by Pete Hobby and Ed Wiseman at 1600 hrs on July 5th 1966. It was watched flying north towards Keyhaven saltings before leaving to the east and was not seen again. Ten days later, on July 15th, Martin Terry saw one flying west off Gilkicker Point which is assumed to be the same bird. Just under a year later two were seen at close range by J. F. M. Fennell and R. Bream as they slowly flew east with other terns off the southern tip of Farlington Marshes at 1530 hrs on April 29th 1967. In the following year another was found in Langstone Harbour when John Smith located one among a flock of Black-headed Gulls and Common Terns off the eastern end of Long Island at 1600 hrs on June 9th 1968. Its identity was confirmed by Bryan Renyard. It disappeared after a few minutes but was relocated at 1625 hrs when it flew off south-east and was not seen again.

Langstone Harbour was again the venue for the next sighting. David Billett spotted one off the southern tip of Farlington Marshes during high tide late on May 31st 1971 and showed it to Bryan Renyard. It flew south down Russell's Lake before perching on a channel marker post about a quarter of a mile away where it spent several minutes preening and resisting displacement by Sandwich Terns. Finally, it flew north again towards the observers, turning east to Baker's and South Binness Islands where it may have landed. It was not seen subsequently. The ninth record for the county was of one seen off Pennington Marshes by Derek Lister at 1015 hrs on May 26th 1972 (not May 27th as in *BoH*). He watched it for a minute as it moved west; it was not seen again. The next was found by David Christie at Hook-with-Warsash on April 29th 1978 when he watched one at close range on the saltmarsh for ten to 15 minutes before it flew off.

Gull-billed Tern, Pennington Marshes, May 31st 2019. Lee Fuller

David Christie found another on May 10th 1982 when he saw one fly down the River Hamble towards the estuary mouth, closely followed by a Sandwich Tern which offered immediate comparison. The next was found by Mike and Pete Combridge at Pennington Marshes at 0745 hrs on July 11th 1982. It was refound at 0830 hrs and seen by ten other observers up to 1200 hrs when it became elusive and was last seen at 1400 hrs.

There was a long gap until the next was found by Pete Combridge at Milford on Sea on May 12th 1994 when he saw one move east at 0855 hrs.

The first for the 21st century was found by Rob Innes and confirmed by C. D. R. Jones as it flew around the South Scrape at Titchfield Haven at 0810 hrs on May 26th 2003. It was watched intermittently until 0850 hrs when it headed west out into the Solent. What was presumably the same bird was seen at Littlehampton, West Sussex between 1600 hrs on 26th and 0630 hrs the next morning but then returned to Hampshire. Doug Robertson saw it flying east in Hayling Bay at 1740 hrs on 27th and the next day Tim Lawman saw it in Chichester Harbour entrance at 1325 hrs and flying west in Hayling Bay at 1650 hrs.

One was seen by Mark Palmer on the shore at Hill Head between 2000 and 2005 hrs on May 23rd 2005 before leaving east. This record has not been submitted to BBRC but is assumed to relate to the individual recorded widely in Kent and Sussex between May 4th and June 6th.

The most recent record was of one found by Lee Fuller at Pennington Marshes at 1545 hrs on May 31st 2019. It arrived from the east and spent 45 minutes preening and sleeping at the back of Fishtail Lagoon before flying off north-west. It was also seen by several other observers who managed to twitch the bird before it departed.

This is one of the most difficult species to catch up with in Hampshire. All were seen on one date only, seven by single observers, a further eight by two or three finders, and only two stayed long enough for some local observers to catch up with them.

One seen at close range by Pete Hobby, Dave Wooldridge and Ed Wiseman flying east off Hurst Beach on May 5th 1970 was harshly considered not proven by BBRC. Four other records were considered not proven by BBRC (see Appendix VIII).

GB status: 52, 1802–1949; 331, 1950–2020. About 3–4 per year; stable. (In Europe, small population breeds N Germany & Denmark. Widespread though local Spain, elsewhere colonies small & isolated. To E, breeds discontinuously from Turkey & SW Russia through Kazakhstan, Mongolia & NW China, with isolated population NE China. European population winters coastal W Africa, S to Gulf of Guinea. Asian populations winter Persian Gulf to Indian subcontinent & SE Asia. Other races occur N & S America.)

Caspian Tern *Hydroprogne caspia*

BBRC

A very rare vagrant. Six records since 1974.

The first for the county was seen by Tony Smith and Lester Mulford off Milford on Sea on June 22nd 1974. Their description was as follows:

The bird was first seen at about 3 p.m. flying in an easterly direction about 200 yards offshore. Two hours later at about 5 p.m. what was presumably the same individual was again noted offshore. It flew in from the west and then turned out to sea and was last seen flying in the direction of the Isle of Wight. Mr. Mulford and I made the following notes:

Structure and Flight. A huge tern, gull-sized and much larger than Sandwich. Long wings, slightly notched tail. No tail streamers. Heavy-headed appearance with large, very deep and heavy bill. Bill held slightly downwards below horizontal in flight. Flight action deliberate, less buoyant than Sandwich with slower, more measured wingbeats.

Colouration. Cap from bill to nape black. Upperside of wings and back grey, darker than Sandwich, especially on primaries. Underparts white contrasting with dark underside of primaries from base. Bill red.

Other species present: Sandwich Tern.

Optical equipment used: 10x40 binoculars and x25 telescope.

Both observers had extensive experience of the species abroad although it was a new British bird for them.

Almost twenty years were to pass before the next was found by Trevor Carpenter at Hill Head on April 24th 1995. Extracts from his description are reproduced below.

At 1645 hrs on April 24th 1995, I stopped at Hill Head to see if anything was moving offshore. After ten minutes without seeing anything, I set off to drive home but first paused at the south-easterly point of the parking area to look at the roosting gulls on the exposed shingle in front of the sailing club. There were 30–40 birds present split between Black-headed Gulls and non-adult Herring Gulls. I quickly noticed a single tern which appeared to be very large. It was almost double the bulk of an adjacent Black-headed Gull and similar in size to a Herring Gull on the other side of it. It had the jizz of a large overweight Common Tern. The head was black to the same extent that Common Tern would be at this time of the year with a very obvious shaggy crest. The bill was very large and orangy/red in colour. There was some black on the primaries but at roost it was difficult to determine how much. I then drove to the car park by the sailing club and viewed the bird from a different angle confirming the observations above. It appeared larger if sleeker than some Herring Gulls and was not overshadowed by any of the closely adjacent birds. I left a message in the Titchfield Haven reserve centre and then drove to the Osborne

View pub where I used the phone to ring other observers. I returned to the car park to find the bird had gone, presumably pushed off by the rising tide. Shortly after Richard Levett drove into the car park; he had received the message and had arrived to see the bird take off at 1715 hrs. He then followed it along the beach and watched it disappear to the west. He was in no doubt it was a Caspian Tern.

Two were seen in 2013. Ian Pibworth was watching from his regular Sunday viewpoint overlooking Testwood Lakes on June 9th when at 1430 hrs he noticed a distant bird high to the south. As it came closer, he realised it was a Caspian Tern, and fortunately it landed on an island below. He was able to watch it fishing and resting at length and realised that it was ringed. It was seen by Simon S. King and Kevin Butcher at 1715 hrs and then by five other observers before it flew off east at 1815 hrs. It was subsequently seen in Dorset on June 11th and 12th. Philip Fawkes had good views of an adult as it moved slowly west off Normandy Marshes towards Oxey at 0845 hrs on July 28th before it was lost to view while he was phoning the news out.

Two were seen in 2020. Ian Pibworth saw one fly south-east over Testwood Lakes on May 31st 2020. It was relocated off Eling Tide Mill by Ian Wells and seen by Andy Lester and David Thelwell. As the tide came in, it flew off down Southampton Water at 1607 hrs and was not seen again. David had the good fortune to see another just under two months later when he, along with Matt Parkes and Emma Butcher, found one fishing at Fishlake Meadows at 1230 hrs on July 27th. It had disappeared by 1315 hrs but returned between 1700 and 1745 hrs and finally between 2005 and 2015 hrs, much to the delight of several birders including the author. It was seen again briefly the next morning.

Caspian Terns: Eling Tide Mill, May 31st 2020 and Fishlake Meadows, July 27th 2020. Lee Fuller and Simon M. King

The 1954 *HBR* includes a report in square-brackets from John Ash, an experienced observer, of two very large terns flying east near Damerham directly overhead at c. 300 feet on April 17th, an overcast day with a strong NNE breeze. He noted:

> *Silent, appeared definitely larger than Sandwich Terns, having markedly heavier build. Impression of larger size and more robust appearance AT ONCE noticed. Definitely terns and not gulls. Watched by my wife and I through 8x30 binoculars (two pairs) for about one minute as they flew steadily on. Long head and neck noticeable ('heavy-headed'), angled, broad wings, relatively short tail which was forked but not deeply. Typical tern flight but less graceful than Sandwich, i.e. stronger, heavier, more powerful.*

Without sight of the birds' bills the record can only be considered as a possible.

More recent sightings of two birds were undoubtedly good records. One was seen by John Bohun and D. M. Jones at Farlington Marshes on July 2nd 1996 and then by Doug Robertson and Chris Cockburn there on July 8th. Doug's description was misplaced by BBRC and when it resurfaced was considered not proven; Chris provided additional notes and yet it was still not accepted. Another was seen by Pete Potts while drinking coffee in his garden at Solent Court at midday on June 27th 2005. He described his sighting as follows:

> *The call drew my attention to the bird. I first saw it and immediately thought blow me down I'm sure that is a Caspian Tern. I dashed into the kitchen, grabbed my bins and watched it for a few minutes circling the garden as it slowly drifted west towards Hook-with-Warsash. It was a large, long-winged, well-built tern, size of Herring Gull, harsh rasping call, black primaries visible against white underwings, dark head and red bill. 100% certain about its identity for sure.*

BBRC did not accept his description. Pete admits that it was too casual, e.g. dark head rather than dark cap, but he stands by the record!

GB status: 23, 1825–1949; 308, 1950–2020. About 4–5 per year, influx 1988 (18); recent decrease, now stable. (European population breeds Baltic coasts of Estonia, Sweden & Finland to head of Gulf of Bothnia. To E, fragmented populations from Black Sea coast of Ukraine across steppe-lake region of C Asia to NW Mongolia & E China. European birds winter W Africa to Gulf of Guinea, Asian birds winter on coasts to S of breeding range. Other populations occur Australia, S Africa & North America.)

Elegant Tern *Thalesseus elegans*

BBRC

A very rare vagrant. One record in 2017.

The only record for Hampshire is of a colour-ringed bird found by Andy Johnson in a high tide roost with eight Sandwich Terns at Fishery Creek, Hayling Island at 1000 hrs on June 7th 2017. Extracts from Andy's account of its discovery, published in the 2017 *HBR* (Johnson 2018) are reproduced opposite.

With high tide at 1115 hrs, I decided to stretch my legs and make my way to Sparkes Marina to check the high tide roosts. Unsurprisingly Black Point had very little – just two Ringed Plovers and two Dunlin trying to shelter from the wind. I moved across to Fishery Creek and could see that there were quite a few gulls and terns on the small promontory on the north side of the creek. I decided I'd better have a scan, just in case there were any roosting waders that I couldn't see with bins. Sheltering in one of the slipways, I trained my scope on the spit. The next thing I knew was that I was looking at an Elegant Tern.

Elegant Tern with Sandwich Terns, Fishery Creek, Hayling Island, June 7th 2017. Andy Johnson

Five other birders managed to get there before it flew off east at 1115 hrs. Andy found it again among a flock of Common and Sandwich Terns off the seafront at nearby Sandy Point at 1700 hrs on June 9th but after 20 minutes it left east, flying past Chichester Harbour entrance and into West Sussex waters. It was located the next day in the tern colony at Pagham Harbour, West Sussex where it remained until June 20th before appearing at Brownsea Island, Dorset on June 21st and 22nd. The colour rings confirm that it is the same individual seen at Dawlish Warren, Devon on May 18th 2002 and subsequently trapped at Banc D'Arguin, France in summer 2003. This has recently been accepted as the first for Britain (Holt *et al.* 2018). For further details of the discovery of this bird and its life history see Johnson (2018).

GB status: 4, 2002–19. Very rare. (Breeds in colonies on Pacific coast of California, USA, & Gulf of California, Mexico. After breeding disperses N along Pacific coast; most winter South America S of Ecuador.)

Sooty Tern *Onychoprion fuscatus*

BBRC

A very rare vagrant. Two records in 1961 and 1969.

The first for the county was found by Dave Wooldridge and Doug Poole while they were surveying the tern colony on Keyhaven Saltings on June 17th 1961. They watched it for ten minutes from 1320 hrs down to 70 yards range. Their description was as follows:

Forehead – white. Whole of upperparts including all of tail – black. Sides of head and whole of underparts – pure white. Bill – black, of Sandwich Tern proportions. Tail – rather deeply forked.

A large tern about Sandwich Tern size but with longer and broader wings probably accentuated by the black and white plumage. Flight was rather slow with shallow wing beats suggesting Common, not Sandwich, Tern. It was vigorously pursued by Black-headed Gulls suggesting it was mistaken for a skua. On returning to our car and checking with the Field Guide it was positively identified as a Sooty Tern.

Later, at 1720 hrs it was seen again, feeding off Hurst Beach in company with one Black and several Common, Little and Sandwich Terns; also Black-headed and Herring Gulls. The closest it came to the shore was c. 120 yards; watched through a 60x scope the following extra points were noted.

Underside of primaries dusky.

The outer margins of outer tail feathers were white. This character, though not normally noticeable, showed well when the tern splayed its tail when swooping to the surface of the water.

Feeding was by (a) picking off surface in manner of a Black Tern and (b) often plunge-diving as does a Common Tern.

It remained off Hurst Beach for at least 90 minutes and was also seen by Jim Williams, who additionally noted the white forehead, not extending behind the eye and black nape.

Dave Wooldridge recently recalled his memories of that red-letter day.

I remember it well. A misty day with a south-west wind. In the morning, my friend Doug Poole and I set off in our canoe through Hawker's Lake to visit the local tern colonies. Out on the open water I saw a flock of 30-40 Common Terns being blown in the mist up the Solent. With them was a tern new to me but I thought I had enough on it to identify it. I did not expect to see it again. However, later as the weather became fine, we sat on the bank in the evening sun watching a flock of terns fishing. Lo and behold – there was the Sooty, so close we could see every feather detail! Even more of a bonus, the tern flock included Black, Common, Sandwich, Little and Rosy Tern. Six species! We watched until the light faded.

The second record involved an adult found dead at Fawley Refinery by Mr. T. E. Stoppel on August 18th 1969. He identified the bird from books and the skin was passed to the Natural History Museum. Alex Bond, Senior Curator of Birds at Tring, has confirmed that they hold the skeleton of the bird under registration number NHMUK S/1973.10.1.

GB status: 11, 1852–1949; 11, 1950–2019. Very rare. (Breeds on oceanic islands in tropical Atlantic, Indian & Pacific Oceans, S Red Sea & throughout Caribbean. Outside breeding season disperses widely throughout tropical oceans.)

Whiskered Tern *Chlidonias hybrida*

BBRC

A rare passage migrant. Twelve records of 15 birds since 1970.

The first for the county was an adult in summer plumage seen by Geoffrey Fisher near Hurst Castle between 1830 and 1900 hrs on May 12th 1970. Extracts from his description are reproduced below.

> *I was on a 17 ft fishing boat returning from Yarmouth to Keyhaven when I saw a very dark bird at 200 yards range whose flight distinguished it quickly from the Black-headed Gulls with which it was fishing. We approached this bird and (having a high-powered motor) were able to follow it for half an hour at ranges from 50–200 ft. It was moving up and down a stream of surface flotsam fishing in a typical marsh tern fashion, i.e. picking off the surface after gentle swoops. It was never observed to plunge in the water or alight on the surface at all although feeding continuously.*
>
> *Description. About the size of a Common Tern. The striking feature was the dark grey of the upper wings, primaries, wing-coverts, mantle and back. The cap was deep black and there was a striking white stripe below the cap across the face incorporating most of the face, the belly was dark grey and this colour was continuous across to the under wing. The under wing was not as dark as the upper wing. The body colour became very pale towards the vent and the undertail was almost white. The tail was short and rather 'square'. The uppertail and rump were almost white. The beak (seen at 30 ft.) was dull red (much darker than Common Tern) also shorter and stubbier. The feet were red. It did not call.*

The next was seen down to 40 yards by Bob Dunn and Ed Wiseman as it flew in off the sea over Hurst Beach with two Common or Arctic Terns and flew on eastwards towards Keyhaven at 1545 hrs on May 13th 1972.

The first in autumn was an adult in summer plumage found on a shingle bar at Hook-with-Warsash by David Christie at 0945 hrs on August 17th 1974. It was accompanied by 44 other terns of four species. After 15 minutes, as the tide rose, all the terns flew off to join a feeding flock of about 300 Common Terns in the Hamble Estuary. It was looked for the next day but not seen again.

The next was also found by David Christie at Hook-with-Warsash. It was an adult in summer plumage which he watched for 25 minutes on a shingle spit with a Common or Arctic Tern from 1235 hrs on May 20th 1978. It then took flight and was not seen again.

The fifth record involved two found by Mike Witherick at Mockbeggar Lake at 1155 hrs on May 14th 1988 and seen by Rod and Pat Allison. They were seen in flight at first but then alighted on wooden posts at 70 yards range where they were watched closely through telescopes for ten minutes. At 1215 hrs the three observers left the site to telephone other local birders from the nearest telephone box but evidently the terns had departed when further observers arrived at 1300 hrs.

The next were not until 2005 when two were found on May 20th. A first-summer was found by Peter Davidson perched on a wooden rail at Titchfield Haven at 1415 hrs. It was seen by at least six other observers down to 30 yards range before flying off west after 15 minutes. It was probably the bird involved in an unconfirmed sighting at Testwood Lakes later that afternoon. The second was an adult found by Brigid Campbell at Tundry Pond at 1540 hrs and still present when she left at 1605 hrs. She was not 100% certain of its identity but on checking reference books at her home in Basingstoke realised it could only be a Whiskered Tern, so she telephoned the author who released the news. Fortunately, it remained at Tundry Pond for several hours enabling many observers to see it, before leaving north-east at 2006 hrs. It was then relocated at Fleet Pond, 4 km to the east, and left north at 2106 hrs. It was found roosting at Moor Green Lakes, Berkshire at 0445 hrs the next morning and subsequently was at Fleet Pond from 0745 to 0915 hrs before moving to Tundry Pond where it remained until 1735 hrs. It returned to roost at Moor Green Lakes at 1755 hrs. Although it roosted there again the following night and was seen there briefly on May 27th it was not seen again in Hampshire but was present for long periods at King George VI and Staines Reservoirs, Surrey. These were part of an influx of around ten into England in May.

The next was in 2008. During a sea-watch from Hurst Beach on May 6th Marc Moody noted three terns flying from Dorset across Milford Bay towards him at 0728 hrs. He was able to identify them as two Black Terns and a Whiskered Tern. By this time Pete Durnell had also seen the birds in his telescope. They passed over the shingle beach just 20 feet from their observation point and continued east into the Solent. Pete was able to phone Marcus Ward enabling him to see the three birds five minutes later as they flew past Oxey Point and along the seaward side of Normandy Marshes.

Almost a year later Ian Pibworth was watching from his viewpoint overlooking Testwood Lakes on May 4th 2009 when at 1920 hrs he found two adults circling overhead before gaining height and after three minutes disappearing over the M27 motorway and pylons to the east.

The earliest spring record was of one found by John Norton perched on a fluorescent marker at least 1 km offshore at Stokes Bay at 1838 hrs on April 14th 2011. John was able to alert other local birders and as a result Dan Houghton was able to get there in time to see it, despite having to walk half a mile due to a road closure to reach the site. At 1923 hrs it was displaced by a Sandwich Tern, flew up and circled a few times before moving off to the east.

The next were two seen by Steve Lavington moving east across Keyhaven Lagoon at 1500 hrs on May 7th 2016. He was able to take confirmatory record shots during the short period they were in view. Finally, one was found by Amy Robjohns fishing over Posbrook Floods at 0722 hrs on May 27th 2020. After five minutes it headed south and was seen at Hill Head by Mark Francis and Richard Carpenter. It spent the day feeding with other terns and gulls offshore and came into the river and scrapes at Titchfield Haven. It was last seen by Andy Collins and Brian Smith early the next morning when it flew off high to the west.

Whiskered Terns: Titchfield Haven. May 20th 2005 and May 27th 2020. Ted Pressey and Lee Fuller

The record of a first-winter seen by Pete Combridge with four Common Terns from Hamble-le-Rice as they drifted down Southampton Water at 0850 hrs on October 2nd 2007 was harshly considered not proven by BBRC. Pete confirmed in 2020 that he stands by the record. Five other records were not accepted by BBRC (see Appendix VIII).

GB status: 19, 1836–1949; 205, 1950–2020. About 4–5 per year, influx 2009 (25); stable. (Breeds in small, scattered colonies through S & E Europe from Iberia to Poland. Numerous & widespread from N Black Sea E to W Kazakhstan, with Volga/Ural River complex holding most of European population. Winters tropical W & C Africa & from Nile Delta to E Africa. Other populations occur Indian subcontinent, E Asia, S Africa & Australia.)

White-winged Black tern *Chlidonias leucopterus*

BBRC to 2005; HOSRP 2006 to date

A rare passage migrant. One record in 1886 and 32 records of 33 birds since 1959.

The first for the county was one shot by Edward Hart at Avon Tyrrell on May 18th 1886. The specimen is now held by the Hampshire Cultural Trust at Chilcomb House, Winchester.

The first modern record was of a juvenile moulting to first-winter seen independently by N. Money and Ed Wiseman over Russell's Lake and Langstone Channel on September 14th 1959. The description submitted to BBRC compiled by Michael Bryant was as follows:

Seen with both 'Comic' and Little Terns. Smaller than 'Comic', larger than Little. EJW thought it was even smaller than Black Tern with quicker wing beats. Forehead white, top of head blackish. Nape white, forming a collar. Wings, paler on primaries which were more greyish than on inner wing, which was darker with a decidedly brownish appearance. Back similar but appeared slightly paler than inner wing. Rump and tail were very pale, the rump being whiter than the tail although the difference was only obvious at very close range. There was no dark shoulder mark as in Black Tern. Underwing surfaces were whitish. Bill rather small and wholly black. Birds seen down to 10 yards.

Ed Wiseman, writing in 2019, said that its back was obviously much darker than the rest of its body (contrary to Michael's description) and that he was still very happy with the record.

Since then, six have been recorded in spring and a further 26 in autumn.

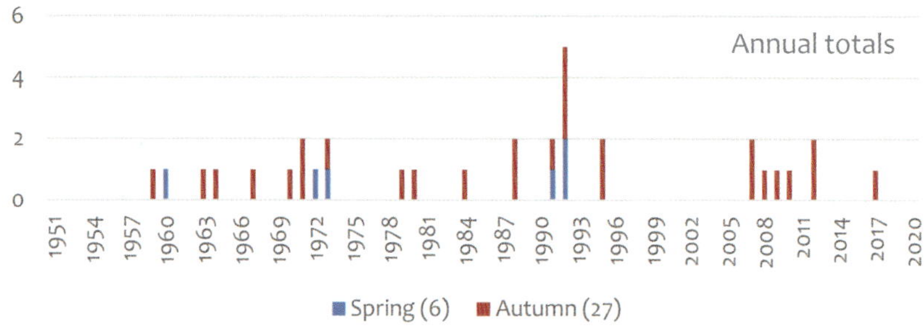

The next was an adult in summer plumage seen by David Price as it moved east with three Black Terns past Lee-on-the-Solent at 1815 hrs on May 11th 1960. This was followed by a moulting adult seen by Jim Williams and Pete Hobby with a flock of 11 Black Terns at Hurst Castle on August 11th 1963. What was assumed to be the same bird was seen at the same place by Jim Williams, Pete Hobby and Gerald Westerhoff on August 18th. A year later Graham Rees saw a juvenile at Black Point, Hayling Island on September 1st 1964. The next was a first-winter found by members of the Portsmouth Group at Farlington Marshes on October 22nd 1967 which remained in the area for several weeks and was last seen on November 15th.

The first inland was a juvenile with nine Black Terns found by Graham Harrison, a visiting birder from the West Midlands, at Fleet Pond on September 8th 1970. He showed it to Dorothy Herlihy, Noel Elms and Richard Millington. The next was a juvenile watched by John Smith feeding over the river at Titchfield Haven between 1915 and 1950 hrs on August 18th 1971. Later that autumn Ed Wiseman and David Peart had brief views of a juvenile flying over Keyhaven Saltings on September 28th; on October 4th what was presumably the same bird was seen by Ed Wiseman and Tom Brice to arrive over the sea from Keyhaven, feed for one or two minutes over a freshwater pool at Pennington Marshes and then continue eastwards.

An adult in summer plumage was found by David Thelwell feeding over a lagoon inside the sea wall at Pennington Marshes at 0710 hrs on May 29th 1972. He was able to watch it for ten minutes before it left unseen. Almost a year later a first-summer was found by Keith R. Grant and Ron Huxford at Farlington Marshes at 1045 hrs on May 26th 1973; it fed over the lake for 30 minutes before continuing eastwards towards Langstone Bridge. This record was not submitted to BBRC at the time but Keith's excellent description and sketches were found in the Langstone Harbour yearbook and finally submitted to BBRC and accepted in 2019. Later that year a juvenile was seen moving slowly east off Pennington Marshes by Bob Dunn, Owen Haisell and Ed Wiseman on August 12th and was seen again the next day by Bob and Ed. The next was not until 1979, when Mike and Pete Combridge saw a juvenile move east off Pennington Marshes during a break in heavy rain at 1555 hrs on August 25th. A sighting of a juvenile at Calshot on September 1st was published in the *HBR* and *BoH* but was never submitted to BBRC. The observer of this record has not been traced although at the time it was assumed to relate to the Pennington bird.

The first of the 1980s was an adult moulting out of summer plumage which Mike and Pete Combridge saw fly west off Pennington Marshes with four Black Terns at 1100 hrs on September 21st. The next was a juvenile which Bob Chapman found with ten Black Terns on the South Scrape at Titchfield Haven at 1735 hrs on August 22nd 1984. He saw it for around 15 seconds before the flock flew off eastwards. Fortunately, the bird returned and was seen later by Dan Powell, David Parsons and Martin Terry. A long-staying moulting adult was first seen by Jimmy Scott at Broadlands Lake on July 6th 1988 where it was seen almost daily until July 28th. Other sightings were made by Jess Pain at Lower Test Marshes on July 26th and August 24th, Brian Small at Eling Great Marsh on August 5th and 6th, Trevor Codlin at Titchfield Haven on 8th, Ron King at Winchester Sewage Farm on 10th and Bob Chapman at Hythe on most evenings between August 14th and September 5th where it came into roost on floating pontoons. Also during this period, on August 22nd, Barry and Tor Duffin saw a juvenile fly over the car in which they were travelling along the coastal road at Titchfield Haven. They subsequently watched it for 20 minutes with Dennis Bright and Richard Levett feeding over the river at Titchfield Haven before it left south-west with Common Terns at 1635 hrs. This record has never previously been published in a *HBR* or in *BoH* due to confusion with the record of the long-staying adult although it was published by BBRC.

There was a glut of records in the 1990s. An adult was found by Adrian Clark at Pennington Marshes during the morning of June 3rd 1991. It disappeared but was refound by Owen Haisell, Ed Wiseman and Adrian at nearby Iley Lane Lake where it was watched at close range for 30 minutes. Later that year a juvenile was found by Ian Prophet at Spinnaker Lake, Blashford on September 22nd where it was subsequently seen by several observers and remained until 25th. Five were seen in 1992. Andy Parfitt was patrolling the Langstone Harbour islands nature reserve by boat when he found an adult perched next to a Black Tern on Round Nap Island at 0935 hrs on May 14th. He was able to watch it down to 50 metres until 0955 hrs and saw it in flight at 1018 hrs as it headed off east over Long Island with the Black Tern. This was no doubt the same individual that was found by Barry Collins roosting with 11 Black Terns on Pilsey Island in the Sussex part of Chichester Harbour at 1300 hrs that day and was seen again the next day. A second-summer was seen by Ed Wiseman as it moved east past Normandy Marshes at 0900 hrs on May 18th and later by Mark Palmer coming in off the sea at Hill Head and moving north up the Meon Valley. A juvenile was found by John Dixon with two Black and nine Common Terns at Fleet Pond on August 22nd; it was subsequently seen by several local observers before leaving high to the south-east. Guy Thompson found an adult moulting into winter plumage at Alresford Pond at 1225 hrs on September 3rd; he watched it until 1335 hrs but had left by the time another observer arrived ten minutes later. Finally, Mike and Pete Combridge saw a juvenile fly west off Keyhaven from 1220 to 1222 hrs on September 6th. This was during a noteworthy passage following south-westerly gales which involved 21 Black Terns, over 300 *Sterna* terns, 19 skuas and 2 Manx Shearwaters. Two juveniles occurred in September 1995. Keith Wills found one at Camp Farm Gravel Pit, Aldershot in mid-afternoon on 8th which was seen by several local observers until dusk. The next morning it was seen briefly at Frimley Gravel Pits, Surrey before being rediscovered by John Collman at Fleet Pond where it remained until 13th when it left high to the south with a Black Tern at 1900 hrs. Another juvenile was found by Barry Duffin as it arrived off the sea at Titchfield Haven at 1035 hrs on 16th and performed in front of the Meon Shore hide before disappearing after a few minutes. It was also seen by Dave Treacher, who submitted a description to BBRC, Paul Heath and Tor Duffin.

White-winged Black Terns: juvenile (left) with Common Tern, Black Point, September 29th 2009 and adult (left) with Common Tern, Black Point, August 17th 2017. Andy Johnson

The next were in 2007 when there were four sightings relating to two individuals. Tim Lawman saw a juvenile fly north through Langstone Harbour entrance at 1950 hrs on August 7th. A minute later it was independently found by Jason Crook in flight off The Kench where it was also seen by Martin Gillingham. Simultaneously they received texts from Tim who joined them to watch the bird which was now perched on a buoy. It was lost from view at 2005 hrs. Later that month Ian McKechar found a juvenile off the Ship Inn, Langstone at 1400 hrs on 19th and the next day presumably the same individual was watched by Jason Crook off West Hayling between 1905 and 2013 hrs. Based on plumage differences, Jason concluded that it was definitely different to The Kench bird. In the following year Tim Lawman was again counting terns flying into Langstone Harbour to roost on September 14th when he found a juvenile at 1930 hrs. In 2009 Andy Johnson found a juvenile in the high tide roost at Black Point from 0847 to 0900 hrs and again at 1002 hrs on September 29th.

An autumn juvenile was found for the fourth consecutive year when Marc Moody located one in Hurst Narrows at 0852 hrs on September 11th 2011. It fed over the tide race with 20 Common and two juvenile Black Terns and rose up and moved east into the Solent at 0924 hrs with the other two marsh terns. The nocturnal roost at Langstone Harbour again proved a reliable location for this species when Tim Lawman located two juveniles flying north through the harbour entrance at dusk on August 24th 2012. The most recent record was of an adult found by Andy Johnson feeding in Chichester Harbour on August 14th 2017. It remained in the area until 17th and was seen by many observers in the vicinity of Black Point where it visited the high tide roost on 16th and 17th.

As shown on the next page, analysis of arrival dates shows the six spring records evenly spread between May 11th and June 3rd. Apart from the long-staying summering bird in 1988 and the lingering first-winter in late 1967, autumn arrivals occurred between August 7th and September 29th. Of 27 at that season, five were adults and 22 juveniles or first-winters.

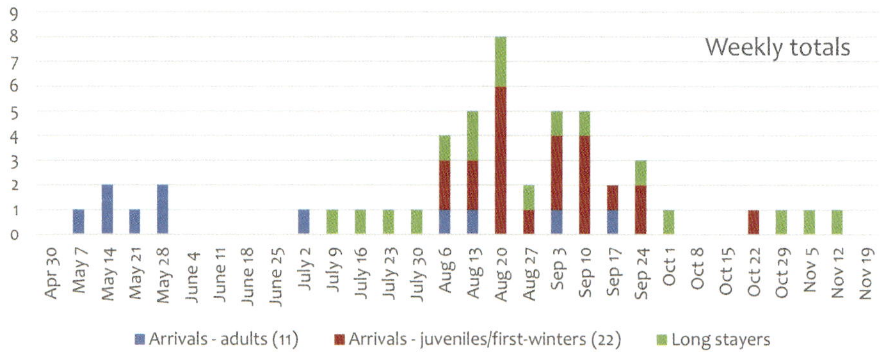

The first accepted modern record was pre-dated by another of one seen by Michael Bryant over The Lake at Farlington Marshes on the evening of August 20th 1959. His description was as follows:

> *Attention was drawn to it by a quacky call, not unlike a Little Tern. Size similar to Little Tern or slightly larger due to broader wings and heavier body. Forehead white, cap black extending to below eye and clear cut, i.e. not streaked, appearing like cap of Black Tern. Upper wing pale grey shading to darker grey along leading edge and on outer primaries, upper surface of rump and tail white. Whole of underparts white with no trace of dark shoulder patch of autumn Black Tern. Bill reddish black, legs dark red. Flight similar to Black Tern, dipping and skimming over surface of water. The bird eventually left eastwards when the wings appeared particularly long as it flew off in contrast to its appearance over the lake. I did not notice that the beak appeared particularly short but it seemed shorter than the beak of Whiskered Tern so far as I can remember. Good views in fair light down to 10 yards. I have previously seen White-winged Black and Whiskered Terns.*

It was not accepted by BBRC but Ed Wiseman (*in litt.*) considers it a good record. Further records harshly considered not proven by BBRC were of a juvenile seen by John M. Jones and Pete Combridge off Pennington Marshes on September 22nd 1983 and one found by Ivor MacPherson and also seen by Dan Houghton at Titchfield Haven on September 26th 2004. This was submitted to BBRC as a first-winter by Dan although he now considers that it was an adult moulting into winter plumage.

Three other genuine records have not been submitted to BBRC. Martin King and Mike Read saw two adults fly south over Mockbeggar Lake on May 3rd 1980. Dan Powell saw a moulting adult resting and later hawking over the South Scrape at Titchfield Haven on the morning of August 6th 1995. Mick and William Scott found a juvenile at Fleet Pond on the evening of September 7th 1999. Mick relocated it at Tundry Pond the next day where he watched and videoed it between 1100 and 1200 hrs. The author has seen the video which confirms the identification but it has now gone missing.

K & M included three sight records – of three at Woolmer Pond in April 1883, 'some' passing down the Avon towards Christchurch on April 30th 1892 and three on a piece of water near Portchester on April 27th 1897. Another anonymous record, overlooked by K & M, was of three flying up and down the River Itchen at Bishopstoke on April 27th 1883 (*The Field*, May 5th 1883). In the absence of specimens, confusion with Black Tern seems likely. A further record of two small parties seen by Mr. Neville Gladstone passing up the Avon at some hours interval on May 29th 1937 was submitted to the *HBR* 1937 by Lt. Col. Cyril Mackworth-Praed but is unacceptable in the absence of details.

GB status: 62, 1853–1949; 1,055, 1950–2019. Stable, mean 19 per year in 2010–19. (Breeds Poland to Hungary, with sporadic breeding to W. Breeds commonly from Belarus, W Russia & Ukraine E to S Siberia, N Kazakhstan, Mongolia, Russian Far East & NE China, but absent from large areas. Winters throughout sub-Saharan Africa, Indian subcontinent, SE Asia & N Australia.)

White-winged Black Tern, moulting adult, South Scrape, Titchfield Haven, August 6th 1995

Long-tailed Skua *Stercorarius longicaudus*

BBRC 1976–79; HOSRP 1980 to date

A very scarce visitor. One record in 1891 and 50 since 1962.

The first for the county was in October 1891. An article in the *Hampshire Independent* of November 7th 1891 referred to a young bird which was shot at Crawley near Winchester and identified by the eminent ornithologist Dr. Philip Sclater. The specimen has not been located. However, Edward Hart, writing in *The Zoologist* for December 1891, described a large invasion of Long-tailed Skuas, Grey Phalaropes and Sabine's Gulls at this time, with hundreds of Grey Phalaropes and a large number of Buffon's (Long-tailed) Skuas, including five in view at once, in Christchurch Harbour between October 22nd and 28th.

Long-tailed Skua, adult, West Hayling Shore, December 5th 1994

The first modern record was of one seen moving east off Gilkicker Point by Martin Terry on May 3rd 1962. This was not accepted by HFCERC at the time but the discovery of acceptable notes in the HOS archives led to its inclusion in *BoH*. Extracts from those notes are reproduced below.

The bird flew east along the Solent at a range of 150 yards from 1255 to 1300 hrs. At once I noticed the size of the body, which was much smaller than an Arctic Skua, and the elongated central tail feathers. The flight was noticeably different from an Arctic Skua, it was buoyant and much more graceful, more like a tern's. It was direct and lacked the characteristic movement from side to side as with Arctic Skua. Plumage description: the bill was noticeably smaller than that of any of the other skuas and was dark in colour. There was a dark cap which extended just below the eye and back to the base of the nape and a distinct white collar contrasting with the cap. The back and rump were a paler brown than an Arctic Skua. The tail was darker brown and the central tail-feathers were remarkably elongated and slender, approximately one third of the overall length of the bird. These feathers were very flexible. The wings were long, very pointed, and slenderer than an Arctic Skua's. They were dark brown in colour; the underwings were also dark. The white primary patches were virtually indistinguishable on the upper surface of the wings and considerably less noticeable than those of an Arctic Skua. The underparts were white but became dusky white on the flanks and undertail-coverts.

Since then, a total of 49 has been recorded.

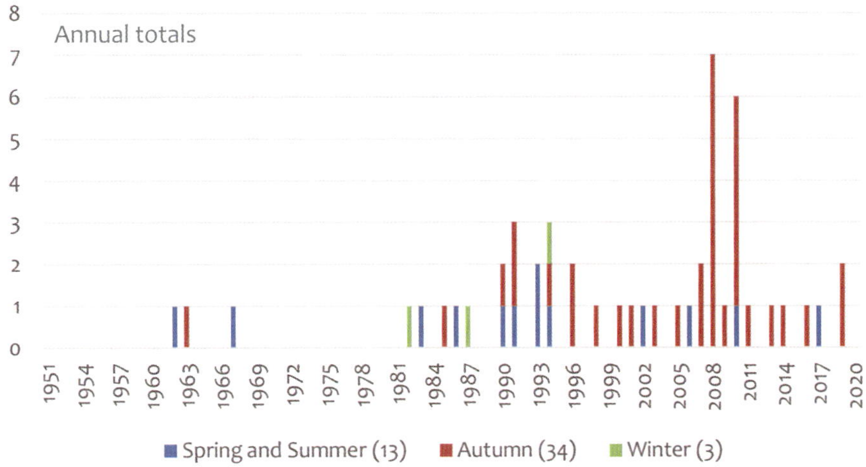

Analysis of arrival dates shows that spring records spanned the period between April 29th and May 28th. The earliest, seen moving east through Hayling Bay by George and Rob Clay on April 29th 1990, was omitted in error from *BoH*. The latest was seen by Phil Cooper moving north over Hale, Avon Valley on May 28th 1983. Other records were coastal and came from Gilkicker Point (May 20th 1967), Hayling Bay/Sandy Point (May 11th 1986), Needs Ore (May 24th 1993), Hurst Beach (May 10th 1994) and Hill Head (May 17th 2002 and May 21st 2006). Two further records involved one which flew east past Hurst Beach at 1821 hrs on May 10th 2010 and was seen later that evening off Haying Bay and another tracked past Hurst at 0832 hrs on May 11th 2017, Hill Head at 0917 hrs and Sandy Point at 1013 hrs.

There were two mid-summer records of adults. Simon Aspinall saw one off Stokes Bay on June 21st 1991 and Ed Wiseman watched another as it soared high to the north-west over Normandy Marshes on June 30th 1993. Possibly these were birds that were summering locally.

Most have been recorded in autumn with records spanning the period between August 11th (in 2016) and November 8th (in 2007 and 2014) and a clear peak in the second half of September. The Hurst Castle to Milford on Sea area was the favoured location with 17 recorded there including at least four (an adult, two dark and one pale juveniles) on October 5th 2008 and a long-staying juvenile from September 10th to 18th 2011 which was probably the bird off Lepe on September 6th. Other coastal records came from Keyhaven/Pennington Marshes (3), Black Point (September 20th and 21st 1998), Sandy Point (October 4th 2008), Hill Head (September 23rd–27th 2007), Dibden Bay (November 4th 1985 and August 26th 1990), Langstone Harbour (October 7th 2001), Stokes Bay (November 8th 2007), Hayling Bay (2 west on October 5th 2008) and Weston Shore (August 11th 2016). Inland a juvenile was at Fleet Pond on September 13th and 14th 1991 and three juveniles paused briefly at Ibsley Water during torrential rain on September 23rd 2010 before moving on southwards.

Long-tailed Skua, juvenile, Milford on Sea, October 26th 2019. Alan Lewis

Three have been recorded in winter: a first-winter off Hurst Beach on January 17th 1982, a sub-adult which circled Titchfield Haven on January 29th 1987 before leaving north-east and an adult which frequented the Hayling Shore of Langstone Harbour and nearby fields on December 4th and 5th 1994.

The weekly distribution of records is shown below.

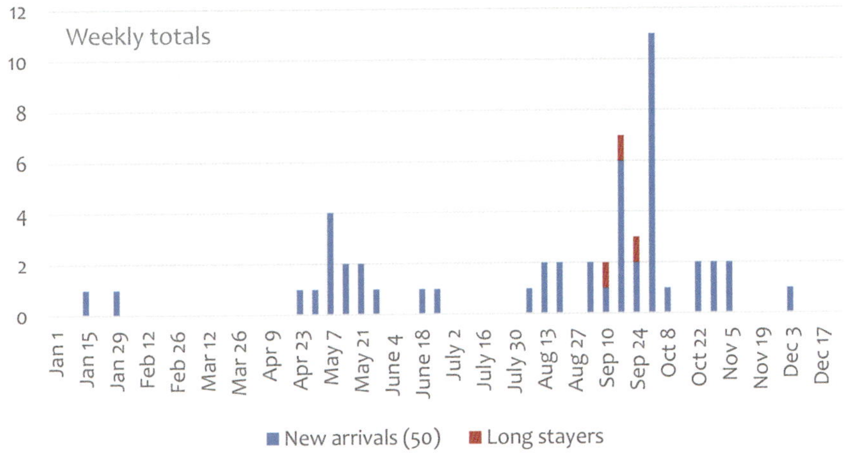

The upsurge in records in the 1990s and 2000s probably reflects a combination of increased abundance and greater observer awareness; the recent small downturn may be a genuine decline. A further 15 claims between 1993 and 2014 were not accepted by HOSRP due to inadequate or no documentation.

GB status: scarce passage migrant, especially in spring off NW Scotland and in autumn in E England. (Breeds in Arctic regions, winters in southern oceans.)

Little Auk *Alle alle*

HOSRP

A rare visitor, usually occurring following autumn gales in the North Sea. Several old records of inland storm-driven birds and around 176 recorded in the modern era.

The first for the county appears to be for spring 1771. Gilbert White, writing to Thomas Pennant in November 1773, said that 'Two years ago last spring, the little auk was found alive and unhurt, but fluttering and unable to rise, in a lane a few miles from Alresford, where there is a great lake; it was kept awhile, but died.' K & M mentioned further inland records including a living specimen brought to Mr. Bell from Le Court, near Greatham, on November 19th 1861, one found by Lord Northesk at Longwood House near Owslebury in November 1884 and specimens from the New Forest and Wield, near Alton, in January 1895 when there was an irruption into the county. K & M described it as an occasional visitor to the coast in winter, sometimes numerous, but did not provide any such records. However, there is a specimen on display in the Science School at Winchester College which is labelled Southampton Water, 1888.

The only records traced for the first half of the 20th century were of one picked up alive at Alresford on October 30th 1930 (*The Field*, December 6th

1930) and another retrieved by the dog of Miss V. Goodwin at Vales Moor Pond, New Forest on November 23rd 1949 and released at the coast. This was given erroneously as November 16th in *British Birds*. One at Hurst Beach on November 27th presumably refers to the same bird. There was a large invasion into Britain in early 1950 but the only record for Hampshire appears to be of one at Barton on Sea during February.

In the 65 winters between 1955/56 and 2020/21, a total of around 176 was recorded in 45 winters. This includes 17 inland, all but one of which was dead or subsequently died, and 27 dead or dying birds at the coast.

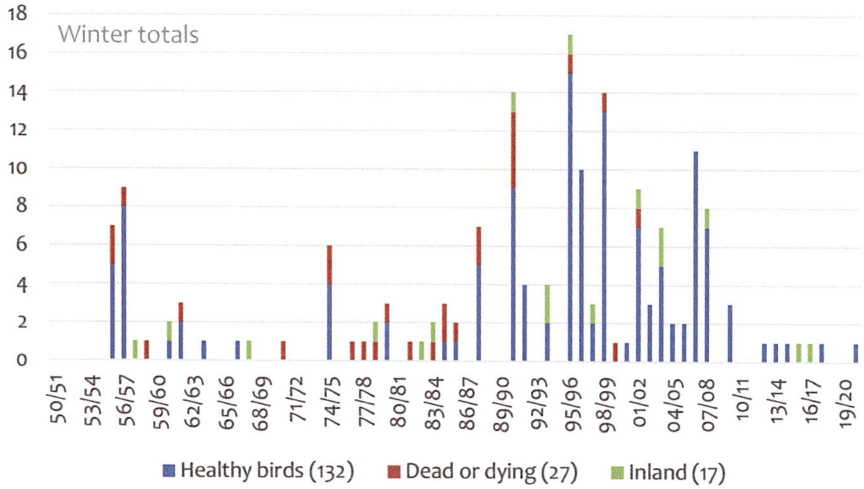

Most Little Auk records are associated with large movements in the North Sea which occur during strong winds or gales, and as a result the numbers recorded vary from year to year. Most occur in late autumn, particularly November, which accounts for 102 of the 176 recorded. The earliest ever was one picked up on a lawn at Overton on October 19th 1997. Only five more have been recorded in that month including the earliest healthy bird off Southsea on 23rd in 1991.

Little Auk, Lower Pennington Lane, December 5th 2009. Simon Vale

As shown opposite, numbers tail off in the early months of the year. The latest ever live birds were two flying east off Hill Head on February 9th 1988 although there are later records of one found dead at Hurst on February 11th 1991 and another found long dead there on April 7th of that year.

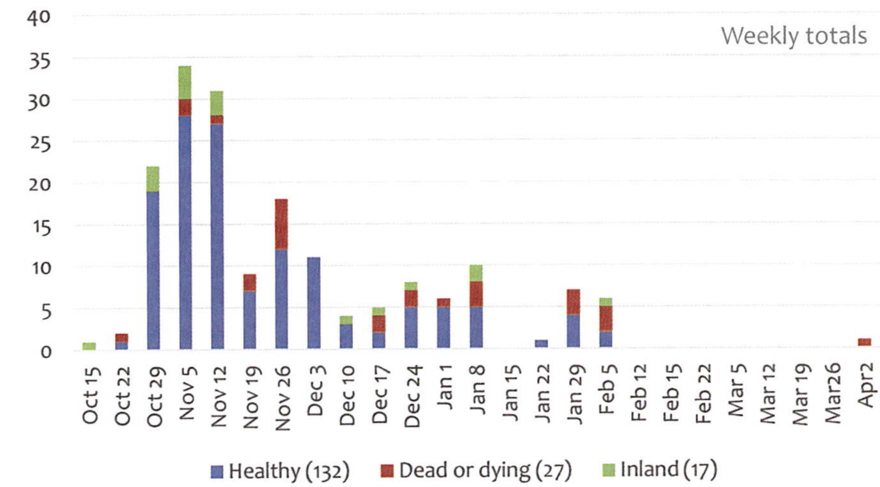

Most records of healthy individuals involve one or two flying past the observation point. Larger numbers were eight seen from the Portsmouth–Ryde ferry on December 1st 1956, seven flying east off Hill Head on November 5th 1998, five off Hurst on November 18th 1995 and three off Needs Ore on November 6th 1998. This is a difficult species to catch up with in the county. Long-staying individuals include those at Black Point from November 20th to 23rd 1985 which relocated to Langstone Harbour entrance from November 24th to 28th, Hilsea Moat from December 29th to 31st 1990 which unfortunately died and IBM Lake, Cosham from January 13th to 25th 1994. Recent records have included those of one seen off Weston Shore on November 10th 2009 and relocated further up the Itchen Estuary at St. Denys the next day and another seen and photographed in a roadside

Little Auk, IBM Lake, January 14th 1994. Finally a live one!

ditch at Lower Pennington Lane on December 5th 2009 which was seen by many birders later that day; it was not seen again.

Well inland, storm-driven birds have been recovered at Headley near Bordon (1957), Old Alresford (1960), Kings Worthy (1967), Winchester (1983), Braishfield (1993), Overton (1997), Farnborough (2003), Lower Froyle (2007) and Whitchurch (2015). Others were nearer the coast at Fareham (1978 and 1984), Havant (1991), Portchester (1993), Totton (1995), Lordshill (2001) and Rowner (2003). The first Fareham bird, picked up on December 12th 1978, was cared for at Titchfield Haven but eventually died on January 12th 1979. The most recent inland record was of an apparently healthy bird which spent two hours on Ibsley Water on November 3rd 2016.

Totals recorded at various coastal sites were as follows: Chichester Harbour including Black Point, 9; Hayling Bay including Sandy Point, 13; Langstone Harbour, 11; Eastney to Southsea, 3; Portsmouth Harbour, 3; Gilkicker Point to Lee-on-the-Solent, 11; Hill Head to Brownwich, 32; Southampton Water, 8; Lepe to Needs Ore, 9; Thorns Beach to Pitts Deep, 3; Lymington to Milford on Sea, 47; Hordle to Barton on Sea, 4.

GB status: Irruptive autumn passage migrant and winter visitor, mostly to east coast. (Breeds in high Arctic from E Baffin Island to E Siberia; winters in northern N Atlantic.)

Black Guillemot *Cepphus grille*

HOSRP

A very rare vagrant. Seven records since 1986.

The first for the county was a first-winter found by Pete Potts off Southsea on January 31st 1986. Extracts from Pete's description are reproduced below.

> *The bird was found while birding along Southsea sea front – I noticed a Slav. Grebe so stopped to look at it and then a second bird appeared – a Tystie in winter plumage diving regularly and heading towards the beach – WOW! Eventually it came within ten feet of the shore but, unfortunately for it, the gulls noticed it and mobbed it continuously causing it to dive frequently while moving along the beach. The bird was watched for 30 minutes off and on between fruitless phone calls. Eventually I left the bird to attend a lecture!*
>
> *Slightly smaller than Guillemot/Razorbill but larger than the nearby Slav. Grebe. Head and neck white with grey mottling. Back mottled black. Wings black with large white secondary coverts/lesser coverts patch. These white feathers had dusky tips making it a first-winter bird. Throat to belly white, tail black, legs (seen when diving) red. Eye dark.*

The next was an adult in summer plumage seen by Marc Moody on May 11th 1989. This record was originally considered not proven by HOSRP so was not included in *BoH*; however, a subsequent review of Marc's notes led to the record being accepted. A further 21 years were to elapse before another occurred, an adult in summer plumage found by Tim Parminter when it flew in from the west and landed on the sea off Hurst Beach at 1314 hrs on May 12th 2010. It remained for a minute before continuing eastwards and may have been the bird seen off Selsey Bill, West Sussex the next day. The next was a first-winter seen by Barry and Geoff Goater and Tim Norriss swimming and diving just off Hayling Island lifeboat station at Sandy Point at 1315 hrs on November 23rd 2017. It left south out to sea after five minutes before local birders could be summonsed. One seen by Philip Fawkes between Lepe and Inchmery in the late afternoon on January 8th 2018 is assumed to be the same individual which was also seen off Seaview, IOW on December 25th. It disappeared before other observers arrived. There was a record for the third year running when Steve Keen saw one fly east past Milford on Sea in poor weather at 0818 hrs on November 26th 2019.

Steve found another moulting out of summer plumage off Barton on Sea on July 19th 2021. This bird proved very obliging, remaining in the area until July 23rd and enabling many birders to see it at close range. It was taken into care after being picked up off Highcliffe, Dorset on July 24th.

Black Guillemot, Barton on Sea, July 22nd 2021. Rob Porter

There is an adult in summer plumage on display in the Science School at Winchester College which is labelled Southampton Water, 1888. K & M did not accept the record, stating, 'We are inclined to doubt the correctness of the locality, as specimens in this plumage would be extremely rare on our coasts, and Mr. Chalkley [the Winchester taxidermist] kindly tells us that he is now unable to find out when and where it was killed, the person who obtained it being dead'.

GB status: Resident coasts from Anglesey and E Scotland northwards, rare visitor further south. (Breeds N Arctic coasts and in N Atlantic to Gulf of Maine and NW British Isles, N birds move S in winter.)

Puffin *Fratercula arctica*

HOSRP

A rare visitor, most frequent in April and May but recorded in all months. At least four old records and 68 birds in the modern era of which 30 were dead or dying.

The first records for the county appear to be of two storm-driven birds in 1892. K & M stated that the Prebendary Gordon wrote to John Kelsall about a specimen that was found at Hawkley, near Petersfield, in November of that year and that another in the Alton Museum was obtained at Froyle in the same month. These were notable as inland records but there may have been earlier coastal occurrences since the species bred on Freshwater Cliffs, IOW and Edward Hart held a specimen obtained on January 29th 1886 which was probably from the vicinity of Christchurch in present-day Dorset.

Wiseman (2017) mentioned a corpse found at Milford on Sea in 1947 but the first dated 20th century record is of one picked up in a ditch at Farringdon, near Alton, on December 15th 1949 which was released at Lymington the next day.

Annual totals from 1951–2020 are shown below.

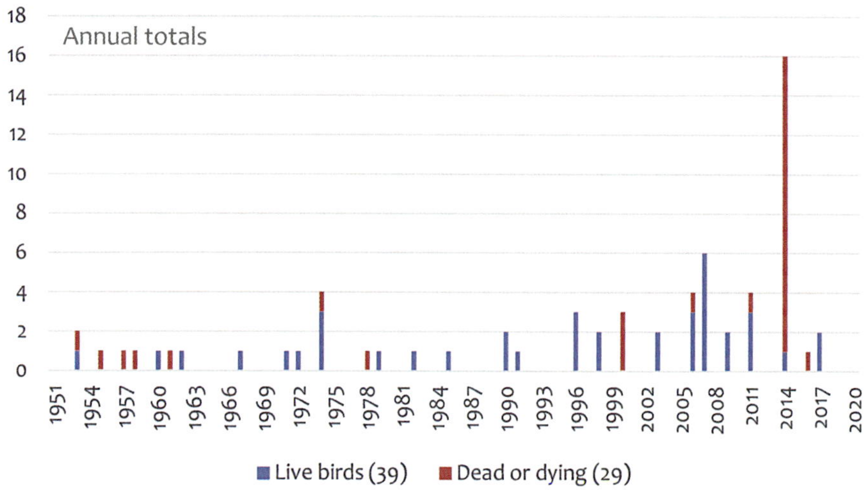

Puffins continued to breed at Freshwater Cliffs in decreasing numbers through the 1950s but most records during this decade referred to inland storm-driven birds. One was picked up at Baughurst on February 14th 1953, a juvenile was found in a field at Crondall on October 3rd 1953 and released on the coast and another juvenile was taken into care from the Basingstoke Canal at Eelmoor on September 19th 1955 but died after five days. A juvenile was found dead at Hill Head on July 14th 1957 and an adult was found dead and oiled at Brownwich on July 31st 1958. Four records in the 1960s were all from the Hurst area: one offshore on October 2nd 1960, a corpse found on October 28th 1961, one on August 6th 1962 and one on July 16th 1967.

Between 1972 and 2020 there were records in 20 years with totals of 35 live birds and 23 picked up dead or dying. Most were recorded at Hurst Beach or nearby with totals of 22 live birds and 16 dead. Others occurred at Hill Head/Brownwich (5 live, 2 dead), Sandy Point (4 live, 1 dead), Lepe/Inchmery (1 taken into care, 2 dead) and single live birds at Calshot (October 11th 1971), Eling (a juvenile picked up on July 12th 1982 which was released off the Needles), Hamble Common (October 25th 1974) and Black Point (January 18th 1996). The only inland record was of a storm-driven bird picked up alive at Winchester Hospital on November 30th 2011 which unfortunately died in care.

Analysis of arrival dates shows that the best chance of seeing a live bird is in spring when a total of 23 has been recorded between April 14th (in 1979) and May 14th (in 1996) with totals of 17 at Hurst, four at Sandy Point and two at Hill Head. A late bird was off Hurst on June 5th 1972. Records of healthy birds in other months are very scarce with only 11 in total. Most storm-driven birds occurred between late November and mid-March. A particularly severe wreck occurred in early 2014 when a series of gales in the English Channel resulted in unprecedented mortality of Puffins and other auks. A total of 15 was found dead on the Hampshire coast between January 30th and March 15th but only one live bird was recorded which flew off east from Hill Head on February 5th.

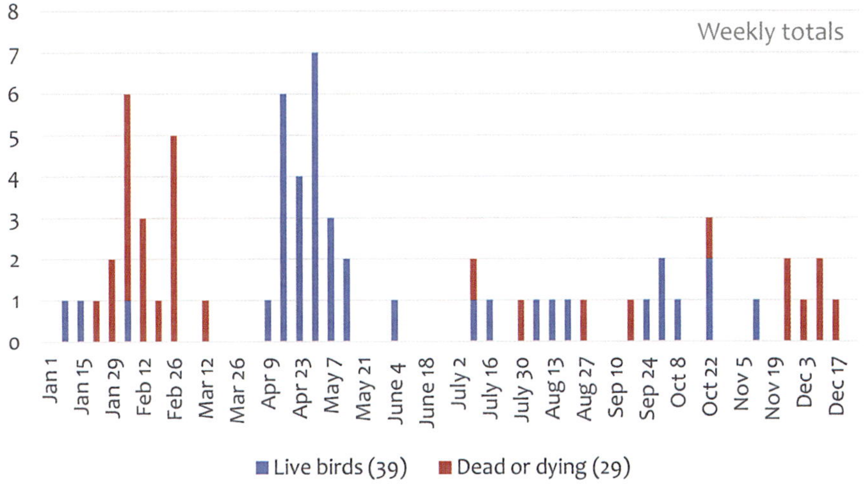

GB status: Summer visitor coasts and offshore islands except SE England; disperses outside breeding season. (Breeds N Atlantic from E USA to Novaya Zemlya, winters at sea S to Mediterranean.)

White-billed Diver *Gavia adamsii*

BBRC up to 2008; HOSRP 2009 to date

A very rare vagrant. One record in 1991.

The only record for Hampshire is of an adult in winter plumage seen by John Garr off Oxey Marsh on February 14th 1991. Extracts from John's description published in the 1993 *HBR* (Garr 1994) are reproduced below.

At about 1015 hrs on February 14th 1991 while watching from the mouth of Oxey Creek I noticed at c. 1,000 m a large bird on the sea. It was observed for about 20 minutes using binoculars and telescope during which time it swam directly towards me to within about 80 m. It was a very large diver with a relatively long, thick, erect, somewhat goose-like neck, large angular head and massive angular bill, with the pronounced angled lower mandible having the upwardly inclined distal part longer than the much less inclined proximal part. This angularity was clearly visible at 1,000 m using 10 x binoculars. The bill was relatively deep at the base and appeared flat-sided rather than curved with a very marked upwardly inclined taper. The tip was blunt rather than pointed. The entire dorsal ridge from the base to the tip was straight with no downward curve at the tip. The bill was entirely light coloured, namely 'dried bone' when seen at long and medium ranges and a slightly warmer ivory colour suffused with just a hint of yellow in the middle part, only visible at close range. There was no dark colouration of any description on any part of either mandible.

It is surprising that this bird was not seen by anyone else in this well-watched area.

K & M included one shot in Emsworth Harbour during the winter of 1895-96 which was in the possession of The Rev. Canon S. G. Scott, Rector of Havant. In a review of records of the species, Witherby (1922) considered the record unacceptable as the specimen had not been traced and insufficient details had been published.

Cohen (1963) mentioned two records. A diver was found dead in Chichester Harbour by Keith Shackleton in about 1953. He initially identified it as Great Northern, but after five years or so, on the basis of the bill shape, re-identified it as White-billed. The bill was transferred to the Natural History Museum where it was examined in the early 1970s and found to be of a Great Northern Diver (Burn & Mather 1974). Six observers, including Mrs E. Miller, Norman Pullen and David Glue, observed a bird which was probably of this species down to 50 yards at Needs Ore on January 13th 1962. They furnished a full description to BBRC. It is listed by Burn & Mather (1974) as an insufficiently substantiated record, but BBRC has never published it as not proven.

GB status: 8, 1829–1949; 711, 1950–2019. Large increase, mean 33 per year in 2010–19. (Breeds along Arctic coasts of European Russia from Yamal Peninsula & Novaya Zemlya

E through Arctic coastal regions of Siberia, N Alaska, & Canada E to Melville Peninsula, Nunavut. Winters at sea in E Atlantic S to S Norway, & ice-free N Pacific S to N British Columbia and N Japan. Some winter inland in North America.)

Storm Petrel *Hydrobates pelagicus*

HOSRP

A rare spring and autumn visitor, usually storm-driven; five records in December and January. Two records in 1835 and 1882 and around 359 birds since 1954.

The first for the county was one attracted to Hurst lighthouse during the night of December 15th/16th 1835. It was taken alive and acquired by Col. Peter Hawker (Payne-Gallwey 1893). This record was overlooked for all the county avifauna until attention was drawn to it by Wiseman (2017). The only other 19th century record was a rather vague claim of one found near Stockbridge by J. H. Willmore about the year 1884 (K & M). However, research in *The Zoologist* for 1882 reveals a note by the same John Willmore which describes a Storm Petrel which had evidently been dead for some days and was found on a hedge not far from here (Stockbridge) on October 23rd. It was lying on its back on the top of the hedge. I suppose it must have been driven inland by a gale and blown against some tree which killed it'.

The next was not until 1954, when Mike Adams saw one in the Solent from Warden Point, IOW on November 28th 1954 (not August 14th as in *BoH*). The first seen from within Hampshire was one watched by Gerald Kinsey off Hythe Pier on October 9th 1955. The 1955 *HBR* states that a full description was given but this has not been traced. Since then, there have been records in 29 years up to 2020. Annual totals have not exceeded 16 apart from exceptional influxes of 138 in autumn 2000 and 128 in spring 2006 which are excluded from the chart.

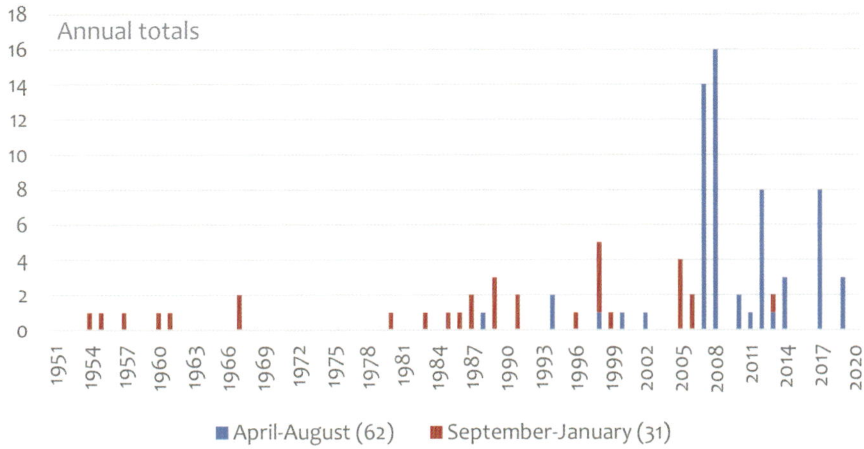

Records until the turn of the century involved 28 birds, usually storm-driven and during the months from September to January. Four were picked up dead or dying. Only four occurred between May and July.

In 2000, one off Hurst Beach on May 15th was the prelude to an unprecedented influx following south-westerly gales and torrential rain during the night of October 29th/30th. Marc Moody and others who made it out to Hurst Castle the next morning logged 55 flying west between 1100 and 1600 hrs. Also on this day, four moved up Southampton Water from Weston Shore and two were off Hill Head.

The following day Lee Marshall saw 18 fly west off Stokes Bay between 0740 and 1242 hrs, five moved west off Hurst between 1140 and 1420 hrs and one flew east in Hayling Bay at 1430 hrs. Some evidently sought shelter in Southampton Water. Three flew north-west off Weston Shore between 0715 and 0755 hrs and then Andy Collins saw 31 do likewise between 1545 and 1645 hrs. On November 1st, a further three moved north-west at Weston between 0707 and 0750 hrs and probably five were seen there between 1410 and 1615 hrs. The destination of these was possibly revealed by the observation of up to three coming in from the south and landing on the sea off Hythe Pier from 1600 to 1620 hrs, suggesting that they were using the sheltered waters as a safe roosting site, despite the continued harassment from Black-headed Gulls. Elsewhere on this date, one moved west off Stokes Bay at 0725 hrs, while the final sightings were on 2nd when five flew west off Stokes Bay between 1005 and 1055 hrs and one was off Lepe.

Storm Petrel, Hill Head, October 30th 2000

In Dorset only 16 were logged over the three-day period at Hengistbury but over 100 moved west past Portland Bill and around 150 were estimated to have been wrecked in the area around Portland Harbour. Very few were noted in counties further east and west; for example, only one was seen in Sussex, suggesting that most were pushed onto the coast in the centre of the track of the low-pressure system, which ran north-eastwards over central southern England.

During the 21st century spring records have predominated. The biggest ever influx occurred between May 19th and 29th 2006 when at least 128 occurred during a period with gales and heavy bands of rain. There were daily sightings between Hurst Castle and Milford on Sea including totals of 15 on May 20th and 19 on 22nd. Just to the west, 185 were logged off Hengistbury Head, Dorset on May 27th. At the eastern end of the county the highest counts off Sandy Point were of 20 at least on May 25th and seven on 27th. A few penetrated the Solent with four off Hill Head between 20th and 24th, singles off Needs Ore and Lepe on 26th and 27th and two west off Eastney on 29th. Noteworthy totals were also recorded in 2007 with 14 between May 10th and July 5th and in 2008 with 17 between May 28th and July 10th. Other than in 2006, many of the recent spring birds off Hurst and Milford occurred in calm conditions when it is thought they were attracted to 'dissolving bait' placed in crab and lobster pots (Marc Moody *in litt.*).

Analysis of arrival dates shows spring records between April 30th (1 off Hurst Beach in 2012) and July 12th (1 off Hurst Beach in 1998). Even if the 2006 influx is discounted, spring sightings show a clear peak between mid-May and early June. Autumn records have been evenly spread between August 6th (1 off Hurst Castle in 2010) and December 5th (1 off Hill Head in 2006) with no more than four in any week apart from the 2000 influx. The only later records have been of singles picked up at Hill Head on December 21st 1989, in Langstone Harbour on December 25th 1989, picked up at Lee-on-the-Solent on January 1st 1957 and feeding over the Frying Pan at Titchfield Haven during a gale on January 21st 1980 which eventually moved off inland.

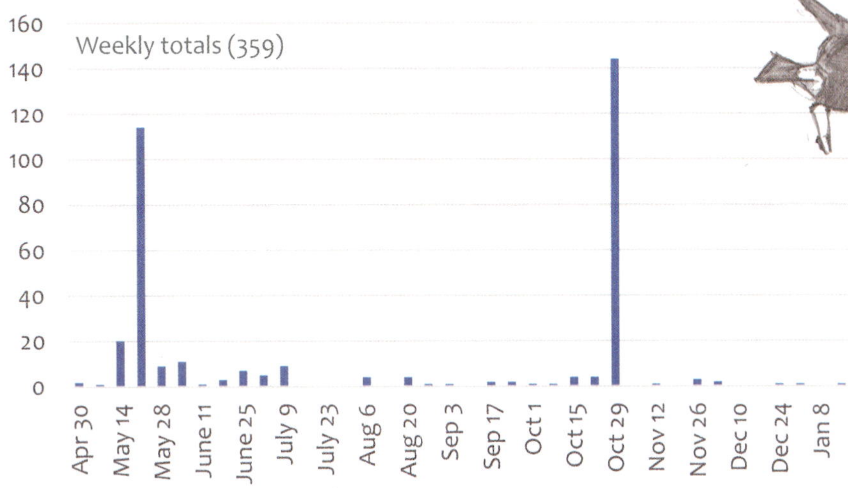

Very few have been recorded away from the open coast. The only other record in Langstone Harbour in addition to the December record above was of one on November 15th 1998. An exhausted bird was found on the container ship *Kowloon Bay* in Southampton Docks after a crossing from Le Havre on November 27th 1986; it was released south of the Isle of Wight the next day. Less fortunate was one found freshly dead in the car park of Southampton Oceanography Centre on November 1st 1996 following overnight gales. In contrast to Leach's Petrel there are no modern records of birds found far inland.

GB status: Breeds on islands from Scilly to Shetland; passage migrant to all coasts. (Breeds from Norway to Canaries and in Mediterranean Sea; winters at sea.)

Leach's Petrel *Oceanodroma leucorhoa*

HOSRP

A rare autumn and winter visitor, usually storm-driven; single records in April and June. Eleven old records and around 325 birds since 1952.

The first for the county was one recorded by Wise, on the authority of Rake, that was picked up near Fordingbridge in November 1859 (Wise 1862). K & M referred to a specimen obtained by Bell 'many years ago' at Alton and several recorded by George Corbin from the neighbourhood of Ringwood in December 1882 and December 1885. Research in *The Zoologist* for March 1882 reveals a note from Corbin which stated:

> *During December three specimens of this petrel at least were met with near Ringwood, one of which I had the pleasure of examining. This specimen weighed scarcely an ounce and measured exactly 18 inches across its expanded wings. The stomach contained only a small quantity of oily matter, although the bird was in very fair condition. It was picked up dead, doubtless driven inland by stress of weather. Another specimen was picked up and partly devoured by a cat.*

Note the error in K & M as the birds were clearly obtained in December 1881, at a time when there was a wreck in many parts of England. *The Zoologist* for May 1886 includes a further note from Corbin. This includes reference to a Brent Goose near Ringwood, which was evidently considered more noteworthy than the petrel, which he described as follows.

> *The Fork-tailed Petrel was caught by a boy on 9th or 10th December last, not half a mile from where the goose was killed. He saw it flying about in a half-dazed state, and, thinking it was a curious swallow, kept it in a cage till it died on 11th after refusing all food.*

There are six records for the first half of the 20th century. One obtained at Portchester in 1921 is on display in the Science School at Winchester College.

Following a south-westerly gale which lasted from November 7th to 9th 1936, one was picked up alive near Beaulieu on 12th and two others, both described as young males, were picked up near West Dean that day and at Hurstbourne Priors on 13th. Further gales in December 1948 accounted for one picked up alive by Tony Norris at Alverstoke, Gosport on 9th and another found dead at Warren Farm, Beaulieu on January 3rd 1949.

The first of the modern recording era was one picked up alive near Portsmouth Cathedral on October 13th 1952. This predates the massive 'wreck' throughout the country involving an estimated 6,700 individuals which occurred in late October and early November 1952 (Boyd 1954). In Hampshire, 17 dead or dying birds were discovered including 15 inland, and five were seen offshore, as shown in the table below.

Offshore	Dead or dying
Hill Head, October 25th	Barton on Sea, October 26th
Lymington, October 30th	Lee-on-the-Solent, October 26th
Hill Head, October 31st	Walkford, New Milton, October 27th
2, Southampton Water, November 7th	2, Laverstoke/Whitchurch, October 31st
	Stoke Charity, October 31st
	Cove, Farnborough, October 31st
	Sutton Scotney. October 31st
	6, Hambledon/Droxford, November 1st
	Andover, November 3rd
	Burntwood near Kings Worthy, November 3rd
	Beaulieu Road, November 16th

Subsequently, as shown on the next page, there were records in just six years between 1956 and 1972 then in 28 between 1982 and 2020. Annual totals did not exceed six prior to 1989 when there was a major wreck involving around 62 individuals. Further influxes occurred in 2003 (13), 2005 (18), 2006 (51) and 2009 (100) but in other years totals did not exceed four.

November 29th 2009

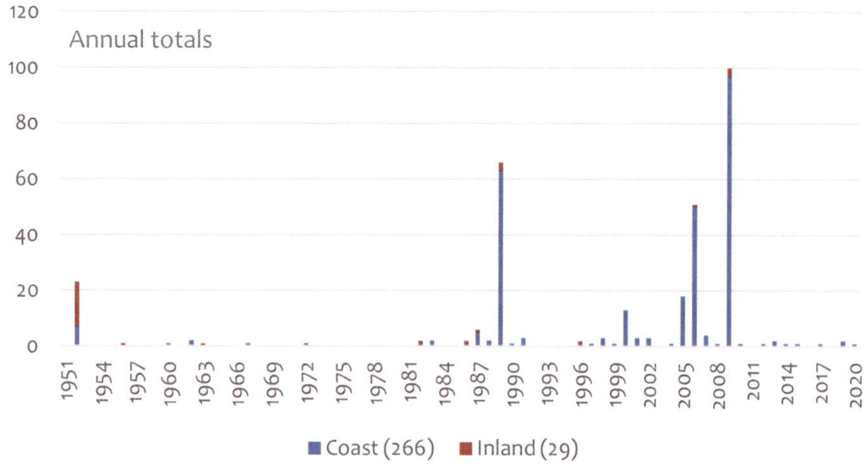

As shown opposite, analysis of arrival dates shows that influxes have occurred between late October and late December and are always associated with both strong winds and heavy rain. Counts exceeding five all occurred in the four years with the largest influxes. In 1989 around 62 occurred between December 16th and 26th with the highest numbers on 23rd when there were ten off Lee-on-the-Solent and 12 in Hurst Narrows and on 25th when at least 15 including eight together were off Hurst Beach. In 2005 there were 18 between October 30th and November 6th with most on November 3rd when there were four east and five west off Hurst Castle and Milford on Sea between 0735 and 1552 hrs and six west and one east off Hill Head between 0935 and 1240 hrs. The 2006 influx involved around 51 between December 5th and 11th. On the first day 16 were logged off Milford on Sea between 0830 hrs and dusk and three were seen elsewhere. There was a lull the next day but following overnight rain more were recorded on 7th with possibly seven in Chichester Harbour, at least 11 off Hill Head between 0820 and 1300 hrs and eight off Milford on Sea between 0839 and 1506 hrs. The most recent and largest influx in 2009 involved around 100 between November 21st and 30th. The highest numbers occurred on 29th when 44 were logged moving east into the Solent from Hurst Castle between 0700 and 1600 hrs, at least eight were seen off Oxey Marsh (including one taken by a Great Skua), 20 were off Hill Head between midday and dusk and at least five were seen from Sandy Point. Later that afternoon three were in nearby Chichester Harbour.

Although most have been from the open coast, particularly the Hurst/Milford area, small numbers have occasionally penetrated Southampton Water with a total of 14 recorded in 1983 (1), 1989 (3), 2000 (1), 2006 (2), 2009 (4), 2013, 2015 and 2020 (1 each). Most were seen from Weston Shore or Hythe but singles were seen further up the Test Estuary off Town Quay on December 21st 1989 and Mayflower Park on December 7th 2006. One found on a cargo ship in the docks on November 29th 1991 was released at Hill Head the next day. There have also been eight within Langstone Harbour in 1972, 1982, 1987, 1989, 2001, 2007, 2019 and 2020. The most recent sightings of singles off Weston Sailing Club at 0820 and 0950 hrs and in Langstone Harbour from 1500 hrs until dusk in calm weather on December 27th 2020 surely referred to the same individual as none was reported anywhere else on the English south coast.

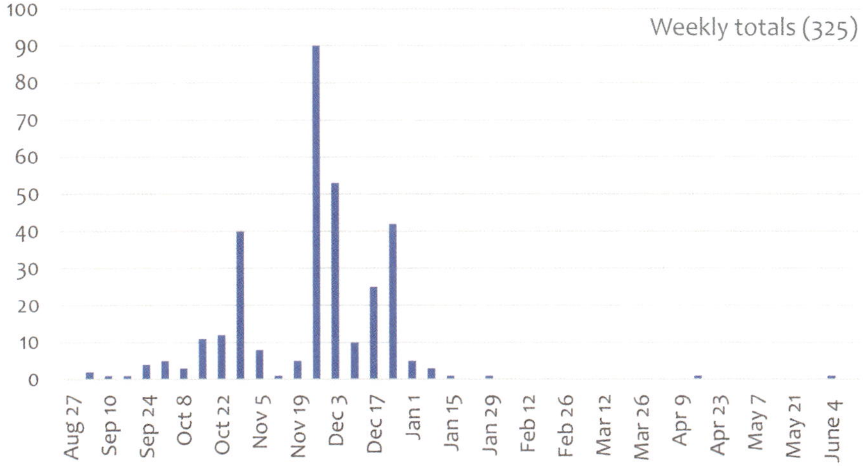

Inland sightings of live birds were of singles picked up at Bisterne on December 11th 1956 and released at Keyhaven the next day, flying over Fullerton on September 27th 1963, picked up at Brambridge on September 26th 1982 which was ringed and released at the coast, flying over Nursling on November 29th 2009 and also on that date one at Ivy Lake and presumably the same later at nearby Ibsley Water which evaded chasing gulls and a Peregrine before leaving south-west. Birds were found dead or dying at Whitchurch on September 9th 1986, Horndean around October 18th 1987, Stubbington on December 25th 1989, Queen Elizabeth Country Park and Hursley the next day, Paper Mill Farm Watercress Bed (now Testbourne Lake) on October 29th 1996, Medstead on December 7th 2006 and Cheriton on November 29th 2009 which was picked up by a dog and later perished.

The earliest in autumn was one seen by Mike and Pete Combridge and John M. Jones which flew east through Hurst Narrows and then back west on September 4th 1983. Only a further seven have been seen in that month. There have been nine in January including the latest one found by Andy Parfitt off West Hayling at 1100 hrs on 20th in 2007 which was last seen in mid-afternoon moving east over Langstone Bridge by one of the attendant year-listers!

There have been three records outside this period. One was located by Jason Crook in Langstone Harbour at 1308 hrs on April 21st 2001. Despite the

Leach's Petrels, chased by Common Gull, Hill Head Harbour, November 29th 2009

attention of mobbing Black-headed Gulls it remained until dusk and was viewed by many birders from the harbour entrance. It occurred at the end of a period of persistent northerly winds. In the following year one was seen off Hill Head in stormy weather by Barry Duffin at 0800 hrs on February 2nd and what was assumed to be the same individual was seen again by Sue and Peter Morrison at 1335 hrs. Perhaps the most extraordinary record was of one seen by Marc Moody off Milford on Sea between 0940 and 0947 hrs on June 8th 2017 during an influx of Storm Petrels. There are very few June records for southern England but it has been suggested that failed breeders may join feeding flocks of Storm Petrels.

GB status: Breeds on islands off NW Scotland; occurs off all coasts particularly following autumn and early winter gales. (Breeds N Atlantic coasts of Europe and N America, Pacific coast of N America and S Africa; winters at sea.

Cory's Shearwater *Calonectris borealis*

BBRC up to 1982: HOSRP 1983 to date

A rare visitor. Seven records of six birds since 1982.

The first for the county was one seen by Dave Wooldridge off Hurst Beach on May 12th 1982. Unfortunately, it has not been possible to trace Dave's description which was accepted by BBRC. In 2020 he commented that he has subsequently seen very many and has no doubt about the identification.

The next was seen by Marc Moody and Martin Cooper flying south off Hurst Castle at 1703 hrs on August 18th 2001. It was watched for about five minutes as it slowly moved south out of the bay, until it was lost from view behind The Needles headland. What is assumed to be the same bird was seen the following day from Milford on Sea by Glyn Horacek-Davis and Ian Williamson, heading back north-west into the bay at 1045 hrs. The next was in 2006, when one was observed by Tim Lawman and Jason Crook in Hayling Bay from 2015 to 2017 hrs on May 23rd. It arrived from the east before heading south-west; its occurrence coincided with a period of stormy weather and an unprecedented influx of Storm Petrels along the Hampshire coast. Later that year, following strong south-west winds and heavy rain overnight, Vicky Hassell saw one arrive from the west and sit on the sea off Hill Head at 1030 hrs on October 20th. After about half an hour, it took off and flew towards the shore, flying within ten metres of the observers, and then continued east over Hill Head harbour and along the shore. In 2007, Andy Johnson watched one off Sandy Point from 1954 to 1959 hrs on July 4th. It was first seen heading towards the observer before veering first west and later moving off in a south-easterly direction.

Cory's Shearwater, Hurst Castle, November 22nd 2012. Marc Moody

The most recent occurrence was of one seen by Marc Moody off Hurst Castle between 1004 and 1010 hrs on November 22nd 2012; it was chased by gulls and left westwards.

An earlier record of one seen by John Walters flying east in Hayling Bay on October 19th 1981 was harshly considered not proven by BBRC.

GB status: 35,699, 1958–2019. No clear trend, mean 940 per year in 2010–19. (Breeds E North Atlantic in Canary Islands, Madeira, Selvagens, Desertas, Azores and other islands. Disperses widely N to S Ireland, S England, and E to coastal NE USA and Canadian Maritime Provinces. Winters S to South Atlantic.)

Sooty Shearwater *Ardenna grisea*

HOSRP

A rare visitor. 33 records of 39 birds since 1974.

The first for the county was a corpse which was found by Tim Lawman and M. Ayres at Hook-with-Warsash on July 22nd 1974 and subsequently identified by Barry Duffin. It was possibly the unidentified bird seen in flight off Hill Head on April 28th of that year. The first live bird was over 20 years later, when Jacquie Hull and Steve Keen watched one offshore at Pennington Marshes for 30 minutes on October 1st 1994 before it left west through Hurst Narrows.

Since then, a further 37 have been recorded with sightings in 14 years.

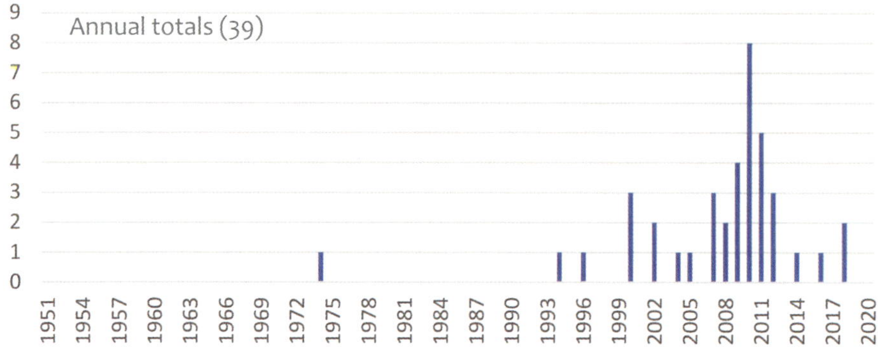

Most were seen from the coast between Hurst and Milford on Sea with 25 there including five west off Milford on September 10th 2011 and three south-east off Hurst Castle on September 11th 2011. At Sandy Point nine have been seen including two offshore on September 9th 2009. Other records were of singles off Stokes Bay on October 21st 2005, Barton on Sea on September 24th 2007, Hill Head on June 13th 2011 and Weston Shore on July 27th 2018. The last two birds were both taken into care but subsequently died. Records spanned the period between April 28th and December 1st with a clear peak in September.

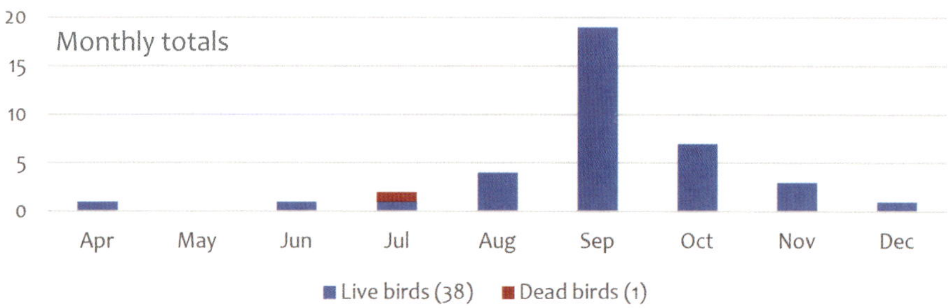

Individual records are tabulated opposite.

Sooty Shearwater, Hill Head, June 13th 2011. Bob Marchant

	Date(s)	Circumstances	Location	Observer(s)
1	1974: July 22nd	1, long dead	Warsash	Tim Lawman and M. Ayres
2	1994: October 2nd	1 offshore, left west	Pennington Marshes	Jacquie Hull and Steve Keen
3	1996: November 3rd	1 offshore 0800–1315 hrs	Hurst Beach	Mike & Pete Combridge, Mike Duffy, Chris Parr, Phil Toye and Ed Wiseman
4	2000: August 26th	1 offshore, left west	Hurst Beach	Marc Moody and Martin Cooper
5	2000: September 5th	1 offshore, left east	Hurst Beach	Marc Moody
6	2000: October 7th	1 offshore	Milford on Sea	Steve Keen
7	2002: April 28th	1 offshore	Hurst Beach	Marc Moody, Tim Parminter and Martin Cooper
8	2002: October 20th	1 offshore	Hayling Bay	Tim Lawman
9	2004: September 18th	1 offshore	Hurst Beach	Tim Parminter
10	2005: October 21st	1 flew east	Stokes Bay	Trevor Codlin

11	2007: September 19th	1 flew east	Hurst Beach	Marc Moody and Colin Allen
12	2007: September 24th	1 offshore	Barton on Sea	Bob Chapman
13	2007: December 1st	1 flew south-west	Sandy Point and Beach-lands	Andy Johnson and Tim Lawman
14	2008: August 13th	1 flew east	Hayling Bay	Tim Lawman
15	2008: September 10th	1 flew west	Sandy Point	Andy Johnson
16–17	2009: September 2nd	2 flew west	Sandy Point	Andy Johnson
18	2009: October 6th	1 east at 0944 hrs and presumed same at 1000 hrs	Milford on Sea	Nigel Jones, Simon Ingram and Steve Piggott
19	2009: October 6th	1 offshore at 1520 hrs	Sandy Point	Andy Johnson
20–24	2010: September 10th	5 flew west	Milford on Sea	Marc Moody
25	2010: September 14th	1 flew east	Milford on Sea	Steve Piggott
26	2010: October 26th	1 flew north-west	Hurst Castle	Marc Moody
27	2010: November 11th	1	Chichester Harbour Entrance	Andy Johnson
28	2011: June 13th	1 found exhausted on tideline, later died	Hill Head	Bob Marchant
29–31	2011: September 11th	3 flew south-east	Hurst Beach	Marc Moody
32	2011: September 12th	1 flew west	Hurst Beach	Graham Barrett and Joe Stockwell
33	2012: August 25th	1 offshore	Milford on Sea	Tim Parminter, Simon Ingram and Colin Allen
34	2012: August 29th	1 offshore	Milford on Sea	Steve Piggott and Colin Allen

35	2012: September 23rd	1 offshore	Hurst Castle	Marc Moody
36	2014: October 26th	1 offshore	Hurst Castle	Marc Moody
37	2016: September 24th	1 flew west	Sandy Point	Andy Johnson
38	2018: July 27th	1 in distress, later picked up and subsequently died	Southampton Water off Weston Shore	Mark Painter
39	2018: November 29th	1 flew east	Milford on Sea	Steve Keen

GB status: regular passage migrant in variable numbers off all coasts. (Breeds on islands in southern oceans; migrates northwards in Pacific and Atlantic Oceans.)

Balearic Shearwater *Puffinus mauretanicus*

HOSRP

A scarce passage migrant. One in 1859 and around 268 recorded since 1971.

The first for the county was one shot by Col. Townsend near The Shingles in Christchurch Bay in August 1859. According to Edward Hart, who mounted the specimen, it was an immature male. It is held at the Horniman Museum, London in a case with a Manx Shearwater obtained on September 15th 1873.

The first modern record was of one seen by Ed and Wendy Wiseman as it flew west off Oxey and Pennington Marshes in calm, sunny conditions on July 21st 1971. Extracts from Ed's description are reproduced below.

> *The bird flew to within 40 yards of our boat moored just west of Oxey Creek. If flew on to the west, just off the saltings, before flying out into the middle of the Solent and on towards Hurst Narrows. A medium-sized shearwater, a little larger than nearby Black-headed Gulls and with longer wings. Not flying in typical shearwater fashion but flapping along just above the water in the calm conditions. The upperparts were distinctly brownish, paler underneath, shading to very pale whitish brown on the belly. There were pale areas on the underside of the wings. The bill was long and horn-coloured.*

The species remained rare in the next two decades with the only records of singles seen off Hurst Beach by June Irvine on August 2nd 1981 and by Dave Wooldridge and Mike Terry on May 11th 1982. There was a gradual increase in the 1990s with five singles off Hurst between 1993 and 1999 and others off Hill Head on August 3rd 1994, Sandy Point from May 28th to 30th 1996 and Pennington Marshes on September 18th 1998 and October 3rd 1999. There was a large influx into the English Channel in August 2001 with at least 41 and possibly as many as 80 of this critically endangered seabird recorded in Hampshire. At least ten were recorded annually from 2006 to 2011 followed by further notable influxes of 41 in 2012 and 28 in 2013. In recent years, numbers have declined with a maximum of 12 in 2016 and none in 2018. Annual totals shown below.

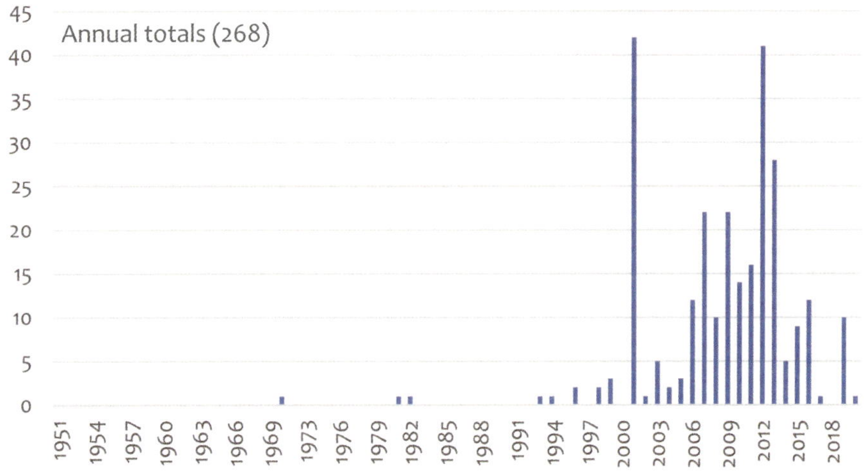

Analysis of arrival dates shows a scattering of spring records, a marked passage between July and September with a peak in August and a few records in October.

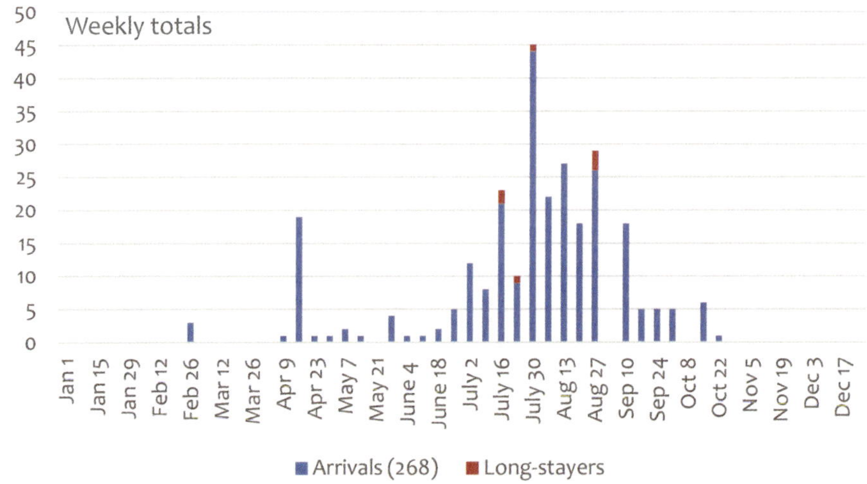

Most were seen from viewpoints between Hurst Castle and Milford on Sea with approximately 220 recorded. Totals of 16 off Barton on Sea and 15 off the sea wall between Oxey Marsh and Keyhaven included some also seen off Hurst. A total of 35 was seen off Sandy Point and in Hayling Bay but the only ones seen in the mid-Solent were the Hill Head bird mentioned above and one which flew east there on April 25th 2012. The most were seen on August 2nd 2001, when there three off Sandy Point at 1400 hrs and seven sightings off Hurst between 0917 and 1601 hrs which totalled 49 with a maximum flock size of 11, a flock of 11 off Barton on Sea at 1300 hrs, and three east and four offshore at Keyhaven at 1740 hrs. The latter were almost certainly the same as the group of eight which went east at Hurst at 1601 hrs as one was seen to come back. The total present off Hurst and Barton that day was assessed as 22 (Wynn 2003). Other high counts were of eight west off Hurst on September 10th 2010, 11 off Milford on Sea between 0721 and 0918 hrs on August 15th 2012, seven off Milford on August 29th 2012, 11 distantly off Sandy Point on August 31st 2013, 12 flying east off Sandy Point on April 21st 2016 and six seen from Milford on Sea just inside the Needles on October 21st 2013.

Balearic Shearwaters: Hill Head, August 3rd 1994 (above) and April 25th 2012

The earliest ever were three off Milford on Sea on March 3rd 2009. This was part of a small influx into the English Channel between early January and mid-March which included at least 20 recorded off Portland Bill, Dorset. Otherwise, the first was one off Milford on April 12th 2015. The latest was off

Milford on October 6th 2009 apart from the group of six mentioned above and a later bird off Milford on October 23rd 2013.

GB status: mostly late summer visitor off SW coasts, c. 4,500 records on Birdguides, 2000-2020. (Breeds W. Mediterranean Sea, disperses to E Mediterranean and off coasts of NW Europe.)

Black Stork *Ciconia nigra*

BBRC

A rare vagrant. 46 records involving around 16 individuals since 1970.

The first for the county was found by Mark Angliss when he flushed it from a stream just to the east of Godshill at 1500 hrs on May 12th 1970. Extracts from his description are reproduced below.

> *It flew up from a stream in front of me and I had a good view through 8.5 x 44 binoculars at 75–100 yards range. I noted large size, long neck and bill, long, broad wings with fingering of primaries, bill, legs and feet were bright red, the feet projecting beyond the tail, completely black above, black neck and head, white body and undersides of wings. Wingbeats slow and deep but mostly soaring.*
> *I watched it for five minutes as it flew off to the north very high.*

What is assumed to be the same individual was relocated by John Oakshatt at Ober Water on May 17th and seen again by Keith R. Grant and John Barker on the evening of 18th and by at least four more observers the next day.

Subsequently, a further 15 have been recorded.

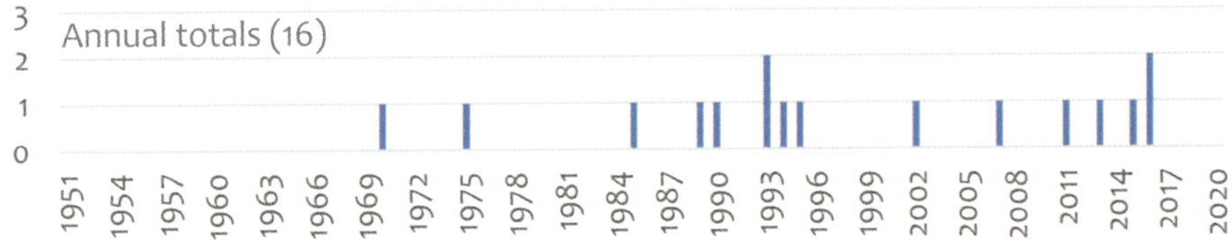

The next was one seen by Kevin Kearns from Farlington Marshes as it lifted off from Baker's Island, Langstone Harbour at 1615 hrs on August 22nd 1975. It soared away to the north and then east over North Hayling and then about 45 minutes later it reappeared from the south and again soared off to the north. The third was seen by R. Dobbs circling overhead at Plastow Green near Kingsclere at 1100 hrs on June 9th 1985 before moving off high to the north-east.

One evidently summered in the New Forest in 1989. It was first seen by Colin and Jenni Tubbs in flight over Burley on May 20th. It was then seen by Granville Pictor over Pound Bottom from within Wiltshire on May 26th. This was presumably the bird at Luccombe Down, IOW on 27th and 28th and reportedly seen over the Island from Hill Head the next day. What is assumed to be the same bird was seen again by Colin and Jenni distantly over Lyndhurst on July 29th and in the Burley area on August 12th and 13th. The Wiltshire and IOW records were accepted by BBRC as the same bird and so although the Hampshire sightings were not submitted, they are included here. In the following year one was first seen by Mike Mockler in water meadows at Kingston between 1700 and 1930 hrs on April 7th before flying off south down the Avon Valley. George Green recalled being invited to Mike's house overlooking the meadows and watching the stork from his lounge window while he watched the FA Cup semi-final between Oldham Athletic and Manchester United which ended in a 3-3 draw after extra time. Mike saw it the next day in the same area when it again flew off south at 1800 hrs. The next sightings were made by Dave Burges who saw it circling over Castle Hill, Woodgreen in the early afternoon on both April 11th and 13th. Two days later Mike and Pete Combridge saw it flying north over Ringwood. It was then seen in flight by Geoffrey Field over Kingston Great Common on April 22nd and by George Green flying west over Ibsley on May 2nd. Several sightings followed from various locations in the New Forest on seven dates between May 5th and June 17th with a further unconfirmed sighting on July 11th.

Black Stork, New Forest, April 15th and 16th 2011. Richard Ford and Ian Williamson overleaf

Mike Wearing was driving north along the A325 between Bordon and Farnham at 0910 hrs on July 2nd 1993 when he saw a large bird soaring above the road which he thought was a stork. He was able to pull over and confirm its identity as a Black Stork. It appeared to drop down in the vicinity of Straits Inclosure, Alice Holt Forest but Mike could not find it. Further enquiries by Mike revealed that it had been found by Laurie Whittle at Straits Inclosure pond the day before. He had returned with Christine Rose the next morning and found the stork perched on a dead tree alongside two Grey Herons. They aged it as first-summer. Mike looked for the bird again in the afternoon of July 2nd but it was not seen again. Later that year another was seen soaring high to the south-east over Yew Tree Bottom, New Forest by Iris and Daniel Powell and Peter Palmer on August 25th. This was presumably the juvenile at Brading Marsh, IOW from August 26th to September 4th.

One was tracked along the south coast on September 17th 1994. It was first reported over Littlehampton Marina, West Sussex at 0720 hrs and then watched as it circled slowly westwards over Pagham Harbour between 0930 and 1000 hrs. In Hampshire it was reported by John Sands over Langstone Harbour at 1110 hrs, Les Williams over Portsdown Hill at 1140 hrs and Brian Cox over Fareham Creek at 1200 hrs. Although the Hampshire sightings were not submitted to BBRC this record is included in the county total based on its acceptance for Sussex. The next was an adult seen by James Andrews as it circled over Tadley Common between 1450 and 1500 hrs on April 15th 1995 before soaring off to the south-west.

The first this century was seen by John Switzer flying east over his garden in the Test Valley at Whitchurch at 1700 hrs on May 17th 2002. Another was seen by Damien Offer flying north over Steep Marsh at 1330 hrs on July 12th 2007.

In 2011 one was first reported by an unknown observer near Fritham, New Forest on April 10th and by D. and J. Hudson just north of Brockenhurst on 11th. It was seen by Nick Hull thermalling over Acres Down at 1012 hrs on April 15th before heading west where it was relocated in Slufters Inclosure by Keith Betton and Martin Cooper. It was then seen by many birders in several localities in the New Forest up to April 22nd as far west as Ibsley Common and, finally, in the east at Denny Wood.

The next was seen on successive days in May 2013. On 18th Rob Souter saw one soaring to the north-east over Acres Down and the next day it was seen by Dan Hoare over the Itchen Navigation at Hockley Viaduct and later by Tim Ball circling between Butser Hill and Oxenbourne Down before moving off SSW. On July 30th 2015 three were seen flying in off the

sea at Southsea by C. and G. Hall. One which they photographed proved to be a Black Stork; although they thought the others were the same BBRC accepted one only in the absence of photographic evidence. Two were seen in 2016; one was noticed by Graham Stephenson and the author as it flew west along the line of the A30 near Yateley Heath Wood at 1035 hrs on July 16th and another was seen by Brian Goddard over Peak Lane, Stubbington at 1245 hrs on September 6th before thermalling off to the north-east.

Analysis of records shows arrival dates between April 7th and September 17th. The chart below shows mini-peaks in spring with three between April 7th and 15th and four between May 12th and 20th.

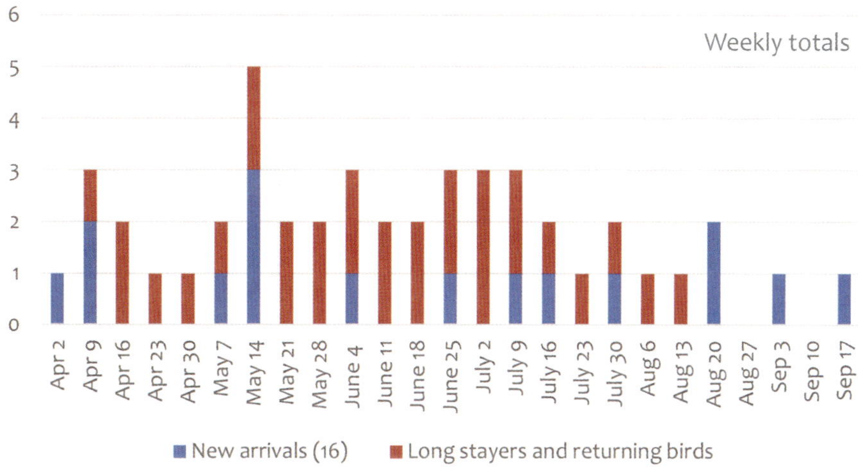

GB status: 22, 1814–1949; 260, 1950–2020. About seven per year; stable or increasing. (Breeds C Iberia & E France through C Europe to Russia & small numbers into N Greece & Turkey. To E, breeds widely in forested temperate regions of Russia & Siberia to Russian Far East. Most are migratory, wintering Africa, S & SE Asia.)

White Stork *Ciconia ciconia*

BBRC up to 1982: HOSRP 1983 to date

A rare vagrant. Five old records and 51 records of 56 birds since 1952.

The first mention of a White Stork in the county involves one killed in 1808 by the gamekeeper of John Guitton of Little Park near Wickham (Yarrell 1843). The next involves a flock of six flying north over Laverstoke on April 23rd 1884 (K & M) which were then seen later that day flying north-east three miles south of Newbury by two independent observers (*The Field*, May 3rd 1884; K & M). Further sightings were of one at Bedhampton in August 1877 and two at Hayling in September 1887, both recorded by W. T. Pearce (K & M). The only record for the first half of the 20th century is somewhat bizarre. It was reported in *British Birds* (25: 362) and is reproduced on the next page.

During the summer of 1931, a White Stork appeared at Southwick Village, and was apparently fired at, as it had a broken leg and a tipped wing. After it had been a week about the fields, a man killed the bird to put it out of its misery, it being very emaciated. Thinking it was a Heron he tied it up in a tree as a scarecrow, where it remained all the winter till March 5th, 1932, when my friend, Mr. J. W. Stares, saw it, and asked for the body, which he brought to me. It may, of course, have been an escape from captivity, but there is no evidence of this. GREGORY M. MATHEWS

The first in the modern era was notified by John Ash in 1952, who said 'Mr. Lush, a farmer near Damerham, reported on July 8th that a White Stork had followed his plough one day during the previous week. His description tallied with this species'.

The first fully documented sighting was in 1960, when Michael Bryant saw one flying north and then west over Cosham at 1900 hrs on May 28th. Extracts from his description to BBRC are reproduced below.

Seen in flight only at a range of c. 150 yards flying over houses and gardens along the Portsmouth to Chichester road at a height of about 150 feet. I had no time to use binoculars but had a very good view with bright sunlight behind me. It was a very large bird, larger than a common heron, with very long neck and head held drooped at an angle below the line of flight, and legs held extended. Wings long and broad, the tips of the primaries separated like fingers. Plumage white except for black primaries and secondaries and part of lower mantle, forming black wing tips and continuous black line across rear edge of wing and back. Beak longer and stouter than herons, tapering, red. Legs pink. Shortly after flying over the road, it turned west and I lost sight of it in the bright evening sun. I have seen this species abroad.

The next was one seen by Michael Dann in a low tree and later photographed in a ploughed field at Martyr Worthy on April 12th 1975 which was still present the next day. In the following year, one was watched by Bob Dunn, Ed Wiseman and Mike and Pete Combridge as it circled over Keyhaven saltings for five minutes before soaring off inland on July 18th and another was seen by Barry Duffin and two other observers as it circled low over Titchfield Haven for several minutes before moving off to the south-west on September 18th. A mid-winter record of one seen by David Billett and Wilf Nelson flying west over Farlington Marshes in freezing conditions on January 1st 1979, later that day over Sway and the next day over Holmsley was not submitted to BBRC. However, an accepted record of one flying west over Goring, West Sussex on December 31st 1978 clearly relates to the same bird and so it can safely be added to the county total (Rogers *et al.* 1979).

HOSRP assumed responsibility for this species from 1983. Subsequently, there were a further 14 records up to the end of the 20th century, and 31 in the new millennium, as shown opposite.

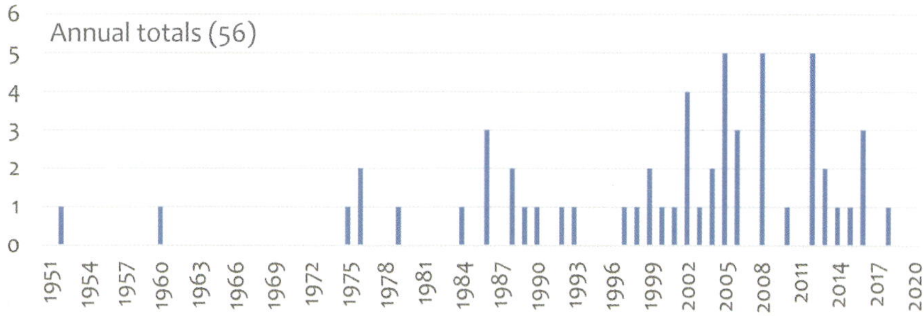

The increased frequency in recent years coincides with the recovery of some populations and reintroduction schemes in other parts of Europe. The ongoing reintroduction scheme centred on Knepp in West Sussex has resulted in several Hampshire sightings of birds thought to have originated from there. Since 2017, records have been accepted only when it is certain they are not colour-ringed birds from Knepp.

White Stork, Hedge End, June 25th 2014. Alan Lewis

Most records were of singles flying over, and in a few cases, what was presumed to be the same individual was seen in different areas over the course of several days or weeks. Only two records have involved more than one bird. A flock of five was seen circling over West Wittering, West Sussex from Sandy Point on May 2nd 2012 and the same group were over Havant on May 4th. Six had been in Oxfordshire in late April but unfortunately one died when it flew into power lines on May 1st. Further records of two over Eldon on May 8th, three over Up Somborne on May 17th and one over South Warnborough on May 19th probably referred to the same birds. The following year two flew NNE over Riverside Park, Southampton on June 2nd. Other series of records presumed to refer to the same individuals include one in the Sopley area from September 2nd to 7th 2002 which was seen over Lepe on 8th and Oliver's Battery, Winchester on 11th, one over Locks Heath on March 21st 2006 and over Needs Ore and Portsmouth Harbour on March 29th, and another seen at various localities in the Avon Valley between July 19th and August 7th 2008 and over Stoney Cross, New Forest on August 14th. Most recently one with a blue colour ring of Dutch origin was seen at various sites in the south-east of the county between April 24th and 30th 2018. It was raised in the wild in Freisland in 2015 having been descended from birds released in the Netherlands in the 1980s and 1990s. At that time, the wild population there had declined to 19 pairs but there are now over 800 pairs.

Areas with the most records are the New Forest (11), Titchfield Haven (8), the Avon Valley (7) and Needs Ore and the Beaulieu Estuary (4).

Apart from the mid-winter record, the earliest is of one at Bickerley Common on March 17th and 18th 2003 which was also seen at Needs Ore on the latter date. The latest was one over Lower Exbury on November 26th 2005. The weekly distribution of records is shown below.

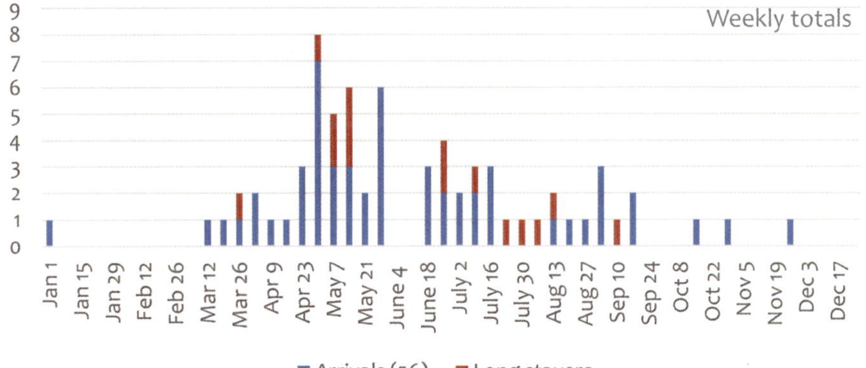

A record of one at Eling Great Marsh on October 26th 1971 published in *BoH* has been confirmed as erroneous. One seen by Rod Ship flying east over Regent's Park, Southampton on September 17th 1978 was harshly considered not proven by BBRC. Another record of one seen by Ron Huxford over Horndean on October 13th 1978 has not been submitted.

GB status: 193, 1416–1949; 1,143, 1950–2015. (Breeds Iberia, C, E and SE Europe, W Asia, N and S Africa; northern populations winter in Mediterranean Basin and sub-Saharan Africa).

Glossy Ibis *Plegadis falcinellus*

BBRC up to 2012; HOSRP 2013 to date

A very scarce visitor. Two old and two modern 20th century records. Increasingly frequent this century with around 60 recorded.

The first for the county was shot by one of Sir H. Mildmay's gamekeepers at Tundry Pond on September 15th 1881. It was exhibited by Mr. Sclater to the Zoological Society of London and was reported in the society's proceedings (1881: 827). The second record was of one killed near Romsey in October 1902 and reported by George Corbin to K & M. Corbin was a reliable observer from Ringwood and there seems no reason to doubt this record.

The first modern record was of one found by Frank Pierce at Balmer Lawn near Brockenhurst on May 24th 1965. It was subsequently seen by many observers and photographed by Norman Orr and remained in the area until at least August 4th. This was possibly the bird which wintered in Devon and Cornwall in 1964/65 and was at Pagham Harbour, West Sussex on April 22nd and 23rd 1965.

Since then, a further 61 have been recorded

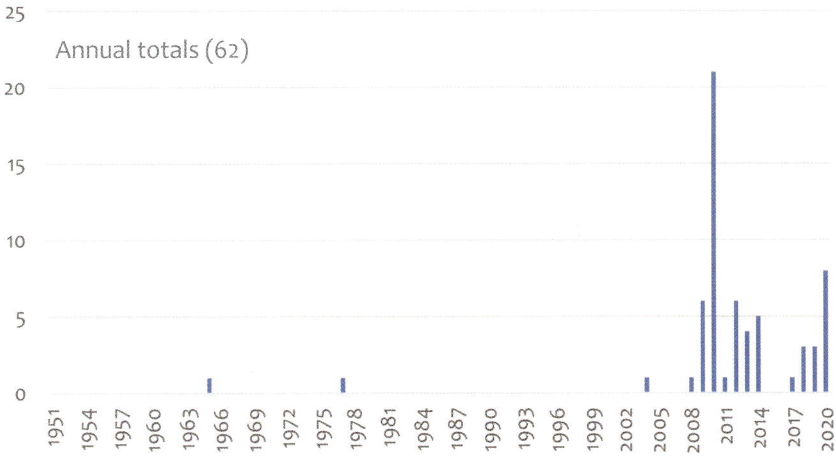

The second was one seen by Keith Douglas as it flew in from the south-west and landed on flooded meadows near Crofton House at Titchfield Haven at 1030 hrs on June 6th 1977.

After a long gap of 27 years, one was found by Ian Morgan-Jones and Christine Emm at Bickerley Common, Ringwood on April 29th 2004. It was present from 1240 to 1600 hrs. This was the individual which arrived at Bowling Green Marsh, Devon as a juvenile in September 2002 and was last seen there on the morning of April 29th. It was further up the Avon Valley at Charlton All Saints, Wiltshire on May 1st and 2nd, then at Otmoor, Oxfordshire from May 2nd to June 20th, before relocating to East Anglia where it remained until June 2005.

A presumed first-summer was found by Tim Parminter at Fishtail Lagoon, Pennington Marshes on May 17th 2008 where it was still present on 19th and six juveniles, also found by Tim, flew east over Butts Lagoon, Pennington Marshes at 1645 hrs on September 19th 2009. This flock was first seen on the Axe Estuary, Devon earlier that day, then at Radipole Lake and Littlesea, Dorset before traversing the Solent; possibly it split as one was at Pagham Harbour until September 28th and five were at Dungeness, Kent from September 21st to October 4th. This movement coincided with an influx into Britain and Ireland during September and October which totalled at least 60 birds. Many of these had been colour-ringed as nestlings in the Coto Doñana National Park, Spain and the Camargue, France in May and June 2009.

A further influx into Britain in autumn 2010 produced several records in Hampshire. One found by David Thelwell at Pennington Marshes on August 14th was still there the next day and had been present at Stanpit Marsh, Dorset on 12th and 14th. This was followed by the astonishing sight of a flock of 19 found by Simon S. King, Phil Toye and Jason Claxton which arrived from the north-east over Pennington Marshes at 1010 hrs on September 13th; they dropped down and circled low over the marsh before leaving to the east after five minutes. The same flock had been seen earlier in the day at Beer Head, Devon and later at Dungeness, Kent. Next was another single found by Nigel Elliott at Ibsley Water on September 18th which then moved to Stanpit where it remained until October 10th, although it seems likely that one seen by Colin Allen flying east over Keyhaven and later at Titchfield Haven by Tony Heath on September 26th was the same individual.

Glossy Ibis:
Bickerley Common, December 8th 2012.
Pennington Marshes, May 13th 2014 and Sturt Pond, February 28th 2018. Lee Fuller and Gareth Rees

Glossy Ibis, Warblington, March 2013. Peter Milinets-Raby

Only one was seen in 2011 – near Lymington on October 17th and at Titchfield Haven from October 21st to 23rd. In 2012 around six were recorded. One left west from the Hamble Estuary on February 11th and two were at Keyhaven on April 7th both of which left north-east separately; presumably one of these was in flight over the Beaulieu Estuary later that day and again at adjacent Needs Ore the next morning before departing. Another flew in over Hurst Beach at 1040 hrs on May 4th and paused briefly at Mount Lake before settling at Lower Pennington Lane floods where it remained until the next day. It had been at Stanpit earlier on 4th. Presumably, the same individual was at Needs Ore on May 13th and 14th, Pennington Marshes on May 27th, Farlington Marshes on 28th, Lower Pennington Lane on June 2nd, Farlington again from June 5th to 16th, Keyhaven Lagoon on 20th and Farlington on July 17th and August 11th. One at Keyhaven Marshes on October 23rd and 24th may have been that on the Avon floods at Bickerley Common from December 2nd until January 15th and then again on January 26th. The following day it was seen moving north over Ibsley Water and presumably the same individual flew into Needs Ore at 1150 hrs where it remained until dusk. A further four were seen later in 2013. One was at Castle Farm, Warblington from February 23rd to March 24th. Another at Fishtail Lagoon, Pennington Marshes on October 15th left east at 1100 hrs and was presumably the same bird seen over Milton Common, Langstone Harbour at 1330 hrs that day. Finally, two flew east over Fishtail Lagoon on November 25th.

Probably five were seen in 2014. One was at Farlington Marshes on January 21st. There were no further sightings until two were found on Butts Lagoon, Pennington Marshes at dawn on May 13th. Later that day, one arrived from the south-east over Titchfield Haven, gained height and left high to the north-east at 1330 hrs. An hour later two, presumed to be the Pennington birds, were on the Bridge Street floods north of Titchfield Haven and departed south at 1530 hrs. One of these (a female) was colour-ringed as a pullus on July 24th 2013 in the Coto Doñana, Spain. The two birds were in the Avon Valley at Cowards Marsh, Dorset from May 6th to 12th and again from 15th to 26th with one remaining in that area until July 21st. Finally, one was at Farlington Marshes from June 14th to 16th.

There were no further records until 2017 when four sightings in the second half of November are assumed to refer to the same bird. Singles left high west from Farlington Marshes at 1030 hrs on 16th, flew south with Cormorants over Avon Tyrrell at 0730 hrs on November 19th, flew south at Avon Causeway in the evening on November 21st and landed in reeds at Avon Floods, Keyhaven at 1100 hrs on November 26th. The following year there were singles at Sturt Pond, Milford on Sea on February 27th and 28th, Testbourne Lake on September 4th and Needs Ore on December 9th which later that day was seen flying west over Pennington Marshes. In 2019 one was at Sturt Pond on February 25th and another was seen flying over Posbrook Floods and later Mill Rythe, Hayling Island on the same day; presumably a third was resident at Fishlake Meadows from March 24th to April 29th before moving to Titchfield Haven on May 1st and 2nd. At least eight occurred in 2020. One was at Posbrook Floods from May 14th to 16th, another flew south over Overton on September 19th and was probably the same bird which flew east over Sparkes Marina, Chichester Harbour the next day, five flew over Pennington Marshes on October 20th and finally one flew over Keyhaven on November 3rd. The latter is no doubt the bird which arrived at Christchurch Harbour, Dorset the next day and remained into 2021.

The monthly distribution of records is shown below.

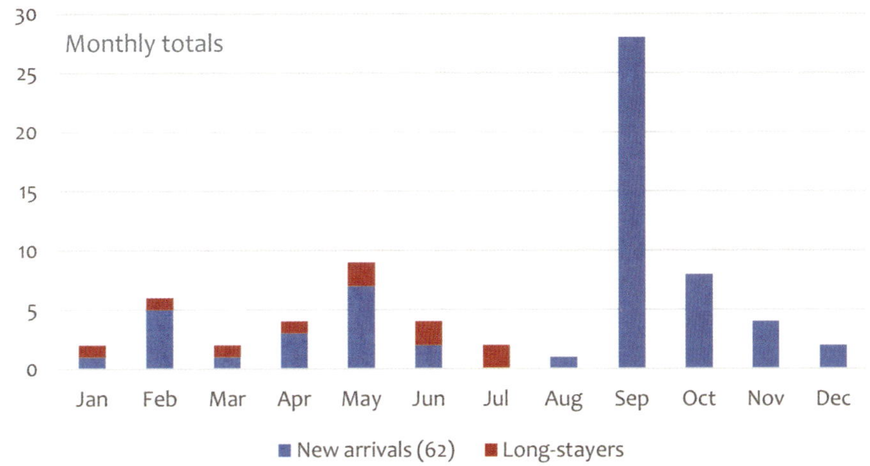

GB status: 193, 1416–1949; 919, 1950–2019. Increasing, mean 76 per year in 2010–19. (Breeds S France & Spain; otherwise, European breeding range centred N & W of Black Sea in Ukraine & Romania, with small, declining population in Balkans. To E, breeds from Volga River to Kazakhstan. Migratory, most wintering E Africa, but W European population winters Morocco & Mediterranean basin. Resident or dispersive populations occur Africa, S Asia, Australia, E USA & Caribbean to N South America.)

American Bittern *Botaurus lentiginosus*

BBRC

A very rare vagrant. Two records in 1876.

The first for the county was shot at Woodside, Lymington in January 1876. The specimen was in the Hart collection and is now held by the Hampshire Cultural Trust at Chilcomb House, Winchester.

Details from Edward Hart's notes give more information:

Shot by I. Bran near Lymington, January 1876, and presented to me by Francis Edwards, Woodside, Lymington. Whilst shot in January 1876 it was not given to me until July 1876, and in the meantime had been mounted on a board by someone in Lymington. The actual place where it was shot was on the Keyhaven Marshes, near the boat house which was used by Col. Hawker, the great punt shooter. Mr Edwards' house overlooked these marshes and was close by.

© Hampshire Cultural Trust

A month later another was shot at Cadnam by Mr. G. R. Nunn of Lyndhurst (*The Field* March 10th 1877). K & M stated that the specimen was sold with Mr. Bidwell's collection in 1904.

GB status: 26, 1804–1949; 11, 1950–2019. (Widespread across Canada, mostly S of 60°N, & throughout USA except southern states. Northern breeders migrate to winter mostly S of 40°N in USA & into C America.)

Little Bittern *Ixobrychus minutus*

BBRC

A rare vagrant. Two records in 1851 and 1869 and 12 of 13 birds since 1953.

The first for the county was shot at Hilsea in 1851. It passed into the collection of P. W. West of Cosham Lodge (*The Naturalist*, Vol. VI. p. 93, January 24th 1856). The second was shot in Bickton Meadows near Fordingbridge on April 15th 1869; it was purchased and stuffed by Mr. Hayden (*The Field*, May 1st 1869).

The first in the modern era was an immature found by Canning Suffern at Titchfield Haven on August 9th 1953. Extracts from his notes are reproduced below.

I watched an immature Little Bittern basking in sunshine at 0800 hrs from the harbour bridge. It was in a little recess in the reeds on the opposite bank of the river. It was hunched up, with neck withdrawn, and looking rather quail-like. It was pale brown, with longitudinal darker stripes above and below, but the stripes above were darker, wider and closer together than those below. There were no transverse stripes or bars. Startled by a nearby Mallard, the Little Bittern instantaneously threw itself into the 'bitternoid' posture of alarm with body drawn up vertically, neck extended and bill pointing skywards. Later it resumed its hunched posture and a Moorhen passed close to it appearing only slightly larger than the bittern.

In 1954 an adult male was flushed at Titchfield Haven by Bill Truckle, Canning Suffern and R. Parsons on June 13th and 14th and a female or immature was found dead on the Eastern Road at Portsmouth by A. F. McQuarrie on August 12th or 13th. Graham Rees confirmed its identity from a coloured sketch drawn by G. Kinns. Also in that year, an adult male was found at Andover and released at Leckford by Terry Jones on an unknown date.

The next was one found by Nigel Wyatt at Titchfield Haven on May 21st 1960; later that day two were seen. Extracts from Canning Suffern's notes submitted to BBRC are reproduced below.

On May 21, 1960, at Titchfield Haven an eleven-year-old boy, Nigel Wyatt, watching for our Great Reed Warbler (which is still present), saw (but did not identify) a Little Bittern adult. He reported to me as follows: 'Heron-shaped bird not much bigger than a thrush hovered over water and then flew to reeds, on which it perched, afterwards standing in reeds – light buff underparts, slate grey upperparts, with conspicuous whitish patch on each wing'.

I identified this at once as a Little Bittern and it has since been seen and identified by a whole host of birdwatchers. Later that morning we saw simultaneously two Little Bitterns, and I thought one was distinctly duller than the other and thought they might be a pair. I got excellent views of one of them and noted its red bill and legs coloured like the dead reeds on which it perched.

Two were still there the next day and one was still present on May 26th.

The next was an adult female found at Winchester Water Meadows by John MacKinnon on May 7th 1964 and subsequently seen by Thomas Sparrow, Robert Cheke and other students from Winchester College. The fifth to be recorded at Titchfield Haven was found by John Cantelo and seen by John Barker and George Preddle on September 3rd 1966. The next was an adult male first seen and identified by David Peart at Titchfield Haven on June 9th 1968 and seen by others including Canning Suffern and Martin Terry. The second for the Langstone Harbour area was an adult male seen by Keith R. Grant and John Barker during a ringing session at Farlington Marshes on May 2nd 1970; it stayed until 6th.

A further eleven years elapsed before the next occurrence, a rather elusive adult male flushed by Barry Duffin at Titchfield Haven at 1755 hrs on May

18th 1981 and seen again briefly by Barry and Tim Lawman at 2055 hrs. The following day it afforded flight views for 40 seconds to six observers but was not seen again. A lack of sleep for the finder led to the discovery of the next bird. Graham Stephenson noted in his description to BBRC that his planned dawn visit to the Hampshire coast was thwarted by his two-year old insomniac twin daughters, so having overslept he arrived at Fleet Pond at 0850 hrs on August 11th 1996 and promptly saw a Little Bittern fly across the pond. Although elusive early in its stay, it remained until August 19th and was seen by hundreds of observers. The final record was of an adult male found by Chris Tabor by his garden pond in Barton on Sea on the evening of May 26th 2003.

Little Bittern, Barton on Sea, May 26th 2003. Chris Tabor

Two labelled Fawley 1845 are on display in the Science School at Winchester College; these were not mentioned by K & M or included in the catalogue of the collection (Fearon 1900). Since there are also two Night-herons and two Squacco Herons also labelled Fawley 1845 the provenance of all must be considered suspect.

GB status: 254, 1773–1949; 281, 1950–2020. About 3–4 per year, occasional breeding; stable. (Widespread but local in Europe N to 53°N. To E, breeds to 60°N in European Russia, & E to Kazakhstan & NW China. Migratory, wintering mainly E Africa, S from Sudan & Ethiopia. Other populations largely resident or dispersive in sub-Saharan Africa & Madagascar.)

Night-heron *Nycticorax nycticorax*

BBRC up to 2001; HOSRP 2002 to date

A rare vagrant. Four old records and 16 since 1961.

The first for the county was shot at Ringwood on July 22nd 1868. It is included in a *List of the Rarer Birds obtained by Mr. Hart in the vicinity of Christchurch, Hants* contributed by J. H. Gurney to *The Zoologist* 1869: 1510–1512. The specimen was part of the Hart collection and is held at Leicester Museum. Another was shot by the River Avon at Ringwood on August 4th 1880 (*The Field,* August 21st 1880). This was not included by K & M and subsequent authors until the reference was discovered by Naylor (1996). K & M also reported one which had been killed at Avon Castle in 1901 and exhibited by Mr. Turner at Bournemouth in that year. Another at New Milton from mid-May 1911 was reported to R. E. Coles who saw it on July 14th 1911. In his description he aged it as an adult (*Brit. Birds* 5: 113).

The first modern record involved an adult seen by Peter Britton near Lyndhurst on the evening of June 14th 1961. Extracts from his description to BBRC are reproduced below.

> *After waiting for some time by a small copse in the hope of seeing a Long-eared Owl, which I had previously heard, an adult Night Heron flew at height of c. 20 ft. from one of the outside trees. I observed it in fairly good light for about a minute at c. 30 yards. It was noticeably smaller than a Grey Heron. It was a compact looking bird with short, broad, very pale grey, rounded wings. It was 'short-bodied' and its head appeared to almost merge with its body, and the black of the crown and mantle coincided to form a large area of black, contrasting sharply with the pale wings. The bill was rather short and thick, reminding me of a Bittern. The legs protruded slightly beyond the short square grey tail. The underparts were white.*

Since then, there have been 15 further records involving 11 adults, one first-summer and three first-winters.

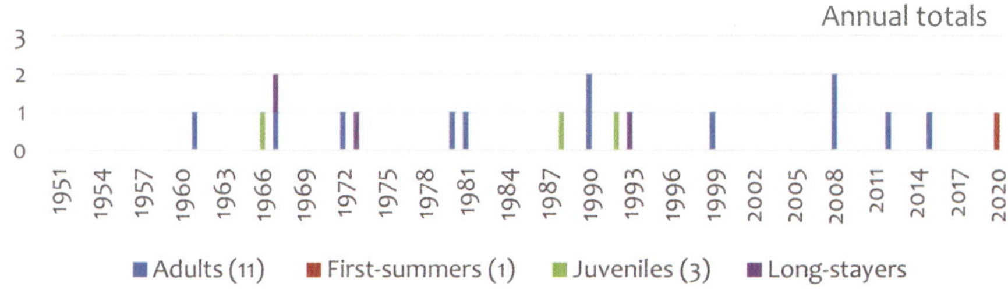

The next record is now assumed to relate to two birds. A juvenile was found by Norman Pullen and Mrs. E. Miller in woodland on the Beaulieu Manor Estate on December 8th 1966 and last seen on March 27th 1967. Another sighting of an adult seen by A. W. G. Smeed at Calmore on June 20th 1967 was included by BBRC as the same bird on the basis of a postcard from Pullen which said that the Cadland (*sic*) record could refer to the same bird as it is only four miles from Beaulieu. But Calmore is north of Totton and given the near three-month gap in sightings it can safely be regarded as a different individual.

The next was an adult which wintered at Sowley Pond between December 1972 and February 1973. It was included in the 1973 *HBR* and in *BoH* with the caveat that it had not been submitted to BBRC. During research for this book notes were discovered for a sighting by A. P. Clark for January 27th 1973 which have now been accepted by BBRC. It was seen by other observers on five dates between December 11th and February 15th.

Night-heron, juvenile, Danes Stream, November 1988.'Graack' and there it was flying over the road like a prehistoric beast, 'graak' and there it was gone into the dusk,

The fifth for the county was an adult flushed by Philip Jones from the bushes area at Farlington Marshes on June 9th 1980 which was watched in flight for several minutes before landing in the reedbed. Another adult was flushed by Peter Milinets-Raby from a small pond near Titchfield Abbey early on May 4th 1981. The first long-staying bird was a juvenile at Milford on Sea from October 30th to November 16th 1988 which was regularly seen flying from its roost site along the Danes Stream at dusk. This record has only recently been accepted by BBRC based on notes submitted by the author.

There was an influx of over 30 into southern England in early 1990 (Rogers *et al.* 1991) which produced at least two in Hampshire. An adult first spotted by John Eyre and seen by the author on floods at Hucklesbrook Water Meadows in the Avon Valley on February 24th remained in the area until at least March 25th. Another adult was flushed by Richard Levett from riverside willows while he was carrying out a Common Bird Census at Upper Titchfield Haven on May 14th. It was later seen by Barry Duffin and has only recently been accepted by BBRC.

Night-heron, juvenile, Woodmill, River Itchen, November 17th 1992. Dennis Bright

Following up on several unconfirmed reports, Dave Unsworth discovered a juvenile roosting in a willow tree by a fishpond near Swaythling Church,

Southampton on November 15th 1992. It remained until at least December 8th. A juvenile found by Richard Mould-Ryan at nearby Itchen Valley Country Park on January 7th 1993 where it was last seen on February 10th is assumed to be the same individual. Later in the 1990s the only record was of an adult flushed by Lee Marshall in the Alver Valley at Gosport at 1130 hrs on April 11th 1999 and seen in flight by Dan Philpott at dusk.

The next were not until 2008 when two were recorded. Kevin Crisp saw and heard an adult flying over his house in North End, Portsmouth at 1810 hrs on April 22nd and Andy Johnson saw an adult fly over Sparkes Marina, Chichester Harbour and continue eastwards until a distant speck at 0850 hrs on September 27th. An adult was found by Brian Baker roosting in a tree by a small pond at Lower Pennington on June 2nd 2012. It was there until 4th, from 12th to 21st, on July 19th and from August 3rd to 19th. Steve Seagrave saw an adult fly west calling over his garden in Dibden Purlieu at 1415 hrs on October 19th 2015. Most recently Joost Brandsma located a first-summer roosting high in a tree along Veal's Lane between Hythe and Marchwood at 0650 hrs on April 19th 2020. He was able to watch it at fairly close range for ten minutes but on his return later in the day it was nowhere to be found.

Analysis of arrival dates shows that there has been records in every month although new arrivals have been confined to February, April–June and September–December. All three juveniles arrived in October–December. Only five birds have stayed for longer than a day.

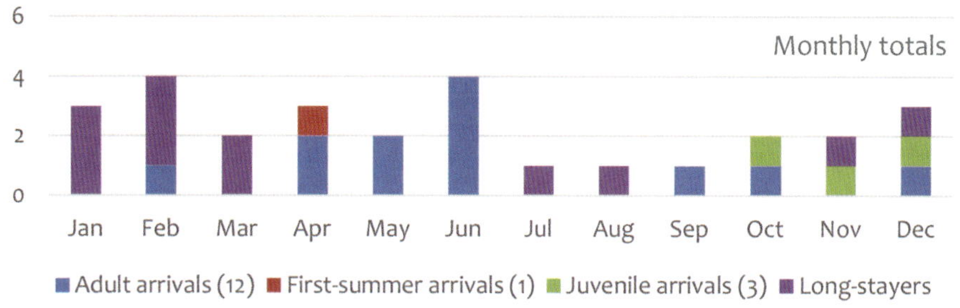

Canning Suffern claimed one at Titchfield Haven on June 27th and 28th 1952. The record was not accepted at the time due to insufficient detail recorded. He watched it for half an hour on the second date. There is a record in the HOS files of an adult at Sowley Pond from February 4th to 20th 1972 (included in HBR but excluded from BoH) which is interesting in view of the bird which was there in the following winter. A third bird recorded in the influx in spring 1990 was a sub-adult seen in flight by Charles Cuthbert at Arlebury Lakes on March 25th. This was recently submitted to BBRC and harshly judged not proven.

Two labelled Fawley 1845 are on display at Winchester College; these were not mentioned by K & M or included in the catalogue of the collection (Fearon 1900).

Night-heron, adult, Lower Pennington Lane, June 20th 2012. Ted Pressey

Since there are also two Little Bitterns and two Squacco Herons also labelled Fawley 1845 the provenance of all must be considered suspect.

GB status: 159, 1782–1949; 690, 1950–2019. No clear trend, mean 14 per year in 2010–19. (Widespread but local throughout Mediterranean basin & N to France, Belgium & occasionally Netherlands. Elsewhere, breeds Morocco, Ukraine, Turkey & throughout Indian subcontinent & SE Asia N to Japan, & locally throughout sub-Saharan Africa. Widespread N & S America.)

Squacco Heron *Ardeola ralloides*

BBRC

A very rare vagrant. Four records since 1982.

The first for the county was recorded on the evening of June 11th 1982. James Venner, the site manager at Needs Ore, was driving slowly past Warren Farm when he observed a bird flying close to his vehicle at a range of no more than 15 metres. Though late in the day the light was good and he noted the colours. He was unfamiliar with the species but was able, without hesitation, to identify it as a Squacco Heron from his field guide. The bird was searched for the next day without success. However, W. Loughborough, who lived at nearby Gins Farmhouse, phoned James to inform him that on June 12th he had observed a Squacco Heron at close range feeding in the fishpond in his garden; he had not been aware of the sighting the previous day. Although included in *BoH* the record was submitted to and accepted by BBRC only as recently as 2016, thanks to the efforts of Ed Wiseman to compile the necessary details.

Squacco Heron, adult, Titchfield Haven, August 27th 1996

The next was one seen briefly twice by Pete Potts during a ringing session at Titchfield Haven on July 7th 1994. Fortunately, it was seen at length during the evening by many observers including Dan Houghton, who provided a detailed description. It remained elusive during the next two days and was seen by Dan and Rosemary Powell circling over their garden around 1700 hrs on July 11th before leaving east. Extracts from Dan Houghton's description are reproduced below.

Having heard that Pete Potts had seen a Squacco Heron earlier in the day, I decided to visit the Haven in the evening where I joined ten or so locals gathered on the seafront. I first saw the bird in flight over the reedbed for a few seconds at 1937 hrs at 500 m. The next sighting was at 1945 hrs when it was seen in flight for a longer period. At 2013 hrs it was located perched on top of a duck trap. The reserve was opened enabling closer views down to 200 m from one of the hides. The face, upper breast and upperparts were a yellowy-brown colour. This colour was most intense on the mantle and had a purplish tint to it. The crown had white feathers with black edges; three of these formed long head plumes which extended on to the mantle. The lores and bill from its base to halfway along was a blue-grey colour and the other half was blackish with a well-defined diagonal demarcation between the two colours. The iris was yellow and the legs a washed-out orange colour. It gave a hunched, squat impression when perched due to its relatively short legs and short neck which it occasionally stretched. It remained on the duck trap for over an hour until the light faded, spending its time preening and standing almost motionless.

Squacco Heron, adult, Posbrook Floods, Titchfield Haven, May 26th 2019. Alan Lewis and Lee Fuller

The third record was also of an adult at Titchfield Haven, found by Peter Marston at 1120 hrs on August 27th 1996. He subsequently informed the author by telephone, who released the news. In the meantime, it was found independently by Michael Clark, a visiting birder from Denbigh, at 1315 hrs. It was seen by many observers during the afternoon and was last seen at 1940 hrs when it flew off south towards the Isle of Wight.

The most recent occurrence was again in the Titchfield Haven area. Mark Francis found one on Posbrook Floods at 1310 hrs on May 24th 2019. It was later seen by Andy Lester flying south and dropping into a reedbed at Titchfield Haven. It was seen by many observers during the next two days. What was assumed to be, the same individual was seen by Ian Pibworth and Kurt Love at Testwood Lakes on June 2nd. It was then refound by Ken Martin at Posbrook Floods at 0713 hrs on June 8th where it remained on view until the evening of the next day. This is presumably the same bird that was found at Pagham Harbour, West Sussex at dusk on June 11th but was not seen again.

Two labelled Fawley 1845 are on display at Winchester College; these were not mentioned by K & M or included in the catalogue of the collection (Fearon 1900). Since there are also two Little Bitterns and two Night-Herons also labelled Fawley 1845 the provenance of all must be considered suspect.

GB status: 64, 1775–1949; 113, 1950–2020. About three per year; stable or increasing. (Breeds Mediterranean basin from S Spain to Black Sea & E to Kazakhstan, with large population in Danube Delta. Northern populations migratory, wintering N tropical Africa. African population largely resident.)

Chinese Pond Heron *Ardeola bacchus*

BBRC

A very rare vagrant. One record in 2004.

Chinese Pond Heron, East Dean, November 13th 2004. Jen Churchill

The only record for Hampshire is of an adult in breeding plumage seen and photographed at close quarters by a garden pond at East Dean on November 13th 2004 by Rob and Jen Churchill. It flew off east after 90 minutes and was not seen again. BBRC considered there was a high probability that it was not of wild origin and it was included in the escapes section of the 2004 *HBR*. In 2016 BOURC admitted the species into category A of the British List based on a record from Hythe, Kent in January 2014. This led to a review of the 2004 record which shortly afterwards was also accepted into category A. It was assumed to be the same bird that was at Eccles on Sea, Norfolk on October 31st 2004. Fuller details are given in Cox (2017) and McInerny & Stoddart (2017).

GB status: 2, 2004 and 2014. (Breeds China from Liaoning and Nei Mongol provinces S to Guangdong, Guangxi, Hainan and Yunnan, S Japan, Taiwan, W to Assam and Andaman Is., India, and S into N Burma and N Vietnam. Winters S China and SE Asia S to Singapore and Borneo.)

Cattle Egret *Bubulcus ibis*

BBRC up to 2008; HOSRP 2009 to date

Formerly a rare vagrant with five records between 1986 and 2001.
An increasingly frequent visitor since 2006; bred for the first time in 2019.

The first for the county was seen briefly by Pete Potts and Barry Duffin as it flew down river at Titchfield Haven while they were ringing Bearded Tits on November 6th 1986. They later relocated it at roost in a clump of sallows. It flew off north the next morning and was found feeding with sheep in a field at Bridge Street, Titchfield, where the farmer, Stephen Caws, reported that he had first seen it on 4th. It continued to feed in this area and later with cattle in a field near Hook Lane; it roosted within Titchfield Haven each evening until it was last seen on January 11th 1987. It seems likely that it succumbed to the very cold weather at the time.

Cattle Egret, Bridge Street floods, Titchfield Haven and Hook Lane, November 1986.
photos Barry Duffin

The second was seen by Chris Patrick in flight over Farlington Marshes early on May 22nd 1992. It was relocated by Ivan Lakin with cattle on the eastern side of the Marshes at 0740 hrs and proved to be in near full summer plumage. It left unseen after 15 minutes. The next was seen by Roger Casalis de Pury as it came into roost at Snails Lake, Blashford on successive evenings on July 14th and 15th 1994. Another was seen by Ron King at Winchester Sewage Farm on the evening of July 15th 1998 but left unseen and was not relocated. The first for the new century was found by Marc Moody as it flew into roost with Little Egrets at Sowley Pond on August 11th 2001. It was seen again on the next two evenings and then was found feeding during the daytime at Needs Ore before transferring to Normandy Marshes where it remained from August 15th to 19th. It was probably the same bird seen in Dorset at Lodmoor on July 26th and in Poole Harbour on July 30th.

No more were recorded until 2006, since when numbers have increased considerably. Between 2006 and 2011, a total of 21 was recorded, mostly in mid-winter, with several making extended stays. One was at Warblington from January 9th to March 3rd 2006. Another was at North End Farm, Harbridge from December 27th 2007 to March 27th 2008. Two were in the Beaulieu area from January 30th to April 11th 2008 with one remaining until May 2nd.

One was in the Pennington Marshes area from August 19th 2009 and was joined by a second on October 1st, both remaining until December 16th. Two were discovered at Ashford, near Fordingbridge on December 31st 2010; they were still there on New Year's Day and then again between February 16th and March 2nd 2011. Three appeared in August 2011. One, first seen at The Kench, Langstone Harbour on August 2nd 2011 moved to Thorney Island, West Sussex before returning to Warblington on December 18th where it remained until February 11th 2012. Others were with cattle in a field at Ladycroft near Alresford on August 8th and 9th 2011 and at Avon Causeway from August 20th to 23rd.

None was seen in 2012 or 2013 and only one in 2014 but subsequently numbers have increased exponentially with at least 250 recorded up to 2020. These included some substantial flocks. A group of 22 at Coward's Marsh, Dorset on August 30th 2015 departed up the Avon Valley at 1755 hrs and were seen flying north over Harbridge at 1820 hrs. One arrived at Warblington on December 7th 2016 with numbers rising to five by 30th and eight on January 2nd and 3rd and the last seen on January 24th. Eight were at Winnall Moors on September 1st 2018 and presumably the same group roosted at Alresford Pond on September 4th with seven on 7th and diminishing numbers until the last was there on 23rd. An unprecedented flock of 56 was feeding with cattle in a field at Manor Farm, Michelmersh on October 24th which left north after they were spooked by a Goshawk; no gatherings of this size were see anywhere else in southern England at that time. Between seven and nine were feeding with cattle in water meadows near Squabb Wood, Romsey between November 27th and December 12th 2018. Up to five were regularly seen at Warblington between February 5th and March 24th 2019 and subsequently breeding was confirmed at nearby Langstone Mill Pond when four pairs raised a total of 11 young (Milinets-Raby 2020). The breeding colony at Langstone Mill Pond expanded in 2020 with eight pairs raising at least 18 young. Several double-figure counts occurred between Emsworth and Titchfield Haven in July and August 2020 including 20 in two groups flying over Emsworth towards Langstone at dusk on July 22nd and 28 doing likewise on August 17th. The next day, 28 including seven juveniles, were at nearby Thorney Island, West Sussex and on August 23rd there were 12 juveniles there.

Cattle Egrets: left, Needs Ore, August 27th 2020 and top with Little Egrets, Langstone Mill Pond, August 27th 2020.
Ian Williamson and Richard Ford

GB status: 3, 1805–1949; 966, 1962–2016. 10 breeding pairs in 2017, increasing. (Expanding N from Iberia, widespread resident through much of Africa, S USA, Mexico, C & S America.)

Purple Heron *Ardea purpurea*

BBRC up to 1982; HOSRP 1983 to date

A rare vagrant. One record in 1875 and 40 since 1964.

The first for the county was an adult male procured by Mr. F. Stevens at Ashley Farm near Stockbridge on June 30th 1875 (K & M). The specimen was part of the Hart collection and is held at Leicester Museum (L.Z252.1982.742.0). The next was not until 1964, when an adult was found by Peter Colston feeding in a creek off Farlington Marshes on August 3rd. It was watched at length later that day by Bryan Renyard. Extracts from Bryan's description are reproduced below.

Watched intermittently from 1530 to 1845 hrs. General appearance dark in contrast to white appearance of nearby Grey Heron which was larger and with a heavier build. Upperparts had a brown-black sheen. Bill yellowish. Throat white. Most noticeable feature snake-like neck with sharp kink and squared grebe-like shape to back of head, an appearance enhanced by two distinct medium length plumes.

Subsequently, a further 39 have been recorded in 26 years.

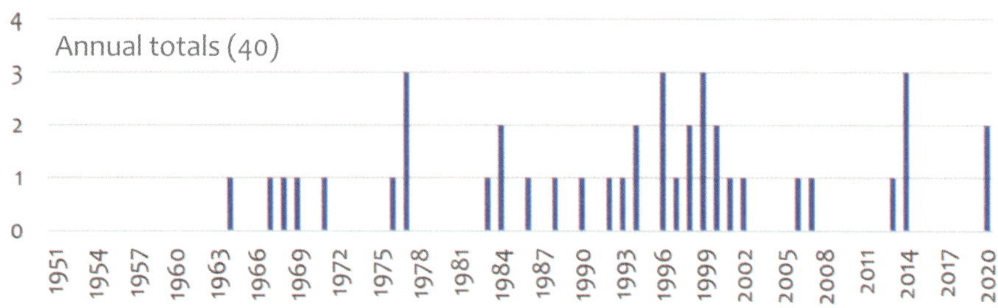

The second in the modern era was a first-summer seen by Michael Bryant when it was disturbed by trespassers in the reedbed at Farlington Marshes on 1100 hrs May 28th 1967. It remained in the area all day and was seen by several observers before leaving high to the north-west in the evening. The next was a first-summer found by Ed Wiseman, accompanied by George Green and Diana Jones (later Westerhoff), as it flew east over Keyhaven on May 19th 1968. An adult was reported by wildfowlers at Cherque Gravel Pit, Gosport in late September 1969 and its presence was confirmed by Richard Carpenter on October 5th. It was later seen by Martin Terry and stayed until October 15th. The third for Langstone Harbour was seen at close range by J. Gallagher at Farlington Marshes at 1120 hrs on August 31st 1971; it flew off to the north over Portsdown Hill. A first-summer was seen flying south-west over Hampton Ridge, New Forest by Gerald Smart on May 1st 1976. Another first-summer was found by John Cantelo, Tim Norriss and David Thelwell at Marsh Court on May 7th 1977. They watched the bird in flight and later saw it walking across a meadow towards the lake. Later that year Nigel Kendall found a juvenile at Sowley Pond on August 29th; presumably the same bird was seen nearby in the New Forest at Bishop's Dyke by Ed and Wendy Wiseman and Diana Westerhoff on September 1st and Tim Norriss on 3rd but the later sightings were not submitted to BBRC. Another juvenile was found by Ron Hunt at the Beaulieu Estuary on October 14th 1977; it was still present two days later when he showed it to Salisbury NHS members including David Paradise.

Purple Herons:
juvenile, Titchfield Haven, August 14th 1994 (above) and adult, The Wildgrounds, Gosport, April 22nd 2020 (right).
Fred Gibson

HOSRP assumed responsibility for the species in 1983. Since then there have been a further 31 records although only 12 have been in the 21st century. Over half have come from coastal localities although most were seen at Fleet Pond, where there were five occurrences between 1998 and 2000. All were seen on one date only apart from a first-summer at Farlington Marshes on April 14th and 17th 2002.

See below for details of all records since 1983.

	Date(s)	Circumstances	Location	Observer(s)
10	1983: May 4th	Sub-adult, left west	Titchfield Haven	Peter Milinets-Raby
11	1984: May 10th	Adult flew off towards IOW	Keyhaven	Bob Dunn
12	1984: May 30th	Sub-adult soaring	Liss	Robin Griffiths
13	1986: August 21st	First-winter	Lower Pennington	Ed Wiseman
14	1988: April 25th	Adult	Needs Ore	Rosemary Cook
15	1990: April 9th	First-summer in flight	Curbridge	Trevor Codlin
16	1992: May 5th	Unaged, flying south	Lower Test Marshes	Mike and Pete Combridge
17	1993: May 4th	Adult or first-summer left reedbed	Lower Test Marshes	Pete Combridge
18	1994: June 11th	One soared off to south	Acres Down	Ron Huxford and Simon Ingram
19	1994: August 14th	First-winter arrived off sea and later seen on scrapes	Titchfield Haven	Ken Martin and Dan Powell
20	1996: April 16th	First-summer left SE	Sinah GP	Tim Lawman *et al.*
21	1996: June 2nd	Sub-adult departed south	Winchester SF	John Cloyne
22	1996: June 16th	Arrived from west and dropped down to south	Romsey/Broadlands Estate	Mike Rafter
23	1997: May 29th	Adult flew north	Lower Test Marshes	Simon S. King
24	1998: May 24th	Adult flew in off sea and moved east	Beaulieu Estuary	John Taverner
25	1998: June 27th	Second-summer/ adult in flight	Fleet Pond	Mick and Will Scott
26	1999: April 26th	Adult flew in off sea	Hayling Bay	Tim Lawman
27	1999: May 8th	One flew ESE	Fleet Pond	Graham Stephenson
28	1999: May 24th	Adult SE over observers' garden	Westover Road, Fleet	Mick, Will and Thomas Scott
29	2000: May 15th	Sub-adult left high to south	Fleet Pond	Mick Scott
30	2000: July 26th	Juvenile left WSW	Fleet Pond	Mick and Will Scott *et al.*
31	2001: May 24th	One unaged seen briefly flying east 0540 hrs	Farlington Marshes	Bob Chapman

32	2002: April 14th and 17th	First-summer	Farlington Marshes	Jason Crook and Bob Chapman
33	2006: May 9th	Adult flew west	Normandy Marsh	Marcus Ward
34	2007: May 2nd	Adult	Titchfield Haven	Barry Duffin
35	2013: April 26th	One flew north	Portsdown Hill	Mike Wearing
36	2014: April 6th	First-summer flew up in front of train	East Dean	Marc Moody
37	2014: June 3rd	First-summer left high east	Farlington Marshes	Tim Doran
38	2014: July 4th	First-summer	Ivy Lake, Blashford	G. Doré
39	2020: April 22nd and 23rd	Adult	Wildgrounds, Gosport	Paul Thurston
40	2020: May 31st	Adult	Farlington Marshes	Jason Crook *et al.*

Purple Herons: first-summer, Ivy Lake, Blashford, July 4th 2014 and adult, Farlington Marshes, May 31st 2020. Geoff Doré and Ian Williamson

As shown on the next page, arrival dates spanned the period between April 6th and October 14th as shown below. Most were in spring with a clear peak between mid-April and late May. Early records involved one seen by Marc Moody from his train driver's cab as it flew up from the River Dun Water Meadows between East Dean and West Dean on April 6th 2014 and a first-winter found by Trevor Codlin at Curbridge on April 9th 1990. None was seen later than September 3rd apart from the 1969 Cherque and 1977 Beaulieu Estuary birds already mentioned.

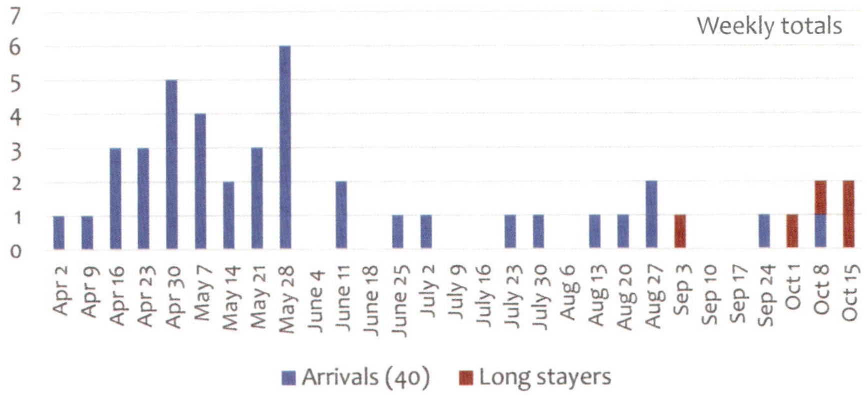

GB status: 89, 1722–1949; 1,158, 1950–2019. No clear trend, mean 26 per year in 2010–19. (Breeds throughout S Europe from S Spain locally N to Netherlands & E to S Russia, Iran, & C Asia to Kazakhstan. Winters sub-Saharan Africa. Other resident races in S & SE Asia.)

Great White Egret *Ardea alba*

BBRC up to 2005; HOSRP 2006–2018

Formerly a rare vagrant with nine records between 1985 and 2000. An increasingly frequent visitor recorded annually since 2003.

The first for the county was one seen by Barry Collins flying from Thorney Island, West Sussex towards Hayling Island on June 15th 1985. The first grounded bird was found by an unknown observer at Blackwater, Needs Ore on July 17th 1992. It was subsequently seen by many birders during the day and photographed by George Green. The author heard about it while at work in West London and was able to set off for Needs Ore rapidly. He saw it on the De L'Orne pool and made the following notes:

> *It was as large as, or slightly larger than, the nearby Grey Heron. It was in summer plumage, with the long dagger-shaped bill being largely black with a yellow base to the upper and lower mandibles. The yellow gape flange extended back behind the eye, thus eliminating the possibility of confusion with Intermediate Egret. The legs were a pale yellow colour and the plumage was, of course, unrelieved white. The bird was feeding at close range but was not seen to fly.*

The next was seen by A. J. and G. C. Tompsett flying north-west near the Old Mill, Damerham at 0815 hrs on January 6th 1996. Rogers *et al.* (1997, 1998) considered this the same individual that was at Bough Beech Reservoir, Kent from January 5th to 14th but given that both were seen early on January 6th this is clearly impossible. However, the Kent bird occurred at various sites in Sussex including Pagham Harbour up to March 31st before moving to Hampshire. Ron Huxford saw it fly out of the reedbed at Farlington Marshes

White egrets:
Cattle, Warblington, January 26th 2018;
Little, Normandy Marsh, March 2nd 2020
and Great White, Edenbrook Country Park, Fleet, July 2nd 2020.
John Hilton, Rob Porter and John Clark

at 1105 hrs on April 7th. He watched it with Bob Chapman and Chris Cockburn feeding on the mud flats before it flew to one of the islands in the harbour, disappearing at 1230 hrs and not being seen again. The third of the year was found by John Lewis at Sadler's Mill, Romsey at noon on December 31st. It was refound by Tony Blunden further up the Test Valley at Horsebridge on February 15th where it was still present two days later. However, these sightings were not submitted to BBRC.

Great White Egret, adult with Little Egret, 11 Acre Scrape, Titchfield Haven, June 10th 2007

Great White Egret, Testwood Lakes, November 25th 2017. John Hilton

One new bird occurred in each of the next four years. A presumed adult was seen in flight by Graham Giddens over the Lymington River at 1945 hrs on August 19th 1997 while he was ringing roosting Swallows. It came in high from the south-east, circled, and then left to the south-west. Another was seen by Mike Rafter as it flew south with a Little Egret over the De L'Orne hide at Needs Ore at 1100 hrs on January 9th 1998. An adult in summer plumage was found by Kevin Crisp on North Binness Island in Langstone Harbour at 1555 hrs on August 13th 1999. It later moved to Farlington Marshes then returned to North Binness before leaving high to the north-east at 1955 hrs. Another was seen by Jason Crook as it flew east over Farlington Marshes at 0853 hrs on May 1st 2000.

No more were recorded until 2003 when there was a series of sightings from July onwards which involved a minimum of six individuals. The most significant of these was a colour-ringed bird found by Lawrie Chappell and Roger Howell at Mockbeggar Lake, Blashford on August 21st which had been ringed as a nestling at Lac de Grand-Lieu, Loire-Atlantique, France on May 4th 2003. This bird, subsequently bestowed the name of Walter, has returned to the Blashford Lakes area for the next 16 winters and last seen in early 2020. It normally arrived in July or August (earliest July 3rd in 2006) and departed in late January or February (latest March 3rd in 2012). Subsequently, sightings of this species increased exponentially so that the overall number of individuals became impossible to judge. Annual totals for 1951–2016 are shown on page 318.

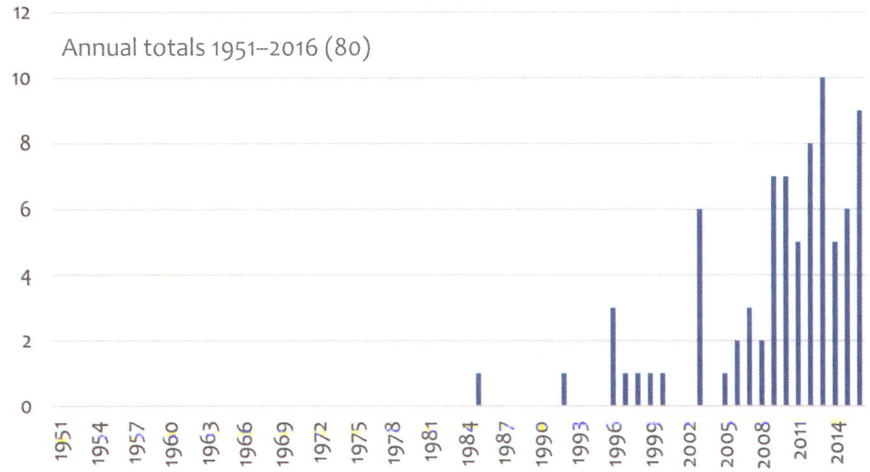

Analysis of arrival dates shows a clear peak in autumn which suggests post-breeding dispersal from elsewhere and a lesser peak in mid-winter which may be connected with cold weather movements.

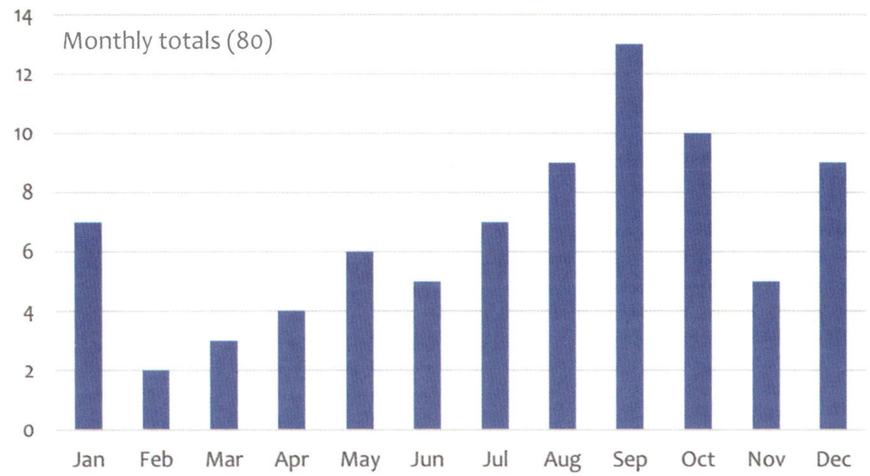

Most sightings were of singles or occasionally two together. The first larger group was of three flying west with three Little Egrets over Sandy Point on November 26th 2016. In 2017 there was a further explosion in numbers. Over 600 records were reported to *HOS*. Numbers were highest in the Test Valley. There were four at Testwood Lakes on April 27th and up to five at nearby Fishlake Meadows between May 1st and 10th. A further influx to Testwood Lakes produced nine in late afternoon on October 27th rising to 11 on 29th and 30th, with four or five remaining until December 14th. At Farlington Marshes, five flew west on September 19th, six south-west on September 24th and seven north at 1130 hrs on October 27th which were presumably part of the group of nine later recorded at Testwood Lakes.

GB status: 8, 1821–1949; 1,923, 1951–2015. 8-12 breeding pairs in 2018, increasing. (E and SE Europe spreading to N, W and SW Europe; other populations Africa, Asia, Australia and Americas.)

Little Egret *Egretta garzetta*

BBRC up to 1990; HOSRP 1990–1992

Formerly a rare vagrant with 33 records between 1957 and 1988. Now a moderately common resident and winter visitor which first bred in 1998.

The first for the county was found by John Simons at Farlington Marshes on June 2nd 1957. The extract which follows is from the Langstone Harbour yearbook.

First seen by JS when he flushed it at c. 80 yds and it flew out towards North Binness flats. Towards high tide it was on North Binness Island and watched for 2 hrs or more by DFB, PFLeB, RHD, GHR, BWR and JS. It was eventually disturbed by people and flew around the north end of the harbour being mobbed by godwits and curlews, finally flying off towards Chichester Harbour. It was a small heron, little larger than a curlew, pure white plumage with black bill and legs. Feet not visible on ground and their colour difficult to judge when in flight as against bright blue sky. However, the godwits which were mobbing it clearly had dark feet which would suggest they were light. Neck very long and slender. Wings rounded.

This was probably the bird seen at Pagham Harbour, West Sussex on June 11th and 12th.

Between 1959 and 1988, a further 32 occurred, as shown below.

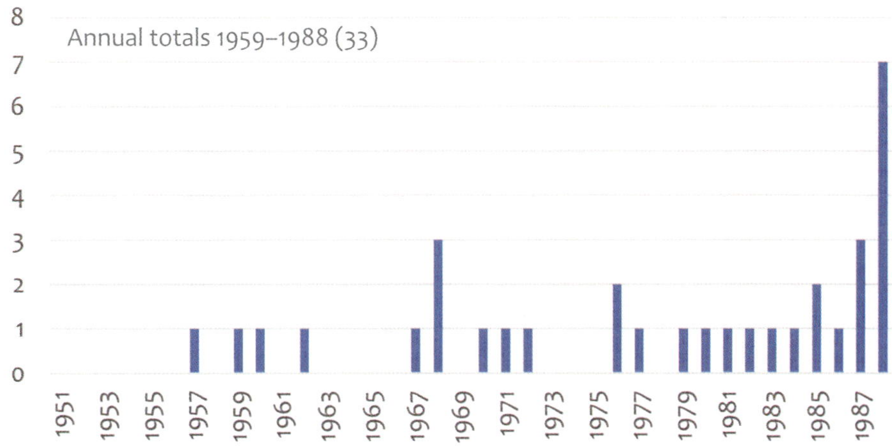

Sightings were made between April 19th and November 18th with a peak in late May and early June.

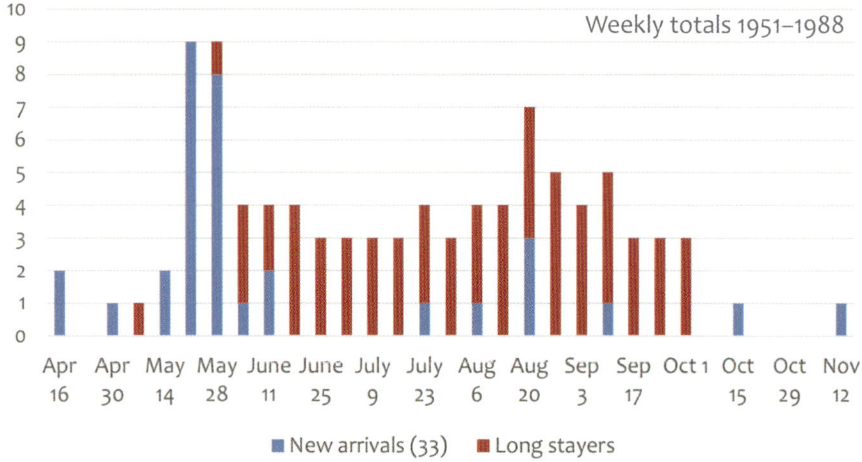

Summering was recorded in 1976 (June 4th–October 3rd), 1984 (June 15th–October 1st) and 1985 (June 11th–October 6th). Most were seen at coastal localities, especially Langstone Harbour, Titchfield Haven and in the north-west Solent area between the Beaulieu Estuary and Hurst but inland singles occurred at Bickton from May 21st to 25th 1962, Bishopstoke on May 18th 1976, Fleet Pond on April 22nd 1982 and Leckford on October 17th 1987. All records were of singles apart from a flock of five which spent the day at Beaulieu Mill Pond on May 26th 1988.

Titchfield Haven, 1984

An influx into southern England from mid-July 1989 involved a total of at least 20 in Hampshire including 17 at Lisle Court, Lymington on August 28th. Around seven remained in the subsequent winter. This pattern recurred in increasing numbers in subsequent years with the first three-figure count of 109 flying east to roost at Thorney Island, West Sussex on December 4th 1995 and the first confirmed breeding in 1998.

GB status: 10, 1826–1949; 702, 1950–1990. Increasingly common, c. 1,500 breeding pairs in 2017. (S and W Europe, S Asia, Africa and Oceania.)

Short-toed Eagle *Circaetus gallicus*

BBRC

A very rare vagrant. One record in 2014.

The only record for Hampshire is of a first-summer found by Simon Colenutt at Bishop's Dyke, New Forest at 0735 hrs on June 8th 2014 and subsequently photographed by Simon and David Cuddon before disappearing at 0955 hrs. This was the bird discovered at Morden Bog, Dorset on May 31st and still present there the next day. On June 10th it was seen in Ashdown Forest, East Sussex but it was back at Bishop's Dyke on June 12th and 13th. Two days later, it had returned to Ashdown Forest where it remained until 29th. It was again in the Bishop's Dyke area from June 30th to July 2nd, when many Hampshire listers were able to catch up with it, and on July 5th. After this, it was seen at both Thursley Common, Surrey and in Ashdown Forest on July 12th and finally reported from Thetford Forest, Norfolk on July 19th and 20th. Extracts from Simon's account of his discovery (Colenutt 2015) are reproduced below.

*At 0600 hrs on June 8th I decided to have a short walk across the heathland at Beaulieu Road Station to Bishop's Dyke. It was a beautiful early summer day, Hobby were hunting between Shatterford Car Park and Denny Wood and Woodlark and Curlew were singing in the distance while Snipe drummed overhead. As I continued south a male Dartford Warbler showed very well singing atop a silver birch sapling while its mate was a little more secretive in the heather nearby. A classic New Forest morning. As I wandered over the Woodfidley Passage I viewed the area of mire, heath, scrub and scattered trees to the west. Two male Cuckoos were very vocal and conspicuous in the silver birch at the transition between the mire and the woodland. As I watched the Cuckoos, their attention switched between chasing one another to chasing a Buzzard which had appeared, at low level, from the woodland and was flying north-east towards the railway line. As I followed the Buzzard and its attendant Cuckoos in my binoculars my attention switched to a large pale bird perched in an oak in the background and as I refocused, there, before me was a massive pale raptor. Immediately my mind rang 'what the **** is that' and at that moment I dropped my scope to the ground to try to get some digi-shots knowing what I was looking at without confirming it conclusively. The bird almost immediately flew and circled the clearing below tree top height and it was evident that it was a Short-toed Eagle. The pale undersides to the wings with very few markings on the underwing coverts with the exception of a string of black 'pearls' on the inner greater coverts and some speckling on the lesser coverts were*

Short-toed Eagle, Bishop's Dyke area, New Forest, June 30th and July 1st 2014. Ian Williamson (left) and Alan Lewis (middle and right)

spot on for the Dorset bird from the week before. The bird landed out of site in a silver birch. I looked around and called the only visible birder, David Cuddon. There was no sign of the eagle from where we were viewing and so I ran, north-eastwards, back down the main path and around the north side of the mire to get the correct angle to see where it had landed. And there it was, the Dorset Short-toed Eagle! I composed myself and took some acceptable digiscoped shots, called David and put the news out on Birdguides. David and I watched the bird for around 40 minutes as it perched in the birch before it took off, circled about 20 m above our heads, literally eyeing us up, and then headed east towards Pig Bush. The first two birders arrived and we managed to put them onto the bird before it became a distant speck. Brilliant, I just wished I had had my SLR, which I had left in the car, when it was circling overhead! I then continued my walk west through the woodlands, mire and heath at Bishop's Dyke but I could hardly focus on the birding at hand such was my excitement in having refound this bird, so after an hour or so I turned and headed back towards the car. I stopped for a scan shortly after crossing back over the Woodfidley Passage and in the distance I could see a hovering raptor, I set up my scope and could see that it was again the Short-toed Eagle hovering over the bog below Pig Bush, it was very distant and so I headed off – elated!

GB status: 4, 1999–2020. (Breeds NW Africa N to N France, & E through C Europe to Estonia & S to Balkans. To E, breeds E through Turkey& Israel to Caucasus Mountains, Iran & India, also European Russia to Urals, S Siberia, C Asia, W China & N Mongolia. Winters Africa S of Sahara, Iran, India & occasionally to Indochina & Malay Peninsula.)

Pallid Harrier *Circus macrourus*

BBRC

A very rare vagrant. One record in 2016.

The only record for Hampshire is of a first-winter found by Bryan Pinchen at Needs Ore while leading a group of visitors around the reserve on November 6th 2016. Nigel Jones takes up the story:

> *I arrived with my wife Sandy at Needs Ore and we were informed that a Pallid Harrier had been found. We walked a short distance to view the foreshore and saw the harrier, on the ground, feeding on the corpse of a Canada Goose which had died a few days previously. We watched as it tore strops of meat from the breast of the goose. After a few minutes, it took flight and quartered the marsh before pitching down into the vegetation. It later reappeared and continued to feed on the goose. The goose was the focus of its attention for the next few days but it began to range over a larger area, hunting over the reedbeds, the salt marsh of the Beaulieu Estuary and straying as far as Park Shore.*

Further details about the identification of the bird and the changing status of the species are found in Jones (2017). Because Needs Ore is a private reserve, news was released only locally. During its stay until November 13th at least 60 birders obtained permits from the Beaulieu Estate Office and were able to view the bird.

Pallid Harrier, first-winter, Needs Ore, November 6th and 7th 2016. Nigel Jones and Lee Fuller

GB status: 2, 1931–1949; 127, 1950–2020. About four per year, influx 2011 (29); recent increase. (Breeds from Ukraine E through Russia to c. 100°E, & S to Kazakhstan & NW China. Occasionally breeds Finland. Migratory, wintering throughout much of E & C Africa, & Indian subcontinent.)

Montagu's Harrier *Circus pygargus*

HOSRP

Formerly a rare summer visitor which last bred in 1997. Now a very scarce passage migrant with around 193 migrants or non-breeders since 1951 including 111 after the cessation of breeding.

Early records are clouded by confusion with Hen Harrier, the two species not being separated until 1802. Gilbert White referred to the blue hawk as a resident species in the Selborne area but it is impossible to know which species he was referring to; in all probability both were involved. He realised that the female was brown but George Montagu questioned this in a letter to him in 1789. William Yarrell later wrongly gave credit for this discovery to George Montagu who had, however, been responsible for realising that there were two distinct species.

The first dated record for the county is of a pair shot at Newton Valence by Mr. Eames in May 1850; K & M noted this from a footnote included in an edition of White's *Natural History of Selborne* published in 1877. K & M suggested that there was a rapid increase between 1860 and 1880 coinciding with a reduction of the 'larger and more powerful Hen Harrier as a breeding species'. They described several instances of successful breeding in the New Forest and Avon Valley in the last three decades of the 19th century but also unfortunately the shooting of a large number by collectors and gamekeepers. They also referred to the north-east of the county where it was said to occur every year and a nest was found at Woolmer Forest in 1878. Nests were found in the south-east of the county in June 1883, which was robbed, and in July 1898, which contained three young. This was possibly on Hayling Island where John Kelso (1912), probably referring to the turn of the century, recorded the species as having nested for several years and been afforded protection. Kelso settled on Hayling in 1909 but by then Montagu's Harriers had ceased to breed there.

It is probable that Montagu's Harriers continued to breed in the New Forest throughout the early 20th century. In the 1930s and 1940s, pairs were found breeding or on territory there in at least ten years. Elsewhere, pairs bred unsuccessfully at Bransbury Common in 1942 and successfully at Liss Forest in 1945.

Between 1950 and 1961 there were reports from six areas in the New Forest, involving at least 22 pairs which were breeding or holding territory with 11 of these known to have reared 35 young. Subsequently, pairs were present in five years up to 1967 but the only one known to have nested (in 1965) was robbed. Between 1968 and 1994 sightings were made in the New Forest in 21 years involving approximately 32 birds. Two to four were present for several weeks in 1979, 1981 and 1985 but breeding was not suspected. However, it was a very pleasant surprise when a pair established a territory at Black Gutter Bottom in 1995 and subsequently bred successfully raising two young. A total of seven was seen on August 19th of that year – the pair and their two young, a first-summer male and two more juveniles from elsewhere. A pair bred in each of the following three years although unfortunately they were only successful in 1997 raising two young; in 1996 the young were predated by a fox and in 1998 the eggs were taken by a Carrion Crow. This happened the day after the pair had been disturbed by the presence of a first-summer male and a second female on May 14th.

Montagu's Harrier, pair, Black Gutter Bottom, New Forest, May 1996

Breeding also occurred in the Damerham/Martin area although persecution was rife. A pair bred successfully in 1952, a female was shot there in 1954, a male was present in 1955 and 1956 and a pair in 1957. Three pairs took up territory in 1958 but unfortunately one pair and a second female were shot. Single pairs were present in the next three years but with no indication of successful breeding. Thereafter, pairs were recorded in 1965, 1966, 1968, 1970, 1977 and 1978. Only those in 1966 and 1978 were known to be successful. In the period up to 2004 birds were seen in every year but six although most sightings involved males which were probably on hunting expeditions from the Sixpenny Handley area in Dorset. Breeding has not been recorded in Dorset since 2004 and subsequently there have been only five spring sightings at Martin Down in 2011–13, 2016 and 2019.

Breeding has also been recorded on the downs to the west of the Test Valley. Pairs bred in 1956 and 1971; the latter had three young in a nest in a grass crop but they died after the female disappeared. Between 1994 and 1997 a pair bred annually raising a total of at least 14 young. Pairs returned in 1998 and 2000 but breeding was not confirmed. Elsewhere in the county, a male or female was seen on several occasions in 1963 hunting over Keyhaven/ Pennington Marshes and carrying food away to the north; in the same year a pair bred successfully in a young conifer plantation in Ringwood Forest and was present in the following year. A pair probably bred in a barley field at Sherborne St. John in 1965.

Records away from breeding areas since 1951 are analysed below. The annual totals in spring and autumn were low in the second half of the 20th century despite the presence of breeding birds in the county. Breeding has not occurred in the 21st century yet numbers, particularly in spring, were initially high but have since shown a downward trend. This presumably reflects the fortunes of the English population which peaked at 17 pairs in 2006 but was only five pairs in 2017 and collapsed to one pair in 2018 following adverse weather on spring migration (RBBP reports, Paul Castle *in litt.*).

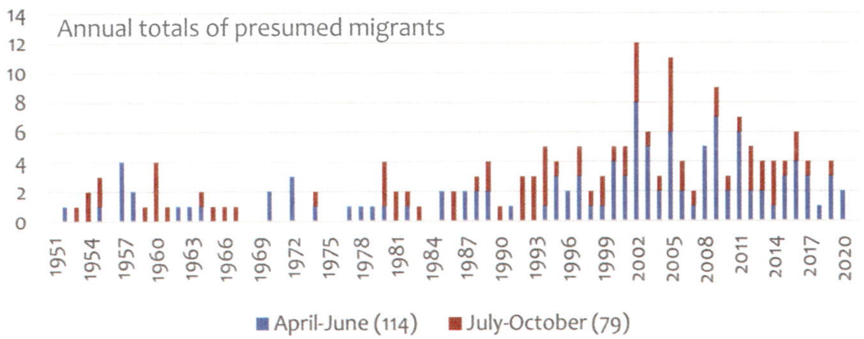

Unsurprisingly coastal locations feature strongly with 28 in the Keyhaven/ Pennington Marshes area, 16 in Langstone Harbour (mostly at Farlington Marshes) and 14 at Needs Ore. These include respectively 12, three and four this century. Other records since 2001 include 26 in the New Forest, 23 on or adjacent to the downs between Cheesefoot Head and Portsdown Hill, 11 in the Test Valley and five on the North Hampshire Downs. All records were of singles on one date only apart from those of a pair which flew high to the north over Hurst Beach on May 2nd 2005 and a first-summer male in the north-west of the New Forest between May 5th and 13th 2008.

The records also include data from the joint satellite tracking project of the Wiltshire Ornithology Society (WOS), the Dutch Montagu's Harrier Foundation, the RSPB and The Sound Approach. Four were recorded in Hampshire air-space although only one was seen by a local birder. A female

'Madge' crossed Hampshire and the Isle of Wight on August 10th 2014 but unfortunately died in Senegal in December 2014; predation by a Verreaux's Eagle Owl *Bubo lacteus* was suspected. A female 'Beatrice' was tracked over Barton on Sea on August 24th 2016; she wintered in Mali, returned to breed successfully in 2017 and again wintered in Mali. Her northbound migration was much delayed by adverse weather in spring 2018 but she was tracked crossing the Hampshire coast on May 27th 2018. She was too late to breed and unfortunately died on her return migration in Mauritania in early October. An adult male seen in the west of the county on April 24th 2017 was confirmed to be 'Mark' who had arrived back in Wiltshire on April 18th and bred successfully from 2014 to 2017. His satellite aerial fell off soon after the Hampshire sighting so it is not known if he returned to his usual wintering area in Senegal. He did not return in 2018. An adult female 'Ruth' roosted overnight near Havant on August 29th 2017 and left across the Solent the next morning. She wintered in Senegal and started her northward migration on March 21st. She made three attempts to cross The Sahara but died in southern Morocco around May 26th. She had originally been ringed as a pullus in Dorset in 2004 and so was 13 years old at the time of her death which is a British longevity record. Many thanks to Paul Castle of *WOS* for supplying this information.

Montagu's Harrier, male, Milkham, New Forest, April 20th 2020. Russell Wynn

As shown opposite, analysis of arrival dates shows the start of spring passage in mid-April with a peak from mid-May until early June. Numbers of males and females were equal at first but females predominated in May. Small numbers including first-summer males continued to be seen through June and July; these were probably wandering non-breeders but there is a remote possibility that a pair was breeding undetected somewhere on the downs. Return passage started in the last week of July and peaked in late August with surprisingly few seen thereafter. Of 79 recorded in autumn, 18 were identified as males and 19 as juveniles with the earliest on July 31st.

The earliest ever in the New Forest were a male at Broomy Plain on April 8th 1979 and a female at Black Gutter Bottom on April 11th 1997 which subsequently bred there. Away from breeding sites, early arrivals involved a male moving east over Itchen Waterworks at 1033 hrs on April 10th 2020 and presumably the same individual continuing east along Portsdown Hill at 1200 hrs the same day, a female in off the sea at Hill Head on April 12th 2016, a female at Burton Common on April 15th 1995 and a male moving east over Warblington on the same date in 2019. There have been only three records after the end of September which involved juveniles at Farlington Marshes on October 2nd 1974, Needs Ore on October 18th 1981 and Micheldever on five dates between October 1st and November 2nd 1960.

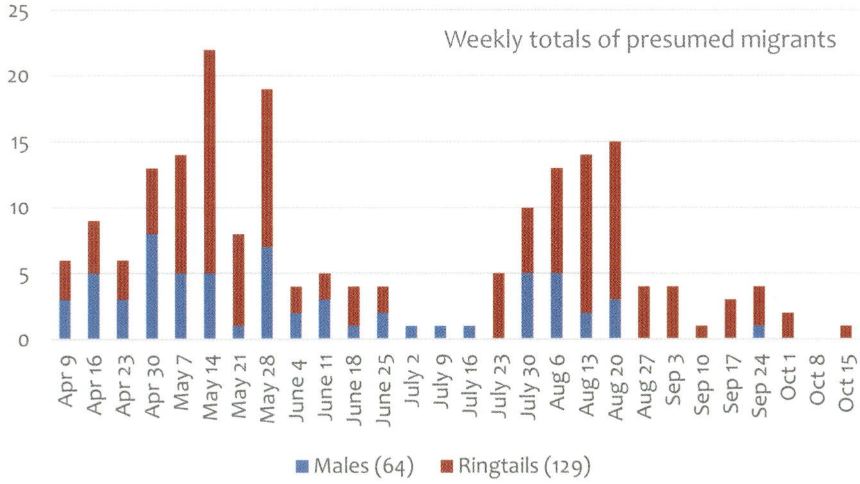

A small number claimed as Montagu's Harrier but accepted only as unidentified harriers by HOSRP have been included above. Initially this was due to possible confusion with Hen Harrier but in recent years the increased frequency of Pallid Harriers in Britain has required observers to consider this species as well.

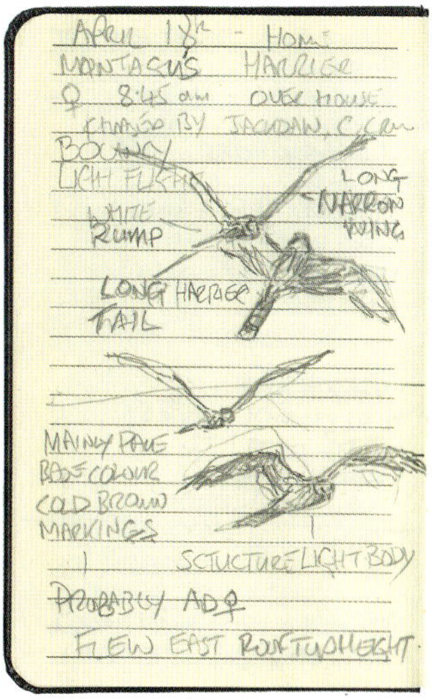

Montagu's Harrier, female, low over our house, Hill Head. Pallid not ruled out! April 18th 2019

There are two records of the rare dark morph. John Ash reported a second-summer or adult female which was shot at Damerham on July 19th 1954. Russell Wynn watched a first-summer female from 1026 to 1030 hrs as it migrated over Pennington Marshes on May 31st 2003. This may well have been the same bird seen circling high over Cley, Norfolk and a few minutes later flying east over Sheringham on June 8th of that year. See Wynn (2003c) for further details. *Birdguides* national database shows seven records of dark morphs since 2003 including one flying west over The Mill Field, Old Basing on September 14th 2016 for which no details have been received.

GB status: rare and declining summer visitor, peak 13 confirmed breeding pairs 2006–07, only 1 pair in 2018. (Breeds from W Europe to C Asia; winters in Sahel Zone of Africa.)

Red Kite *Milvus milvus*

BBRC to 1962; HOSRP 1963–1992

Historically a common resident but exterminated by the mid-19th century; thence a rare visitor until the late 20th century when reintroduced to southern England; now once again a common resident in most of the county.

The first dated record for the county appears to be of a pair with a nest and three eggs on April 17th 1768 referred to in a note by Gilbert White. Boyes (1774) wrote that Kites frequently pursue Starlings roosting at Alresford Pond. W. P. Heathcote, writing about the birds of prey in the parish of Hursley in *The Zoologist* in 1844, said that 'it [the Red Kite] was formerly very common, but of late has much decreased in numbers ... it is not uncommon to see it wheeling in circles or drifting away on the winds. It does not often breed with us.' K & M, writing in 1905, commented that it was evidently at one time a plentiful resident but that it is now of the rarest occurrence. They quote a record of eggs being taken in the New Forest in 1850 and describe the last known nest in the county, from which the clutch of three eggs was taken, near Broughton in 1864. Their only subsequent record was of a very fine specimen from Lower Lanham near Alresford in January 1890 which was displayed in taxidermist William Chalkley's shop in Winchester.

Red Kites: juvenile, Timsbury, October 27th 2017 (left) and adult, Mottisfont, June 29th 2021. Richard Jacobs

The first in the 20th century was seen by Ian Prestt at Old Winchester Hill from August 15th to 17th 1956. The species was considered by BBRC from 1959 to 1962 and during that period there was just one claim at Adhurst St. Mary on July 14th 1960 which was adjudged not proven. The description in the HOS files is brief although a deeply forked tail was noted. There were no further

sightings until Ewart Jones saw one between Preshaw and Dur Wood on August 23rd 1970. Subsequently, there were records in every year but five with increasing frequency from the late 1980 onwards. One found dead at Freefolk Wood near Whitchurch in January 1978 had been ringed as a pullus in eastern Germany in June 1976. A long-staying bird at Lower Whitehill Farm, Overton from January 12th to 24th 1990 offered the first opportunity for many birders to add the species to their county lists. In late 1991 the first wing-tagged birds from the JNCC reintroduction scheme were recorded. Sightings increased rapidly and culminated in the first successful breeding for well over 100 years when a pair bred near Overton in 1995, raising two young. Red Kites are now, in 2020, a common sight over much of the county.

GB status: Remnant Welsh population had recovered to 102 pairs in 1992 when first successful breeding by birds introduced to England and Scotland occurred; now 2,000+ breeding pairs. (Breeds British Isles, from Belarus and S Sweden SW to Iberia, and Italy. Winters south to N Africa.)

Black Kite *Milvus migrans*

BBRC up to 2005; HOSRP 2006 to date

A rare vagrant. 46 recorded since 1980.

The first for the county was found by R. H. Lawrence at Emery Down, New Forest at 1550 hrs on July 13th 1980. It was also seen by Mike Witherick and Royston Coles. Extracts from R. H. Lawrence's description are set out below.

> *On the afternoon of Sunday July 13th 1980 MEW, RKC and I were on a flat-topped ridge near Emery Down watching for raptors. We were looking west over wooded ridges of mixed deciduous and coniferous timber. The weather was poor – light drizzle and occasional breaks with 8/8 cloud and low cloud-base (200 to 300 feet), mild with a light south-westerly wind. At about 1550 hrs I happened to glance*

behind us towards the east and saw a large black bird approaching us which at first I took to be a Carrion Crow. However, its tail appeared too long. When I looked at it through binoculars its identity was immediately obvious – a Black Kite. By now it was level with us, at which point I drew my companions' attention to the bird. They too agreed instantly upon its identification. At this time, it was no more than 150 yards away from us about 30 feet up. Although identification was positive, I dismissed similar species with which Black Kite could be confused. Its tail was too angled at the corner for a Marsh Harrier moulting central tail feathers. It glided on flat wings and showed no white on the leading edge of the wing or head. It was too black and the tail was not forked deeply enough for a Red Kite. There were no pale patches beneath or above the wing on the primaries and the only feature which relieved the all dark plumage was a diagonal paler bar across the wing caused by paler coverts. Its flight was rather 'loose' and floppy yet purposeful, with slow, deep wingbeats interspersed with glides when the wings were held flat, slightly raised to the carpal joint with primaries depressed. It had a distinctly head down attitude as if searching for food. When gliding it twisted its tail from side to side and opened and closed it, being almost square when open and slightly forked when closed. Circled in distance, again on flat wings.

This is presumed to be the same individual present at Washington, West Sussex on July 12th and early on 13th and seen moving west over Pagham Harbour at 1210 hrs that day (Rogers *et al.* 1981).

Subsequently, a further 45 have been recorded.

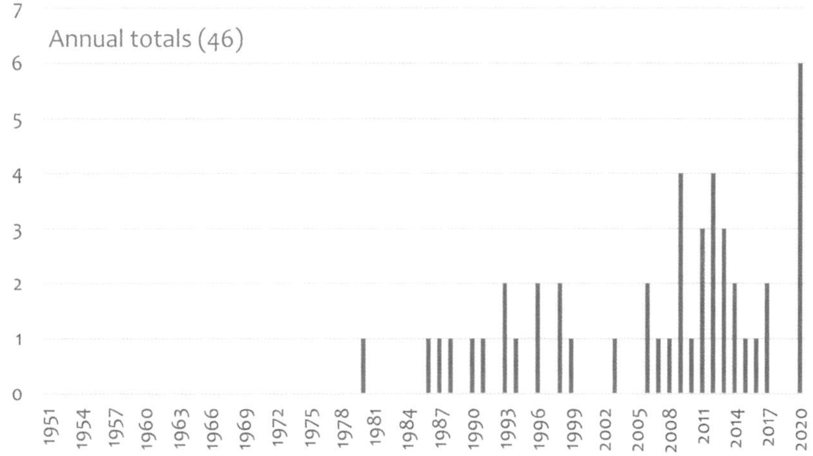

The next was one watched by Mike Rafter which flew south from Wiltshire over Ashley Walk, New Forest at 1500 hrs on June 12th 1986. In the following year Brian Gee and John Norton saw one arrive at Hurst Beach from the direction of The Needles on May 4th and then fly off west towards Dorset, and the third in consecutive years was identified by Gilbert Rowland as it picked up a fish from the surface of Fleet Pond on April 23rd 1988 as he led a HOS outdoor meeting.

The first of the 1990s was watched by Dave Burges over Hawkhill Inclosure, New Forest on May 9th 1990. A year later Rob and George Clay saw one flying north-west over Langstone Harbour on April 27th. There were four sightings in spring 1993 which are assumed to relate to only two birds. Richard Levett saw one drifting north up the Hamble Estuary and circling over Wendleholme between 1235 and 1245 hrs on April 28th. Two days later David Christie saw one flying NNE over nearby Harefield, on the east side of Southampton, at 1400 hrs. Another was seen by Wayne Percy over Bishop's Dyke, New Forest at 1115 hrs on May 7th and Mrs. J. Corley saw a Black Kite over Mill Lawn, Burley at 1700 hrs on 9th. The next was exactly a year later when Paul Pearson and Gavin Maclean watched one drift westwards over Pig Bush, New Forest between 1435 and 1445 hrs on May 9th 1994. The first (and only) multiple occurrence was on May 25th 1996 when Graham Stephenson watched two over Fleet Pond between 1457 and 1500 hrs which were refound at 1510 hrs as they drifted off ENE. The next was one which Barry Collins observed as it flew south-west from Thorney Island, West Sussex across Chichester Harbour to East Hayling on May 11th 1998. Another was watched by A. W. Shakespeare as it flew over Martin Down and into Wiltshire on May 17th of that year. The first opportunity for most birders to add this species to their Hampshire lists was in 1999 when a wide-ranging juvenile was in the Dunberry Hill/Whitsbury Down area between August 26th and September 10th. It had previously been seen flying south over Redlynch in nearby Wiltshire on August 20th.

Black Kite, Titchfield Haven, April 30th 1998

The next was not until 2003 when Mike Rafter saw one flying south-west over Millersford Bottom, New Forest between 1535 and 1539 hrs on May 6th.

Several other records in the period up to 2005 are considered acceptable by the author. Some of these were harshly adjudged as not proven by BBRC despite the descriptions being better than others which were deemed proven. Others were of birds seen by experienced observers most of whom were reluctant to submit descriptions particularly if they had not seen the upperwing pattern well, which BBRC had indicated was necessary for acceptance.

Records harshly considered not proven by BBRC were of singles seen by Graham Glombek over Chilbolton on June 21st 1987, Graham Huxley over Well on May 7th 1990, Graham Stephenson over Fleet Pond on July 3rd 1993, Trevor Codlin over Rownhams on May 29th 1995, Alan Snook over Eastleigh on September 16th 1996, Mark Simmonds at East Stratton on May 4th 1997, Dan Powell and Richard Carpenter over Titchfield Haven on April 30th 1998 and Mick Hay at Beacon Hill, Warnford on July 5th 1998.

The most controversial not proven record involved the wintering bird found by Mike Rafter at Cheesefoot Head on January 19th 2003 which stayed until at least March 5th. This was seen by many birders often in the company of up to three Red Kites. The waters were undoubtedly muddied by the fact that one of the Red Kites showed some features of the rarer species and was misidentified as Black Kite by some observers. The identity of this individual was discussed in detail by Simon Woolley in the 2004 *HBR* (Woolley 2005). Further claims of a Black Kite in the same general area on July 1st, September 5th, October 3rd, 4th and 21st and December 21st are assumed to refer to the same individual. It was stated in the introduction to the 2008 BBRC report that an article on the assessment of this bird as not proven was in preparation (Hudson *et al.* 2009) but it was never published. Attempts to establish the precise reasons for non-acceptance have not succeeded as paperwork relating to not proven records was not retained by BBRC when their old files were digitized. However, Paul French, BBRC Chair, did reveal that a comment on the database stated that it was 'an *interesting* bird with anomalies for Black Kite. Another bird was also present which may have been a Black Kite x Red Kite hybrid. S. Woolley says it was present 'in 2002' and was also seen later in 2003 but no documentation to support any of these extended earlier or later dates was received.'

Non-submitted records included those of birds seen by Dennis Bright over Titchfield Haven on April 28th 1993 (no doubt the bird seen 20 minutes later over the Hamble Estuary), Mick Scott over Fleet Pond on May 5th 1996, May 26th 1996 (no doubt one of the two birds seen there the previous day), and May 29th 1998, Mick and Will Scott and Graham Stephenson over Fleet Pond on April 13th 2001 and presumably the same bird by Mick two days later, and Paul Sterry over St. Catherine's Hill on June 14th 2005.

A further nine records were considered not proven by BBRC (see Appendix VIII) and 27 were not submitted.

In 2006 HOSRP took over responsibility for this species. Accepted records for that year were of singles seen by Peter Milinets-Raby over Bridgmary, Gosport at 1440 hrs on May 4th and Brian Rickwood and Tony Polley flying south-west over Anton Lakes, Andover at 0930 hrs on June 24th. In the following year Mark Palmer saw one over his garden at Hill Head at 1121 hrs on June 6th and was able to text Dan Houghton who saw it drifting north over Stubbington a few minutes later. 2008 produced the earliest ever spring record when John Eyre saw one flying high to the north over Long Valley near Fleet on April 1st. There were four records in 2009. One was watched by Peter Milinets-Raby over his garden in Havant between 1204 and 1227 hrs on May 31st and on the same day but at the opposite end of the county Lawrie Chappell was at Ibsley Water when he located one over the Somerley Estate which he watched between 1445 and 1455 hrs. The next was an adult spotted by Nigel Jones over the Straight Mile, Ampfield at 0820 hrs on June 23rd which drifted off east. Later in the year Barry Goater watched an adult circling over Keyhaven reed-bed before drifting off east at midday on September 21st.

A total of 17 was recorded in the 2010s up to 2017. In 2010 one was found by Mark Palmer over Ranvilles Lane, Stubbington at 1215 hrs on September 12th which he watched for three minutes before it slowly drifted off south-east towards Gosport. In the following year three were seen in spring. Steve Piggott and Mark Halligan saw one fly north-east over Ibsley North Lake on April 25th, G. Lyons saw one with a Red Kite over East Tisted on April 28th and Mike Rafter claimed one near Broadlands, Romsey on May 29th. There was evidently a small influx in May 2012 when Mark Halligan recorded one moving north over Ibsley at 1610 hrs on 5th, Barry Goater had one moving north over Needs Ore at 1346 hrs on 7th, Pete Campbell saw one moving east over Hatchet Moor at 1015 hrs the next day (possibly the same bird) and Joe Stockwell had one over Goodworth Clatford near Andover at 1310 hrs also on 8th. In 2013 Martin Terry saw one over Stubbington on April 26th, Rob Hume had one over Ocknell Inclosure, New Forest on May 17th and Marcus Ward saw one over Lower Pennington two days later. The pattern of spring sightings continued to dominate in the next four years with singles recorded by Gavin Maclean over Fulflood, Winchester from 1510 to 1515 hrs on May 16th 2014, Graham Stephenson over Miles Hill, Bourley from 0905 to 0920 hrs on May 18th in that year, Doug Robertson flying low ENE over Hambledon at 0700 hrs on May 15th 2015, Alan Lewis over Ibsley Water on April 30th 2016, Ed Wiseman moving south-west over East Boldre at 1000 hrs on May 18th 2017 and Graham Stephenson flying south-west over Yateley Common at 0905 hrs on May 26th of that year.

The largest ever influx occurred in spring 2020 with six confirmed sightings between April 13th and May 10th. At least 25 were reported in England during that period. John Eyre saw one over Long Valley on 13th, Carol Earnshaw videoed one over New Milton on the same day, Andy Lester photographed one moving north-east over Fishlake Meadows on 25th and Richard Ford photographed one drifting east over Woolmer Forest on 30th.

Black Kites: Cheesefoot Head, February 20th 2003 and Ibsley, April 30th 2016. Simon Woolley and Alan Lewis

Pete Gammage and Tim Doran independently witnessed an amazing eastward movement of at least 45 Red Kites along Portsdown Hill during the late morning of May 10th. Pete detected a Black Kite among its commoner congeners and was able to alert Tim to its presence by text. Just after noon on the same day Graham Barrett saw one flying north over Itchen Valley Country Park which is approximately 18 km north-west of Portsdown Hill. Based on the timings this is assumed to be a different bird.

Analysis of arrival dates shows the majority in spring and early summer between April 1st and June 24th with a clear peak in the last week of April and May, then just four records up to September 21st including the long-staying bird at Dunberry Hill in 1999.

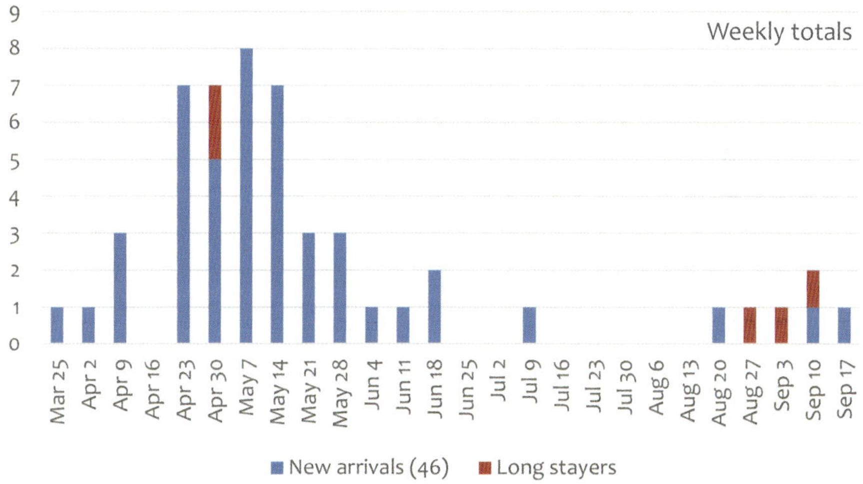

During the period between 2006 and 2020 there have been a further 58 claims of Black Kite. At least 31 of these originated via *Birdguides*, *Rare Bird Alert*, *Going Birding* or *Birdtrack* and were without documentation. Multiple attempts were made by the County Recorders to obtain adequate evidence for these sightings but in most cases little was forthcoming. The remaining 27 were considered not proven by HOSRP. While in earlier years Marsh Harrier was considered a major pitfall in correctly identifying this species, clearly the explosion of Red Kites in Hampshire in the 21st century presents a likely source of error. Several claims have been from moving cars but clearly such non-binocular views of a rare bird are not acceptable!

GB status: 5, 1866–1949; 682, 1950–2019. Increasing, mean 23 per year in 2010–19. (Nominate race breeds throughout Mediterranean basin & continental Europe except maritime NW & Scandinavia, with most in Spain, France & Germany. To E, breeds European Russia to W Kazakhstan. Most winter Africa. Other races in C & E Asia, Indian subcontinent, E & SE Asia & Australia.)

White-tailed Eagle *Haliaeetus albicilla*

BBRC to 1999; HOSRP 2000 to date

A rare winter visitor. At least 11 old records and four since 2007.

The first dated record for the county appears to be of one seen by Col. Peter Hawker at Keyhaven on February 10th 1827. He saw others on December 28th 1829, January 9th and 15th 1837, and January 22nd and 23rd 1841 but failed to kill one (Payne-Gallwey 1893). The next was shot in Somerley Park (Ringwood) on December 28th 1861 which was identified as a Spotted Eagle at the time (Wise 1862). The specimen was later examined at Somerley House by Lt. Col. Cyril Mackworth-Praed and proved to be of this species (Cohen 1963). John Levell (*in litt.*) examined the specimen in 2020; it is an immature White-tailed Eagle and unfortunately is in a very poor state of repair. Most of the feathers have now disappeared. It resides in a crate in the stables. Among several vague records quoted by K & M were the following which were definitely in Hampshire: Edmund Meade-Waldo saw an adult over Wilverley Plain (New Forest) in July 1885 (which roosted throughout the summer of that year at Hengistbury Head, now in Dorset) and an immature near Micheldever Station in December 1895. One was shot at Cadland in about 1886 and another was there for two or three months in the winter of 1888-89.

The only 20th century records were of one at Highclere on December 18th 1927 which was seen later in the month at Combe, Berkshire (*Oxford Ornithological Society Report*) and an immature at Somerley Park on January 6th 1947 (Cohen 1963). Unfortunately, Cohen does not provide any further detail of the Somerley bird which is not included in the 1947 *HBR*. John Cloyne has traced further records of a probable White-tailed Eagle seen by Maj. T. Phelps flying downstream at Brambridge in the Itchen Valley on May 1st and June 11th 1942 which are reported in *The Field* for June 13th and August 1st 1942. The latter also includes a report of presumably the same bird seen flying over Lainston near Sparsholt towards the Itchen Valley on May 1st.

White-tailed Eagle, juvenile, Downton, January 1st 2011. Keith Maycock

A juvenile was discovered by Henry Edmunds sitting in a field on the Cholderton Estate on November 18th 2007. It remained in the area until January 13th. Many thanks are due to Henry, the owner of the estate, for setting up a viewing area which enabled hundreds of birders to enjoy this magnificent bird often at close quarters. It also ranged north to Shipton Bellinger and east to Quarley Hill. It evidently relocated to the Porton Down/Over Wallop area where it was reported intermittently between January 23rd and April 9th. The bird was colour-ringed as a chick and proved to have originated from a nest in Finnish Lapland in 2007. Earlier sightings near Braintree, Essex on November 11th and 15th are presumed to refer to this individual. This was only the second to be recorded in south central England in the modern era following the famous wintering individual at Brill on the Buckinghamshire/ Oxfordshire border from November 1983 until February 1984.

Another juvenile was seen by Andy Johnson flying west just north of Sandy Point at 1326 hrs on December 12th 2010, subsequently being seen over Southsea and Titchfield Haven. It had been present at Amberley Wild Brooks, West Sussex, the previous day but departed in the early afternoon of the 12th. It then went missing until January 1st when it was discovered by Russell Wynn on farmland at Downton near Hordle. It remained in the area until February 12th before disappearing again, only to be refound by Nigel Bell on farmland between Old Basing and Newnham on February 23rd. It was last seen in this area on March 24th, having also been seen at Sherfield on Loddon on March 4th and 12th. The same individual was then seen in Lincolnshire, Norfolk and Suffolk from April 3rd and finally settled at Ruckland, Lincolnshire where it remained from April 23rd until August 7th.

A third juvenile was present during the 2018/19 winter. It was first seen at Milkham Bottom, New Forest on December 16th. It remained in this area until January 11th but was also seen at Ibsley Water on December 18th and 21st. In the following winter, Judy Hunt videoed one at Hasley Inclosure, New Forest on February 13th which stayed until at least March 18th and ventured to Ibsley Water on March 1st. Presumably the same individual was also seen over the Hawk Conservancy at Weyhill on March 31st.

White-tailed Eagle, juvenile, Brockenhurst railway crossing, the late afternoon albicilla *express, with a train of 100s of corvids. Reward on the homeward journey after a fruitless six hour vigil at Downton looking for this bird, January 22nd 2011*

A project to re-introduce White-tailed Eagles to southern England was initiated by the Roy Dennis Wildlife Foundation in 2019 and in August of that year six juveniles were released on the Isle of Wight of which four survived into 2020. A further seven juveniles were released in 2020. They all carry satellite transmitters and unsurprisingly several of them have been tracked over Hampshire since their release. Thanks to satellite tracking data it is known that none of the sightings in early 2020 match any journey made by Isle of Wight birds. Examination of Judy Hunt's video confirmed that the eagle was the same as one photographed in Wiltshire which was thought to be a second-winter bird from Sweden. Possibly it was the 2018/19 juvenile returning.

GB status: 490, 1732–1950; 26, 1951–98. Formerly resident, last bred Skye 1916. Reintroduced Scotland, first successful breeding 1985, increased to 111 breeding pairs 2018. Increasing frequent winter visitor to England as Netherlands, Germany and Fennoscandia population recovers. (Breeds Greenland and W Europe to Japan, moves S in winter.)

Rough-legged Buzzard *Buteo lagopus*

HOSRP

A rare winter visitor and passage migrant. 13 old records and around 56 since 1955.

The first for the county was a male shot by Edward Hart at Bramsgrove (Bransgore) on November 25th 1875. The specimen is now held by the Horniman Museum, London.

© Horniman Museum

A further eight or nine were recorded in the second half of the 19th century. One was shot near Ringwood on December 13th 1879 and reported by George Corbin in *The Field* for December 27th of that year. This was overlooked by K & M. Another was shot near Stockbridge following the great blizzard in mid-January 1881 (WC) and one was trapped near Faccombe around the same time and kept in captivity until at least 1894 (K & M). The next was shot near Ringwood in December 1884 (Corbin). Another was found caught in a gin-trap in a rabbit warren near Winchester by Sutton Davies, a student at Winchester College, on March 13th 1892. The bird was taken into care but its eventual fate was not recorded. A large hawk, presumed to be the same bird, had been seen in the area on November 17th 1891 and February 14th 1892 (*Zool.* 1892: 147, 1893: 8–9). K & M did not know about one shot by Captain Bacon's keeper at Burghclere in 1892 which 'had been taking a heavy toll of rabbits' (Cohen 1963). Corbin described one that was shot near Ringwood in January 1897 that was much whiter than specimens seen in 1884 and 1894 and was probably an old male (*Zool.* 1898: 124). An older record was quoted by Wise (1862) who said: 'Mr. Rake informs me that a specimen was trapped near Fordingbridge in the summer of 1857. It is, however, more generally noticed later in the year'. This was erroneously given in K & M as 1887. This would be the first Hampshire record but given the unseasonal date and lack of a specimen it is unacceptable.

The only records for the first half of the 20th century are of one trapped near Stockbridge in January 1916 (Cohen 1963), another seen by D. A. Phillips at Faccombe on December 18th 1927 (*Brit. Birds* 21: 203), a cock reported by Col.

W. A. Payn as trapped near Andover in 1940 but no date was given (*HBR*) and one seen at Stockbridge on an unspecified date in 1945 (Cohen 1963).

The first in the modern era was one found by Eric Jones and Gerald Kinsey at Brockley Warren on January 30th 1955. In the *HBR* it stated that a full description and sketch was submitted but it has not been possible to trace the original notes. The record was subsequently excluded by Cohen (1963) who was under the impression that the species could only be safely identified if the feathered tarsi had been observed. Eric and Gerald were reliable observers and so the record was accepted for *BoH*.

Subsequently, there were records in 25 winters up to 2019/20 involving around 55 individuals.

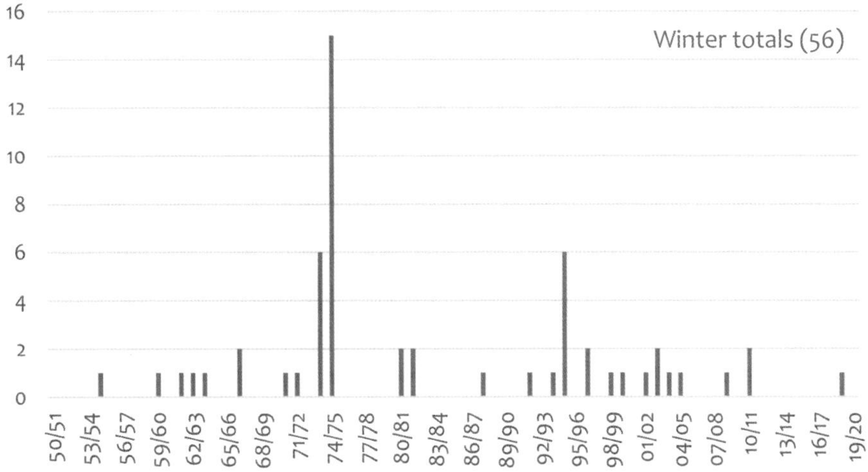

The second was one seen by Canning Suffern at Titchfield Haven on November 15th 1959 (Duffin 1984, 1991). This was followed by one seen by Norman Orr over Rockford on April 7th 1962 which was well described (*HBR*); he also mentioned that one was seen around the same time by Colin Tubbs. The author contacted Jenni Tubbs in 2021 and she kindly transcribed the following note from Colin's journal:

> *Rough Legged Buzzard. Picked up a buzzard beating steadily out of the eye of the sun over Fritham Plain, moving N, 1300 hrs, April 13th. Even against the light I thought it looked remarkably grey on the upper wing and pale beneath, but as it worked round so that the light was behind me my first impression was confirmed. White tail with black terminal bar very conspicuous when it turned into a brief soar. Upper side of wings showing very grey, with what appeared to be white fleckings on the coverts; primaries and secondaries brown. Altogether a very pale bird. After I had been watching it for a few seconds I suddenly caught sight of a Buzzard poised above it, about to stoop. Down it came in a breath-taking dive,*

wings completely closed, and a short twisting chase followed, the buzzard strafing the RLB for several minutes afterwards before finally giving up. Both birds finally lost to view in the haze over Coopers Hill.

The next was one seen by David Bolton over Beacon Hill, Highclere on April 26th 1963. Another was seen near Beaulieu by Edwin Cohen on January 13th 1964 and by David Glue five days later on January 18th.

Between 1966 and 1981 there were influxes into Britain in five winters including at least 57 wintering, mostly in south-eastern England, in 1966/67 (Scott 1968), 85 in 1973/74 and 100 in 1974/75 (Scott 1978). Around 27 occurred in Hampshire during this period accounting for almost 50% of those seen in the modern era. In 1966/67, one wintered in the Cheesefoot Head area where it was recorded by Bill Norton and David Pumfrett between December 10th and February 26th. Another was seen by John Ash between Boveridge and Martin on the Dorset border on February 19th 1967. One was seen by David Christie at Crockford Bridge, New Forest on October 18th 1970 and in the 1971/72 winter one in the Beacon Hill/Great Litchfield Down area between December 15th and February 15th was recorded by John Taverner, Barry Goater, Phil Gregory and others. The influx in autumn 1973 was signalled by singles seen by Rob Fuller at Great Litchfield Down on October 21st and Colin Tubbs at Selborne Common on October 26th. Wintering birds were recorded at Cheesefoot Head between November 17th and January 12th and at Great Litchfield Down between December 2nd and January 1st. Later records were of singles seen by Ted and Penny Raynor over Harewood Forest on January 19th and by Derek Lister moving east over Old Winchester Hill on February 24th. In autumn 1974, two were found by George Johnson at Beacon Hill on October 26th, two by Ron King near Cheesefoot Head and one by Derek Lister at Old Winchester Hill on 30th and one by Barry Duffin at Titchfield Haven on 31st. Subsequently, three, possibly four, wintered in the Beacon Hill/Great Litchfield Down area, at least two at Cheesefoot Head and one at Old Winchester Hill. Two or three were found by A. V. Lee in the Whitchurch/Tufton area in early January and up to three were seen by Molly Tucker and others in the Faccombe/Pilot Hill area between March 16th and 23rd with one still present on April 6th. In the New Forest singles were seen by Mick Hay at Bishop's Dyke and Mike and Pete Combridge at Ashley Walk on December 28th, with the latter seen again on January 5th. No more were recorded until the 1980/81 winter when one or two wide-ranging birds were seen by B. J. Wyatt, Ian Weston, Steve Smith and others in the Faccombe/Pilot Hill area and adjacent areas of Berkshire between December 2nd and March 20th.

Since then, all records have been of singles on one date only apart from two wintering birds. Three in the 1980s comprised singles seen by Nigel Cleere at Faccombe on December 6th 1981, by Keith Godfrey as it flew north over

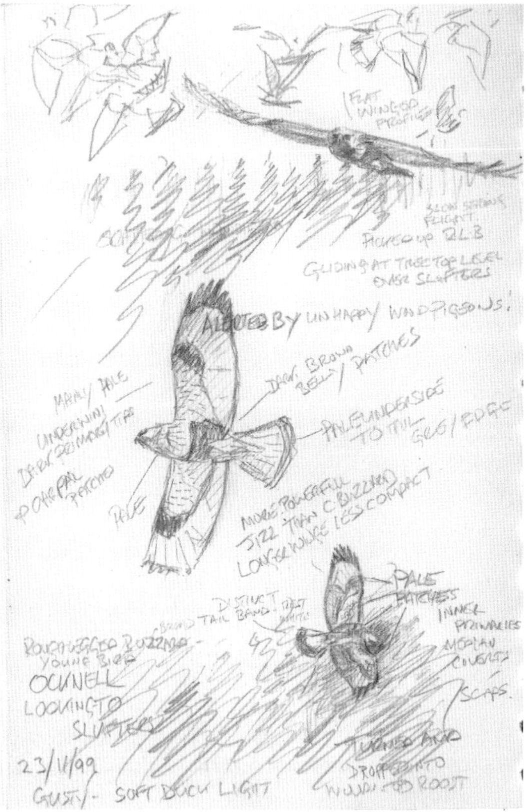

Rough-legged Buzzard, Slufters Inclosure, New Forest, November 23rd 1999

the Royal South Hants Hospital, Southampton on December 12th 1981 and by David Billett at Harvest Slade Bottom, New Forest on April 20th 1988.

Mick Hay and Keith Maycock recorded one flying north-east over Old Winchester Hill on November 2nd 1991 and David Christie saw a juvenile fly in off the sea at Hook-with-Warsash on October 18th 1993. A national influx in the 1994/95 winter produced six in Hampshire but the only wintering individual was recorded by N. W. Bissett, Nigel Cleere and David Barker in the Pilot Hill area between November 7th and March 18th. Other records were of one seen by Mick Scott as it flew west over Fleet Pond on October 18th and another watched by Dave Unsworth, Barry Goater and other members of the NFOC at Needs Ore on the same date, an adult seen by Mike Baker as it flew east over Staunton Country Park, Havant on January 30th, another seen by Mike at Wheely Down on March 16th and presumably the same individual seen by Mick Hay at nearby Beacon Hill on April 9th and a late bird seen by Ian Weston and Jan Legg over Lower Woodcott Down on April 23rd. One was seen by Ian Waddington at Cadnam on January 10th 1996 and later that year Norman Orr saw one over Holmsley Ridge on March 21st. The next was one seen by Mick Scott as it drifted east over Fleet Pond on October 5th 1998. This is the earliest ever autumn record. The last for the century was one watched by Dan and Rosie Powell near Slufters Inclosure on November 23rd 1999.

The species became even scarcer in the 21st century with only nine recorded. Mick Hay and Bryan Pinchen saw one fly south-west over Needs Ore on April 5th 2002, Dan Houghton noted a first-winter flying east over Titchfield Haven on February 8th 2003 and almost a month later Barry Goater and NFOC members saw one at Needs Ore on March 4th, David and Rosemary Billett saw one fly east over Braishfield on March 31st 2004 and Marc Moody saw an adult female flying north over Hurst Castle on October 17th 2005.

A second-winter male was found by an unknown observer north of Faccombe on February 14th 2009. It remained in the area on both sides of the Berkshire boundary until at least March 8th and was at times attracted to a raptor feeding station at Pilot Hill where many birders enjoyed excellent views. Two spring migrants occurred in 2011: one seen by Andy Lester over Milkham Bottom on April 25th 2011 and an adult found by Joe Stockwell at the Harrow Way north of Overton on May 3rd 2011 which was subsequently seen by several observers and was reported the next morning. This is the latest ever spring record. The only record since was of a juvenile located by Tim Doran and Jason Crook as it flew west over Farlington Marshes on October 29th 2018.

The monthly totals suggest that some birds recorded in October moved through the county before further arrivals in December and January. The data also indicate spring passage with possibly 13 new arrivals from March onwards. The latest date for a presumed wintering bird was April 6th 1975 at Pilot Hill but a further seven have occurred later than this.

Rough-legged Buzzards: Faccombe, February 14th 2009 (top and middle) and Overton, May 3rd 2011 (below). Jerry O'Brien and Joe Stockwell

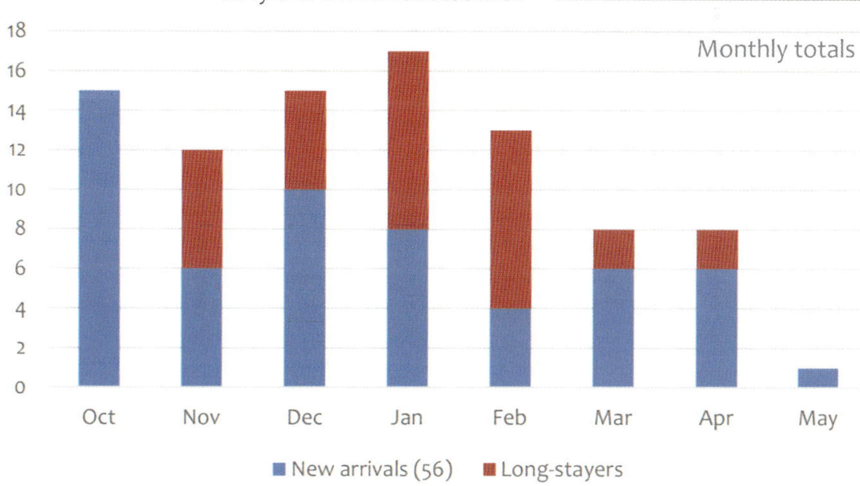

This species presents a significant identification challenge even to experienced observers due to the risk of confusion with pale Common Buzzards, especially if they show whitish upper-tail coverts. Some forty records in the HOS files for the period from the mid-1960s to the 1980s lacked convincing descriptions. Although some were published in *HBR*s all of those from non-invasion years were excluded from *BoH.* Many of these related to claims of wintering birds at Cheesefoot Head which other observers considered were Common Buzzards. Subsequently, at least 15 further records, most of which had minimal documentation, were considered not proven by HOSRP and details of ten sightings originating from bird information services were not forthcoming.

GB status: 2,167, 1986–2019. No clear trend, mean 81 per year in 2010–19. (Circumpolar low Arctic breeding range from mountains of S Scandinavia, E through N Fennoscandia, Russia E & S to Kamchatka, Alaska & N Canada E to Newfoundland. Winters to S of breeding range.)

Scops Owl *Otus scops*

BBRC

A very rare vagrant. Four old records and three since 1968.

© Leicester Museum

The first for the county was a female which was shot by the gamekeeper Mr. Perkins at Wilverley Lodge in July 1866. The second was a male shot by the gamekeeper Mr. Coles at Holmsley on October 29th 1884. Both specimens passed into the Hart collection and are now held at Leicester Museum.

The finding of another in 1884 was recounted by K & M as follows.

> *Mr. Walter Jeffrey, taxidermist, of Gosport, a relative of Mr. William Jeffery, the well-known naturalist of Ratham in Sussex, had a female brought to his house on June 10th 1884 which had apparently only just reached our shores. The following particulars were kindly supplied by Mr. Jeffery and Mr. W. T. Pearce, of the Portsmouth NHS. The bird was picked up on the shore at Stokes Bay by some boys who had been bathing; it appeared to be quite exhausted; and died in their hand, but it was still warm when Mrs. Jeffery bought it off them. It was in good plumage, but very emaciated, weighing barely two ounces, and contained only the remains of a few beetles in its stomach. Its length was seven inches, expanse of wings one foot eight inches. Mr. Pearce possesses the sternum and the skin is in Manchester.*

One at Totton in 1950 had probably escaped after being brought in by a seaman (Cohen 1963).

The first modern record was of one found dead at Highclere on May 11th 1968 and sent to John Ash for identification. The skin subsequently passed into the collection of Dr. James M. Harrison.

A singing male was identified by Ted Raynor at Dummer on May 12th 1980. Details of this bird were initially withheld for fear of upsetting the locals but it was subsequently seen by hundreds of observers during its long stay until July 20th. The landlord of the village pub was very grateful for the extra business that came his way as a result! Richard Millington, in his book *A Twitcher's Diary* (1981), described the events of the evening of July 3rd as follows:

Scored at last! On the fifth visit, crippling views of the Scops Owl, perched on edge of a hole in a tree-trunk close to the road. Watched here for about 20 minutes, spotlighted by car headlights, and seen to be a small, intricately patterned grey-brown owl with extended ear-tufts and penetrating yellow eyes set in dark centred facial discs, the latter also edged with a blackish line. After staring at us, apparently fascinated by the curious twitching antics, it flew up into the branches of a tree, perching on a small branch, then off into the gardens where it commenced calling.

Pete Wheeler   *Off down the pub*

This became a very controversial record with strong rumours that the bird had escaped from local captivity, having initially been taken to Dummer as part of an illegal consignment of birds which arrived at Heathrow Airport. Mike Rogers, secretary of BBRC, subsequently exchanged copious correspondence with various witnesses but on review the committee did not consider there was any corroborated evidence to overturn their decision to accept it as a wild bird.

The most recent record was of one picked up freshly dead by P. J. White at Dean Lane, Bishop's Waltham on April 2nd 2004. The skin was passed to Paul Roper of the Rye Meads Ringing Group.

GB status: 42, 1802–1949; 43, 1950–2020. Rare, about one every 1–2 years. (Breeds N Africa & S Europe, from Iberia N to C France & E to Greece. Also breeds Ukraine, S Russia & S Siberia to W Mongolia, Kazakhstan & Iran. Most winter N equatorial Africa, but some remain S Europe.)

Snowy Owl *Bubo scandiacus*

BBRC

A very rare vagrant. Three records in 1848, 1946 and 1965.

The first for the county appears to be one which was shot at Burley Wood in the winter of 1848. The specimen was acquired by the Hon. Grantley Berkeley before passing into the Hart collection. It is now held at Leicester Museum.

Almost a century was to elapse before the next was seen by F. J. Goddard near Southampton on January 22nd 1946. The notes below are from *Brit. Birds* 39: 216 which also describes another seen in Cornwall five days earlier.

> *At about 5 p.m. on January 22nd, 1946, at Woodlands, near Southampton, during a period of very cold weather, my attention was drawn to a very large white owl, which flew at some fifteen feet from the ground and about 50 feet ahead of me. It settled in an oak tree and I estimated its length as about 20 inches or slightly more. It remained perched for about 20 seconds and then flew off at right angles to me, passing within 20 feet, when I noticed some brown spots or irregular marks on the shoulder region. Apart from this the bird was almost pure white, but I could not observe the colour of the beak, or the eye, as the light was failing. The wing-beats were very powerful, quite different from those of the Barn Owl, with which I am very familiar, and it was also about twice the size of that species. It was clearly a Snowy Owl. I searched the vicinity on subsequent evenings, but saw no more of the bird, and local enquiries failed to elicit that it had been seen by anyone else. The locality is one of open fields surrounded by small copses and woods on the edge of the New Forest, with many open spaces of gorse and heather.*

The only modern record involved one seen by J. M. and Mrs. A. D. Crocker at Pennington Marshes on March 14th 1965. This was presumably the same bird as that seen at Brading Marsh, IOW on March 28th. Extracts from J. M. Crocker's description are reproduced below.

> *Date and Time: 14th March 1965. 1700–1800 hours.*
>
> *Optical Aids: Binoculars, x9, x8, and telescope x20.*
>
> *The bird was first seen sitting on a low bank among brambles in the open marsh. The general impression was of a large pale owl, overall dirty grey and whiter on the face. It was sitting down on the tarsi, inclined forward at about 60 degrees, with*

the body feathers reaching to the ground like a 'skirt'. At close range, the overall colouring was dirty white, all the feathers tipped with broad bands of dark brown. The upperparts formed a uniform area, greyish white banded with broad brown tips. The folded wings hung at the sides, all the coverts and flight feathers again broadly tipped dark brown, but with a prominent white patch at the carpal joint. The primaries were whiter and each was tipped with brown. The breast, flanks and belly were whitish with narrower brown tips which extended as fine bars across the body. The tail was short, brownish white and faintly barred with brown.

The head was large and round with a white facial area which appeared yellowish at certain angles. The eyes were large with amber-yellow irises and large pupils. The eyelids were prominent and white. The bill was obscured by white feathers. Brownish flecks edged the facial discs and extended as a 'beard' beneath the bill. The rest of the head was heavily flecked dark brown extending over the facial discs like eyebrows and down to the side of the neck and nape. At the sides of the neck, when seen in profile, there was a prominent black patch. The hind part of the crown was without flecking giving a whitish 'bald patch' appearance. There were a number of large blackish feathers on top of the crown. From directly in front and behind two very small ear tufts could be seen.

At the approach of a herd of cows it became agitated and watched them over its back with its head turned through 180 degrees, before flying off. The legs were trailed behind showing dirty white feathering to the toes before being retracted. The wing beats were strong and heavy but not deep. The tail was fanned and seen to be indistinctly barred. The bird landed close by and was mobbed by three Black-headed Gulls avoiding their swoops by ducking. It then flew to an earth bank where it remained for about 20 minutes, and when it finally flew off it was carrying a starling-sized bird which had one wing dangling.

GB status: 193, 1811–1949; 240, 1950–2020. About 3–4 per year; stable. (Breeds N Scandinavia & Iceland, depending on availability of small mammals. Outside Europe, erratic circumpolar breeder across tundra & N islands of Arctic Russia, Siberia, Alaska, Canada & N Greenland. Most disperse S in winter but some resident or nomadic if food available.)

Hoopoe *Upupa epops*

A very scarce passage migrant which has bred, most recently in 1959. At least 47 old records and 259 in the modern era.

The first mention of this species for Hampshire is in *Pinax rerum naturalium Brittanicarum*, an alphabetical list of flora, fauna and minerals found in England published by Christopher Merrett in 1666. It mentions its occurrence in the New Forest, *sed raro invenitur*. A century later, Gilbert White, in Letter XI to Thomas Pennant, dated September 9th 1767, wrote 'The most unusual birds I have observed in these parts were a pair of Hoopoes which came several years ago in the summer and frequented an ornamental piece of ground, which joins to my garden, for several weeks'. The earliest dated record is of one shot at Liphook on April 22nd 1845 (Newman 1849). K & M listed a further 27 records in the period up to 1905. These include reports of breeding in the New Forest in 1897 and 1898 and near Winchester in 1900 and 1902, the latter successful, eight records with exact dates including one shot by J. Barnes at Cranesmoor on May 3rd 1887 which passed into the Hart collection and is now held by the Horniman Museum and the latest of one seen by Harry Witherby at Burley Lawn on April 25th 1903, six records with vague dates including two of pairs in 1861 and 1888 and nine undated specimen records. To these can be added a record of one heard and seen at St. Cross on May 26th 1894 by John Walpole-Bond, a student at Winchester College. In the years up to 1950 there were 18 records in 1907–13 (4), 1920–26 (3), 1931–36 (3) and 1940–49 (8). The latter included one picked up dead at Swanmore in severe weather on January 9th 1940.

In the modern era there have been records in every year except 1992. Single pairs bred near Sherfield-on-Loddon in 1953, Southwick in 1954 (H. E. Woods *in litt.*), Middle Wallop in 1956 (Taverner 1956), Knowle from 1956–59 (H. E. Woods *in litt.*) and near Lyndhurst in 1959 (Ed Wiseman *in litt.*). All were successful apart from the pair at Knowle in 1958. Other than breeding birds, around 259 have occurred with records in every year except 1992.

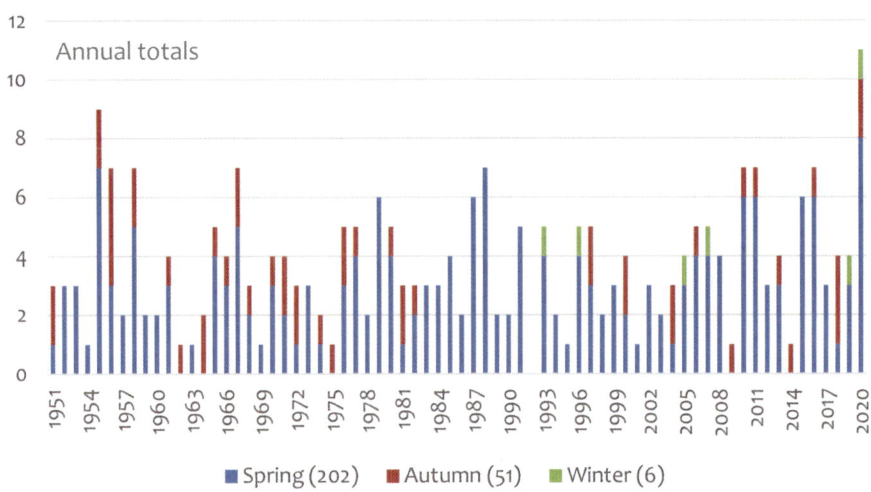

Most occurred in spring when the highest totals were eight in 2020, seven in 1955 and 1988 and six in six years including four between 2010 and 2016. Autumn totals have not exceeded two except for in 1956 and three in 2018. There are six winter records.

As shown on page 352, analysis of arrival dates shows spring passage from mid-March onwards with a peak between early April and mid-May. The earliest were both in 1979: singles at Chalton from February 24th to 28th which may have wintered and at Yateley from March 13th to 20th.

Hoopoes: Calshot, April 25th 2007 (top), Sandy Point, April 11th 2020 (top right), Fawley, February 3rd 2020 (above) and Winchester, April 24th 2015 (right). Kit Day, Andy Johnson, Rob Porter and John Hilton

Other early birds were at Crow on March 15th 1981, Farlington Marshes on March 17th 1990 and Wimpson Lane, Southampton on March 17th 1993. Most move quickly on although stays of up to a week are not uncommon. Individuals may remain for several weeks in summer although even at this time some are only seen briefly. Long-staying birds include singles at Penton Mewsey from May 16th to June 5th 1988 when it was taken by a cat and at South Wonston from around May 15th to July 27th 2020. Unfortunately, there has been no suggestion of breeding since the series of records in the 1950s. Post-breeding long-stayers include singles at Keyhaven from July 15th to August 3rd 1958, Martin Down from July 10th to August 19th 1997 and Redenham Park from June 26th to July 7th 2019. The earliest presumed autumn migrant was at Petersfield on August 4th 2020. From that date records are evenly spread through until mid-October with a small peak in late August and early September. As in spring, most are seen on one date only but stays of up to eight days have occurred. Two were together on just one occasion, at Godshill on August 29th 1964. Late birds may linger such as those at Upton Grey from October 11th to 24th 1976, Bentley from October 25th to 27th 2018 and Broxhead Common on October 30th 2007.

The last was presumably that discovered nearby at Kingsley Common on December 14th where it remained until at least January 14th 2008. Other records of wintering involved birds at Flexford from November 11th to December 31st 1993, when it perished in cold weather, one heard by two experienced observers at Titchfield Haven on December 26th and 27th 1996 and one which flew west past Sandy Point and Beachlands, Hayling Island on November 14th 2005 and was reported later that day in Walpole Park, Gosport. This seems likely to have been the bird located in nearby Felicia Park on January 7th 2006 where it remained until March 15th. Finally, one was at Needs Ore from November 22nd to December 27th 2019 and was presumably the same bird found at Badminston on February 1st where it stayed until March 15th 2020. Another seen at New Copse Inclosure, New Forest on February 17th 2020 is assumed different to the Badminston individual which was also recorded that day. It may have been an early spring migrant.

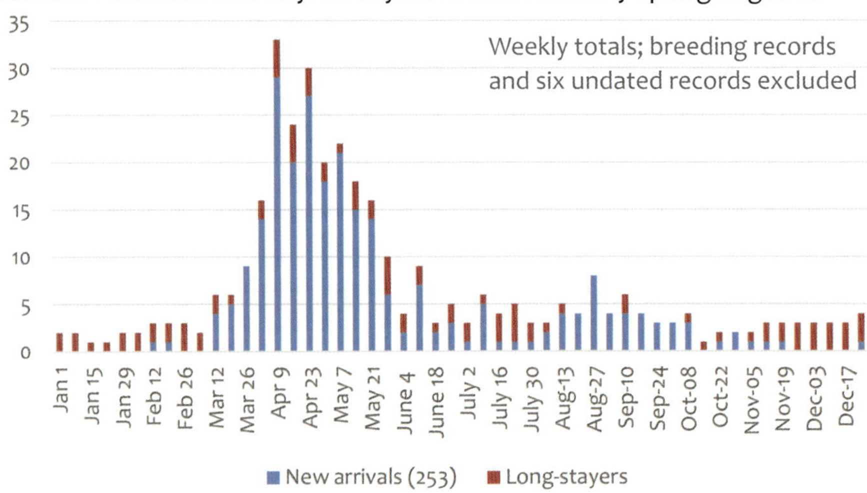

Hoopoes have been found all over the county. Downland locations predominate with around 70 birds including six at Martin Down but curiously none at the well-watched Old Winchester Hill. Approximately 30 have occurred in the New Forest. Totals in coastal areas include 20 on Hayling Island, seven in Langstone Harbour including five at Farlington Marshes, ten around the margins of Portsmouth Harbour, 23 between Gilkicker Point and Hook-with-Warsash including eight in the vicinity of Titchfield Haven and 13 between Lymington and Milford on Sea.

GB status: 6,061, 1968–2019. No clear trend, mean 120 per year in 2010–19. (Breeds widely warm & temperate regions of Europe & Asia. Some winter in Mediterranean basin & N Africa, but most winter in sub-Saharan Africa and S & SE Asia.)

Roller *Coracias garrulus*

BBRC

A very rare vagrant. Four old records and four since 1955.

The first for the county was an adult male shot by the gamekeeper G. Barnes at Allenworth, Hinton near Christchurch in July 1874. The specimen was mounted by Edward Hart and is now held by the Horniman Museum, London.

© *Horniman Museum*

In the first half of the 20th century there were records of one seen near Ringwood in July 1904 (*Zool.* 1904: 458), a female shot at New Milton on September 23rd 1918 (*Brit. Birds* 12: 140) and a juvenile near Fritham, New Forest (not Fareham as in *BoH*) from September 15th to 21st 1947 (*Brit. Birds* 41: 119–120).

The next was one found by Lt. R. G. S. Champion at Butser Combe on May 30th 1955. Ed Wiseman recently recounted the story behind this sighting, as follows:

> *'We met Lt. Champion at Farlington Marshes. He asked where he might be able to see one or two species, including Stone-curlew, so we suggested he go to Butser. A few hours later he returned to FM to say he'd found a Roller, and did we wish to try for it. He drove a small sports car, but managed to squeeze in DFB, GHR, JKB & EJW. On arrival we all saw it except Taff (who failed to get on to it), but only in flight, as it flew past, below us, at close range. That was the only view we had, though Champion had prolonged views earlier.'*

Over 30 years were to elapse before the next was found in the New Forest on May 30th 1987. Keith Maycock and Mick Hay were seated at Furzy Brow overlooking the Rowbarrow area. At 1005 hrs Keith saw a medium-sized bird fly low over the heath and pass within about 20 yards. He lifted his binoculars and immediately identified the bird as a Roller by its unmistakeable pattern of brilliant colours. Keith alerted Mick who was onto the bird in seconds as it flew towards Penny Moor, about 500 yards away, where it was lost from view. They were joined by Dick Senior and the bird was finally relocated by Mick around 1145 hrs; subsequently they obtained prolonged views until leaving the area after three hours. It remained until the next day and was seen by hundreds of observers. Subsequently, it came to light that it had

first been seen by Andy Clayton, a railway worker, who saw it perched on wires at Setley Plain alongside the Brockenhurst to Lymington branch line on May 28th. Another was found in the New Forest by Ed Wiseman along the Crockford Stream at Norley Inclosure at dusk on May 25th 1997. Ed initially flushed the bird and it flew off into the setting sun without its identification being confirmed. Fortunately, he was able to relocate it and obtained good views for ten minutes at ranges of 50–250 metres as it made frequent sallies from the tops of small Scots pines. The bird was settled when Ed left but it was nowhere to be found the next day.

Roller, Pig Bush, New Forest, May 31st 1987

Most recently one was seen briefly by Jim Smith at Broxhead Common on May 23rd 2013. He was unsure of its identification and returned on May 25th when he refound it. He alerted Andy James, Denise Murray and Steve Mansfield who located it at 1140 hrs. It was subsequently seen by hundreds of birders later that day and on 26th. On its final day it flew off at 0910 hrs and paused at Thursley Common, Surrey before continuing north-east.

Roller, Broxhead Common, May 26th 2013. Jason Ward and Mike Trew

GB status: 174, 1664–1949; 127, 1950–2020. About 1–2 per year; stable. (NW Africa & S Europe from Portugal to Greece, & locally through Balkan countries & E Poland to Estonia & E to Ukraine. More numerous from Turkey & S Russia to S Urals, SW Siberia & C Asia to W China & N Pakistan. Some winter equatorial W Africa but most winter E Africa from Kenya to Zimbabwe.)

Blue-cheeked Bee-eater *Merops persicus*

BBRC

A very rare vagrant. One record in 2009.

Blue-cheeked Bee-eater, Needs Ore, June 21st 2009. Alan and Caroline Dawson

The only record for Hampshire is of one seen and photographed by Alan and Caroline Dawson at Needs Ore at 1710 hrs on June 21st 2009. After satisfying themselves as to the identity of the species the finders alerted other wardens at Needs Ore but it was not relocated that evening or the following day. Extracts from the submission to BBRC compiled by Dave Unsworth are reproduced below.

Alan and Caroline Dawson are voluntary wardens at North Solent (near Beaulieu). They were leaving Needs Ore at around 1720 hrs when Alan noticed a bird sitting on top of a large (dying) tree about 40 m out from the track in an adjacent field. This caught his attention and they stopped to look at it. Their immediate reaction was that they thought it was a European Bee-eater. They watched it, both perched and in flight and took several photos. Caroline tried ringing me while watching the bird although I missed the call, and after about 15 minutes they left the bird in the same place. I picked up Caroline's message at about 1840 hrs (I was in south-east Dorset at the time) in which she said they had just been watching a Bee-eater. I was keen to see it, so we diverted towards Needs Ore on the way home (I am also a voluntary warden at Needs Ore).

En route, Caroline rang back. They had now arrived home and had checked their photos and consulted a field guide. My wife took the call, and you can imagine my surprise and panic when I heard her say 'Blue-cheeked, oh, that's nice!' We got to Needs Ore at about 1930 hrs and a quick look at the tree revealed that it wasn't there. We then spent about another hour looking for it, both in the immediate vicinity and elsewhere around Needs Ore, but to no avail. I returned the following morning and spent about three hours looking for it, both at Needs Ore and nearby Park Shore, again without success.

Later that evening a bee-eater, identified at the time as European (*M. apiaster*), was seen about 10 km to the south-west over the car park at Lower Pennington Lane at about 2000 hrs, and may have been the Blue-cheeked. Then back at Needs Ore, Mike and Peter Oram, unaware of the previous sighting, had brief views of a bee-eater flying east at 1810 hrs on 23rd which they described as having blue-green upperparts. However, Mike felt their views were too brief to claim either species. Subsequently, Blue-cheeked Bee-eaters were found at Braunton Burrows, Devon on June 30th and Bockhill, Kent on July 22nd. BBRC concluded that the Hampshire and Devon records involved the same individual and that the Kent bird was different.

GB status: 1, 1921; 10, 1951–2019. Very rare. (Nominate persicus *breeds Egypt & Middle East, N to S Turkey, Armenia, Uzbekistan & S Kazakhstan, E through Iran to NW India. Race* chrysocercus *breeds SE Morocco & Algeria, also W Africa. Winters Africa S of Sahara.)*

Bee-eater *Merops apiaster*

BBRC to 1990; HOSRP 1991 to date

A rare vagrant. 29 records involving 41 birds since 1979.

The first for the county was one seen by Ron King as it flew east over his garden at Oliver's Battery, Winchester on August 19th 1979. Extracts from Ron's description are reproduced below.

> *Whilst working in my garden at 1940 hrs on August 19th I was distracted by birds calling. One call was much louder and more distinct – a sort of repetitive queep – queep. The call was unusual but recently familiar and I immediately looked up and saw a small group of House Martins and a much larger bird flying overhead. Unbelieving and without seeing any colours, the call, shape and flight was sufficient for me to identify this as a Bee-eater!! The wings were pointed, thick at base and forward in flight. The tail was fairly long with two central tail feathers protruding from the centre. The bird was flying with a deep undulating flight (one or two flaps up and gliding down) in an easterly direction and disappeared after a few seconds over the houses. I have seen lots of Bee-eaters before and at the time had just returned from the Camargue – hence my certainty of the bird being a Bee-eater.*

Subsequently, there were ten further records of singles up to 2002 followed by a clear increase from 2005 onwards with 18 records involving 30 birds up to 2020, as shown opposite.

The second was seen by David Christie as it flew fairly low (about the height of 1½ telegraph poles) eastwards over the River Hamble north of Warsash on August 29th 1981. The next was watched by Ian Brown over his house at Yateley as he left for work on July 27th 1989. It was initially considered not proven by BBRC but accepted after recirculation in 2020. The first multiple sighting was of two seen by James Andrews over Hatchet Pond on August

5th 1995. Two singles were seen in 1998. The first was found by Richard Carpenter at Cowes Lane, Hook-with-Warsash at 1630 hrs on April 26th; it disappeared but was then relocated at Hook Park where it stayed until dusk and was seen by at least 20 observers. The second was seen and heard by Tim Cowley as it moved north over Worthy Down at 1400 hrs on October 23rd.

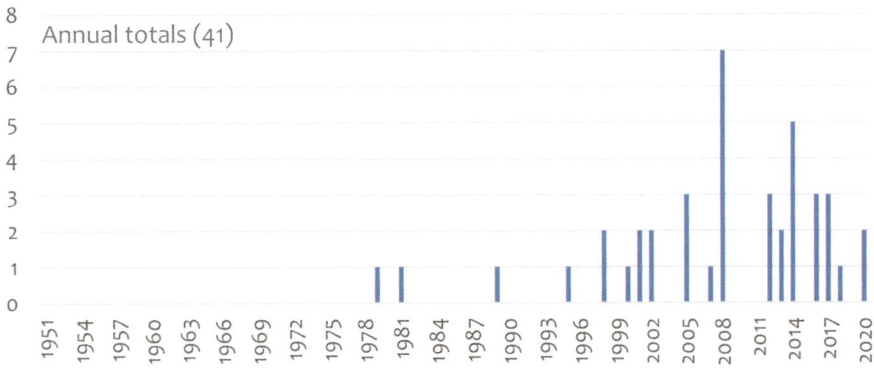

Bee-eater, Hook-with-Warsash, April 26th 1998

One was found by David Ball at Sandy Point at 1045 hrs on May 2nd 2000. This may well have been the individual seen at Selsey Bill, West Sussex the previous day. John Cloyne saw one which was calling continuously and loudly as it flew low to the east over his house in Fulflood, Winchester at 0830 hrs on May 31st 2001 and later that year Ivor McPherson saw one over Titchfield Haven on July 26th. The following year M. Jones heard and saw one over Longstock at 1855 hrs on August 15th. It was relocated hawking for insects over the village early the next day and seen again later that morning by Barry Goater and other observers. A few days later, P. Smith heard and saw one over his garden in Oak Road, Ashley on August 26th. In 2005 Brian Goddard and R. Pitty saw two circling with Swifts over the scrapes at Titchfield Haven at 1130 hrs on June 3rd before leaving north. These were probably the same birds seen at Shoreham, West Sussex on June 1st and 2nd. Andy Johnson watched an adult over the northern edge of Sandy Point between 1804 and 1811 hrs on August 3rd before it drifted off west. Andy had the good fortune to find another there on August 9th 2007. He first heard the bird and then saw it loosely associating with a flock of 14 Swifts which moved east at 1310 hrs. In 2008 P. Davies found a party of five around Cadman's Pool car park in the New Forest on May 11th and he reported that at least two were still there on 13th. The next day Martin Bennett had brief views of a flock of five probables which flew north over Brogenslade Bottom at 1945 hrs. Subsequently, Simon Hartill and David Holland saw one over Denny Wood on the afternoon of May 23rd and Alistair Martin saw another over West Hayling Shore at 1450 hrs on July 14th.

The next were in 2012 when Paul Winter saw one over Lower Test Marshes on May 5th, Tim Lawman had one fly north over South Hayling on May 17th and Ed Bennett saw another fly west over Titchfield Haven at 0948 hrs on July 28th and presumably the same bird fly north there at 1026 hrs the next day. In 2013 there were two records of singles heard by Simon Ingram over Keyhaven on April 28th and by Keith Vokes over Marchwood on May 18th. In the following year Alan Lewis heard one calling over Acres Down and picked it up gliding high to the south-west over the viewpoint at 1115 hrs on May 18th. It was seen by at least nine observers who had gathered there to look for raptors. Tim Parminter found a group of four circling low over Pig Bush at 1420 hrs on May 31st which remained in the area for four minutes before flying north. Two pairs bred on the Isle of Wight in 2014 and conceivably a probable sighting of five seen from a moving vehicle over the A27 at Emsworth on June 3rd involved these birds.

In 2016, S. Walker saw three fly south-east over his garden in Fareham at 1800 hrs on August 14th. The following year Andy Rhodes photographed three at Needs Ore Point on May 21st and in 2018 Bryan Pinchen photographed one perched on a wire there from 0615 to 0620 hrs on May 11th before it left north-east. Most recently one was seen by Andy Lester over Hut Wood on May 10th 2020 and another was seen by Mark Anderson flying south-west over Fishlake Meadows on July 18th.

Bee-eater, Needs Ore, May 21st 2017. Andy Rhodes

Analysis of arrival dates shows that 25 occurred in spring between April 26th and June 3rd and 15 after the breeding season between July 14th and August 29th apart from the late bird on October 23rd.

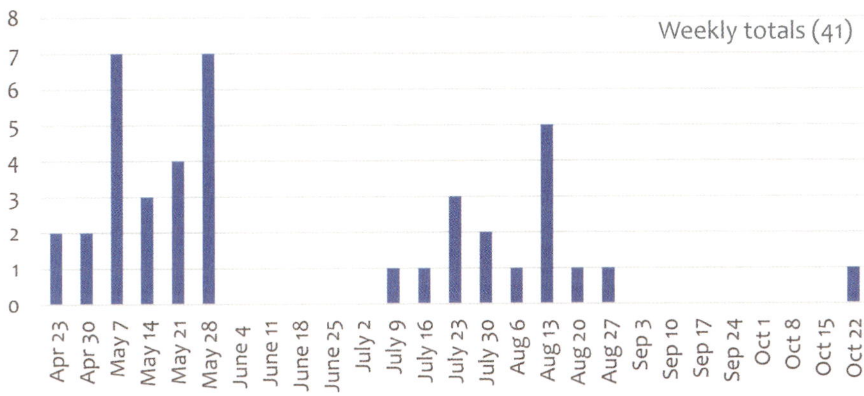

K & M mentioned that Edward Hart informed them of two specimens obtained in the New Forest district in the 19th century but give no details of dates or locations.

GB status: 139, 1793–1949; 1,896, 1950–2019. Increase, mean 71 per year in 2010–19. (Breeds NW Africa & Mediterranean Europe, N to C France & N Ukraine, occasionally N to S Sweden & Netherlands, E through C Asia to E Kazakhstan. Winters throughout sub-Saharan Africa, & locally India S to Sri Lanka, and S Arabian Peninsula.)

Wryneck *Jynx torquilla*

Formerly a moderately common summer visitor which last bred in 1975. Now a very scarce passage migrant with around 478 recorded since 1952.

The species was known to Gilbert White in the second half of the 18th century. He mentioned that it sometimes arrived in mid-March while K & M stated that White recorded it as early as March 5th. In the late 19th century K & M described it as 'a summer visitor to all parts of the county' and quoted extreme dates of March 19th 1903 at Emsworth and December 5th 1861 in the New Forest. It bred on heathlands and in parkland, large gardens, orchards and other rural situations.

In common with other parts of the country, a major decline took place in the first half of the 20th century although coverage was poor during this period. Published records for the 1930s include nesting at Wellow, presence at Ampfield, Farnborough, Havant, Linwood, Medstead, Otterbourne, Petersfield, Ringwood, Wickham and Winchester and the earliest arrival at Brockenhurst on March 15th 1936 (*SEBRs*). In the 1940s, pairs bred near Hinton Admiral in 1940, Crondall in 1944, Petersfield in 1946 and Kingsley in 1949 (*HBRs*), an early arrival was at Lymington from March 18th to 26th 1946 (Wiseman 2017) and one on the shore at Emsworth on August 30th 1945 was the only record of an autumn migrant (*HBR*).

In the 1950s breeding was confirmed at Aldershot and Kingsley in 1950 and Drayton, Portsmouth in 1954, with six other pairs possibly breeding between 1954 and 1958 (Monk 1963). Subsequently, one summered at Woolmer Forest in 1966 and 1972 and pairs attempted breeding at Cranesmoor in 1968 and Four Marks in 1971. Finally, a pair was seen feeding young in a nest at Newnham in 1975 (Norman Clayden *in litt.*). The only subsequent confirmed breeding in England was in Buckinghamshire in 1985 although the species continued to breed sporadically in Scotland until 2002 (Holling *et al.* 2012).

Excluding breeding birds, 478 have been recorded since 1951 with 89 in spring (up to the end of July) and 389 in autumn.

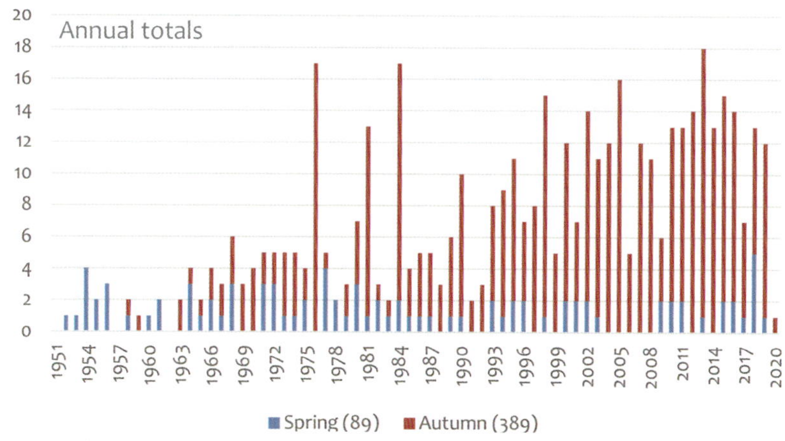

During the 1950s and 1960s numbers were low with spring records predominating. This reflects records of unmated males at former breeding sites, some of which stayed for several weeks, and rather poor coverage of most coastal sites in autumn. Subsequently, the numbers recorded in autumn increased. As many as 16 were found in 1976 and this total was not equalled again until 2013. The trend during the 21st century has been stable with occasional poor years including the most recent in 2020, when only one

Wryneck, Hook-with-Warsash. September 5th 2015.
Pennington Marshes, September 13th 2012 (overleaf). John Hilton

was found. This presumably reflects the frequency of timely easterly and north-easterly winds to drift migrants from Scandinavia which is believed to be the main source of our autumn migrants. Spring records have remained rare with none in some years. However, in 2017 one was at Noar Hill on June 1st, in 2018 five were seen including one which held territory near Burley in the New Forest from May 18th to June 3rd and another which flew in off the sea at Lee-on-the-Solent on July 7th and in 2019 a wide-ranging singing male was at Ocknell Plain, New Forest on June 19th but not seen again.

As shown below, analysis of arrival dates shows the earliest records of two at East Tytherley on April 2nd 1961 and one at Backley Plain, New Forest on April 3rd 2010 (omitted in error from the 2010 *HBR*) with records thereafter evenly spread through until late May. Most were seen on one date only although from mid-May longer stays mostly relate to presumed unmated males in the 1950s and 1960s apart from the 2018 bird mentioned below.

One in a garden at Weston, Southampton on July 28th 1982 was probably an early autumn migrant but otherwise the earliest were at Amport and Farlington Marshes on August 8th 1994. There is a clear peak in late August and early September with gradually declining numbers until late October. There have been several exceptionally late birds in recent years: singles at Calshot on October 25th 2015, Keyhaven on October 27th and 28th 2013, Fawley Power Station from October 31st to November 2nd 2004, Fishery Creek, Hayling Island on November 5th and 6th 2011 and Old Winchester Hill on November 9th 2013.

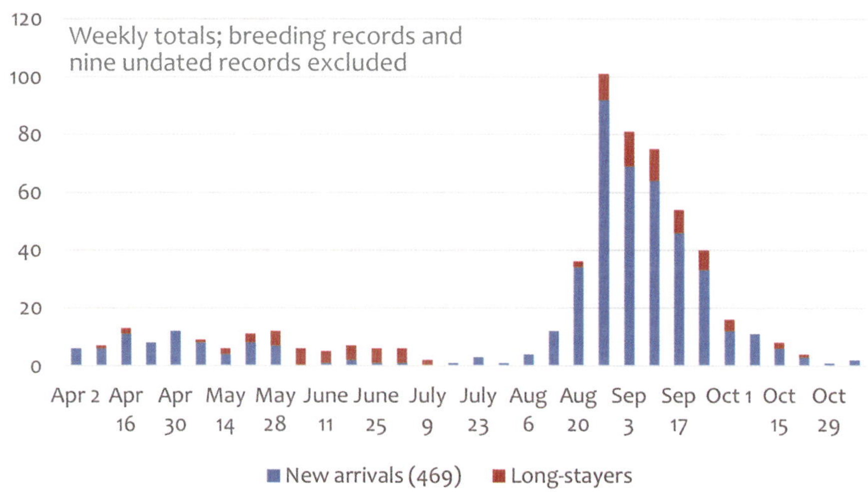

Most spring records were at inland locations with 25 in the New Forest and a scattering elsewhere. Very few occurred at coastal sites with three at Titchfield Haven (May 8th 1981, April 13th 2016 and April 20th 2018), two at Lower Test Marshes (April 18th 1996 and April 17th 2010) and singles at

Wryneck, Hook-with-Warsash, September 4th 2015. Dave Ryves

Ashlett (May 1st 1977), Sinah Common (April 20th and 21st 1983), Farlington Marshes (April 30th 1986) and Hook-with-Warsash (April 26th 2009).

Coastal records predominate in autumn. Favoured locations include Farlington Marshes (61), the Keyhaven area (40), Hayling Island (36 including 13 at Sandy Point, six at Northney and four each in the Sinah area and at West Hayling), Hook-with-Warsash and Chilling (15), Titchfield Haven and Brownwich (14), Fawley to Calshot (11), Needs Ore (7), IBM Lake, Cosham (5) and Southampton Western Docks (5). The records from Hayling Island include four occasions when Andy Johnson discovered the distinctive droppings of the species; twice a bird was subsequently discovered: a day later at Sandy Point on August 31st 2008 and three days later at West Hayling on September 11th 2010. Most records are of singles but two have been seen together on six occasions at Calshot on September 2nd 1984, Southampton Western Docks on September 26th 1993, Hook-with-Warsash on September 9th 1998 and September 8th 2010, Farlington Marshes on August 27th and 28th 2000 and Needs Ore on September 15th 2017. Three were together at Fawley on September 9th 1981. Birds often remain in one area for a week or longer as shown by the numbers of long-stayers in the chart above. Examples of long stays include singles at Hook-with-Warsash from September 5th to 20th 1998, Farlington Marshes from September 22nd to October 1st 2001, Sandy Point from September 21st to 29th 2005, Iley Point, Keyhaven from September 23rd to October 3rd 2017 and Titchfield Haven from October 15th to 22nd 2019.

Inland at least 18 have been seen in the New Forest, an almost unbelievable total of 11 at one site, The Mill Field, Old Basing and seven at Old Winchester Hill. Many have occurred in gardens where ants have been a popular food source. Such records include those of two at Nomansland from August 15th to September 7th 1981 and singles at Springvale, Winchester from August 22nd to September 2nd 1976, Eastleigh from September 15th to 22nd 1981, Church Crookham from September 15th to 22nd 1986, Whitenap from October 12th to 18th 1990, Ringwood from September 10th to 19th 1994, Woodlands, Totton from August 26th to 30th 2002 (omitted in error from the 2002 *HBR*), Broughton from October 2nd to 4th 2004 and Hinton Ampner on August 20th and 21st 2018.

GB status: 10,495, 1986–2019. Increase, mean 366 per year in 2010– 19. (Breeds throughout Europe E to Japan. Some winter in Mediterranean basin, otherwise winters widely throughout Africa and S & SE Asia.)

Red-footed Falcon *Falco vespertinus*

BBRC to 2005; HOSRP to date

A rare non-breeding summer visitor and passage migrant. Two old records and around 61 individuals since 1959.

The first for the county was a female shot by Mr. H. B. Leggatt near Titchfield in October 1875 which was reported in *The Field* for October 30th 1875. The specimen had been sent to the editor who confirmed its identity. The next was a male seen at close range by Roy Pitman, an experienced observer, and others at Martin Down on August 24th 1937 (*SEBR*).

The first modern record involved a party of five consisting of an adult male, an adult female, a first-summer male and two first-summer females which were found by Colin Tubbs at Bishop's Dyke, New Forest on May 16th 1959. Up to four remained until May 24th and the last sighting was of the adult male on June 6th. Extracts from the description submitted to BBRC by Mick Goodman are reproduced below.

A total of five Red-footed Falcons were watched hawking insects at a height generally over 100 feet and at times much higher. Four were often seen in the air at once and five were accounted for on one occasion. Two were adult males and three were adult females or possibly in some stage of immature plumage. In shape and flight they resembled Hobby (one or two of which were present for a short time) but tail slightly longer and wings narrower and more flexible. Slaty grey upperparts of males with russet undertail-coverts, red feet and cere; wings with pale silvery grey area on upper surface of outer secondaries and/or inner primaries. Underparts including wings grey. Three remaining birds were essentially alike but may have varied in colour to some extent. Description of one bird as follows. Paler appearance than male with pale honey-coloured crown, dark moustache and pale creamy underparts with faint streaks; upperparts sandy buff barred darker, wing quills dull brownish; rump and tail grey above, tail barred, underwing finely barred.

Ron Johns, also present on that day, told this story in Moss Taylor's book *Guardian Spirit of the East Bank: A celebration of the life of R. A. Richardson* (Taylor 2002).

Along with Richard [Richardson] and Mick Goodman, I went to the New Forest in mid-May 1959, where we had arranged to see Howard Medhurst and some others who were camping there. Shortly after we arrived, Colin Tubbs appeared in a rather agitated state and took Howard to one side. After a lengthy discussion, he explained that they had found a party of very rare birds not far away which may possibly stay and breed in the area. If everybody agreed not to mention them to anyone, Colin was prepared to take us to see them, which we all agreed. After a rather circuitous journey we ended up at the bottom of Bishop's Dyke, where we were shown a party of five Red-footed Falcons. Everyone was sworn to secrecy.

After leaving, the three of us went on to Portland and then to see Horace G Alexander at home in Swanage for tea. Almost the first thing Richard said was 'you'll never guess what we've been watching'!

Horace Alexander saw three there the next day!

Since then, there have been records in 25 years involving a minimum of 56 individuals of which 28 occurred in the New Forest.

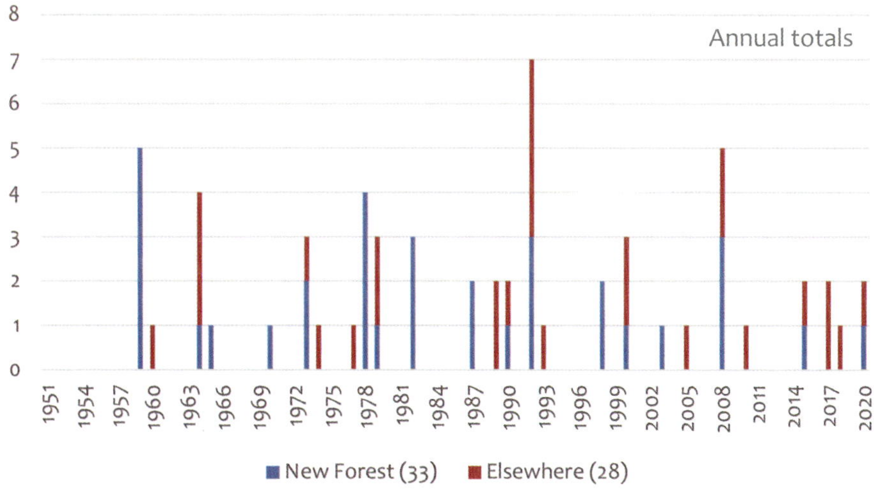

Six were seen in the 1960s. A first-summer male was seen by David Billett, John Simons and Ed Wiseman flying north-west over Farlington Marshes on May 29th 1960. A small influx in the spring of 1964 produced three: a first-summer male found by David Glue at Nursling Gravel Pit on May 30th which was seen by many observers until June 9th when a second bird, an adult female, was found there by Norman Pratt, and a first-summer male found by Graham Rees and Bill Truckle at Ashley Walk on May 31st. Later that year an adult male was seen by Mr. and Mrs. I. R. Beames at Milford on Sea on October 2nd. In 1965 a first-summer male was found by M. T. Barnes, C. A. E. Kirtland and O. R. Marks at Rhinefield Walk on June 7th.

Numbers increased in the 1970s with a total of 12 seen in five different years. The first was an adult male seen by A. J. L. Smith at Bishop's Dyke on May 9th 1970. There was a series of records from the Broomy Walk and Slufters Inclosure area of the New Forest in 1973. A first-summer male on May 25th was found by an unknown observer. John Boys discovered an adult female on June 7th which was subsequently seen by many observers until June 21st. On June 16th Len Mummery found a first-summer male which is assumed to be the earlier bird. Later that day both birds were seen by David and Rosemary Billett. Away from the New Forest, Andrew Williamson found an adult female at Farlington Marshes at 1600 hrs on June 5th which was also seen by David Billett and remained until dusk the next day when it left west. In 1974 the only record was of an adult male found by John Oakshatt, who was leading a group of 36 members of the Thomas Hardye School Natural History Society, at Bickton on May 5th. It was watched from 1305 to 1314 hrs and seen perched on a telegraph wire for some of this time. The record was published in error in the 1973 *HBR*. In 1977 John Taverner found an adult male at Winnall Moors at 1800 hrs on June 2nd which was seen the next day by an unknown observer. A further series of records from the New Forest in 1978 involved at least four individuals. The first sighting was of a male seen independently by John Taverner and Dave Smith at Bishop's Dyke on May 17th; Dave also suspected a female was present. The male was still present on 17th and both a male and a female were seen by Dave Smith and R. G. Mayall on 20th. Johnny and Anne Bowers found a male near Burley on May 30th which was also seen by Mike and Pete Combridge. Terry Martin found a first-summer female at Bishop's Dyke on June 17th which he showed to John Taverner. Finally, Mike and Pete Combridge found a male at Sway on June 24th which was different to the bird they saw near Burley; it was still present the next day. In 1979 Ed and Wendy Wiseman watched an adult female at Salterns Marsh, Lymington for an hour on the evening of June 4th, and later in that month Ed found an adult male at Bishop's Dyke on 20th which was seen independently by several other observers and was still present the next day. The third that year was an adult male spotted by Johnny Bowers while driving along the A34 at Cole Henley, Whitchurch on July 30th. He was able to stop and get excellent views.

The next were in 1982 when three were seen at Bishop's Dyke. An adult male was found there by Eric Baker and Graham Brandejs on the afternoon of April 28th but not seen again. Just over a month later, on the afternoon of May 30th, a first-summer male was found there by M. A. Hallett which he showed to several birders. The next day it was refound by Dave Nurney, Dave Harris, Jonathon Bowley who also saw an adult female in the same area. There was a five-year gap until the next were seen in 1987, again in the New Forest, with an adult female found by Pete Combridge at Stoney Cross Airfield on July 2nd and a first-summer female found by C. Heyward and also seen by Kevin Ilsley at Bishop's Dyke on July 5th. 1989 produced two records away from the forest, the first of a first-summer female found by David Thelwell perched on wires as he drove north through Shipton Bellinger on June 2nd and subsequently seen by many observers up to June 14th and an adult male found by John Lucas at Mortimer West End on June 4th.

In 1990 a female was seen by Graham Stephenson flying west over Farlington Marshes on May 4th and a first-summer male was found by Mark Hardwick at Bishop's Dyke on May 28th and subsequently seen by Colin Savage, Martin Terry and others up to 30th. At least seven were seen in 1992 including three

Red-footed Falcon, first-summer female, Shipton Bellinger. Loafing, watching the watchers, June 3rd 1989

or more in the New Forest where the first reports were on May 17th when an adult male was located at Bishop's Dyke by Ken Arber, Pete Gammage and P. Whitfield and a first-summer male was found in the same area by Peter Hutchins and Andrew Pink. Presumably, the same adult male was found by Mike Baker at Ocknell Plain on 22nd and at Ashley Walk by Mike and Pete Combridge and independently by Graham Stephenson and Andy Hanby on 25th, where it was still present the next day. A first-summer female was located by Norman Orr at Whitten Bottom near Burley on June 10th where it stayed until 15th and presumably the same bird was found by Mike and Pete Combridge at Picket Post on 21st and by Tim Doran at Fritham Plain on 27th. Elsewhere, David Christie saw a first-summer male over Southampton Central Station on May 19th, Andy Hanby and Graham Stephenson found an adult male and a second-summer male together at Horsebridge on May 24th and the adult was seen by many observers there the next day, and nearby John Grafton and Tim Norriss saw a first-summer female at Kings Somborne on June 21st and 22nd (not a first-summer male from June 19th to 21st as in *BoH*). Occurrences were sparse in the mid-1990s. A female was found by Pete Combridge and Simon S. King at Lower Test Marshes on June 21st 1993. In 1998 two first-summer males were found together by an unknown observer at Beaulieu Heath West on May 24th. They were subsequently seen by hundreds of observers; two were still present on 31st and one was last recorded on June 4th.

First-summer male, Beaulieu Heath, May 27th 1998

In 2000 three first-summer females were reported. Mike Rafter saw one fly in off the sea and up the Beaulieu River on May 9th, another first reported by D. Smith, B. F. Forbes and C. Wright delighted hundreds of observers in the Bishop's Dyke area from May 21st until June 13th and finally one seen independently by John Eyre and L. Smith perched on a road sign on the eastbound carriageway of the A31 at the eastern end of Alton by-pass on June 16th. The next was a female watched by Simon Buckell and Wayne Percy hawking insects over Acres Down from 0900 to 0930 hrs before moving off south on May 31st 2003. In 2005 Ian Ellis found and Andrew Self located a first-summer male at Martin Down at 1500 hrs on June 18th; it was found independently by Alan and John Greensmith at 1845 hrs. It roosted in a tree at nearby Bokerley Dyke but was not seen the next day.

Records of a female seen by Ed Wiseman, Colin Tubbs and Gerald Westerhoff at Verely Hill on May 17th 1964, an adult male found by Keith R. Grant at Farlington Marshes on May 12th 1980 and a male which Pete Durnell saw fly in off the sea at Gutner Point, Hayling Island on May 5th 1997 were harshly considered not proven by BBRC. A further 16 records including ten from the New Forest were considered not proven by BBRC (see Appendix VIII).

Unsubmitted records include those of a female seen by David Christie at Bishop's Dyke on May 11th 1971, an adult female found by Ed Wiseman at

Bishop's Dyke on May 28th 1976 and males seen by David Billett at Bishop's Dyke on June 3rd 1978 and Cranesmoor on June 15th 1978.

HOSRP assumed responsibility for this species in 2006.

An influx of around 60 into England in May 2008 produced a minimum of five, and possibly seven, individuals in Hampshire. The first was a first-summer female which Wayne Percy saw flying south-west over Ridley Plain, New Forest on May 10th. Three days later Barry Friend and Barry Goater saw an adult male with four Hobbies at Beaulieu Road and Andy Page and Wayne Percy found a first-summer male at Holm Hill, New Forest which stayed until 15th. On 18th Lawrie Chappell watched a first-summer male drift north over Ibsley Water and on 22nd Damien Offer found two, an adult and a first-summer male with nine Hobbies at Beaulieu Road where the first-summer male remained until June 1st and was enjoyed by many birders. The last two may well have been those recorded earlier in the month at Beaulieu Road and Holm Hill. Later that year Glyn Horacek-Davis watched a female on his local patch at Moorgreen Farm from 1205 to 1230 hrs on September 20th before it flew off north-west towards Eastleigh. The next was a juvenile found and videoed by Marcus Hebbes at Hawkhill Inclosure, New Forest from 1300 to 1320 hrs on October 10th 2010. A five-year gap ensued before the next occurred. Tim Parminter watched a first-summer male for five minutes near Pig Bush, New Forest on June 6th 2015 and V. Boult photographed a first-summer female on private land near Fordingbridge

on June 20th which remained in the area until 27th. The next was a female which Andy Collins watched fly in off the sea at Brownwich at 0710 hrs on May 6th 2017; a week later on 13th another female was found and photographed by Gareth Rees at Fishlake Meadows at 1500 hrs but not seen again. A year later David Mills saw a male flying north over Noar Hill on May 13th 2018. Finally, two were seen in 2020: a female watched by Amy Robjohns circling over Titchfield on June 11th and another female seen by Simon Colenutt and one unknown observer at Acres Down the next day. There was an influx of at least 30 into Britain between mid-May and mid-June 2020.

Analysis of arrivals dates shows that most occurred in spring between April 28th and July 5th with a clear peak between mid-May and mid-June. Later occurrences were on July 30th, September 20th, October 2nd and 10th.

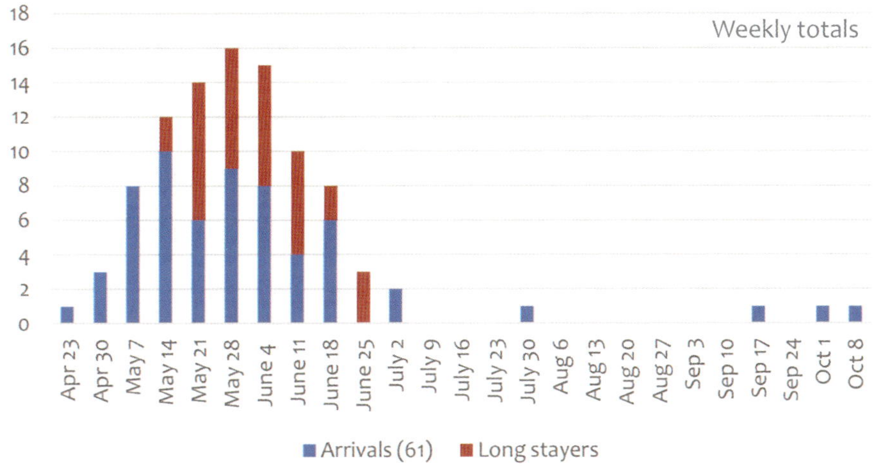

A note by Murray A. Mathew in *The Zoologist* for January 1877 reported that Mr. Hayden informed him that a male was killed close to Fordingbridge in December 1876. This record was mentioned by Mathew in *The Birds of Devon* published in 1892 although it was said to have been shot in January 1877. In the absence of a specimen, such a mid-winter record is unacceptable. Another was reported by V. E. Stillwell at Yateley on April 28th 1917 (*The Field*, May 19th 1917). The description mentions that it was carrying what looked like a young Moorhen, with long, limp little legs dangling done. It seems likely to have been a male Sparrowhawk.

Red-footed Falcon, first-summer male, Beaulieu Road, May 19th 2008. Mike Darling

GB status: 68, 1830–1949; 937, 1950–2019. No clear trend, mean 17 per year in 2010–19. (Breeds in forested steppes of E Europe from E Hungary to temperate Russia, E to Baikal region. Migratory, wintering SW Africa.)

Brown Shrike *Lanius cristatus*

BBRC

A very rare vagrant. One record in 2013.

The only record for Hampshire is of a juvenile moulting to first-winter plumage found by Bob Marchant near Chilling sub-station, Hook-with-Warsash on the morning of September 20th 2013. It remained in the area all day and was seen by many observers. Extracts from Bob's description of events given below are taken from Marchant (2014).

On the morning of Friday September 20th, I was on my local patch just after 0800 hours hoping for some migrant activity in the calm and sunshine following the previous afternoon's high winds and rain. Chiffchaffs were numerous and hirundines were assembling on overhead wires. A few Yellow Wagtails, Meadow Pipits, other common warblers and a Spotted Flycatcher were around and a fine Hobby perched up until mobbed by a Jay. As I approached Chilling sub-station, which borders the eastern edge of Hook-with-Warsash at 0955 hours, I noted a couple of Common Whitethroats dodging around the bramble bushes and there perched up was a SHRIKE! This was a first for me on the patch in 20 years and having missed the 2001 Red-backed Shrike I thought at last I had got it back. The bird was some 50 m away and with a brief sighting through the scope I assumed I had at last ticked Red-backed Shrike. The light was poor with the scope just catching the sun and in a brief two-minute view I noted the dark mask, vermiculated underparts and brownish upperparts and tail. It looked good for a juvenile but it disappeared for some 20 minutes before reappearing further away but in cover alongside the Solent Way footpath. I texted a few locals and put out the news of a probable juvenile Red-backed Shrike. The bird then flew and disappeared over the bramble patch and out of view. I moved up to the Solent Way footpath for a better view overlooking the brambles and paddock area and thankfully refound the bird along the fence line running perpendicular to the footpath and adjacent to the bramble bushes. The shrike was now some 80 to 100 m away but in better light. As more detail could be made out it was becoming apparent that the bird was showing the features of a first-winter Brown Shrike, and revised news was put out as a probable first-winter of that species.

Brown Shrike, juvenile, Hook-with-Warsash, September 20th 2013. Bob Marchant and Lee Fuller

The salient features included chestnut crown, brown upperparts (not noticeably vermiculated at that range), dark-centred tertials with pale edgings, long brown tail with reddish base/rump, a short primary projection reaching only to the lower end of the reddish patch, and buff-coloured underparts with noticeable vermiculation (when not in bright sunlight). The dark mask also continued in front of the eye, though to a much lesser extent and tapered down towards the bill. The tail, though having a pale tip, had no white edging along the length (as in Red-backed Shrike) and the bare parts appeared grey, with a dark tip to the heavy-looking bill. All these points were agreed with birders gathering on the scene and with reference to the second edition of Collins Bird Guide and so confirmatory news was put out that the bird was a first-winter Brown Shrike.

The bird fed frequently from its favoured fence and a nearby hawthorn, mainly catching insects on the ground. At one point it was noted to have bathed as it returned to its perch in a somewhat bedraggled state to preen. Habitually it would cock its tail. The bird continued to range at 80 to 100 m distance and initially was difficult to photograph. However, some excellent images, many of which appeared on internet sites, supported the identification of the bird as a Brown Shrike in moult from juvenile to first-winter plumage.

GB status: 39, 1985–2020. Rare, about one per year; recent increase. (Breeds Siberia from Ob River basin E to Sea of Okhotsk & Kamchatka, N Japan & throughout much of E China. Winters Indian subcontinent & throughout SE Asia to Philippines, N Borneo & Indonesian islands.)

Red-backed Shrike *Lanius collurio*

HOSRP

Formerly a moderately common summer visitor which last bred in 1984. Now a rare passage migrant with around 102 recorded since 1954.

The first for the county was one procured by Gilbert White, presumably at Selborne, on May 21st 1768. In Letter XX to Thomas Pennant, dated October 3rd 1768, he wrote, 'it might easily have escaped notice had not the outcries and chatterings of the whitethroats and other small birds drawn attention to the bush where it was'. He evidently considered it a rare bird as he mentions only one other – shot at Selborne about the year 1773.

K & M, writing in 1905, described the Red-backed Shrike as 'found in most parts of the county, but usually not very plentifully .... of late years, its range has greatly extended'. During the 1930s, it was evidently still widespread with nesting reported at Burley, Church Crookham, Danebury Hill, Gorley, Havant, Hill Head, Hurstbourne Priors, Leckford, Linwood, Medstead, Norman Court, Oakhanger, Penton Mewsey, Petersfield, Portsdown Hill and several sites in the Winchester area. Its subsequent decline as a breeding species in Britain has been well documented (Peakall 1962, Bibby 1973). In the 1950s the New Forest was one of its last strongholds in England and it was widely, if thinly, spread through the rest of the county. The map in *BoH* shows 99 tetrads, just under 10% of the county's total, where breeding occurred after 1950. Only 15 were occupied between 1971 and 1984 when breeding was last recorded by Chris Packham who found a pair with two fledged young near Overton.

Red-backed Shrike, male, Titchfield Haven, June 5th 1996

Analysis of records from coastal locations and inland sites where breeding has ceased reveals a total of 102 birds in 46 years between 1954 and 2020.

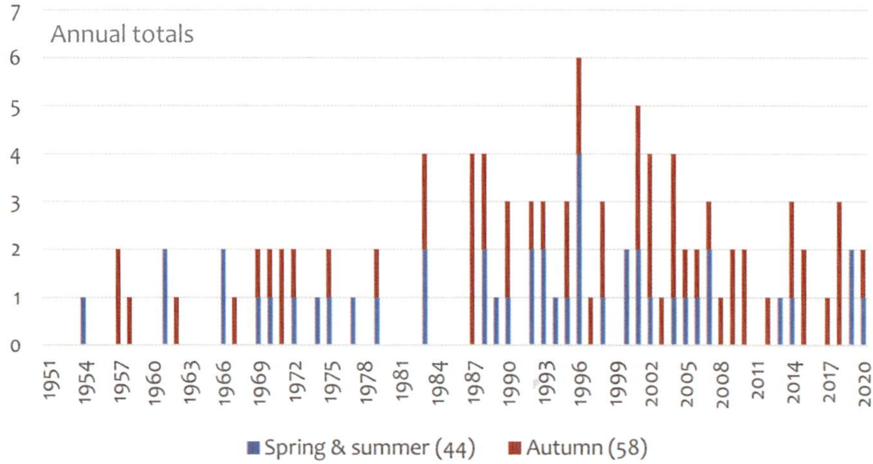

The earliest were males at Park Shore, Needs Ore on April 15th 1961, Bishopstoke on April 24th 1975 and Avon Water, Keyhaven on April 29th 1961. Other spring records fell between May 10th (1998, female, Titchfield Haven)

and July 12th (1979, male, Bramley Camp). Records came from Farlington Marshes (6), the New Forest (5), Martin Down, Titchfield Haven and Yateley Common (4 each), Fleet Pond, Normandy Marshes and Silchester Common (2 each) and Bramley Camp, Bricksbury Hill, Charity Down Farm, Dibden Bay, Eelmoor, Itchen Abbas, Kings Worthy, Mapledurwell, Needs Ore, Old Winchester Hill and Perham Down (1 each). Most were recorded on one or two dates only apart from males at Silchester Common from May 15th to July 13th 1969 and May 19th to June 14th 1970, Yateley Common from July 5th to 14th 1996 and Pig Bush, New Forest from June 15th to July 5th 2014.

Red-backed Shrikes: juvenile, Pennington Marshes, September 29th 2008, male, Cheriton, July 26th 2012 and juvenile, Farlington Marshes, September 21st 2018.
Linda Miller, Ian Williamson and Lee Fuller

Autumn records fell between July 25th (2015, Beaulieu Heath West) and October 10th (2001, Farlington Marshes, until 15th) apart from a late bird at Farlington Marshes on October 22nd 1967. Favoured locations were Farlington Marshes (14), Hayling Island (8) and Keyhaven/Pennington Marshes (7). Other records involved 14 at nine coastal locations and 15 at 12 inland sites mostly in the southern half of the county. Stays of up to a week were not infrequent but longer sojourns involved singles at Farlington Marshes from August 10th to 29th 1957, Keyhaven from September 17th to October 2nd 1983 and September 20th to October 3rd 1992, West Hayling from September 25th to October 2nd 2001, Nursling and Lower Test Marshes from September 5th to 13th 2004, Pennington Marshes from September 27th to October 11th 2008 and Sandy Point from September 4th to 28th 2014. All records involved singles although at Farlington Marshes a first-winter was present from September 9th to 16th 1987 and a presumed adult female from September 11th to 15th; both were seen in the same bush on 15th.

The weekly distribution of records is shown below.

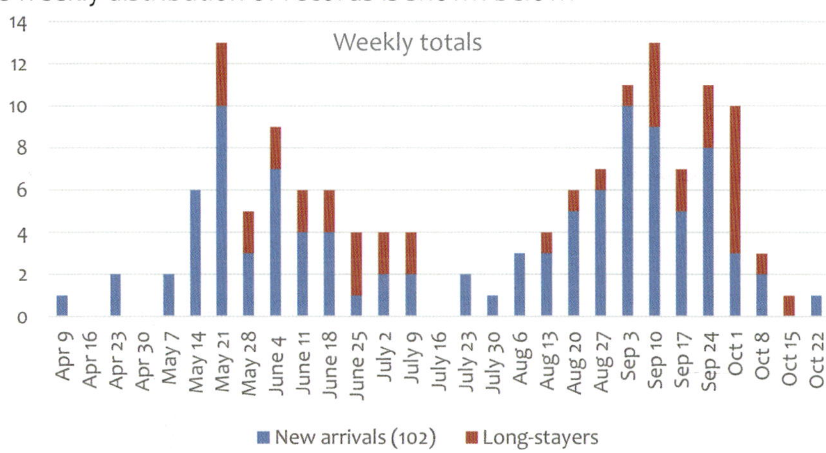

GB status: 7,087, 1986–2019. No clear trend, mean 173 per year, 2010–-19. (Widespread breeder from N Spain E to Greece & N through Europe to C Fennoscandia, Russia E to Ob River in C Siberia, Turkey & Caucasus region S to NW Iran. Winters equatorial & S Africa.)

Daurian Shrike *Lanius isabellinus*

BBRC

A very rare vagrant. One record in 2010.

The only record for Hampshire is of an adult female found by John Norton at Grove Road Recreation Ground adjacent to Forton Lake, Gosport at 1645 hrs on September 10th 2010. He provisionally identified it as an Isabelline-type shrike; it was still present the following day when identification was confirmed and it was seen and photographed by many observers. Extracts from John's account of its discovery and subsequent identification from the 2010 *HBR* (Norton 2011) are reproduced below.

After giving up on work for the day I met up with my partner Debbie Allan at 1645 hrs to go for a short walk over to the Felicia Farm and Priddy's Hard area on the north side of Forton Lake. Even though it was overcast with heavy rain clouds and with a strong wind still blowing I was nonetheless optimistic of finding something unusual. However, by the time we returned to the recreation ground three-quarters of an hour later, we had not seen anything of note and were more intent on getting home as it had started to rain. As Debbie headed off to the local shops, I noticed a bird drop down to the ground and fly back up again near the base of a thick Blackthorn hedge near the shore. It seemed quite pale and my birdwatching intuition told me immediately it was something interesting – perhaps a Redstart? But when I looked through my binoculars, instead of a flickering tail I found myself looking at the unmistakable (or so I thought) outline of a Red-backed Shrike! By now Debbie was nearly at the other end of the park. So I phoned her mobile and waved frantically for her to come over.

We watched the shrike initially at 70 m or so and I pointed out the salient features of broad, flattish head, dark face mask and short, stubby bill. We slowly approached and were able to reach the two wooden benches on the shore of the creek, situated only 20–25 m away from where the bird was perched. During this time it had remained motionless, perched on a buckled section of chain link fence about a metre above the ground and nicely protected from the rain. However, even at this close distance it was not easy to make out features in the gloom and because it was facing us its upperparts, wing and tail features could not be seen. We took some record shots with our compact digital cameras but after watching it for a further ten minutes it disappeared into a dark hole in the hedge and I wondered whether it had gone to roost. It was only then that I started to question some of the details I had seen – the characteristic chevron markings on the sides of the breast seemed very faint for a typical juvenile Red-backed Shrike and the crown seemed greyish brown instead of rufous-toned, as possibly did the nape (which could not be seen well at this stage). While the bird was out of view I attempted to phone or text a few local birdwatchers to let them know I had found a Red-backed Shrike. I spoke to Ian Calderwood who was eating his tea but said he would come over when he had finished. Whilst I was waiting, the shrike reappeared, flew up to the top of the hedge, made one sally for an insect, then went back down again into the hole. It was further away but in better light so rather than check features through binoculars I took a few more record shots just in case it decided to disappear again for good. I did manage to briefly see the upperside of the tail whilst the bird was in flight, which appeared to be the typical dark rufous-brown of a Red-backed Shrike; however, I thought I caught a glimpse of some much warmer orange-brown at the base or near the rump – which started me thinking. Ian eventually arrived just after 1800 hrs armed with telescope and video camera. The shrike thankfully came out again on the fence allowing us both better views, but face on again. We watched it for a further 15 minutes and contacted more local birders, none of whom was able to make it over. I optimistically suggested to Ian that it could be an Isabelline Shrike but he was not too sure. We had still not gained decent views of the upperparts or tail, but I returned home to have some tea of my own and take a detailed look at my photos. It was only then that the penny finally dropped that it clearly was an Isabelline Shrike! I posted a message on Hoslist just after 1900 hrs to alert other Hampshire birders, followed a bit later by a few of the photographs. I was confident that the bird would not move off in the rainy conditions overnight, and it was with much relief that it was still present the next day, showing fairly well on and off, though disappearing for long periods within the Blackthorn hedge or a nearby Hawthorn tree. Several hundred birders from Hampshire and a few from further afield turned up to see it during the day. During the morning it was seen catching spiders inside the hedge and was obviously feeding up quite well. Too well as it turned out, as the next evening it departed under a starry sky.

Daurian Shrike, female, Gosport, September 11th 2010. Keith Maycock

This bird was initially accepted by BBRC as an Isabelline Shrike showing the characteristics of the race *isabellinus*. In January 2018, BOURC adopted IOC taxonomy and split Isabelline Shrike into Daurian Shrike and Turkestan Shrike (BOURC 2018). Since then, a review of post-2002 adult female Isabelline Shrikes has been undertaken, resulting in the acceptance as Daurian Shrike of a further six individuals including the Hampshire bird. There are now ten accepted records of Daurian Shrike for Britain (Stoddart & Hudson 2021).

GB status: 10, 2002–2019. About one per year; stable. (Breeds S Russia, N & W China & Mongolia. Winters Arabian Peninsula, NE Africa & Indian subcontinent.)

Daurian/Turkestan Shrike *Lanius isabellinus/phoenicuroides*

BBRC

A very rare vagrant. One record in 2011.

A first-winter Isabelline-type shrike was found by R. Reeves and T. Dempsey at Moonhills, New Forest on October 30th. Identification was confirmed the next day by photographs taken by David Reeves; it stayed until November 2nd although news was not released.

Daurian/Turkistan Shrike, first-winter, Moonhills, New Forest, November 2nd 2011. Brian Dempsey

As mentioned under Daurian Shrike, in January 2018, BOURC adopted IOC taxonomy and split Isabelline Shrike into Daurian Shrike and Turkestan Shrike (BOURC 2018). Adults of the two taxa can be distinctive but first-winters, which form the bulk of British records, are – on current knowledge – not safely assigned to either species (Holt *et al.* 2019). Thus the Moonhills bird is placed in the wider either/or category.

An earlier record of a putative first-winter Isabelline-type shrike at Farlington Marshes from October 7th to 9th 2001 has not been submitted to BBRC.

GB status: 108, 1950–2020. About three per year; stable. (Daurian Shrike breeds S Russia, N & W China & Mongolia. Winters Arabian Peninsula, NE Africa & Indian subcontinent. Turkestan Shrike breeds C Asia from N Kazakhstan to Afghanistan & Pakistan. Winters SW Asia, Arabian Peninsula & E Africa.)

Lesser Grey Shrike *Lanius minor*

BBRC

A very rare vagrant. Three records in 1900, 1967 and 1973.

The first for the county was an adult male shot by S. Head, the keeper, at Poors Common near Bransgore on June 2nd 1900 and mounted by Edward Hart. The specimen is now held by the Hampshire Cultural Trust at Chilcomb House, Winchester. In K & M it was described as having been procured at Poors Common, Bournemouth. The record was thus overlooked for *Birds of*

© Hampshire Cultural Trust

Hampshire as it was assumed it had occurred in present day Dorset. A sighting of two near Andover on April 5th 1917 was recorded in *The Field* for April 21st 1917. This was included by Naylor (1996) but is surely suspect.

The first modern record involved one found dead in a porch at Holt Pound, Alice Holt Forest, on May 16th 1967. The corpse was passed to Tim Winter who arranged for the skin to be preserved and submitted the record to BBRC. The specimen is now in Haslemere Educational Museum. The most recent occurrence involved one found by 13-year-old N. B. Ansell in scrub adjacent to Fort Purbrook on July 15th 1973. He initially reported it to the RSPB and was advised by Peter Holden to submit a description to BBRC. He wrote:

I should like to report the occurrence of what I believe to have been a Lesser Grey Shrike at about ten o'clock on Sunday 15th July in some hawthorn and blackthorn scrub on Portsdown Hill near Portsmouth. It had been raining for most of the morning but had just stopped when I first saw it, perched on the topmost branch of a nearby hawthorn bush. It remained so near that I did not find it necessary to use my binoculars much. Its pose and build were reminiscent of a Red-backed Shrike, and it was not much larger than this bird, which I have seen several times. Its colour and patterning were similar to a Great Grey Shrike. The bird's most conspicuous features were its two black eyestripes which were joined across its forehead to form a continuous band. The scapulars were the same colour as the rest of the back, and the underparts had a slight pinkish tint. The bird had a white bar on each wing and white outer tail feathers. The bill was black and the upper mandible had a sharp, downcurved tip, and the legs and feet were black. In flight, it swooped down low and then up to the top of the next bush. No call-note was heard. I was alone when I saw this bird.

GB status: 20, 1842–1949; 186, 1950–2020. About 2–3 per year; stable. (Breeds Balkans to E Poland, with small numbers W through N Mediterranean to S France & NE Spain. To E, breeds locally from Black Sea coasts, across S Russia & Kazakhstan to NW China & SW Siberia. Migrates through E Africa to winter S Africa, from Namibia to S Mozambique & N South Africa.)

Woodchat Shrike
Lanius senator
BBRC to 1990; HOSRP 1991 to date

A rare vagrant. One record in 1875 and nine since 1955.

© Horniman Museum

The first for the county was a female shot by Edward Hart at Bransgore on June 1st 1875. The specimen is now held by the Horniman Museum, London and is housed in a case with a male, also shot by Hart, at Somerford, now Dorset, on May 29th 1880.

The first modern records were in 1955. A male was watched by Mr. and Mrs. L. M. Seth-Smith from their car at 20 to 100 yards range at Widden Bottom, New Forest for 30 minutes on May 19th. They watched it for a further 90 minutes the next day but it was not present on the following three days. Unfortunately, it has not been possible to trace a description of this bird. Another male was found by Graham Rees by the side of the railway line running through the copse to the west of Mason's Crossing near Farlington Marshes on June 11th. Graham's description, extracted from the Langstone Harbour yearbook, is reproduced below.

About the size of a Red-backed Shrike with longer tail. Chestnut crown and nape, black band through eye (coming well above bill on forehead), grey black bill, white on rest of face and all underparts. Conspicuous white patches on scapulars, and white bars across wing (when seen in flight). White rump. Back and tail black. Hunted in the same manner as Red-backed Shrike, with short sallies from various perches, flying and gliding low over the ground after prey. It caught a bumble-bee and a large beetle on 11th.

It was still present the next day and was seen by five other observers. Graham commented in 2020 that it was able to remain undisturbed in the area for two days because its visit coincided with a railway strike!

The next was an adult found by Ron Cooke at Dibden Bay on June 14th 1981. He was able to watch it for several minutes dropping to the ground to catch prey before he left to contact a friend, but unfortunately it could not be found on their return an hour later.

There were no further records until the new century. The first was a female found by Mark Palmer at Hammond's Bridge along the Titchfield Canal Path on the evening of June 24th 2002 (not June 14th as in the 2002 *HBR*). It was also seen by Peter Milinets-Raby and several other observers before dusk but was not relocated the next day. In 2006 an obliging first-summer female was found by Peter Milinets-Raby along the lane south of the Bat and Ball public house near Hambledon on May 7th. It remained until the 9th and was much appreciated by hundreds of birders and was the first truly twitchable bird in the county. Later that year a juvenile was found by Darren Fry at South Moor Lane near the north shore of Langstone Harbour on August 28th and confirmed by Jason Crook. It remained until dusk and was seen by many observers but not relocated the following morning.

Woodchat Shrike, female, canal path, Titchfield Haven, June 24th 2002

The next were in 2012. A first-summer female was found by Alan Lewis at Vidle Van Farm, Keyhaven on May 12th and seen by many observers during the day and on 13th. Amazingly Alan found another individual (with a different face pattern and tertial moult) which was thought to be a first-summer male at Martin Down on May 20th where it was still present the next day.

Woodchat Shrikes: juvenile, Southmoor Lane, Langstone Harbour, August 28th 2006 (left), first-summer male, Martin Down, May 20th and 21st 2012 (centre) and first- summer female, Hambledon, May 8th 2006 (right). Richard Ford, Alan Lewis, Mike Trew and Kit Day

Finally, a first-summer female was found and photographed by Steve Lavington at Rowbarrow Pond, New Forest at 1645 hrs on May 29th 2021. It was still present at 1915 hrs but could not be relocated later that evening and was not present the next day.

GB status: 70, 1829–1949; 1,191, 1950–2019. Small increase, mean 27 per year in 2010–19. (Breeds NW Africa & Mediterranean Europe N to C France & S Germany, E through S Turkey to W Iran & S to Israel. Winters in N & C equatorial Africa.)

Golden Oriole *Oriolus oriolus*

HOSRP

A rare summer visitor which has bred. At least 16 old records including one of breeding; in the modern era, 87 records including eight of breeding.

The first for the county was described by Yarrell (1843) from an immature male taken at Aldershot in April 1824. The specimen is still housed in Eton College Museum. K & M listed further specimens procured near Portsmouth and in the New Forest in about 1845, at Woolston in August 1889 and at Avington in 1897. The latter was received by the Winchester taxidermist William Chalkley on December 16th (Smith 1898). A record of breeding at Mottisfont in 1897 attributed to Meinertzhagen is now discounted because it is known that he falsified many of his ornithological records.

As with many species, records were sparse and somewhat vague during the first half of the 20th century. A male was in a field at Milford on Sea on March 19th 1901 (Kelsall & Coles 1913), one was near Chandlers Ford on May 1st 1912

(Munn 1919), one was at Winchester Water Meadows on June 17th 1917 (*WC*), a pair nested unsuccessfully near Owslebury in about 1920 (*WC*), one was shot at Chilcomb in December 1929 (*WC*), one was at Itchen Abbas in May 1930 (Cheke 1962), a pair was seen in Avington Park sometime in 1934 (*WC*), a male was seen well by three persons at Fritham on May 11th 1935 (*SEBR*), one was seen on a fence post at Chilland on May 10th 1936 (*HBR*), one was seen at Embley on May 1st 1937 (*SEBR*), a male was at Alton on April 20th 1943 (*HBR*), a singing male was in an oak copse at Pennington Marshes on May 10th 1945 (*Brit. Birds* 39: 128), a male was at Highclere on December 1st 1947 (*Brit. Birds* 42: 53) and a pair was at Priors Barton, Winchester on June 7th 1948 (*WC*).

Males, toing and froing, East Hyden Wood, June 1986

There were several breeding records between 1969 and 1987. In 1969 a pair nested successfully at Wickham. A male returned to the nest site on May 22nd in the following year but did not stay. In the New Forest, a pair probably bred in Frame Wood in 1975 and a year later a pair was suspected of breeding at Fletchers Thorns Inclosure. Also in 1976, up to three including a juvenile were at Chilland in the Itchen Valley from July 12th to August 3rd; this indicates local breeding but in 1977 the only record from there was of one heard on July 7th. In 1984, a pair nested at Bushy Copse near Longwood Warren, raising two young, and in 1985, a probable juvenile was seen there on July 3rd. In the first year of the HOS Tetrad Atlas Survey in 1986, Golden Orioles were present in East Hyden Wood from May 27th to August 25th. Two males, two females and one other bird were seen on May 31st, though only one pair bred, raising three young. In the following year, a male was first heard on May 25th and a pair was seen on May 30th. Subsequently, two juveniles fledged and the last record was on August 29th. In the adjacent Combe Wood, an unmated male was present during June. Anecdotal evidence suggests that Orioles had bred in these woods since 1971 or even earlier (John Brame, Campbell Burnett *in litt.*). Unfortunately, many trees were uprooted during the great storm of October 1987, and this, combined with disturbance caused by the removal of those trees during the 1988 breeding season, probably discouraged the orioles from nesting. In 1988 the only records were of one heard in East Hyden Wood on May 10th, a female or immature male there from May 30th to June 3rd and an adult male at Combe Wood on July 5th. None have been recorded in the area since.

Excluding breeding birds, 79 have been recorded in 44 years since 1951 including 54 between 1970 and 2003, as shown below.

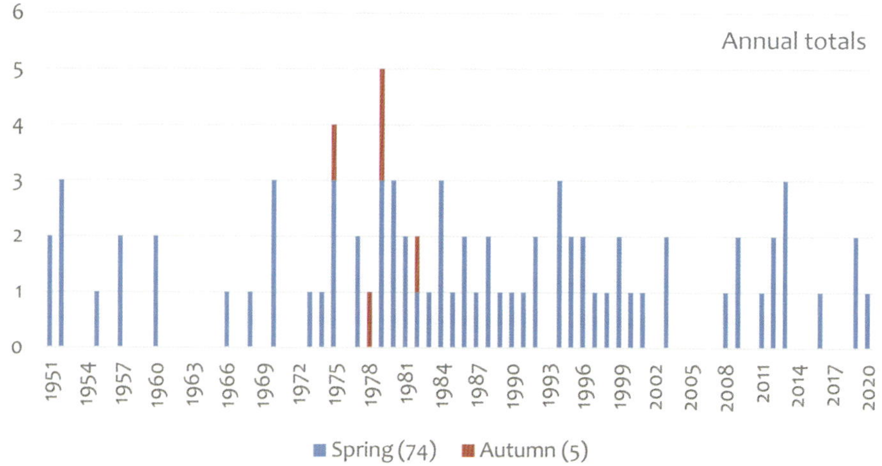

Analysis of arrival dates shows that most occurred in spring.

Golden Oriole, male, Pennington Marshes, April 28th and 29th 2013.
Mike Wearing and David Cuddon bottom right

There was an exceptionally early bird watched by Mr. D Freeman through binoculars in his garden on March 15th 1998. HOSRP accepted the record but felt that the escape possibility could not be discounted. A second-hand report received by Edwin Cohen of one at Holmsley on April 5th 1957 seems very early. The earliest unambiguous record is of one picked up dead at Mansbridge on April 19th 1970. Most sightings occurred from May onwards with a peak in the last week of the month and newly discovered birds continuing to be found through June. Records have been widely scattered across the county although woodlands of the New Forest have predominated with 19 seen and/or heard. Other areas with more than one record are the Avon Valley (6), Hayling Island (4) and Lymington–Hurst,Sowley–Lepe and Titchfield Haven (3 each). Coastal records mostly involved brief sightings apart from an adult male found by Steve Piggott at Pennington Marshes on April 27th 2013 which stayed until 29th and was enjoyed by many visiting birders. Apart from breeding birds there are no spring sightings after July 8th.

Five later records involved singles at Ellisfield on September 10th 1975, Alton Golf Course on August 7th 1978, Denny Lodge Inclosure on August 12th 1979, Heckfield on September 29th 1979 and Fareham on August 31st 1982.

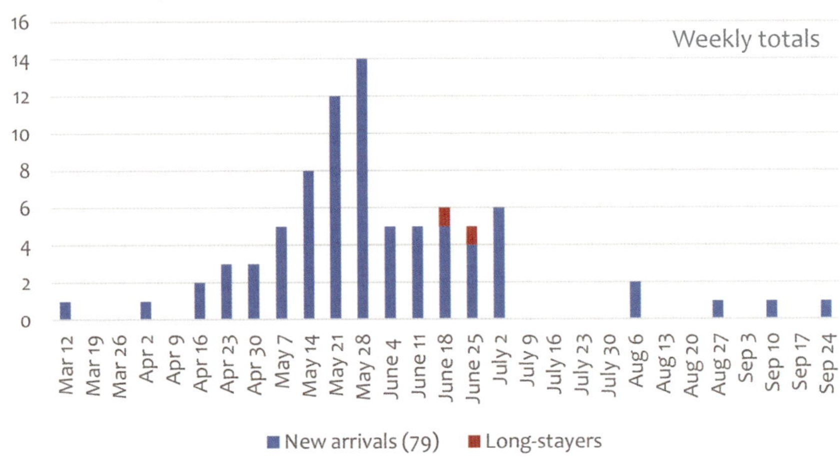

This species is confused with Green Woodpecker especially by inexperienced observers obtaining brief views. Several records, both old and modern, on dates in March, the first half of April and December have been previously published and are included here. However, they must be considered suspect. Since 1987 HOSRP has assessed 16 claims of this species as not proven, all but four of them in the 21st century. A number of these involved birds seen from moving vehicles, either in flight or perched on fence posts.

GB status: 70, 1829–1949; 61, 1950–57; 4,361, 1968–2019. Declining, mean 77 per year in 2010–19. Breeding population peaked at 40 pairs, 1990; now extinct. (Breeds NW Africa & widely throughout Europe from Mediterranean N to S Sweden & S Finland, Russia, Turkey & Caucasus to NW Iran. Winters equatorial & southern Africa.)

Nutcracker *Nucifraga caryocatactes*

BBRC

A very rare vagrant. Two old records and five since 1968.

© *Horniman Museum*

The first for the county appears to be an adult female shot in the New Forest in October 1846. Edward Hart, writing on 6 April 1927, said that the specimen was from the H. Treasure Jenkins collection. The record was omitted by K & M but the specimen is now held by the Horniman Museum, London. K & M did include one first notified in *The Zoologist* of April 1886. Mr. A. G. Renshaw wrote as follows:

> *On February 8th 1886 at Exbury, New Forest, I saw a Nutcracker. It was close to the house where I was staying and I recognised the note (with which one gets very familiar in Switzerland and the Alps generally) as I lay in bed. I got up at once and saw my friend busy amongst the cones of a large fir tree near the windows of my room. Perhaps the bird is commoner in this country than I suppose, but I never saw or heard one here before.*

This record is probably acceptable despite the 'What's hit is history, what's missed is mystery' phrase which was applied to most claims during that era. There was an unprecedented influx into Britain in 1968 involving some 315 birds (Hollyer 1970) but only one was recorded in Hampshire. It was first seen in a garden at Fordingbridge on December 6th and subsequently confirmed by John Ash. It ranged about the district for up to a mile and was found dead on December 28th. However, information has recently come to light of one picked up in a weak condition by a beater during a pheasant shoot at Quarley Park on November 17th. It was taken into care but unfortunately died two days later. The specimen was mounted by a local taxidermist and has been retained by Henry Edmunds (*in litt.*). Another was reported by Brig. Clive Simson in a garden at Hurstbourne Tarrant in late September 1969 where it remained until mid-October. It was relocated by Miss Pentland in her garden at nearby Stoke on November 4th and confirmed by Linda Wren on 15th. It was still present on 24th when it was killed by a cat. This was presumably a straggler from the previous year's invasion.

The next was found by Clive Chatters and the landowner, Mrs. Clifford, at Copythorne Common on October 22nd 1991. News was not released although the local Nature Conservancy Council office was informed. Simon Aspinall, who worked for NCC, relocated it on October 26th when he watched it at close range for 15 minutes.

Nutcracker, Hurstbourne Tarrant, October 1969. © Holmes & Sons

The most recent occurrence was of one watched by Dan Houghton perched in the top of a tree at Denny Wood for 30 seconds on December 29th 1996. Unfortunately, it flew off with Redwings before he was able to show it to other members of his local bird club who had been eating their lunches. It was not thought to have flown far but could not be relocated in the next hour or during the following days.

GB status: 57, 1753–1949; 355, 1950–1998 with most in 1968. (Eurasia from Scandinavia and the Alps to Kamchatka and China).

Chough *Pyrrhocorax pyrrhocorax*

HOSRP

A very rare vagrant. Three records in 1941, 1963 and 2007.

Cohen (1963) gives details of one which stayed for two or three weeks at Cheesefoot Head about 1941; it was seen by a shepherd and a farm bailiff and used to stay around the sheep pen.

The next record was of one at Hordle Cliff on May 26th 1963. This was included in Cohen & Taverner (1972) with the citation Riddick *in litt.* It was not published in the 1963 *HBR* and was overlooked for *BoH.* Ernest Riddick's correspondence with Edwin Cohen describing several sightings has now been discovered in the HOS archives. He stated, 'one Chough on Hordle Cliff close to the main road. Three people were already watching it from a car when I arrived but unfortunately moved off before I could speak to them. The scarlet beak and legs could be seen quite clearly'. No date was given. Edwin evidently asked for further details as in a second letter Ernest stated, 'The Chough I had no doubt about at all. This bird was on the ground about 20 yards from the main cliff road. No mistaking the slightly curved red bill and red legs. I had never seen anything like it before. After about 3 minutes it flew back to the cliff top with an up and down twisty sort of flight. The date May 26th'. Also attached is a note which says, 'May well be reliable but short of meeting observer and seeing him perform in the field, would tend to ignore his records'. This is initialled by four HFCERC members, including John Taverner who subsequently included it in Cohen & Taverner (1972).

In 2007 Henry Edmunds heard one calling over Cholderton Park on the Wiltshire border at 1400 hrs on April 10th. He located the Chough circling in the same thermal as a male Sparrowhawk before it headed off west. Following the re-establishment of the species as a very scarce resident in Cornwall, there have been several sightings further east in Somerset, Dorset and this individual at least in Wiltshire. In 2007 three birds were present in Somerset but somewhat surprisingly colour-rings identified the birds as of Welsh origin; the last was seen on April 6th, four days before the Hampshire/Wiltshire record.

GB status: c. 350 breeding pairs, 2018. Increasing in SW England. (Resident, British Isles, Iberia and Morocco east through S Europe to Mongolia and China.)

Hooded Crow *Corvus cornix*

HOSRP

A rare visitor, usually in winter. Apparently common in the 19th century, but very few in the first half of the 20th century and 50 since 1951.

The first dated record for the county was one recorded by Gilbert White as it flew over Farringdon Heath on November 5th 1773. White considered it rare in the Selborne area but mentions two further sightings of birds flying over his garden on November 16th 1773 and November 29th 1775 and another shot near the village on January 9th 1784. K & M described it as 'of regular and fairly common occurrence on all our coasts but only sparingly and occasionally in some inland localities'. George Corbin, writing in *The Zoologist* in 1882, noticed an unusual abundance in the New Forest in the winter of 1881/82 but did not give any specific records. Mr. Stares recorded about ten along the coast at Browndown on October 10th 1897 (Smith 1898). The only dated records for the first half of the 20th century are of singles at Bullington Cross on January 7th 1934 (*SEBR*), Beaulieu on December 20th 1936 (*HBR*),

West Dean on December 21st 1938 (*SEBR*), Horsebridge on April 2nd 1946 (*HBR*) and Keyhaven on October 25th 1946 (*HBR*).

Between 1951 and 2000 there were sightings in 27 years with a total of 48 recorded but since then only two have been seen.

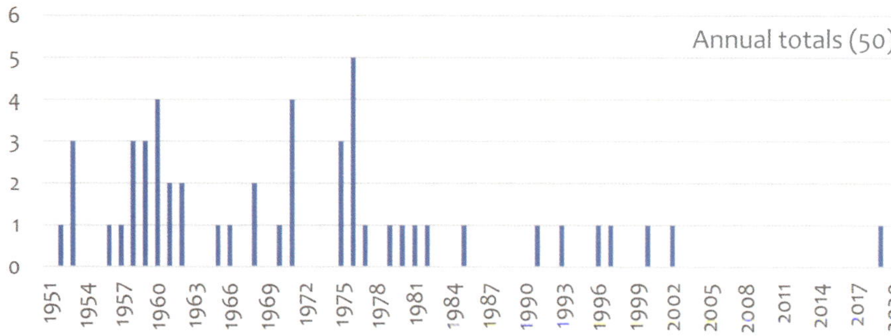

Although 27 out of 50 were at localities at or adjacent to the coast, records have been for all parts of the county, with rubbish tips being especially favoured. Most were seen on one date only but stays of a month or longer have occurred. Well-watched areas with records include Langstone Harbour/ Farlington Marshes (March 1st 1953, October 26th 1956, March 12th to 27th 1965 and October 30th 1971), Titchfield Haven (November 10th 1958 (2) and January 11th 1960), Brownwich (January 9th 1959 (2), November 18th 1979 and April 15th 1980), Woolmer Forest (November 11th 1961 and December 31st 1976), Keyhaven/Pennington Marshes and the adjacent Efford Rubbish Tip (March 1st–22nd 1968, January 26th 1985, April 29th 1996 and June 16th and July 6th 2000), Needs Ore (February 5th 1966 and April 18th 1971), Ibsley (April 23rd 1968), Newtown Rubbish Tip (January 11th 1975, December 20th 1976 and March 12th 1982) and Bramshill Rubbish Tip (January 26th–February 6th and March 20th 1977).

The only records this century have been of one reported by Sir Charles Pringle at Appleyards, Fordingbridge on January 16th 2002 and another found by Paul Heath feeding in fields at Portsdown Hill on March 8th 2019. It remained there until March 17th enabling many birders to add this rare visitor to their Hampshire lists.

Hooded Crow, Portsdown Hill, March 8th 2019. Andy Johnson

Analysis of arrival dates shows records between October 17th and May 26th apart from one at Efford Rubbish Tip on June 16th and July 6th 2000 mentioned above.

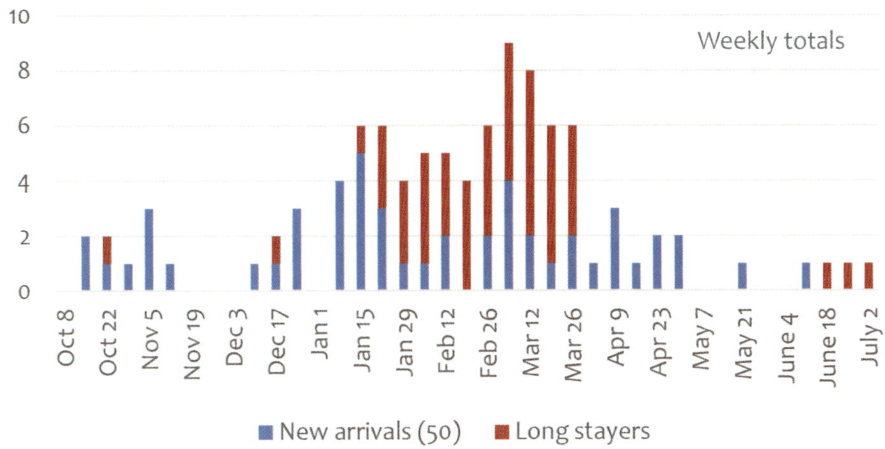

One first seen at Fair Oak in May 1970, paired with a Carrion Crow in 1972 and raised two hybrid young. A hybrid was reported there in late March 1973.

GB status: resident in N and W Scotland; rare winter visitor to E coast of England. (Breeds Ireland, Scotland, Fennoscandia, E Europe and Turkey east to W Siberia and Mongolia; some move south in winter.)

Waxwing *Bombycilla garrulus*

A rare winter visitor and passage migrant which has occurred more frequently in the 21st century.

The first for the county was one mentioned by Gilbert White in Letter XII to Thomas Pennant as being shot in his neighbourhood (Selborne) in January 1767. Longcroft (1856) recorded one at Emsworth during the winter of 1779–80. K & M included records for six years between 1850 and 1895 – all of singles apart from two in 1895. Records for the first half of the 20th century are equally sparse. One was seen in Garnier Road, Winchester on several dates in March 1917 but no more were recorded until 1935 when two were at Swanmore on March 15th and one was at Romsey on April 28th. In the period up to 1950 there were sightings in seven winters with a total of 21 records involving around 90 birds. Most of these were in 1946/47 when there was a large invasion to Britain estimated at 6,000 birds, mostly in Scotland (Gibb 1948). Some 75 were recorded in Hampshire including six at Alton on November 2nd, when very few had reached southern England, 16 at Portchester on January 4th and 20 at Portswood on February 5th.

In the modern era there were records in 40 out of 70 winters. Until the turn of the century, numbers were generally low with significant influxes in 1965–

66 (202), 1970–71 (134) and 1995–96 (215). There were double-figure totals of up to 51 in five winters and single-figure totals in 17. The first two decades of the 21st century produced more frequent and larger invasions than before. By far the highest numbers were recorded in 2004–05 and 2010–11 with estimates of 2,000 and 3,000 respectively. Full details of these invasions are given by Clark (2007, 2013). Other sizeable totals were of 220 in 2008–09 and 586 in 2012–13. There were records in ten other winters with totals reaching double-figures in six of them.

Attacking the berries. January 29th 2011

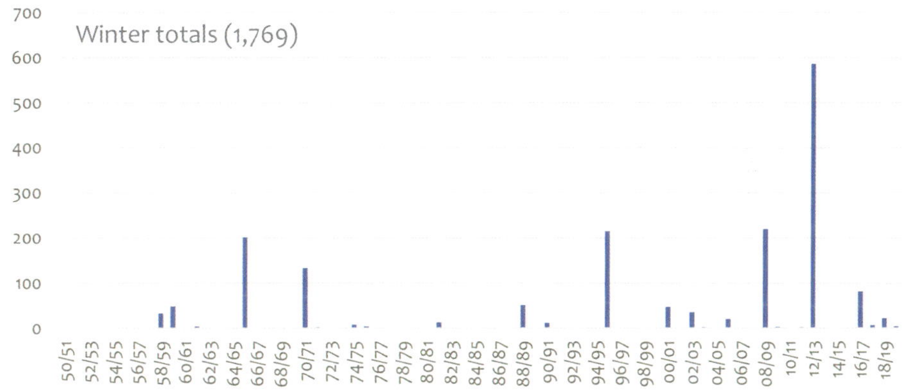

The earliest ever was one heard by Marc Moody in misty conditions at Sowley Shore on October 25th 2007. Few have been recorded in November with the earliest two at Butser Hill on 5th in 1961 and one flying west over Gosport on 6th in 1988. Comparison of the two largest invasions shows different arrival patterns, as shown on page 394. In 2004/05 the arrival was mostly in the first half of January with peak numbers at the end of that month followed by a gradual departure but with significant numbers still present in April. In contrast in 2010/11 numbers increased rapidly from early December and peaked in the first week of January. Many had moved on by the end of the month and few were seen after mid-March. A flock of 40 at Meon Valley Golf Course on April 28th 2011 were possibly migrants. The only later records were of one at Romsey on May 3rd 2013 and in 2005 eight at St. Mary's, Southampton on May 1st, 18 in Church Crookham on 2nd and the latest ever seen by Marc Moody in his garden in Lymington on 8th.

Waxwings, Romsey, December 17th 2010 and January 9th 2011. Martin Bennett and Ian Williamson

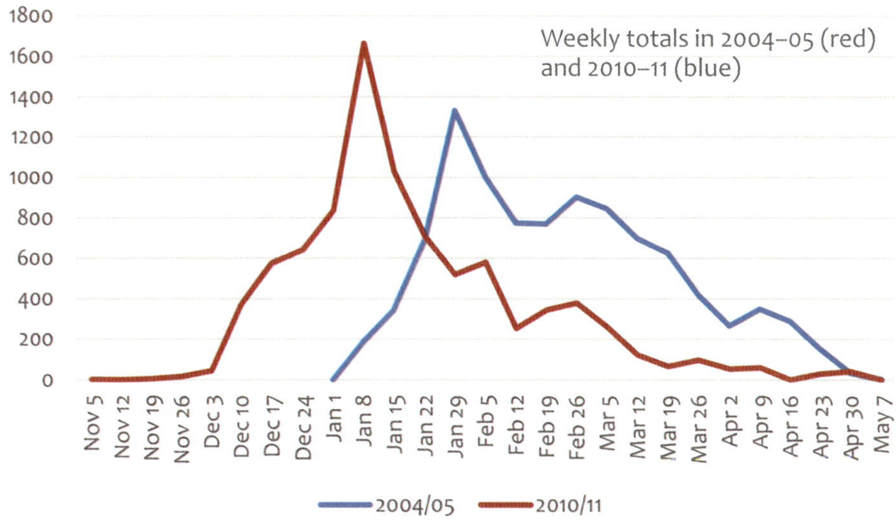

Most Waxwings are found where berries such as rowan and cotoneaster are available and so trading estates, retail parks, supermarket car parks and modern housing estates have proved to be popular venues. Flocks exceeding 200 have occurred on four occasions. In February 2005, a flock centred on the Ordnance Survey HQ in Maybush, Southampton was confirmed to hold 352 when counted by Alan Lewis from a video grab. In January 2011 up to 259 were counted in Beverley and Rimington Gardens, Romsey with the peak count on 7th, an astonishing 773 was counted from a digital photograph taken by the author on the Hampshire/Surrey border at North Camp Station on 14th and 300 were at Hounsdown near Totton on 26th.

GB status: irruptive winter visitor. (Breeds in taiga zone throughout Eurasia and N America, moves S in winter.)

Penduline Tit *Remiz pendulinus*

BBRC to 2014; HOSRP to date

A rare late autumn and winter visitor. Twelve records of 25 birds since 1989.

The first for the county occurred in 1989 when Barry Duffin saw and heard two flying high to the west over Titchfield Haven on October 28th. However, he was not able to identify them until he heard the calls given by the next which was a juvenile trapped at Titchfield Haven on November 9th 1989. Extracts from Barry's description are reproduced below.

> *At approximately 1500 hrs my attention was drawn to a call that was not familiar to me. On looking up two small passerines were overflying the reserve in a westerly direction at some considerable height. The same call was uttered on several more occasions as the birds were kept in view. They were small, the size of Blue Tit,*

> *with a very similar flight to that species. I was not able to get any plumage details during the 1.5 – 2 minutes they were in view. The calls uttered were a clear, sweet 'see-oow' or 'seeow'. When I heard the call uttered by the Penduline Tit trapped on November 9th, I immediately recognised it as that made by the two overflying birds on October 28th. I have absolutely no doubt that they were Penduline Tits.*

Penduline Tit, Titchfield Haven, November 9th 1989. Barry Duffin

Subsequently, a further 22 have been recorded.

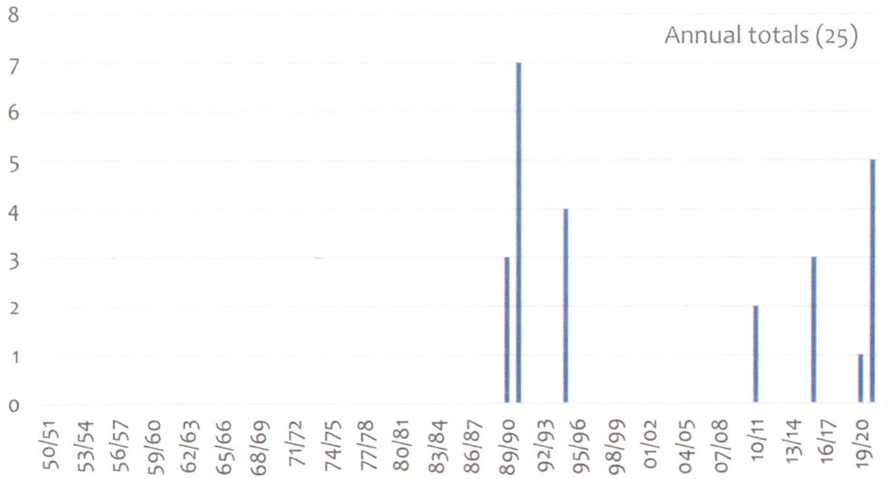

At 1140 hrs on October 23rd 1990, during a ringing session at Titchfield Haven, Barry Duffin heard Penduline Tits calling nearby and on investigation found a group of five juveniles feeding on reed mace. Shortly after they left and gained height as they flew away to the east. The next were two males found by unknown observers at Farlington Marshes on January 23rd 1991 and subsequently seen by many admirers during their stay until March 23rd. A party of four including two males was found by P. Phillips feeding on seed heads of reed mace at Titchfield Haven at 1030 hrs on November 4th 1994 and independently identified shortly after by Sue Morrison. They were later watched by seven other birders before departing high to the west at 1245 hrs.

There was a long gap before the next record of two juveniles found by Barry Duffin at Titchfield Haven on October 12th 2010 which departed east at 1040 hrs. This site maintained its almost complete monopoly on this species in 2015 when Ian Calderwood found a male and two females there on December 7th which remained until 14th. They were then relocated by Tim Doran ten km to

Penduline Tits: male, one of two, Farlington Marshes, January 1991 and Titchfield Haven, February 23rd 2016

the north-east at IBM Lake, Cosham on December 15th where they stayed until the next day. They had returned to Titchfield Haven by December 29th and remained until at least February 3rd 2016 with two to March 6th and a male until March 30th.

There was a four-year gap until the next when one was found by Amy Robjohns at the north end of Titchfield Canal Path at 0825 hrs on February 18th 2020. It flew away south and was not relocated. Later that year a group of two adults and three juveniles was found and photographed by Matthew Dryden from the Spurgin Hide at Titchfield Haven on October 28th. They were still present the next day but only one adult and one juvenile were seen until November 2nd. Subsequently, an adult and a juvenile were found by Nigel Jones in a small reedbed at Needs Ore at 1500 hrs on November 4th; he was able to show them to Chris Button before they flew off west at 1530 hrs. Finally, an adult was found from the Spurgin Hide by Shaun Webb on December 15th which was also seen after an anxious wait by Billy Rodger and Andrew Friend. The last two sightings are assumed to refer to individuals from the original group of five.

As shown opposite, analysis of the records confirms October as the main month for arrivals in the county.

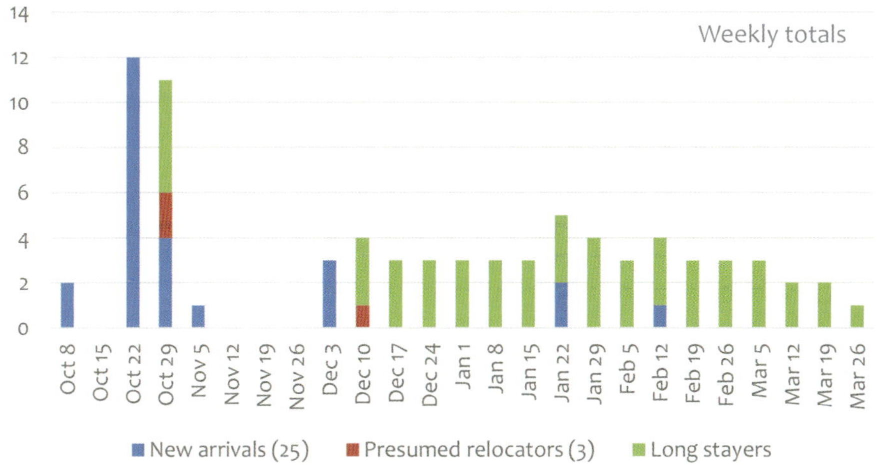

Penduline Tit, male, Titchfield Haven, December 11th 2015. Alan Lewis

GB status: 328, 1966–2019. Stable, around ten per year. (Widely but locally distributed throughout C & E Europe, from Denmark, Germany & Italy NE to C Sweden & Estonia. Absent from much of NW Europe but locally numerous Spain. To E, breeds Turkey, Caucasus region, N Iran, S Russia and W & N Kazakhstan. Largely resident or dispersive in Europe.)

Bearded Tit *Panurus biarmicus*

BBRC to 1962; HOSRP 1963–78

Formerly a rare visitor. Several old records and three records of seven birds between 1956 and 1962. Now a scarce resident, passage migrant and winter visitor which first bred in 1966.

K & M described the Bearded Tit as 'an accidental visitor' but gave two records of breeding, one undated and the other in 1902; neither of them is particularly convincing. Hart, writing on April 6th 1927, described a taxidermy case containing five Bearded Tits obtained on the River Avon at Winkton (now Dorset) and Sopley. The earliest, an adult male, was taken on January 27th 1847 and three others in the period up to 1854. These were mounted by the taxidermist Barron and were originally in the Hon. Grantley Berkeley's collection. The final bird was killed on January 27th 1906 near Splash House, Sopley and passed by the owner Miss H. Penruddock to Edward Hart who mounted it. The specimens are now held at the Horniman Museum, London. The only other records in the first half of the 20th century were of singles at Titchfield Haven on July 3rd 1911 and February 1st 1922.

Bearded Tit, male, Farlington Marshes, January 1st 2020. Rob Porter

Bearded Tits, males, Titchfield Haven, October 25th 2015

In the modern recording era, the first was a female or first-winter male at St. Cross Water Meadows from November 3rd to 5th 1956. It was found by an un-named observer on the evening of 3rd who returned to the site with Stephen Martin on the morning of 5th. Martin provided a full description which was included in the 1956 *HBR*. It was watched for 25 minutes before rising in spirals and leaving high to the north-west. With the formation of BBRC with effect from 1959, records of this species were adjudicated by that body. It accepted records of a male and two females found by Shaun White at Titchfield Haven on January 15th 1960 which he refound the next day and showed to Canning Suffern; they were also found independently by Martin Terry on 16th. Subsequently, a male, female and first-winter male were seen there by Canning Suffern, Michael Bryant and others on February 14th which remained until March 27th with one at least still present on April 3rd. Suffern saw both groups of birds and stated that the first-winter male was not present in the original trio. In 1961 Norman Pratt found a pair at Fleet Pond on December 9th; these presumably overwintered there as Norman refound them on February 10th and 24th 1962 and Malcolm Harvey and Ewart Jones saw them on four dates between February 27th and March 10th.

The situation changed in the winter of 1964/65, when 20 were recorded, including flocks of five at Nursling Gravel Pit on October 24th, Titchfield Haven on October 31st and Farlington Marshes on November 1st. In the following winter, 62 were seen including flocks of 16 flying west at Lower Test Marshes on October 10th and 17 there on November 2nd; up to ten wintered at Titchfield Haven and Farlington Marshes. A pair bred successfully at Titchfield Haven in 1966 and since then the species has established a small breeding population and become a regular winter visitor in variable numbers.

GB status: Formerly rare resident confined to East Anglia; gradual expansion from mid-1960s, now c. 700 breeding pairs, increasing. (Spain, England through C and S Europe to C Asia. Some winter to south)

Shore Lark *Eremophila alpestris*

HOSRP

A rare late autumn migrant and winter visitor. 26 records of 43–48 individuals since 1955.

The first for the county were five found by Gerald Kinsey at Pennington Marshes on October 31st 1955. It has not been possible to trace notes on this sighting although the 1955 *HBR* does state that satisfactory details were received.

Totals for each winter are shown below.

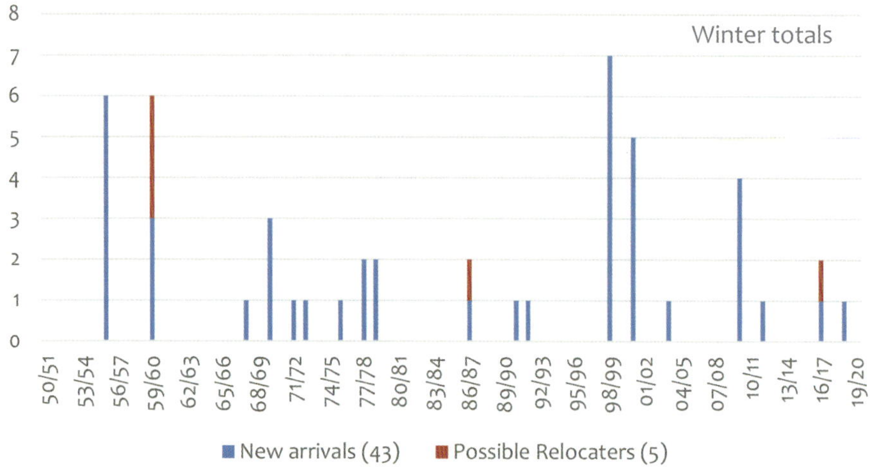

There was another in 1955 when Roy Dennis found one at Hook-with-Warsash on November 9th. The next were three found by Tom Brice at Black Point on November 23rd 1959 and presumably the same group located by David Price at nearby Sinah Common on December 29th of that year. David Evans only recently submitted a description of one which he found at Southsea on December 11th 1967. The first long-staying group was a party of two males and one female found by an unknown observer along Hurst Beach on December 10th 1969 which were last seen together near Sturt Pond on April 7th 1970. The female remained in the area until April 16th.

Further records during the 1970s involved singles at Black Point on January 10th 1971 and Farlington Marshes on October 12th 1972 and October 12th 1975, two at Needs Ore on October 27th 1977, one at Ashley Walk, New Forest on November 4th 1978 and another at Keyhaven on December 14th 1978.

The only records in the 1980s were of singles at Browndown on January 15th 1987 and Weston Shore on January 21st of that year which probably related to the same individual.

The 1990s started with singles at Hook-with-Warsash on December 10th and 19th 1990 and Hayling Oyster Beds on November 23rd and 24th 1991. In 1998 a flock of seven, the largest recorded in the county, was found by Barry Goater and Alan Moody along Hurst Beach on November 12th; sightings of two there the next day and one at nearby Pennington Marshes on 15th are assumed to relate to the same birds.

Shore Lark, Hayling Oyster Beds, January 22nd 2012. Freezing as always.

The first record in the 2000s was of a flock of five that Melvyn Brickwood saw fly in off the sea and land briefly on the beach at Hill Head on November 9th 2000 before departing to the north-west. The next was one in the Hurst Beach area from October 16th to 21st 2003. The next were not until 2009 when there were two at Pennington Marshes on October 13th and singles at Sandy Point on November 5th and Needs Ore from November 17th to 19th.

The only records in the 2010s were of singles at Hayling Oyster Beds from January 15th to February 17th 2012, Eastoke, Hayling Island from October 28th to 30th 2016 and presumably the same at Southsea Castle on November 6th, and at Ibsley Water on October 23rd 2018.

Shore Lark, Eastoke, Hayling Island, October 28th and 29th 2016. Barry Stalker and John Dodds

Analysis of arrival dates shows the species' principal status as an autumn passage migrant with all records between October 12th and December 14th apart from three singles in January. The only records after February 17th have been of the group of three which wintered at Hurst Beach until April 1970.

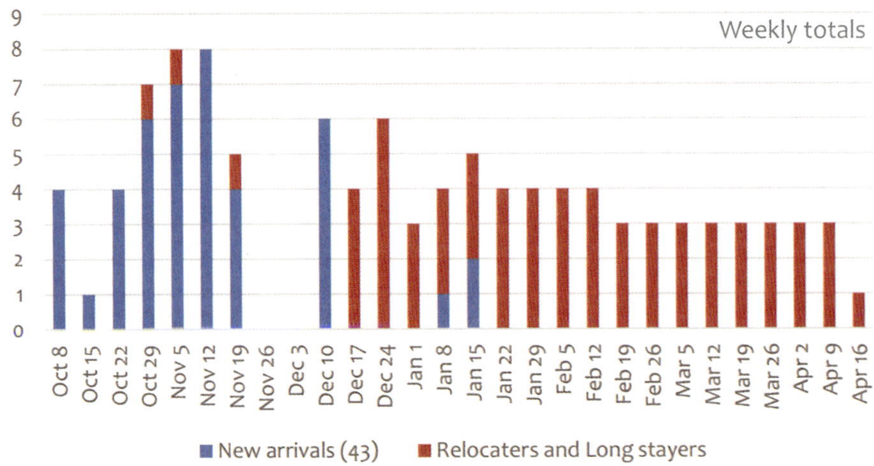

The record of one grounded at Eversley Gravel Pits in poor weather on April 10th and 11th 1998 which was published in the 1998 *HBR* is now excluded as it was only seen on the Berkshire side of the county boundary.

GB status: Passage migrant and winter visitor to coastal E England; total typically c. 300. (Breeds throughout tundra zone of Eurasia and N America. Winters to S.)

Short-toed Lark *Calandrella brachydactyla*

BBRC to 1993; HOSRP 1994 to date

A very rare vagrant. Five records since 1983.

The first for the county was one found by Philip Fawkes and jointly identified by him and Mike Witherick along the shore between Calshot and Ower at midday on October 9th 1983. It was subsequently seen by many observers during its stay until October 19th. The record was accepted by BBRC but the original description has not been traced. Philip kindly penned the account which follows in early 2021.

Mike Witherick and I were co-leading the first of the winter season walks on 'Bird Watching for Beginners', an adult education course based at Calshot Activities Centre. Towards the end of the two-hour session, we happened upon a small, pale lark feeding along the coastal path between Ower Creek and the Spit Road, at a distance of 100 yards. Coming from the other direction and independent of our group, was Dr. Jimmy Scott, who was unaware of its presence and who, despite our shouts and gesticulations, unwittingly flushed the bird. Fortunately, it flew towards us calling several times and settling a matter of metres from where we were standing. It remained a subject of much argument, never to be resolved, as to who first 'called' Short-toed Lark, but it was clear from the bemused expressions on their faces that the party of twelve did not immediately share our joint elation! We pointed out the dark patch on the side of the breast as one of several distinguishing features from Skylark, but we may have been guilty of glossing over some of the other salient points, as we were keen to scribble out a field description. Anyway, I am not convinced that the enormity of the occasion was fully appreciated by the members of the group, as was the case three weeks later when we pointed out Richard's Pipit along the same stretch of coastline. All far too much, too early!

Short-toed Lark, Normandy Lagoon, May 18th 2010. Marcus Ward and Simon Ingram

The next was not until 2010, when Marcus Ward found one at Normandy Lagoon at 0800 hrs on May 18th. It remained all day and was seen by many observers. Further records ensued in the same area. Mike Witherick was leading a group of four birders when he found one at the edge of the old rubbish tip at Lower Pennington Lane which they watched for ten minutes on May 19th 2013. Mike and Peter Oram and Janet and Cyril Shelley saw a bird which they did not recognise on the Ancient Highway near Lower Pennington Lane on August 29th 2017. Fortunately, they were able to photograph it before it flew off east after ten minutes and it was subsequently identified as this species.

Finally, one was picked up by Graham Stephenson and also seen by the author as it flew low west over Tweseldown Hill just behind a group of nine migrating Redwings at 0952 hrs on October 17th 2021.

A further record of one seen well by John Collman, John Eyre, Gilbert Rowland and the author as it flew north-east low over their view-point, calling, near Beaulieu Road Station on May 6th 1989 was harshly considered not proven by BBRC.

K & M mentioned two records. The Rev. J. Pemberton Bartlett, writing from the Exbury Parsonage to *The Zoologist* in February 1862, tells the story of a Short-toed Lark which was caught by a bird-catcher near Southampton during the previous winter and kept in his aviary. Dr. A. G. Butler's *British Birds and their nests and eggs* (1896) mentions one caught near Portsmouth and exhibited at the Crystal Palace Bird Show in 1890. Given the vague details and absence of specimens both of these records are unacceptable as the first for the county.

GB status: 19, 1841–1949; 956, 1950–2019. Declining, mean 20 per year in 2010–19. (Breeds NW Africa & European Mediterranean basin to Black Sea region of Turkey & S Russia, E through C Asia to Mongolia & NW China. Winters along S edge of Sahara from Senegal to Sudan, Arabian Peninsula & N Indian subcontinent.)

Red-rumped Swallow *Cecropis daurica*

BBRC to 2005; HOSRP to date

A rare vagrant. 18 records of 17 birds since 1987.

The first for the county was one seen by Rob Clay over Farlington Marshes on November 14th 1987. Extracts from his description are reproduced below.

Whilst watching two Fieldfares and three Redwings in bushes at Farlington Marshes I noticed what I at first thought was a House Martin flying towards me from the lake. On looking again, I soon realised the bird was a Red-rumped Swallow. As the bird hawked over the ponds, in good light, I noted the warm buffish rump along with the buff-orange nape and supercilia contrasting with the dark upperparts which in the bright light appeared glossy. From overhead the bird showed a buffish wash to the underparts and dark undertail-coverts. The wings appeared rather short – probably due to blunt tips and broad base – and the tail streamers were noticeably thicker than Swallow and considerably incurved. Flight was more like a House Martin – slower and more deliberate than Swallow. After a short time, the bird stopped hawking and flew off north over the A27.

Subsequently, a further 16 have occurred, all since 2002.

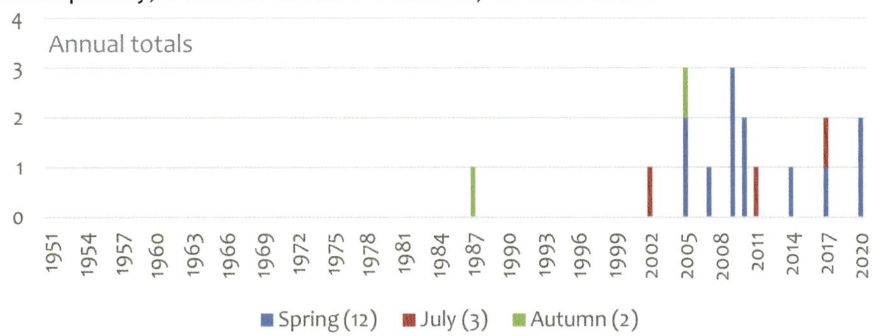

The next was not until 2002 when Mike Rafter observed one at close range hawking insects over the heathland at Pitts Wood Inclosure with House

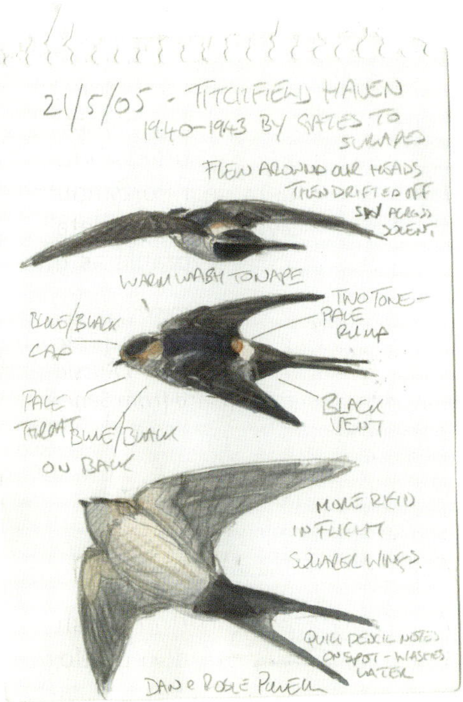

Red-rumped Swallow, Titchfield Haven, May 21st 2005

Martins from 1500 to 1503 hrs on July 10th; it then left south-west. The first in spring was seen by Jason Crook at close range for 15–20 seconds as it migrated north over Farlington Marshes in association with a House Martin at 0913 hrs on May 10th 2005. Later that month Dan and Rosemary Powell saw one over Titchfield Haven on May 21st.

In the autumn of that year Tim Parminter watched a juvenile with a large flock of hirundines over Keyhaven Harbour and reedbed for one minute at 1710 hrs on October 10th. Mike Rafter had the good fortune to find his second for the county when he saw one fly east over Needs Ore at 1130 hrs on May 1st 2007. The best spring to date was in 2009 when there were four records, presumably of three different individuals. Joe Stockwell watched one from 1845 to 1852 hrs over Rooksbury Mill Lake, Andover on April 21st. Simon Ingram saw one as it flew south over his local patch at Lakeside Country Park, Eastleigh between 0942 and 0945 hrs on May 23rd and presumably the same individual was seen by Tristan Norton a little later as it bathed at a fishpond at nearby Bishopstoke between 1000 and 1002 hrs before leaving north-east. The third was found by Peter Milinets-Raby feeding with five Swallows over cereal fields at Swiveltown Lane, Portsdown Hill between 1640 and 1647 hrs on June 2nd. Despite prompt release of the news, it had disappeared by the time other observers arrived.

Two were found in 2010. One was seen by Graham Barrett over Portchester Castle on May 5th and another was found by Paul Winter at Nursling on May 12th. In the following year, Andy Johnson found one feeding and moving slowly west over Sandy Point on July 10th.

In 2014, Simon Boswell found one perched on wires near the entrance to Hurst View caravan site in Lower Pennington Lane at 0915 hrs on April 18th before it flew off south after ten minutes. It was reported over Fishtail Lagoon that evening and was refound with other hirundines by Lee Fuller in the original area at 0920 hrs the following day before drifting off north; it returned briefly before disappearing for good.

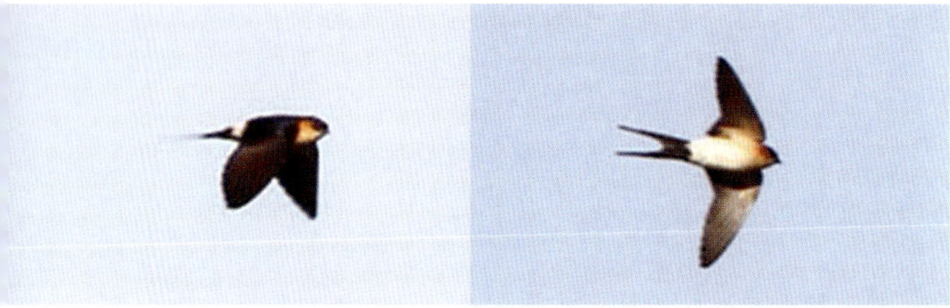

Red-rumped Swallow, Lower Pennington Lane, April 19th 2014. Lee Fuller

The first twitchable bird for the county was initially found by Barry and Tor Duffin flying over the East Meadows at Titchfield Haven at 0905 hrs on May 3rd 2017. It remained in the area north towards Titchfield village until midday and was seen by many observers. Later that year another was seen by Amy Robjohns as it flew south-west over Titchfield Haven at 0925 hrs on July 5th. The Titchfield Haven area's recent monopoly on this species continued in 2020 when Amy found one with a small group of hirundines feeding on newly-emerged midges over Titchfield Canal Path and the adjacent Posbrook Floods at 0847 hrs on April 5th. Other locals on their Coronavirus lockdown exercise were able to catch up with it before it disappeared by 1015 hrs. Later that year Philip Hack saw one fly over his garden at Upton near Romsey on June 16th.

Analysis of arrival dates shows 15 in spring and summer with three in April from 5th, seven in May, two in June on 2nd and 16th and three in July between 5th and 10th, and singles in autumn on October 10th and November 14th.

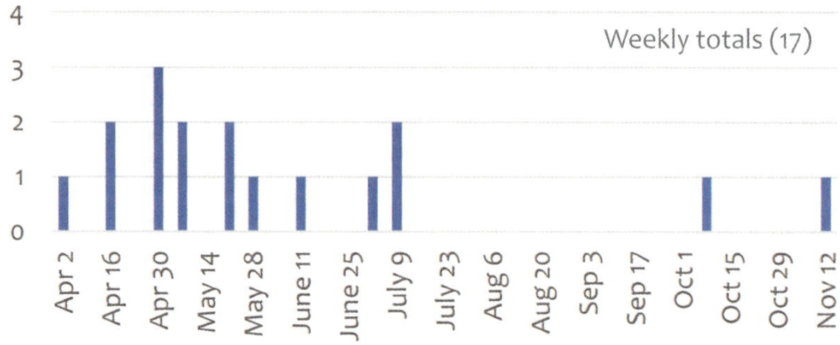

Several other records have not been included. One at Black Gutter Bottom on April 24th 1987 considered not proven by BBRC would have been the first for the county. One seen by Trevor Codlin with Swallows over Titchfield Haven on September 19th 1988 was harshly considered not proven by BBRC. Seven further records between 2010 and 2017 comprised five brief encounters which HOSRP considered not proven and two other claims for which no documentation was received.

GB status: 2, 1906–1949; 1,056, 1950–2019. Increasing, mean 41 per year in 2010–19. (Widespread & locally common NE Africa, Iberian Peninsula, Balkans & Greece but uncommon C Mediterranean to S France. To E, breeds discontinuously W & S Turkey &

Middle East. Wintering area of European population unknown but assumed to lie in N equatorial Africa. Other races, including nominate daurica, *breed E Asia from S Siberia to S China & Japan, Indian subcontinent & locally equatorial Africa.)*

American Cliff Swallow *Petrochelidon pyrrhonota*

BBRC

A very rare vagrant. One record in 2000.

The only record for Hampshire is of a juvenile found by Dan Powell with large numbers of Swallows at Titchfield Haven on October 1st 2000. It was possibly the same as the well-watched individual at Portland, Dorset on September 29th and 30th although BBRC considered it to be a different bird, the ninth for Britain. Dan told the story of an exceptional morning at Titchfield Haven in his paper in the 2000 *HBR* (Powell 2002), extracts from which are reproduced here.

It had been a memorable day for the Titchfield Haven ringing team to say the least. As a birder or ringer you are content to work with familiar birds but you can't help musing on discovering something a little bit special, maybe a White-rumped Sandpiper amongst a flock of Dunlin or a Wryneck laying in a pocket of the nets. So the idea of finding a Paddyfield Warbler looking up at us from the nets was as likely as finding an Andalusian Hemipode to our minds. And then to cap it all a Richard's Pipit flew over us as we furled the nets at the end of the session. We were feeling understandably satisfied by the years of effort we had put into the site.

As we stood in the car park discussing the morning's events it was apparent that there was an extraordinary easterly movement of hirundines unfolding. Barry Duffin decided to carry on ringing by opening the nets by the river to try and catch a few of them – this was at about 10 am. A tape lure was put on and, as the nets were unfurled, I became aware that something very distinctive was flying in among the hundreds of Swallows that were attracted to the tape. At first, I managed a few brief views and slowly built up a jigsaw of features. I was pretty sure of what it was but just couldn't convince myself until I had a clear and prolonged view.

At last it gave itself up by making two or three low passes over the river to drink. I simply bellowed 'Cliff Swallow' and as soon as I said that the bird flew off behind some sallows. The others were by my side in nanoseconds (BSD, Rosemary Powell and Ken Martin) and, for what seemed an age, nothing happened then, much to my relief, the bird re-appeared over the river before drifting towards the scrapes and into the assembled masses. Soon after this a Hobby flew by and the hirundines did what hirundines do under these circumstances, panic and try to put themselves high above the problem. This they achieved with some speed and the Cliff Swallow was lost, never to be seen again.

A record of one at Needs Ore on June 3rd 2016 was considered not proven by BBRC.

GB status: 13, 1983–2019. Rare, about one every three years. (Breeds North America from W Alaska, E across Canada to Nova Scotia, & throughout USA except SE & S, to S Mexico. Migrates to winter S America from Brazil to Chile & Argentina.)

Cetti's Warbler *Cettia cetti*

BBRC to 1976; HOSRP 1977–1982

Formerly a very rare vagrant first recorded in 1961. Annual records from 1972; now a moderately common resident which first bred in 1979.

The first Cetti's Warbler to be recorded in Britain was at Titchfield Haven from March 4th to April 10th 1961. Extracts from the account of the bird's discovery and subsequent identification (Suffern & Ferguson-Lees 1964) are reproduced below.

On March 4th 1961 Canning Suffern heard a rich, fruity song-phrase from a bird which remained hidden in the reeds; it was strange to him and, although it was repeated a few times at intervals of two or three minutes, he was unable to identify it. The next day the bird was still singing irregularly and it continued to be very skulking, but CS glimpsed it briefly and so did D. Price, M. H. Terry and S. L. White among others. During the following fortnight it was seen briefly or heard by countless observers, including H. G. and W. B. Alexander, Dr. J. S. Ash, D. F. Billett, R. H. Charlwood, D. D. Harber, P. A. D. Hollom, D. I. M. Wallace and I. J. Ferguson-Lees, and it became increasingly clear that it was a Cetti's Warbler, a species with which several of those concerned were familiar abroad. Any remaining doubts were finally resolved on March 19th when the bird was trapped in a mist-net by M. J. Carter and P. R. Colston, ringed by J. A. Miller and seen in the hand by about 25 people.

The above account reads like a Who's Who of British ornithology in the early 1960s. At the time, much controversy surrounded the occurrence. Initially, it was considered possibly to be a Moustached Warbler and characteristics of that species were 'seen' by some observers. On March 19th it was trapped enabling its identity to be confirmed. Subsequently, several heated letters discussing the events were published (*Brit. Birds* 58: 225–227, 516–520).

Cetti's Warbler, Ivy Lake, Blashford, March 5th 2019. Mike Read

The next were not until the autumn of 1972, when one was trapped by Andrew Williamson at Farlington Marshes on October 25th and another was seen and heard calling by David Thelwell near the River Test at Timsbury on October 28th. Later that day it was trapped by David and John Taverner. It was not relocated subsequently, despite being searched for. In 1973 Richard Millington found one in a reedbed at Farlington Marshes at 1515 hrs on February 25th which was seen by several other observers that day but not subsequently. In 1974 George Green found a singing male at Keyhaven on March 24th which stayed until at least April 21st and Richard Carpenter discovered another at Alresford Pond on May 5th which remained until June 30th. In 1975 Steve Buckland found one at what is now Itchen Valley Country Park on February 19th; he showed it to Simon Delany the next day and it was still there on 22nd. Another in song was found by Richard and Spike Millington and Tim Henderson at Fleet Pond on June 24th which stayed until the next day. Later that year, Barry Duffin and Bob Gomes trapped one at Titchfield Haven on October 11th and then Barry and Bob Dunn found two on December 10th which were unringed; singles were seen by other observers on December 12th and 20th. In 1976 Hugh Insley and Mike Gibbons found a singing male at Twyford on June 13th which they trapped on 19th. Another located by its song by Mike and Pete Combridge and Ed Wiseman at Keyhaven on October 31st was seen by three more observers on November 3rd.

In spring 1977, single males held territory in the Itchen Valley near Winchester and in the Avon Valley at Hale. In the autumn of that year, a small influx occurred, with 13 recorded, including five in the Itchen Valley and six at Titchfield Haven which arrived between October 14th and 19th. In 1978, at least four singing males were present in the Avon Valley, and in the following year breeding was confirmed there for the first time. The breeding population in the county increased steadily during the 1980s, despite several cold winters. A full breeding survey in 1990 revealed 125–127 singing males. Further consolidation occurred subsequently with occasional setbacks following cold winters such that the population was estimated at 460 singing males in 2016.

GB status: Moderately common but localised resident in England & Wales. (Breeds W Europe and N Africa through Mediterranean Basin to Kazakhstan & Afghanistan.)

Hume's Warbler *Phylloscopus humei*

BBRC

A very rare vagrant. One record in 2014.

The only record for Hampshire is of one seen, heard and photographed by Bryan Pinchen in his garden in Pennington village on November 3rd 2014. Extracts from his account of the events (Pinchen 2015) are reproduced below.

> *On the afternoon of Monday November 3rd 2014 at 1435 hrs I happened to look up from my desk and noticed several passerines around the garden. Alongside three Blue Tits were a stunning male Blackcap and something smaller flitting around the neighbours' sycamore that immediately drew my attention. At first glance it could have been a crest; both Gold- and Firecrest are regular garden birds here so it was worth a second look. Something about this was different and thoughts turned to Yellow-browed or, because of the date, Pallas's Warbler. With binoculars, both crests were quickly ruled out, and it was clearly not bright enough, nor did it have a pale crown stripe, thus ruling out Pallas's. My first impression of the bird was of a rather pale, dull and washed-out Yellow-browed. Setting up my telescope and with a copy of the Collins Bird Guide on my knee I was quickly starting to think this was a Hume's as I started to check off the relevant features: washed-out looking, dull grey-green mantle and back, only a single obvious wing-bar, pale grey tertials and coverts. With the telescope the dark bill, pale cheek and dark legs were obvious. All-in, it was nowhere near as brightly coloured as any of the Yellow-browed I've seen before. Trying to concentrate on watching the bird and gathering enough detail of the identification features I only managed to grab one half-decent record shot as it continued to feed amongst the leaves, constantly flitting about, occasionally stopping to pick at food, but almost always on the move in typical leaf-warbler fashion. Throughout the duration of the sighting (1437–1450 hrs) I was lucky that I was sitting upstairs with the bird barely ten yards away at eye-level, instead of the 'looking up its backside type views' that I would have obtained had I been out in the garden with it above my head. All relevant identification features were clearly to be seen. The afternoon light was excellent from the south-west, with the bird to the south-east of me, giving perfect viewing conditions, albeit through double-glazing, with no glare or heat haze. At 1447 hrs a Magpie landed above it in the sycamore and the warbler was temporarily lost to view. Realising it wasn't going to stay around much longer, I dashed downstairs and into the garden to see if I could see it further or get the clinching call. Sure enough as I raced to the sycamore a clear, short, disyllabic, almost Pied Wagtail-like call could be heard (higher pitched in the first part, dropping down a pitch in the second – as a multi-instrumentalist such things are obvious). The warbler continued flitting around, calling all the while, getting deeper inside the sycamore until eventually disappearing into the trees of the neighbouring copse until it was out of sight and earshot at 1450 hrs. It was only a couple of days later that it dawned on me that in becoming my 105th 'window species' this could also be a county first and the record was duly submitted to the BBRC.*

GB status: 177, 1966–2020. About five per year, influx 2003 (28); stable. (Nominate humei breeds Altai Mountains to W Mongolia, S through Tien Shan & Pamirs to NE Afghanistan, NW Himalayas & mountains of NW China. Winters S Afghanistan to N India, E to W Bengal. Race mandellii breeds C China from Hebei to S Yunnan, W to lower slopes of Tibetan Plateau.)

Yellow-browed Warbler *Phylloscopus inornatus*

BBRC to 1962; HOSRP to date

A rare but increasing autumn passage migrant and winter visitor. 177 records since 1959.

The first for the county was one found by D. W. L. Hubble at Avington on November 30th 1959. The record was accepted by BBRC but unfortunately it has not been possible to trace the description.

Only two were recorded during the next two decades. Norman Pullen trapped one at East Boldre on October 10th 1967 and Barry Duffin trapped and photographed one at Titchfield Haven on December 1st 1978 which was also seen by Bob Chapman and others and was last recorded on January 1st 1979, probably having succumbed to the very cold weather at the time.

The 1980s produced eight. Graham Stephenson found one in a garden adjoining Fleet Pond on October 19th 1981. It was still present the next day when it was seen by several observers. Singles were located by Piers Reid Hart along the footpath between Mill Hill and The Dean, Alresford on October 9th 1984 and later that year by Dave Bishop at Sinah Warren on October 27th. Ralph Simpson found one in a tit party at Wellington Country Park on November 30th 1986 which was seen by the author on December 1st. Andy Welch trapped a juvenile at Bitterne Park, Southampton on October 25th 1988 and Stuart Sporne did likewise on his smallholding at Sway on October 3rd of that year. In 1989, singles were located by Keith Turner at Sinah Gravel Pit on October 1st and Mick Scott at Fleet Pond on October 24th.

A further 11 were recorded in the 1990s including three over-wintering birds. Andy Parfitt found one at Sinah Warren on October 7th 1990 and Alan Claxton saw another in his garden at Cromarty Road, Southampton on September 25th 1993, the earliest record at the time, and again on 28th. The early date was eclipsed in the following year when Jason Crook found one at Farlington Marshes on September 18th. Jason found another near Budds Farm Sewage Treatment Works on December 29th where it remained until at least April 5th 1995 and was seen by many observers. In the autumn of that year Maurice Opie found one along the Titchfield Canal Path on October 11th and Tim Lawman saw two together there on October 15th. Tim Doran discovered one at Budds Farm on December 2nd and later that month possibly two were seen by Peter Le Brocq on 10th and by Ted and Penny Raynor on 26th. The

Yellow-browed Warbler, Newlands Farm, December 19th 2008

presence of two over-wintering birds was confirmed in January 1996 when they were seen together by Kevin Crisp and Alan Snook on 10th and Terry Jennings on 16th. One of these was trapped by Trevor Codlin on February 7th and last recorded singing by Jason Crook on April 12th. In the autumn of that year Mick Scott found singles at Fleet Pond on September 30th, which he relocated the next day, and on November 13th. The final record of the decade involved a rather washed-out looking individual found by Marc Moody at Lower Pennington Lane on December 1st 1999. This raised the possibility of it being a Hume's Warbler but its call confirmed its identity; it was also seen by Ed Wiseman and Adrian Clark and was last seen on 4th.

The 21st century started with singles in each of the first four years. Bob Chapman and Jason Crook found one at Farlington Marshes on October 14th 2001, Will Walmsley saw one with Goldcrests at Shedfield Common on October 24th 2002, Mark Edgeller watched one with five Chiffchaffs in a hedgerow at Brownwich on October 5th 2003 and Tim Doran located one in trees in the car park by IBM Lake, Cosham on October 16th 2004. Since then, the species has occurred annually in increasing numbers. Autumn totals exceeded five in 12 years with maxima of 19 in 2014 and 27 in 2016.

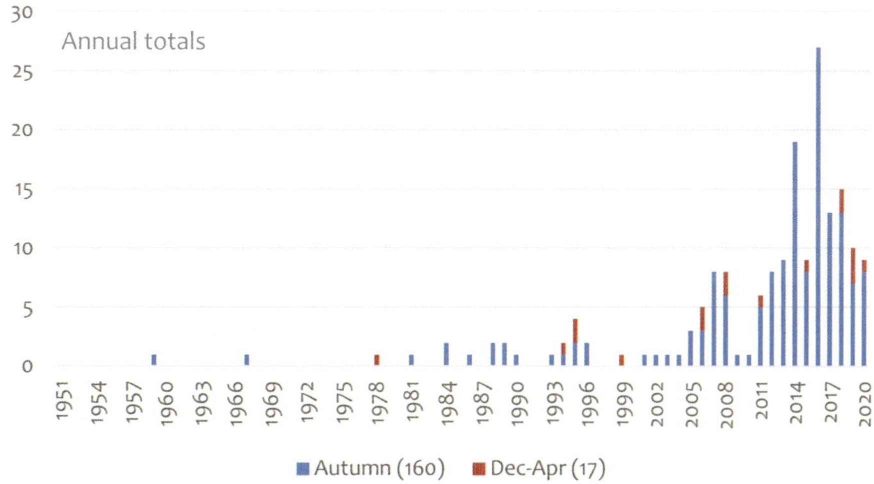

Analysis of the records shows that autumn arrivals have occurred in September in eight years, the earliest at Farlington Marshes on 18th in 1994 as mentioned above. The peak passage is in October, especially the middle two weeks, with numbers rapidly dropping off in November. Most are recorded on one or two dates only although stays of up to seven days, assumed to involve the same individual, have been recorded. A total of 13 has been recorded in December with seven of these staying into the following year. A further five have been found in January and February of which three stayed for several weeks and two were seen on one date only. In addition to the wintering birds at Titchfield Haven and Budds Farm mentioned above, long-staying individuals have been at Hawley Meadows from December 28th 2006 to March 16th 2007, Rooksbury Mill from February 5th to March 25th 2007, Eastleigh Sewage Treatment Works from December 20th 2015 to February 10th 2016 and from February 10th to 25th 2019 and Ivy Lake, Blashford from January 4th to 30th 2019 and from December 11th 2020 into 2021. Although one found at Brownwich on February 26th 2012 is the latest new arrival, two wintering birds at Budds Farm mentioned above stayed into April with the latest date on 12th in 1996.

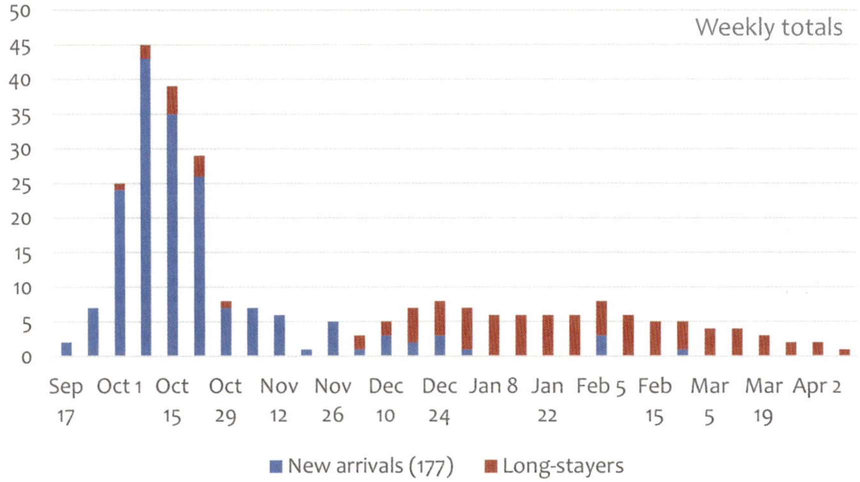

Most have occurred close to the coast with totals of 20 on Hayling Island including 11 at Sandy Point and eight in the Sinah area, 19 in the Portsmouth area including seven at IBM Lake, 34 between Hook-with-Warsash and Gosport including 18 at Titchfield Haven, 23 in the Southampton area including six on Southampton Common and eight between Marchwood and Fishlake Meadows, five between Lepe and Park Shore and 19 between Lymington and Barton on Sea including nine between Normandy and Keyhaven and six at Milford on Sea. Around 57 have occurred inland including 18 in the Itchen Valley, 12 in the north-east including four at Fleet Pond and four at The Mill Field, five in the New Forest, four at Blashford Lakes and three at Old Winchester Hill.

Yellow-browed Warblers: Titchfield Haven, October 17th 2014 and Ivy Lake, Blashford, December 28th 2020. Trevor Codlin and Matthew Barfield

A further 33 records have been excluded from the above account. Of eight between 1994 and 2005, when the species was still rare, two were considered not proven by HOSRP and the remainder were from bird information services where attempts to obtain documentation failed. These included an interesting record of one singing at Itchen Valley Country Park on April 16th 2004. Of 25 since then, nine were considered not proven but for the rest the observers' names could not be traced.

The increasing number of records of this Siberian breeder in Hampshire has mirrored the national trend; greater observer activity and favourable weather conditions have no doubt contributed.

Possibly the first for the county was one seen by Mr. A. Ponchaud in his garden at Ringwood on October 24th 1943 (*HBR*). In a review of the 1943 *HBR* in *British Birds* (38: 160) it described views at three yards range with wing-bar and superciliary stripe seen but said that more detailed evidence was desirable. Where this description came from is unknown as it is not in the *HBR*. It was not accepted by Naylor (1996).

GB status: c. 30,700, 1968–2017. Large recent increase. (Breeds Russia from Urals to NE Siberia, S to N Mongolia & NE China. Winters NE India, S China, Indochina & Malay Peninsula S to Singapore.)

Pallas's Warbler *Phylloscopus proregulus*

BBRC to 1990; HOSRP to date

A very rare vagrant. Nine records since 1974.

The first for the county was found by John Sylvester at Milford on Sea on November 23rd 1974. His description to BBRC follows.

> *On November 23rd 1974 my brother, his friend and I (all keen ornithologists) spotted a Pallas's Leaf Warbler in the Pleasure Gardens, Milford on Sea. In the Readers Digest and AA book of British Birds there is a column naming the rarest British Birds (or vagrants) seen in the British Isles between 1964 and 1971. The warbler is one of those mentioned and therefore I thought you may be interested to hear about it. The Warbler was about the size of a Goldcrest, but not as plump, much more streamlined in appearance. The back and wings were green with whitish lower parts. A streak of black passed through each eye with two streaks of yellow on the crown. The rump was yellow and very prominent in flight, as he flitted quickly from branch to branch finding insects. The flight appeared to be very short and quick and the movements were the same, very quick. It never appeared to be still for any amount of time. He was feeding low in the trees, not above 12 feet. Long-tailed Tits were feeding nearby.*

It was a long wait before the first twitchable bird was found by Dan Powell in Thatcher's Copse, Brownwich on October 26th 1988. It remained until 28th and was seen by many observers during that time.

Pallas's Warbler, Thatcher's Copse, October 26th 1988. Autumn colour and a tiny gem.

The 2000s produced a flurry of records with five recorded during the decade. Tim Lawman watched one at Sandy Point from 1130 to 1150 hrs and again with a small group of observers from 1320 to 1400 hrs on October 30th 2004. The next was seen by Graham Giddens in his garden at Lymington for two minutes from 1355 hrs on January 15th 2005. This was followed by a singing male found by Pete Combridge on private land at Totton on March 1st 2008; it remained on territory until April 3rd and was shown to five other observers. Another singing male was discovered by Andy Johnson at Sandy Point on May 8th of that year; it showed well during the rest of the day but was not found the following morning. An unprecedented year for the species ended when Robert Cheke trapped one in a garden in Bramley in north Hampshire on December 28th. It was also seen and photographed by others.

Pallas's Warblers: River Blackwater, Eversley Gravel Pit, January 15th 2013 and Sandy Point, May 8th 2008. Dave Perrett and Richard Ford

The first of the 2010s was a more expected autumn record when Tim Lawman found one with a tit flock at Sandy Point on October 24th 2011. Although elusive during its two day stay it was eventually seen well by many observers. The most recent record involved one, initially identified as a Yellow-browed Warbler, found at Moor Green Lakes, Berkshire on January 3rd 2013. It was subsequently refound 500 m away along the River Blackwater, which straddles the county boundary, on 6th. Its identity was confirmed on January 12th and it continued to frequent the bushes on both sides of the river until April 18th. At times it gave fantastic close-up views and was often heard singing towards the end of its stay.

GB status: 1, 1896; 2,803, 1951–2019. Decreasing after peak in 1990–2009. Mean 56 per year in 2010–19. (Breeds S Siberia from Baikal region E to Ussuriland, NE China & S to N Mongolia. Winters E China to S of Yangtze River & throughout Indochina S to C Thailand.)

Radde's Warbler *Phylloscopus schwarzi*

BBRC to 2005; HOSRP to date

A very rare vagrant. Three records in 2005 and 2013 (2).

The first for the county was found by Andrew Pink on rough ground near Sherborne St. John on December 7th 2005. Its identity was confirmed by Nick Montegriffo and news released selectively before the end of the day. Unfortunately, it was not present the following day. Extracts from Nick's description to BBRC are provided below.

Mike Duffy

At 1045 hrs I received a phone call from Andy Pink who was visiting an industrial area a few hundred metres from my home in Sherborne St. John. Andy had found a Phylloscopus *warbler which was calling with a hard 'chack', but he didn't have binoculars. I grabbed two pairs of bins and joined him on site. The bird was in a very overgrown area approximately the size of a football pitch, covered in nettles and obviously used as an informal dumping ground for old farm machinery and paving tiles. After an hour searching, Andy found it on top of the embankment surrounding the small car park. We stalked the bird and it eventually popped into view. One glance at the long supercilium and apricot-coloured undertail-coverts was enough: it was clearly a Radde's Warbler! For the next few minutes, the bird showed very well down to five metres and the hard 'chack' call was also heard. Andy had to leave but I was able to return home to fetch my video camera, by which time other local observers arrived and watched the bird intermittently until dusk.*

Raddes Warbler, Titchfield Haven, October 26th 2013. Barry Duffin

The next was found by Joe Stockwell at Anton Lakes, Andover on October 13th 2013. It was subsequently seen by many observers in the next two days but tragically Tim Lawman collapsed and died while running to see the bird on the first day. Later that autumn, one was trapped by Barry Duffin, Duncan Bell and Trevor Codlin at Titchfield Haven on October 26th.

GB status: 1, 1898; 451, 1961–2019. No clear trend, mean 12 per year in 2010–19. (Breeds S Siberia from Ob River region E to Ussuriland & NE China. Migrates through E China to winter N Burma & Indochina S to C Thailand.)

Dusky Warbler *Phylloscopus fuscatus*

BBRC to 2005; HOSRP to date

A very rare vagrant. Two records in 2006 and 2013.

The first for the county was found by Andy Johnson at Sandy Point on November 2nd 2006. It was subsequently seen by many observers. Extracts from Andy's account of his discovery (Johnson 2007) are given below.

The first round of my recording area completed, I headed towards the northern end of the reserve when, at 0856 hrs, I heard a repetitive 'tacking' in front of me. My brain hardly had time to think 'that sounds like a ...' before I was watching a Dusky Warbler in the nearest bramble bush! I didn't have to endure any of the stresses of locating a 'skulker' deep in impenetrable vegetation, or wondering whether I'd heard the call rightly; here was the most obliging Dusky Warbler I've ever seen! By around 1030 hrs it settled into a patch of bramble scrub on the edge of the reserve where it stayed for the rest of the day, being last seen in fading light at 1648 hrs. I was in position before first light the next morning but its chosen patch was silent and it was obvious that the bird had departed in clear conditions overnight.

The second was also found by Andy at Sandy Point on October 8th 2013. It was enjoyed by many observers during its stay until October 10th.

Dusky Warbler, Sandy Point, November 2nd 2006. George Spraggs

GB status: 1, 1913; 607, 1961–2019. Increasing, mean 26 per year in 2010–19. (Breeds Siberia from Ob River N to c. 60°N, E to Sea of Okhotsk, S to Russian Altai, N Mongolia & Ussuriland through NE China. Winters Nepal to S China & SE Asia to Singapore.)

Siberian Chiffchaff *Phylloscopus collybita tristis*

HOSRP

A very scarce winter visitor and passage migrant. Around 63 records since 1982.

The first record for the county of this race of Chiffchaff was of one trapped by David Murdoch, a member of the Lower Test Ringing Group, at Southampton Container Berth on October 31st 1982. Extracts from his description are given below.

> *The bird was watched and heard calling several times during the morning and was eventually trapped and photographed. He described a bird with uniform brownish upperparts with no green, buff eye-ring and well-marked supercilium, buff cheeks and upper breast gradually became dirty cream underbelly. Yellow only on shoulder of wing and axillaries. Call an extremely distinctive, monosyllabic 'weet', quite different from the 'hoeet' of* collybita *with a much harder attack to it. Carried far, often repeated when disturbed.*

Awareness of the eastern forms of Chiffchaff rose from the mid-1980s and several records of birds showing the characteristics of either *abietinus* (from Scandinavia and European Russia) or *tristis* were submitted. Subsequently, knowledge increased and it was realised that *tristis* had distinctive calls. Claims of definite *tristis* included singles reported by Graham Stephenson at Sinah Common on November 7th and 15th 1992, Jonty Denton at Liss on February 2nd 1993, Ivan Lakin at Titchfield Haven on February 21st 1993, Jason Crook at Budds Farm Sewage Treatment Works on January 31st 1996 (possibly two) and Mick Scott at Fleet Pond on November 24th 1996. Subsequently, between two and five were identified in the six winters from 1998/99 to 2003/04. None was claimed in the following three winters although as many as 17 *abietinus/tristis* types were trapped at Eastleigh Sewage Treatment Works in November and December 2005.

In 2007 *British Birds* set up a '*tristis* panel' to establish the status of Siberian Chiffchaff in Britain. Stoddart (2008) summarised the criteria for identifying *tristis* and requested that all records during 2008 should be submitted to county records panels for onward transmission to BBRC if they met the criteria. There were five such claims submitted to HOSRP although it is unclear if all were submitted to BBRC; only one was published as accepted in the subsequent paper reviewing the results (Dean *et al.* 2010). This was of one found by Graham Stephenson at Darby Green which was present between December 19th 2007 and February 28th 2008. There were two main conclusions from the review: firstly that *tristis* should embrace both the east Siberian form *tristis* (breeding eastwards from the Yenisey river) and the west Siberian form *fulvescens* (breeding eastwards from the Ural mountains to the Yenisey river) – the latter really a variant of the former but possibly showing traces of yellow in its otherwise '*tristis*' morphology, and secondly to inform observers that acceptance of a suspected *tristis* requires documentation of both voice and a very careful appraisal of plumage and bare part colour. Subsequently, HOSRP has applied these criteria when assessing records although since 2017 claims of silent birds have also been published in *HBRs* with a suitable caveat. Some birds trapped recently at Eastleigh Sewage Treatment Works have been further confirmed as *tristis* by DNA sampling. The chart below shows totals of claims of definite *tristis* and *abietinus/tristis* types in each winter.

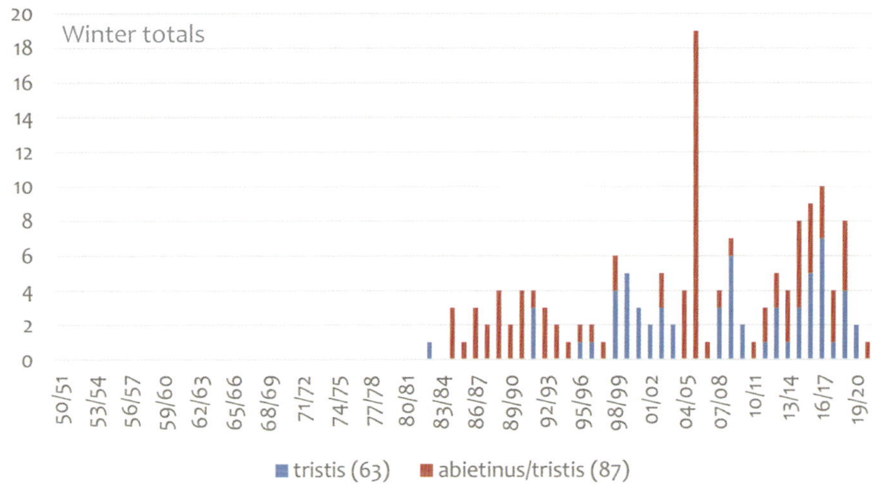

Most records came from four main areas: Eastleigh Sewage Treatment Works, 56 (17 *tristis*), the Blackwater Valley at Darby Green (adjacent to Sandhurst Sewage Treatment Works, Berkshire), 18 (10 *tristis*); Langstone Harbour north shore including Budds Farm Sewage Treatment Works, 14 (6 *tristis*); and Titchfield Haven, 10 (9 *tristis*). Elsewhere, a total of 52 (21 *tristis*) has been recorded at 40 different sites.

Siberian Chiffchaff, right, and Chiffchaff, Fishlake Meadows, November 29th 2014. Simon Ingram

Siberian Chiffchaff, Darby Green, February 26th 2017. Jerry Bird

Analysis of arrival dates shows a clear peak in December with a gradual decline in numbers through to mid-March.

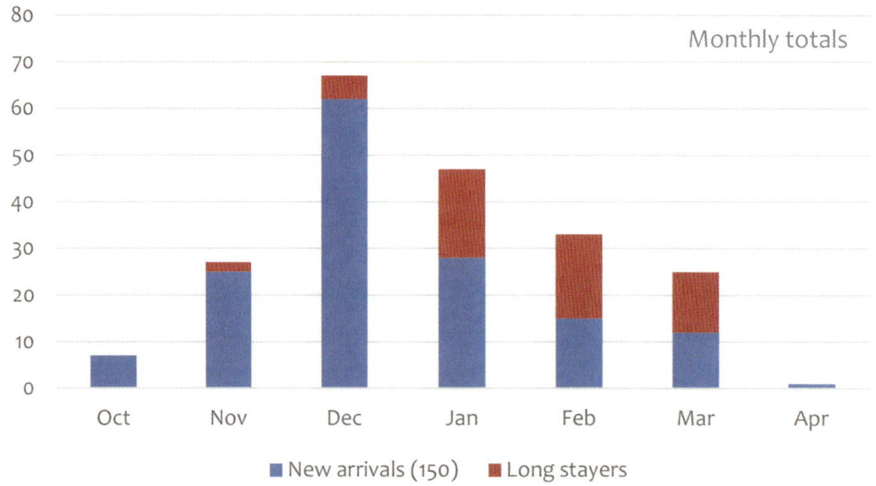

Of seven records in October by far the earliest was a claimed *tristis* seen by Chris Parr in his back garden at Woolston on October 1st 1998. Another *tristis* was seen and heard by Paul Winter adjacent to Lower Test Marshes on October 18th 2003 which had reportedly been present since the 15th. The other records were all between October 28th and 31st. Overwintering has occurred in most recent winters, particularly at the favoured sites, and while it is assumed that the same individuals are present for several weeks, ringing activities have inevitably demonstrated that more birds are involved. At Eastleigh Sewage Treatment Works four were ringed between December 27th 2018 and February 3rd 2019 and at least one unringed bird was present until March 20th. Other long-staying individuals include one caught at Fishlake Meadows on October 28th 2014 which was retrapped there on November 29th of that year, and singles at Newlands Farm, Fareham from December 17th 2011 to March 17th 2012, Seafarers Walk near Sandy Point from January 15th to March 11th 2013 and Upper Titchfield Haven from November 22nd 2015 to January 17th 2016. Most have departed by mid-March; the latest records of confirmed *tristis* were of singles at Budds Farm Sewage Treatment Works on March 20th 2000 and March 23rd 2016. Later records of indeterminate birds involved singles at Darby Green on March 25th 2007 and Bickton on April 25th 1986. The latter was well described by Paul Bowman, an experienced observer.

Siberian Chiffchaff and Chiffchaff, Newlands Farm, January 13th 2012

GB status: 2,968, 2008–19. Increasing, mean 276 per year in 2010–19. (Breeds Russia from Urals E to NE Siberia & S to N Mongolia. Winters Iran to N Indian subcontinent.)

Iberian Chiffchaff *Phylloscopus ibericus*

BBRC

A very rare vagrant. One record in 2009.

The only record for Hampshire is of a singing male at New Copse Inclosure, New Forest from May 10th to 17th 2009. A *Phylloscopus* warbler singing an unusual song was found by Dan Houghton, Mark Rolfe and Alan Butler. Dan videoed it singing and Nigel Jones obtained photographs and further recordings of its song. Detailed documentation of its plumage and an analysis of sonograms facilitated its acceptance by BBRC. Further details are given in Dan and Nigel's paper in the 2010 *HBR* (Houghton & Jones 2011), extracts from which are reproduced below.

Nigel takes up the story: 'On the afternoon of May 12th, after reading comments on Hoslist about the 'iffy chiffy', I decided to try to see the bird. I found it perched high up in an ash tree and singing a mixed song. Very much like the well-known sketch with Morecambe & Wise and André Previn, 'It was playing all the right notes but not necessarily in the right order'. The sequence of the notes was jumbled from what I would describe as pure Iberian and more classic 'chiffchaff' calls were being uttered in response to local collybita *Chiffchaffs. However, the plumage and structure looked fine for Iberian. I played* collybita *Chiffchaff song to the bird and got a lukewarm response. However, when I changed the song to Iberian Chiffchaff, and placed the player on the ground, the bird danced over the speaker within inches, singing incessantly and fluttering its wings in display. I took some photos and made some recordings. I knew the key feature to record was the down-slurred 'piu' but even though it was heard I could not trace it on my recordings. Luckily,* Dan Houghton *had made some video recordings and captured this call, which is described as diagnostic of the species.*

Iberian Chiffchaff, New Copse Inclosure, New Forest, May 12th 2009. Nigel Jones

Several less than enthusiastic observers will regret not going to see the bird until another is found!

A possible hybrid Iberian x Common Chiffchaff with a 'mixed' song held territory at Sandy Point from March 18th to July 2nd 2014.

GB status: 83, 1972–2020. About six per year; recent increase. (Breeds locally French Pyrenees & S throughout W Iberia. N African range restricted to NW Morocco & N Algeria to NW Tunisia. Wintering range poorly known.)

Great Reed Warbler *Acrocephalus arundinaceus*

BBRC

A very rare vagrant. One record in 1884 and five since 1960.

The first for the county was shot near Ringwood on June 3rd 1884 although the record was overlooked for *BoH*. The article in *The Zoologist* for August 1884 states:

> *Your readers will be interested to hear that on the 3rd June last a fine specimen of the Great Reed Warbler was shot by Mr. G. F. Hart, who takes charge of my fishery at Ringwood. He found it by the river side in a rank growth of sedge, reed, water hemlock and willow. It has been preserved for me, and proved on dissection to be a male – T. J. Mann, The Grange, Bishop's Stortford, Herts. [We have been favoured with a sight of this bird, which Mr. Mann very kindly bought for our inspection, and there is no doubt that it has been correctly identified – Ed.]*

The first modern record was of a singing male found by Shaun White at Titchfield Haven on May 18th 1960 and subsequently seen by Canning Suffern, Tony Norris and others during its stay until May 24th. Extracts from Canning's notes to BBRC are reproduced below.

> *On 18th May 1960 S. L. White phoned me at 0615 hrs GMT to say that he had found what he thought must be a Great Reed Warbler. By 0620 hrs I reached the indicated spot and watched it sing while perched high up on the dead stem of a reed. I easily confirmed SLW's identification. The song consisted of a few loud harsh notes oft repeated. The bird's stance while singing was almost upright – i.e. head up, tail down, vertically – and the bird did not move about as an ordinary reed warbler would. Its size was estimated as that of a small song thrush and half as big again as a reed warbler. Its plumage resembled that of a reed warbler but it had also a conspicuous pale superciliary stripe which continued backwards from the eye towards the ear.*

The next three records were all of males singing from the same reedbed at Fleet Pond. The first, on May 24th 1970, was found by Richard and Spike Millington and subsequently seen by Trevor Price, the author and others. The second, present from May 20th to June 1st 1975, was found by Spike Millington and seen by many observers. Finally, Tim Doran found one on May 26th 1980 which stayed until June 3rd and was widely viewed.

This species at last became available to Hampshire birders in 2021 when Mark Edgeller found one singing at Itchen Valley Country Park early on May 23rd 2021. He takes up the story:

> *At 0830 hrs I was an hour into my final BTO Breeding Waders of Wet Meadows survey of the season. As I walked into a new field I heard a short burst of distinctive*

song from a nearby patch of bramble and hawthorn. The bird gave a couple more short bursts and I felt sure it was a Great Reed Warbler. I made a recording on my phone and sent it to a few people to check that I wasn't hearing things and making an identification error! At about 0845 hrs it gave a longer burst of song which confirmed the identification for me. The deep guttural 'grr... grr... ' notes followed by 'guttuck' notes were distinctive and loud. Around 0900 hrs it moved into a larger bank of scrub and finally showed well at 0930hrs just as Nigel Jones and Dave Unsworth arrived. It was a large, almost thrush-like Acrocephalus *warbler; long-winged, strong-billed and with a slightly rounded tail, it moved with hefty jumps through the scrub. Its plumage was very similar to Reed Warbler with uniform brown upperparts, a white chin and throat contrasting with buffy-brown flanks. A distinct white supercilium extended just beyond the eye and was broadest in front. This contrasted with a dark eye stripe and lores. The deep and relatively blunted ended bill showed a dark culmen and tip contrasting with a paler cutting edge and lower mandible. Its shape and size added to the thrush-like feel of the bird. The iris was dark orange with a black pupil. The weather deteriorated from late morning and the bird became elusive and sang less frequently. By mid-evening only three observers remained and they were treated to a final burst of song as it sat out in the open. It was not seen the next day.*

Richard Levett heard one singing at Titchfield Haven at 1955 hrs on May 12th 2001 while undertaking a common bird census. He located the bird and obtained good views over the next five minutes but it was not relocated. The record was harshly considered not proven by BBRC. Richard has recently confirmed that he remains 100% confident of his identification.

GB status: 8, 1847–1949; 310, 1950–2020. About six per year; stable. (Nominate race breeds throughout much of continental Europe from Iberia to Greece, N to S Sweden & S Finland, & E across S Russia, Turkey & Caucasus to W Siberia. C Asian race zarudnyi *breeds from Volga to NW China & W Mongolia. Winters throughout C & S Africa.)*

Aquatic Warbler *Acrocephalus paludicola*

BBRC to 1982; HOSRP 1983–2014; BBRC 2015 to date

A rare autumn passage migrant. 88 records since 1956.

The first for the county was found by John Ash at Latchmore Bottom, New Forest on September 15th 1956. It was followed the next day by another seen by Roy Dennis, Graham Rees and David Billett and eight others at Farlington Marshes. The 1956 *HBR* records that the Latchmore bird was watched at close range under perfect conditions, while the Farlington individual was seen feeding in a ditch with full details noted, including streaked rump. Further birds were found at Farlington Marshes by Alan Walker on August 29th 1957 and Graham Rees on September 23rd 1957.

Annual totals are shown opposite.

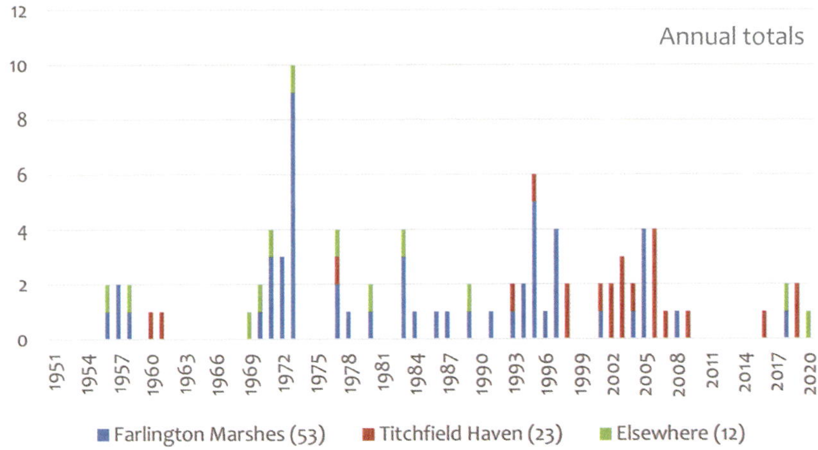

Records were considered by BBRC from 1958. Those of singles seen by John Simons at Farlington Marshes on August 6th 1958, Johnny Bowers at Ashlett on September 6th 1958 and Michael Bryant and Martin Terry at Titchfield Haven on September 25th 1960 and September 24th 1961 respectively were all accepted.

Between 1969 and 2009 there were records in 32 years involving 79 birds, of which 64 were trapped. Subsequently, there was a crash in numbers nationally and the species was restored to the BBRC list. The only recent records are of first-winters trapped by Barry Duffin at Titchfield Haven on August 13th 2016 and September 1st 2019, one photographed by J. W. Burnett at Farlington Marshes on August 20th 2018 and singles seen by Philip Fawkes at Stansore Point, Lepe the next day, Amy Robjohns along the Titchfield Canal Path on September 15th 2019 and Martin Pellett and others at Hook-with-Warsash on September 11th 2020.

Of 88 recorded, 53 were at Farlington Marshes. Of these, 44 were trapped between 1970 and 2008 with a peak of nine in 1973. All were aged as first-winters apart from four adults. A further 23 occurred at Titchfield Haven. Of these, 17 were trapped, one in 1977 and the remainder between 1998 and 2019. All were aged as first-winters including one caught on August 19th 2006 which was controlled 394 km to the south in Loire-Atlantique in western France on August 28th 2006. This is one of only three foreign recoveries of British-ringed birds. Others occurred at Keyhaven/Pennington Marshes (4 – 1 trapped) and Anton Lakes, Ashlett, Hook-with-Warsash, Latchmore Bottom, Lepe, Lower Test Marshes (trapped), Paulsgrove Reclamation and Tournerbury.

As shown on the next page, analysis of arrival dates shows a clear peak in the first two weeks of August with gradually declining numbers through to mid-September. The extreme dates were July 24th (1977, Anton Lakes) and

October 3rd (1971, Tournerbury). All were recorded on one date only apart from one ringed at Farlington Marshes on September 9th 1973 and retrapped on 16th and another ringed there on August 18th 1995 and retrapped on 20th, and sight records of singles at Anton Lakes on July 24th and 25th 1977 and Paulsgrove Reclamation from August 19th to 21st 1983.

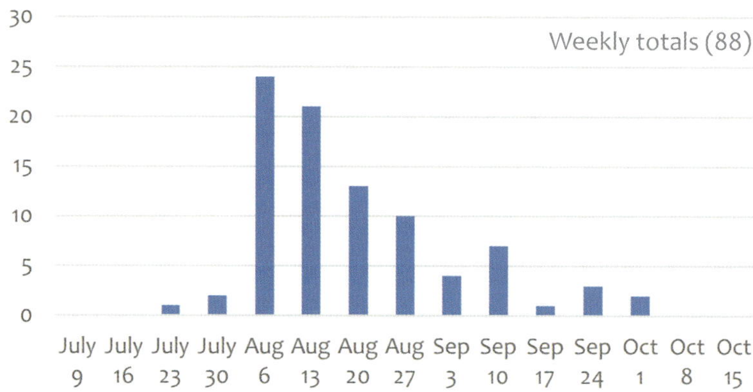

The overall trend has been downward since the turn of the century which is in line with the decline in breeding populations in eastern Europe. The numbers recorded each autumn must be influenced by the incidence of easterly winds to drift migrants from their breeding grounds and the intensity of ringing activity (which has accounted for 71.6% of all records). In the latter part of the 20th century most were trapped at Farlington Marshes but increased levels of ringing at Titchfield Haven since then account for the domination of that site in recent years.

Aquatic Warblers: Titchfield Haven, September 7th 2003 and August 13th 2016.
Dan Powell and Trevor Codlin

Three further records should be mentioned. Roy Dennis saw one at Titchfield Haven on October 3rd 1960 which was apparently submitted to BBRC but no decision was published. Wiseman (2017) included two singles at Keyhaven Marshes seen by Bob Dunn on September 7th 1980 and Mike Stewart and Bob Dunn on September 20th 1981, both experienced observers. The first was not submitted to BBRC and the second was harshly considered not proven.

Aquatic Warbler, Farlington Marshes, August 20th 2018. John Burnett

One at Farlington Marshes was reported in *The Field* for October 9th 1897 and subsequently in *The Zoologist* for October 1897. Rev. S. G. Scott was quoted as follows:

My neighbour, Mr. Richards, of Farlington, sent me the other day a small bird that had been killed accidentally by his fox-terrier in Farlington Marshes on September 20th. Neither of us could identify the species, so I sent it to Mr. Pratt, of Brighton, who pronounced it a male specimen of the Aquatic Warbler. It agrees with the coloured plate in Borer's Birds of Sussex.

The fate of the specimen was not mentioned and therefore the record must be considered questionable.

GB status: 40, 1853–1949; 1,364, 1950–2020. About ten per year, recent decrease. (Breeds locally NE Germany, Poland, Belarus & Russia E to S Urals, with isolated population in Hungary. Entire population migrates along North Sea and Atlantic coasts to winter in N tropics of W Africa.)

Paddyfield Warbler *Acrocephalus agricola*

BBRC

A very rare vagrant. One record in 2000.

The only record for Hampshire is of a juvenile trapped by Duncan Bell, Barry Duffin and Trevor Codlin at Titchfield Haven on October 1st 2000. It was also seen by Ken Martin and Dan and Rosemary Powell. Extracts from Dan's account of events that day (Powell 2002) are reproduced below.

Paddyfield Warbler

At 0700 hrs a team of six Haven ringers was checking the nets soon after first light in anticipation of catching the 400th Sedge Warbler of the season. None was caught but instead a similar sized sandy warbler with a conspicuous cream supercilium was in one of the nets set between mixed scrub and an extensive reedbed close to the river. To the utter surprise of the finders, they discovered that it was a Paddyfield Warbler. A full description was taken before it was ringed, photographed and released into the habitat where it was first found.

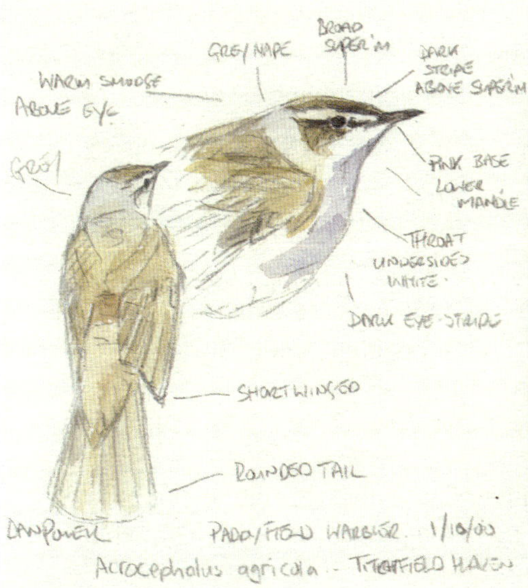

Paddyfield Warbler, Titchfield Haven, October 1st 2000. Trevor Codlin

GB status: 1, 1925; 118, 1953–2020. About 3–4 per year; increasing. (Breeds Black Sea coast from N Bulgaria & Danube Delta E to Ukraine. To E, breeds widely across steppes of S Russia & SW Siberia, Kazakhstan, NW China & W Mongolia, S to Uzbekistan & N Pakistan. Winters throughout Indian subcontinent.)

Marsh Warbler *Acrocephalus palustris*

HOSRP

A rare summer visitor and autumn passage migrant. Five old records and at least 43 since 1975.

K & M recorded a nest with eggs taken by Mr. W. H. Smith at Alresford Pond in June 1863 but no further details have been traced. A transcript of the diaries of John Walpole-Bond, who was a student at Winchester College in the mid-1890s, was acquired by John Cloyne in 2020. He shared the entry for June 25th 1896 which states:

> *Marsh-warbler's nest (c/6 almost hatching) nr. Winchester, in nettles and meadow-sweet in practically dry ditch near Itchen (river). No-one knew what it was then, but I saved 1 egg which told me the tale when I became acquainted with the species.*

The Zoologist for November 1907 contains a lengthy article by Harry Beeston recording the discovery of two nests with eggs in a reedbed near Havant in June 1907. Beeston is at pains to acknowledge that reedbeds are not the normal habitat of Marsh Warbler. The eggs were sent to Dr. Bowdler Sharpe at the Natural History Museum who confirmed that they were of this species. Dr. Alex Bond, Senior Curator in charge of birds in 2020, has been unable to trace any reference to the eggs. The only other records in the first half of the 20th century were of singing males reported by J. C. Boyes at Ringwood on

May 11th 1934 (*SEBR*) and Ken Smith at Keyhaven on May 19th 1935 (*HBR*). Ken became a member of BBRC soon after its establishment in 1959.

None of the above records are entirely convincing. Walpole-Bond went on to become a highly respected field ornithologist and wrote *A History of Sussex Birds* published in 1938; his appears to be the most reliable observation.

It is difficult to be certain of the first modern record. A singing male found by Norman Orr in the Avon Valley on June 2nd 1963 has now, following reference to his diaries, been confirmed as being in present-day Dorset (Stuart Orr *in litt.*). The 1975 *HBR* refers to two, including a singing male, at a southern locality (Lower Test Marshes) on May 3rd and 4th which is unacceptably early for this late migrant. However, an adult was trapped there by Jess Pain on July 11th 1975. In the following year, juveniles were trapped at the same site on August 11th and 18th. There were no further records until 1979, when a male, apparently unmated, was found during survey work for the RSPB in the Avon Valley at Bickton on June 5th and was still holding territory on June 25th.

Subsequently, most records have been from Titchfield Haven and Lower Test Marshes, as shown overleaf. Some of these were not published in *HBRs*.

Marsh Warbler, Titchfield Haven, May 28th 1992

Titchfield Haven: singing males on June 16th 1987 (Duffin 2009), May 26th, June 6th and June 14th 1988 (Duffin 2009, Trevor Codlin), May 29th and June 2nd 1989 (Duffin 2009, Mike Rafter), May 30th and 31st 1990 (Mike Cannings), between May 25th and June 22nd 1992 including two on May 26th and June 9th (Barry Duffin *in litt.*), from May 28th to June 19th 1994 (Mike Cannings, Martin Terry and Barry Duffin), on June 16th and 20th 2004 (Barry Duffin and Richard Carpenter), June 12th 2017 (Amy Robjohns) and May 27th 2020 (Amy Robjohns). Despite some birds holding territory for several weeks, evidence of successful breeding has not been obtained (Barry Duffin *in litt.*).

Lower Test Marshes: singing males on June 14th 1991, June 11th 1992 (not 17th as in *HBR*), from June 7th to 20th 1993 with food-carrying and alarm-calling noted on June 15th, in two areas on May 24th and June 8th 1995, in three areas on June 5th, 10th and 13th 1996, in two areas on May 24th and June 2nd 1997, on May 24th 1999 (Jess Pain *in litt.*) and May 19th 2018 (Simon S. King). The observation of food-carrying in 1993 is the only evidence of breeding in the modern era.

Marsh Warbler, Ibsley, June 6th 2012. Martin Bennett

Other spring records were of singing males recorded at Nursling substation by Tom Dougall on May 21st 1983, Bishopstoke by Dave Unsworth from June 14th to 25th 1986, Ellingham Lake by Mike Read from June 29th to July 11th 1986, Itchen Valley Country Park by Dave Unsworth from June 14th to 21st 1987, Sinah Common by Pete Durnell and Andy Parfitt on June 13th 1989, John Lewis at Timsbury Manor on June 5th 1994, Mike Rafter at Needs Ore on June 3rd 1996, Pete Potts at Solent Court on June 3rd 1997, Tony Blunden, John Norton and Peter Milinets-Raby at Fareham Hill Coppice on May 28th 1999, Mike Rafter at Lee Lane, Nursling on May 19th 2002, Alan Thompson at Ibsley Bridge on June 5th 2012 where it remained until 10th, and Olly Frampton at Waterditch on the Dorset border on June 9th and 10th 2018. The Ibsley bird was the first ever at an accessible site and enabled many Hampshire birders to add it to their county lists.

Autumn occurrences provide challenging identification problems. HOSRP accepted records of singles found at Keyhaven by Pete Combridge on October 1st 1989, Mike and Pete Combridge on August 31st 1992 and Steve Keen on September 7th 2003 and a first-winter found by Nigel Jones at Titchfield Haven on September 2nd 2007. The last was watched through binoculars as close as 15 feet. A juvenile was trapped by Barry Duffin at Titchfield Haven on September 4th 2010 and retrapped the next day.

GB status: 1,586, 1986–2019. No clear trend, mean 48 per year in 2010–2019. 74 pairs breeding 1982; steady decline since with c. 10 pairs in 2018. (Breeds temperate Europe from France to C Fennoscandia, E to Caucasus & NE Turkey, & Russia & Siberia E to Ob River. Winters equatorial E and SE Africa.)

Booted Warbler *Iduna caligata*

BBRC

A very rare vagrant. One record in 2010.

The only record for Hampshire is of one found by Dave Unsworth at Needs Ore on the afternoon of September 12th 2010 which remained until September 15th. Extracts from Dave's submission to BBRC are reproduced below; further details are given in Unsworth (2011).

I had finished my monthly WeBS count at the Beaulieu Estuary and decided to check out an area of low scrub – predominantly bramble, hawthorn and gorse along the seaward side of Needs Ore, hoping to find any migrants. Apart from several Chiffchaffs, there did not seem to be much about. As I skirted the edge of some low gorse, I was following a Chiffchaff and it flew over to join another bird which looked the size and shape of Chiffchaff but was noticeably sandier in colour. First views were brief and it had me puzzled. The strong wind made viewing conditions difficult but over a period of about half an hour I managed to obtain some views which suggested this was something interesting. It had been many years since I had seen a Booted Warbler and my memory had faded but the feature that got me thinking that it might be this species was that it showed a slight dark line above the supercilium, which I recalled from the illustration in the excellent MacMillan guide. (I had forgotten about the classic milky-tea colour of Booted Warbler which on hindsight this bird clearly showed).

I had no ID material with me, so took some notes on the bird. I was aware of the possibility of Sykes Warbler but again was not up to speed on how to identify that recent split. After a while, the bird completely vanished and I could not relocate it. So I headed for home to check the literature. My main references were the MacMillan and Collins Guides, plus the various articles in Birding World concerning the identification of Sykes Warbler. I concluded it was a Booted Warbler but I wanted to be sure so went back for further views the next day.

Luckily, it was still there and it proved to be much showier despite the poor weather and before long I was completely happy that it was a Booted Warbler.

Its behaviour was remarkable. It spent a lot of time on the ground, either moving along under the gorse (which was suffering die-back and was therefore quite open underneath), or out in the open feeding amongst the dead thrift heads on the grassy turf. On the 15th, when the weather was much better, it seemed to have got quite used to me being there and was happy to feed as close as 15 feet away without being the slightest bit bothered about my presence. It was not seen to pump its tail. It would feed by either taking items purposefully from the ground, often with its head cocked (in the style of a Thrush) or lunge with a flurry of wings and tail at passing insects. It was quite entertaining to watch!

Booted Warbler, Needs Ore, September 10th 2015. Simon Ingram

GB status: 1, 1936; 177, 1966–2020. About five per year; stable. (Breeds E Finland, E to C Russia & W Siberia to Yenisey valley, C & N Kazakhstan to W Mongolia & W Xinjiang province, China. Winters N & peninsular India, S to Karnataka.)

Eastern Olivaceous Warbler *Iduna pallida*

BBRC

A very rare vagrant. One record in 2019.

The only record for Hampshire is of one found by Pete Gammage at Farlington Marshes on September 14th 2019 and confirmed by Jason Crook, Andy Johnson and Tim Doran. It showed very well to many observers and stayed until 19th. Extracts from Pete's account of its discovery and identification are reproduced below; it first appeared on the *Birdguides* website and was published in the 2019 *HBR* (Gammage 2020).

I was late into the field, arriving a little after 9 am, and took a route through the western section of 'the bushes', an area of mostly hawthorn and bramble scrub which often holds passerine migrants. There were plenty of Blackcaps present along with a few Whitethroats. Pretty standard fare, but then I had a Garden Warbler and a glimpse of a Spotted Flycatcher. Perhaps there were new birds in after all!

I skirted along the south side of the 'old sea wall', a land-locked bank covered in scrub, notching up a few more Sylvia warblers, and then crossed the gap towards the east section of the bushes. It was just before 10 am as I approached a tall bramble-smothered hawthorn and picked up a pale warbler, seemingly an 'Acro', close to ground level. I was immediately struck by how pale and plain-looking it was and thought I could hear a 'tack' call coming from the bird. I moved forward a few paces so that I could determine whether it was the pale warbler that was calling. It was and as the bird was pumping its tail downwards as it called, the adrenaline was already starting to kick in. I was convinced it was something good but realised I had seen very little actual detail. The only features noted were a plain face, whitish underparts, uniform grey-brown upperparts, a rather long looking pale bill and just a suggestion of a square-ended tail.

The warbler had quickly moved to the top of the bush and flicked over out of view. I moved around only to find a walker wandering past exactly where the warbler had disappeared and

Farlington Marshes, September 18th 2019

thought I caught it double-back. I waited a few minutes but it seemed I had lost touch. At this point I realised I needed help so I decided to phone a friend. I first called Jason Crook who I thought would be on the marsh, but unfortunately he was not. I left him with the news that I thought it could be a Blyth's Reed Warbler but couldn't rule out a grey 'Hippo'. Next, I tried Tim Doran, another Farlington regular. He was not on the marsh but only a few miles away. Ten minutes later he arrived and we split up to try and relocate the bird. The day was already warming up and bird activity was reducing. I was already beginning to regret making those calls.

Fortunately, Jason had contacted Andy Johnson and they both arrived to join the search at around 10.30. It was just as I was explaining that I had not seen the bird since the initial brief sighting when Andy lifted his bins and announced, 'that's got to be the bird'. He and Jason caught the 'tail dipping' warbler just before it flicked into the original bush. We quickly moved to the sunny side and backed off to give the bird room. I beckoned Tim over and the bird started to show. The head shape and broad based, long yellowish bill were apparent and it was lacking the long undertail-coverts of an Acrocephalus. *The tail appeared relatively long and squared off and the legs greyish and sturdy. Now with a decent view through a telescope it was immediately clear we were looking at either a* Hippolais *or* Iduna.

The question now was: which one? Having read a few articles and finders' accounts, I was aware that there were several possibilities that needed to be considered. We quickly excluded Melodious and Icterine Warblers on the lack of green and yellow tones in the plumage. I don't recall that we properly considered Upcher's Warbler as a possibility, but that could have been eliminated as our bird had a long tapering bill, shallow tapering forehead and lacked both the prominent wing panel and tail-swaying habits of that species. That left the Iduna *complex.*

The bird seemed too long-billed and bulky to be a Booted Warbler and the upperpart colouration was more on the grey scale than 'milky tea'. Eastern or Western Olivaceous Warblers were then tabled. Having all had some previous experience of one or other of these two Olivaceous Warbler species, they did immediately fit. However, I was aware that Sykes's Warbler could also be a contender and would need to be eliminated.

Tim checked his Cornell Merlin App and came up with some key features. Tail dipping – a good feature for Eastern; prominent and staggered primary projection – Sykes has a short, even primary projection; the faint presence of a loral line between bill and eye – Western should show plain pale lores. We were all agreed that Eastern Olivaceous seemed the best fit. It was approaching 11 am and the consensus was to start putting news out now that we felt more confident about the identity.

Up until this point the bird had remained faithful to the same clump of bushes. Typically, just as the first visitors started to arrive, the warbler flew strongly west towards a large and dense area of bushes and was lost to view. A tense hour passed before the trail was picked up again. Before long, the Olivaceous Warbler had returned to its favoured bramble-covered hawthorn. The bird remained in place for most of the week performing well at times, especially in the hot sunshine, and was seen by hundreds of observers. To my knowledge it was last seen on the evening of 19th September.

GB status: 23, 1967–2019. Rare, about one every 1–2 years. (Race elaeica *breeds Balkans & Greece E to Turkey, S Caucasus, S Kazakhstan, Uzbekistan, Iraq, Iran & N Afghanistan. Migrates through Middle East to winter E Africa. Other races breed North Africa, some wintering to S of breeding range.)*

Eastern Olivaceous Warbler, Farlington Marshes, September 18th 2019. Gareth Rees

Melodious Warbler *Hippolais polyglotta*

BBRC to 1962; HOSRP to date

A rare passage migrant. 18 records since 1961.

The first for the county was found by Dave Wooldridge and Don Westerhoff at Keyhaven on September 2nd 1961; it was still present the next day. Extracts from Dave's notes are reproduced below.

> *General field characters: silver-blue legs, steep forehead and 'tufted' appearance of crown, and large heavy bill at once distinguished this bird from the other single* Phylloscopus *in an adjoining bush. It was watched several times on the 2.9.61 at distances down to 12 yds in good light. The most noticeable feature was the short wings giving it a rather squat appearance. At no time under any light condition could any sign of a pale wing patch be seen. Size: only slightly larger than Chiffchaff. Description: forehead, crown, nape and whole of upperparts – olive brown. Primaries sometimes seemed to be slightly darker. No wing patch. No superciliary stripe, colour of crown extending down to the eye. Chin/throat: conspicuous lemon yellow. Breast and belly: the rather bright yellow on the throat gradually shaded to pale yellow on the belly. Bill: longer and heavier than* Phylloscopus*; quite distinctive. Upper mandible pale brown. Lower mandible much paler than upper, appearing yellowish especially where it met the yellow of the throat. Legs: silvery blue, but slightly brownish above tarsal joint.*

Since then a further 17 have been recorded.

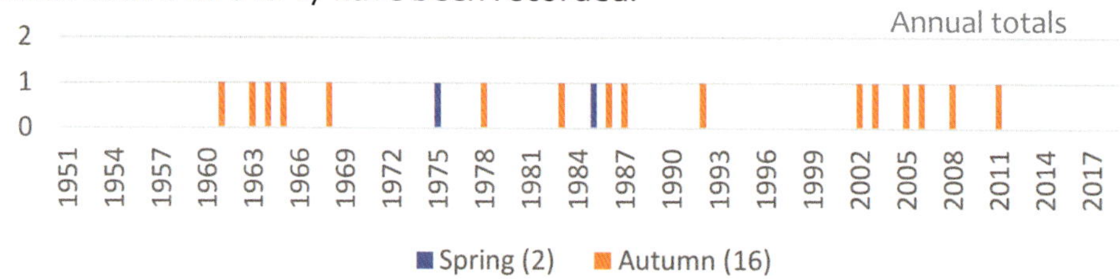

The second was discovered by Johnny Bowers and Graham Rees at Titchfield Haven on September 2nd 1963. Graham also found the next at Needs Ore on October 17th 1964. In the following year another was found by Len Mummery and Jim Williams at Hurst Castle on September 25th which was still present on the following day. In 1968 Bryan Renyard found the first for Farlington Marshes on September 14th.

There then followed a long gap before the next record of a singing male found by David Christie at Culverley, New Forest on June 8th 1975. The only other bird in the 1970s was located by Barry Duffin at Titchfield Haven on August 12th 1978 and later seen by Peter Milinets-Raby and John Norton.

The first of the 1980s was found by George Clay at Farlington Marshes on September 24th 1983 and was well described by Geoff Farwell; it was still present the next day when Dave Unsworth and others saw it. The second in spring was a singing male found by George Green at Ludshott Common on June 10th 1985 while doing research for his book *Where to Watch Birds in Dorset, Hampshire and the Isle of Wight*. Unfortunately, it was not relocated the next day. The next was discovered by Philip Fawkes at Eling on September 26th 1986 which was omitted in error from *BoH*. A good decade for this species was completed when Terry Martin found one at Farlington Marshes on September 29th 1987.

The only record in the 1990s involved a long-staying first-winter at Farlington Marshes which was found by Jason Crook and Geoff Farwell on September 12th 1992 and seen by many observers during its stay until September 27th.

The next was not until the new century, when Mick Hay found one at Park Shore, Needs Ore on August 24th 2002. This was followed in 2003 by one found by Steve Keen and Sue Wood with a flock of *Phylloscopus* warblers at Beckton Bunny, Barton on Sea at 0910 hrs on August 31st. It was seen later in the day by several other observers but was not relocated subsequently. In 2005 one was watched at close range by Tim Doran in bushes adjacent to IBM Lake, Cosham between 0850 and 0910 hrs on September 17th. It then moved 50 m to another clump of bushes but could not be relocated. In the following year, an elusive first-winter was found by Peter Milinets-Raby and John Norton at Gilkicker Point on September 8th which showed intermittently in the area until 11th. The fifth of the 2000s was found and photographed by Bob Chapman at Barton on Sea on October 8th 2008.

Tim Doran found his second at IBM Lake when he located one in hawthorn scrub there on August 13th 2011.

Melodious Warbler, Barton on Sea, October 8th 2008. Richard Ford

Analysis of arrival dates shows two in spring in early June and a protracted autumn passage between August 12th and October 17th with most in September.

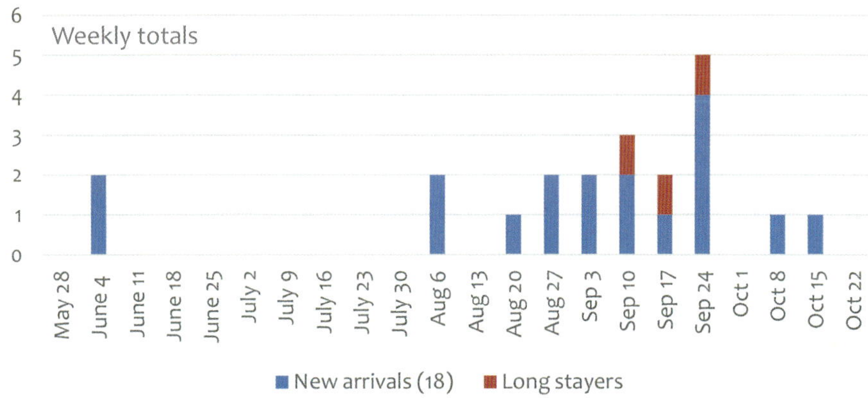

One at Sinah Common on September 12th 1983 which was published in the *HBR* was considered not proven after review for *BoH*.

In addition, *Hippolais* warblers seen at Church Crookham on June 24th 1961 and Farlington Marshes on September 3rd 1963 were almost certainly of this species.

GB status: 2, 1905–1913; 1.491, 1951–2019. Declining, mean 20 per year in 2010–19. (Breeds NW Africa & SW Europe from S Spain to SE Netherlands & E to Italy. Winters W Africa N of equator.)

Icterine Warbler *Hippolais icterina*

BBRC to 1962; HOSRP to date

A rare passage migrant. Twelve records since 1981.

The first for the county was an adult found by David Christie at Hook-with-Warsash on August 8th 1981. Extracts from his description are reproduced below.

In bushes by lake. Skulking but showed itself in open (in fair light in calm conditions) for about 10 seconds. Adult still showing very green and yellow plumage; upperparts greenish with very distinct bright yellow supercilium extending from bill to rear of ear-coverts, and pale eye-ring; primaries browner; inner secondaries and tertials with pale edges, quite prominent when bird turned sideways; underparts noticeably bright yellow; general bright green and yellow plumage striking. Wings long and overlapping tail when folded. Bill long, distinctly longer than Willow Warblers seen nearby. Legs bluish-grey, noticeable. A large, rather clumsy warbler occasionally stretched head and neck upwards to take food from branches. I have seen species many times in eastern Europe and Melodious in west.

Subsequently, a further eleven have occurred with four in spring and summer between May 13th and July 4th and seven in autumn between August 18th and October 11th.

The second was a probable adult trapped by Stuart Sporne at Sway on August 22nd 1983. This was followed by one found by Richard Coomber in his garden at New Milton July 4th 1985. Two years later Mark Simmonds found one at Farlington Marshes on August 30th 1987 which was also seen by Geoff Farwell.

Five were recorded during the 1990s. The first was a juvenile found by Pete Durnell at Sandy Point on September 16th 1992 which was seen by many observers up to 19th. The next was one watched by Jess Pain and Simon S. King at Lower Test Marshes for 15 minutes on August 18th 1993. The following year Andy Welch and Clay Jones trapped a first-winter male at Southampton Common on October 11th. It was sexed as a male based on a wing length of 82 mm. In 1995 Steve Keen watched an exhausted bird for ten minutes at close range on the ground in a small patch of woodland at Barton on Sea on September 7th. It then disappeared into dense brambles and could not be relocated. One occurred for the fifth consecutive year when Simon S. King located a singing male at Lower Test Marshes at 0915 hrs on June 12th 1996. It was seen later by Alan Claxton, Jess Pain and Pete Combridge before it left north at 1800 hrs.

Icterine Warbler, Sandy Point, May 13th 2009. A rich variety of song

There was a long wait for the next, a singing male found by Pete Gammage at Sandy Point on May 13th 2009 and seen by many observers during the day. Ten more years passed before another appeared. A singing male was found independently by Francis Buner and Lewis Hooper in the grounds of the Church of St Peter ad Vincula at Colemore near East Tisted on June 18th 2019. Lewis released the news enabling many people to add this species to their county lists. It remained throughout the following day but was elusive at times and often ventured up to 100 m away from the churchyard. The first in autumn since 1995 was found by Matt Phelps on the cliffs opposite Western Avenue at Barton on Sea at 0800 hrs on September 1st 2020. It was seen by several observers intermittently during the day and was still present the next morning.

Icterine Warblers: Sandy Point, May 13th 2009 and Barton on Sea, September 1st 2020. Richard Ford and Matthew Barfield

One at Farlington Marshes on September 12th 1987 which was published in the *HBR* was considered not proven after review for *BoH*.

GB status: 48, 1848–1949; 5,132, 1950–2019. Declining, mean 75 per year in 2010–19. (Breeds W & C Europe N to C Fennoscandia & E to Black Sea, European Russia & Siberia E to region of Ob River, & N Kazakhstan. Winters throughout Africa S of equator.)

Pallas's Grasshopper Warbler *Helopsaltes certhiola*

BBRC

A very rare vagrant. One record in 2014.

The only record for Hampshire is of a juvenile trapped at Titchfield Haven by Barry Duffin, Duncan Bell and Trevor Codlin on September 6th 2014. Extracts from Barry's description of the events (Duffin 2015) are reproduced below.

> *At approximately 0600 hrs on September 6th 2014, I was extracting birds from a series of mist nets during an early morning ringing session when I came across an unfamiliar passerine in a lower net pocket. Earlier that morning I had extracted several juvenile Grasshopper Warblers, many of which had shown bright yellow underparts as indeed this bird was showing. However, the yellow underparts of this bird were richer than that I had previously seen on Grasshopper Warblers. Of over four and a half thousand Grasshopper Warblers handled at Titchfield Haven in the last ten years a little over 50% had shown predominantly yellow underparts. Above all I was struck by how the bright yellow underparts contrasted with the rufous undertail-coverts and the dark tail feathers with pale tips. The thought of having caught a Pallas's Grasshopper Warbler came to mind. Having extracted the bird from the net I then showed it to fellow ringers Trevor Codlin and Duncan Bell, and we quickly concluded that we were looking at Hampshire's first Pallas's Grasshopper Warbler.*

Trevor Codlin

Of the 62 British records, 45 have been in Shetland, and all of them in September or October. The only previous record from southern England was in Dorset, when a bird was trapped on almost the same date in 1996.

GB status: 1, 1949; 61, 1956–2020. About 1-2 per year; stable. (Five races breed Siberia from Irtysh River E to Yakutia & Sea of Okhotsk, & SW Siberia & NE Kazakhstan through Mongolia to Ussuriland & N & NE China. Winters Sri Lanka & NE India to S China, & S throughout SE Asia.)

Lanceolated Warbler *Locustella lanceolata*

BBRC

A very rare vagrant. One record in 1979.

The only record for Hampshire is of a first-winter trapped at Damerham by Tony Taylor on September 23rd 1979. Extracts from Tony's submission to BBRC are reproduced below.

Tony Taylor

At about 7 am I was checking and furling mist-nets at Damerham Watercress Beds because it had started raining. One net spanned the entrance to a narrow disused bed, about 30 feet wide with a low, thick growth of reeds etc in it and willow and thick undergrowth on each side. There was a small bird caught in the top shelf well above the vegetation. I saw its rounded tail and streaked undertail-coverts and realised it was a Locustella *warbler. I was struck by its small size, strongly streaked breast and drab colouration and was immediately reminded of two Lanceolated Warblers I had handled on Fair Isle in 1976. I was joined by Paul Toynton a few minutes later, and a full description and measurements were taken before the bird was ringed, photographed and released. It flew into some willows and was not seen again.*

This is the only record of this species for southern England apart from one on Annet, Isles of Scilly on September 22nd and 23rd 2002.

GB status: 7, 1908–1938; 160, 1957–2020. About four per year; stable. (Singing males regular SE Finland. To E, discontinuously from C Urals E to Kamchatka, Kuril Islands, Hokkaido & NE China. Winters from Nepal E through NE India to SE Asia & Philippines.)

Savi's Warbler *Locustella luscinioides*

BBRC to 1982; HOSRP 1983–98; BBRC 1999 to date

A rare summer visitor which has bred. At least 43 records since 1969.

The first for the county was a singing male identified by Richard Carpenter on private land adjacent to Upper Titchfield Haven on May 11th 1969. The record was not sent to BBRC at the time but in his recently accepted submission Richard told the story of its discovery as follows:

I was a regular visitor to Upper Titchfield Haven to do CBC and other survey work. The reserve had resident Grasshopper Warblers and I was keeping an eye on them to see if they established territories and bred. Someone I met had said there was a Grasshopper Warbler singing in the reedbed as well as in the usual scrubby area. This sounded odd so I investigated and heard what was clearly a Savi's Warbler, a species I'd heard the previous week in Kent. I eventually managed to see the bird at close range, often perched out on top of reed stems. I was certain it was this species and not a Grasshopper Warbler or Sedge Warbler. It was closest in appearance to Reed Warbler, being unstreaked and with a white throat patch. The bird was also seen and heard by David Billett, Michael Bryant, Canning Suffern and others who were all in agreement with the identification. The key ID feature was the song, different enough from nearby Grasshopper Warblers to make it certain. It subsequently moved a short distance downstream and held territory until June 21st.

Subsequently, singing males were recorded in 22 springs between 1973 and 2020 on dates between April 11th and July 30th. There is also one autumn record.

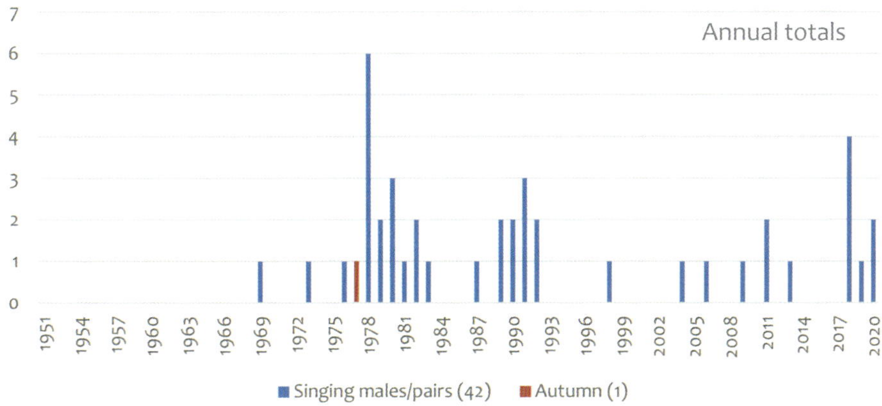

There was four-year gap until the next was found by Andrew Williamson at Farlington Marshes on June 14th 1973 and heard again three days later.

In 1976 a singing male was found by Tim Lawman, D. Pearce and Barry Duffin at Titchfield Haven on the evening of May 27th. It remained in the area until June 5th at least. In the following year another was heard in the same area between June 25th and 29th but there was no evidence they were mated in either year. In 1978 six singing males were there of which three held territories for more than a month. The other three were located at different sites but their periods of song were very short. At two of the main sites, including the 1976 site, food-carrying was noted in the first week of July. During ringing activities at the same time, a male and female were trapped on 1st, the female showing a full brood patch. Later a newly fledged bird was trapped on 6th. Successful breeding at the third of the main sites was not proven but was considered probable.

Details of breeding at Titchfield Haven from 1978 to 1998 are tabulated below. For further details see Duffin (2003).

Year	First date	Last date	Singing males	Probable pairs breeding	Confirmed pairs breeding
1978	May 5th	July 18th	6	3	2
1979	April 27th	July 4th	2	1	
1980	April 22nd	July 6th	3	1	
1981	April 13th	June 16th	1	1	
1982	April 14th	June 26th	2	1	1
1983	April 27th	July 6th	1	1	1
1987	April 24th	May 8th	1		
1989	May 10th	May 29th	1	1	
1990	April 27th	July 13th	2	1	
1991	April 27th	July 30th	3	2	1
1992	April 26th	June 23rd	2	2	1
1998	April 11th	May 18th	1		

Other records during this period involved one trapped by Ron Toft at Farlington Marshes on August 21st 1977, the only autumn record, and a singing male heard by Mike and Pete Combridge and Mike Stewart at Keyhaven on May 6th 1989.

Savi's Warbler, Fishlake Meadows, May 13th 2009. Nigel Jones

Subsequent records were of singing birds at Titchfield Haven on July 4th 2004, April 15th and 24th 2011, May 15th and 16th 2013, May 6th and 7th 2018 and July 16th to 26th 2019, Hook Lake on May 5th and 6th 2006, Fishlake Meadows from May 1st to June 27th 2009, April 21st to June 2nd 2018, April 16th to May 20th 2020 and April 18th to 25th 2020 and Farlington Marshes from April 18th to May 3rd and April 21st to 29th 2018.

Records of singing males at Farlington Marshes from May 26th to June 1st 1974, on April 15th 2011 and from April 21st to 29th 2019, at Blackwater, Needs Ore on April 20th 2018 and Edenbrook Country Park, Fleet on May 31st and June 1st 2018 were not submitted to BBRC.

An autumn record of one seen at Farlington Marshes on September 29th 2001 was harshly considered not proven by BBRC. Another at Keyhaven on September 20th 2003 was also considered not proven (M. J. Rogers *in litt.*) although not published as such.

GB status: 11, 1819–1908; 846, 1954–2020. About ten per year, including breeders; recent decrease, now stable. (Nominate luscinioides *breeds Iberia N to S Scandinavia, & E through E Europe, Russia & Ukraine to Black Sea coasts, & winters W Africa from Senegal to N Nigeria. Eastern race* fusca *breeds C Asia to NW China & W Mongolia, & winters NE Africa.)*

Barred Warbler *Curruca nisoria*

HOSRP

A rare autumn passage migrant. 14 records since 1965.

The first for the county was a first-winter found by Bryan Renyard at Farlington Marshes on October 16th 1965. Extracts from Bryan's description are reproduced below.

> *First seen at 1130 hrs in hawthorn/blackberry bushes. In sight for ten minutes at 30 yds range in good light. In general, a heavy, long-tailed warbler with large bill and no conspicuous markings. Larger than Whitethroat. Upperparts uniform grey-brown with a few lighter grey feathers on mantle and rump. Tail long with end square cut. No wing-bars or noticeable white edgings to flight feathers. No head markings. Underparts white with grey tinge, being darker on throat and lighter undertail. Some faint, darker grey flecking on sides of throat and upper breast. Legs and feet strong-looking, light (grey) in bright light. Bill – a noticeable feature – thicker and longer than in Whitethroat or Garden Warbler. Light in colour. Head sloped sharply back from bill. Eye dark. Bird moved along edge of bushes just inside vegetation. Actions slow and deliberate. Now and again would snap at an insect or blackberry. It finally flew low to another bush but was not seen again.*

The next was not until 1986, when a first-winter was found by Brian Small and Paul Castle at Farlington Marshes at 1130 hrs on August 22nd 1986. It then rained heavily until 1300 hrs when they refound it and were able to watch it eating elderberries and preening in a dead hawthorn before disappearing. It was refound by Pete Potts late in the evening but was not seen again.

The first of four in the 1990s was a first-winter found by Mike and Pete Combridge at Keyhaven on October 7th 1990. Another first-winter was seen by Dan Powell in his garden at Stubbington on November 30th 1993.

1997 produced two records of first-winters found by Pete Combridge at Pennington Marshes on September 24th and Peter Strangeman in Victoria Park, Portsmouth on November 11th.

The 2000s also produced four records. An often elusive first-winter bird was found by Jason Crook at Farlington Marshes on October 15th 2000 and seen by persistent observers during its stay until 22nd. Another first-winter found by Jason in scrub along the north-east sea wall at Farlington Marshes on September 20th 2003 provided good views to many observers but was not relocated the next day. In 2005 a first-winter was seen briefly but well by Pete Potts in his garden at Solent Court on September 18th. Yet another first-winter was located by Simon Wright at Sinah Gravel Pit at 1040 hrs on October 29th 2007; it was subsequently seen by several observers during the afternoon.

Barred Warbler,
Titchfield Haven, December 6th 2017

In 2015 a first-winter was trapped by Barry Duffin at Titchfield Haven on September 8th and retrapped on 10th. In the following year a first-winter was found by Graham Stephenson in tamarisks by the Hayling Yacht Club at Sinah Common on the morning of August 27th. Graham saw it well but it disappeared and was not relocated by other birders during the day. In 2017 a first-winter was found by Tony Heath eating berries in scrub along the sea-front adjacent to Titchfield Haven visitor centre on November 19th. It was subsequently viewed and photographed by many observers at close range later that afternoon but was not seen the next day. However, it was refound by John Shillitoe in the same area on December 3rd. It remained until 20th when possibly it succumbed to the frosty conditions at the time.

The most recent record involved a first-winter found by Pete Gammage at Fort Widley at 0815 hrs on September 15th 2020. It showed well then flew across the moat into a bush. Seconds later a Sparrowhawk crashed into the

Barred Warblers:
Sinah Gravel Pit,
October 29th 2007 (left). Richard Ford
Titchfield Haven, December 2017
(top to bottom). John Hilton,
Martin Bennett and John Dodds

same bush. A trilling alarm call was heard before the raptor flew off unseen. The Barred Warbler was not seen again.

There is also a record of an adult which was ringed at Le Torri near Ravenna, Italy on July 24th 1943 and found freshly dead in the Woolston/Bitterne area on October 20th of that year. However, reservations have been expressed about this recovery because autumn adults are so rare in Britain and the ring was not available for inspection.

GB status: 171, 1884–1949; 157, 1950–57; 8,296, 1968–2019. Increasing, mean 220 per year in 2010–19. (Breeds C & E Europe N to S Finland & S to S Black Sea, Russia E to Yenisey River region of Siberia, & N Kazakhstan. Winters NE & E equatorial Africa.)

Siberian/Central Asian Lesser Whitethroat
Curruca curruca blythi/halimodendri

BBRC

A very rare vagrant. One record under consideration.

A Lesser Whitethroat photographed by Nigel Jones at Barton on Sea on November 19th 2008 was identified by him as belonging to one of the eastern races *blythi* or *halimodendri*. This record along with 16 other claims of these races remain under consideration by BBRC.

Lesser Whitethroat, Barton on Sea, November 19th 2008. Nigel Jones

There have been 13 other records of birds in Hampshire during the months of November to March since 1979. Long-staying birds overwintering in gardens may have been race *blythi* including those at Highfield, Southampton from January 29th to April 19th 1987, Fareham from November 9th 2005 to March 31st 2006, Bursledon from January 23rd to March 7th 2010, Bishop's Waltham from February 28th to March 27th 2013 and Eastney from January 3rd to 31st 2017.

GB status: Rare, about one every three years. (Breeds arid regions of C Asia from NE Caspian Sea, through Kyzyl Kum Desert in Turkmenistan, Uzbekistan & S Kazakhstan to SE Russia Altai & NW Mongolia. Winters SE Iran, S Afghanistan, Pakistan & NW Punjab, India.)

Western Subalpine Warbler *Curruca iberiae*

BBRC to 2005; HOSRP 2006–18; BBRC 2019 to date

A very rare vagrant. Five records since 1984.

The first for the county was a singing male found by Doug Radford along the Hayling Billy footpath just south of the Hayling Oyster Beds at 1025 hrs on May 23rd 1984. He was able to fetch John Walters and they relocated the bird about 400 m from the original site. News was released promptly thus allowing many observers to see it during the day. It was not relocated the next morning. Extracts from Doug's description and John Walters' sketches are reproduced below.

Whilst walking past some blackthorn and hawthorn scrub along the old railway line, my attention was drawn by a short burst of Whitethroat-like song. Immediately the singer (instantly recognisable as a Sylvia *warbler) came into view about 10 m away where it was watched through binoculars for about ten seconds. I obtained further views through a telescope and was able to note the slate-grey head, nape and back and brick-red throat and breast, these being separated by a neat white moustachial stripe. The red iris and orbital ring were also prominent. I fetched John Walters and we relocated the bird about 400 m from the original site.*

This was followed by a singing male found by Jason Crook at Farlington Marshes on May 21st 1995 which stayed all day enabling many birders to see it.

Western Subalpine Warbler, first-summer male, The Hawthorns, Southampton, April 24th 2007. Andy Welch

Yet another singing male was located by Mick Hay at Old Winchester Hill on May 20th 2000. It was also seen by four other local birders during its stay until May 23rd but due to the sensitive nature of the site and limited parking facilities news was not released. Two records in quick succession in 2007 are assumed to relate to different individuals. A male was seen by Mike Oram at Itchen Valley Country Park on April 18th and Andy Welch trapped a first-summer male at The Hawthorns, Southampton Common on April 24th.

The above records were all accepted as Subalpine Warbler *Sylvia cantillans* but subsequently the species has been split into Western *S. iberiae* and Eastern Subalpine Warblers *S. cantillans* (BOURC 2021) and Moltoni's Warbler *S. subalpina* (BOURC 2014). The descriptions and/or photos suggest that all the Hampshire birds are of the Western species. Currently BBRC only accepts ten British records of Western Subalpine Warbler but a review will be undertaken to attempt to assign historical records to a particular species (Stoddart & Hudson 2021).

GB status: 4, 1894–1949; 877, 1950–2020 (all species). Status uncertain. (Race iberiae *breeds Iberian Peninsula, S France & NW Italy; race* inornata *breeds Morocco, N Algeria, Tunisia & NW Libya.)*

Spectacled Warbler *Curruca conspicillata*

BBRC

A very rare vagrant. One record in 2011.

The only record for Hampshire is of one found by Phil Hack at Needs Ore on October 29th 2011 and confirmed by Nigel Jones on October 30th. Extracts from Philip and Nigel's account of events in the 2011 *HBR* (Hack & Jones 2013) are reproduced below.

Day 1 (Philip Hack) In the early afternoon of Saturday October 29th 2011, I visited Needs Ore Point NR. After birdwatching around the Point and seeing several migrant Meadow Pipits amongst the wintering Rock Pipits, I re-parked my car near the Shore Hide and proceeded to walk along the hedge-lined path to the other hides which overlook Blackwater. Upon reaching the end, I noticed a small-sized Sylvia type warbler species land very briefly on the barbed wire fence just to the left of the footpath gate. The bird was only present for a few seconds before flying off behind the field hedge to the left but I had a very clear near head-on view at 4–5 metres. As it was so close, even without binoculars, I knew immediately what it was – a male Spectacled Warbler! The main diagnostic features in my initial brief view were as follows: • slate grey head: • white throat; • pale off-white/pink underparts; • bright yellow/orange legs; • bright uniform/'solid' rufous wings.

I then decided that someone else should support my sighting but rather unusually for a Saturday afternoon nobody else was around! So I telephoned the Reserve Warden (Adam Wells), as is the procedure when finding rare birds on the Reserve.

After a short wait, Adam arrived with a colleague. We quickly relocated the bird low down in the hedge and he agreed with my identification. On returning home I telephoned Mike Rafter but with no success. He returned my call at lunch-time the following day and then departed to the Reserve. There he met Nigel Jones, who relocated the bird.

Based on photos by Nigel Jones

Day 2 (Nigel Jones) My wife Sandy and I visited Needs Ore on Sunday October 30th at around midday. I looked in the logbook and chuckled when I saw that someone had claimed a 'male Spectacled Warbler' the day before! There are often bird species written in the logbook which have to be taken with a pinch of salt. However, I spent the next 20 minutes looking in the area with no success, so we drove to the eastern point of the reserve and had a coffee and hoped the cold, drizzly conditions might improve. I phoned a couple of friends to see if they could shed any light on the sighting, but they also laughed – but asked to be kept informed, just in case! Another half hour search produced nothing so we went to the hides. On returning we bumped into Mike Rafter who said he had spoken with the finder and felt sure that it was a good record. However, Mike had also drawn a blank in relocating it and he went off to the hides. Sandy and I spent another 20 minutes looking and then fine drizzle set in, so she went back to the car. After ten more minutes I was walking back and decided to get my trusty i-pod out and play Spectacled Warbler song. After a minute of playing a small bird perched in the top of a Blackthorn, next to the path, about 5 m away – a male Spectacled Warbler! I watched it for a minute noting the key features, then followed it down

October 30th 2011. Nigel Jones

the path and watched it feed. I ran to the hide and found Mike and we came back and relocated the bird in one of the hedges. Sandy also returned and we watched it for five minutes as it fed and preened. The weather was cold and overcast still and it was now getting into late afternoon. I phoned a few people who I knew were around that day and made sure we had someone with some photographic gear as I knew photos of the bird would be a good idea to make sure the record was accepted. Within 40 minutes a small group had assembled and we all got brief views and took descriptions, along with some poor photo record shots, before dusk settled and we left the bird to go to its roost. We all noted the large rufous wing panel, the small dark centres to the tertials, the short primary extension, yellowy legs, slate grey head, dark lores, obvious pale eye-ring, white throat and pinkish suffused underparts. It was a striking bird and was definitely a male (looking at the literature all birds have undergone a moult by the end of October, so whether it was an adult I cannot be sure). Unfortunately, next morning the bird could not be refound and we presumed it had succumbed to the cold of the night, been predated, or departed.

This record is pre-dated by one of a male which Brian Small saw fly in off the sea at Lepe at 1135 hrs on April 15th 1989. He then watched it at close range during the next 30 minutes before it flew off inland. He did not release the news as it had flown on to private land but he searched for it again over the next two days without success. Brian submitted field sketches and a detailed description to BBRC. The record was accepted by BOURC as the first for Britain but on return to BBRC was narrowly considered not proven. Brian's annotated sketches are reproduced below. He informed the author in 2020 that he remained 100% certain that it was a Spectacled Warbler; the record is currently being reconsidered by BBRC.

Lepe, April 15th 1989. Brian Small

GB status: 9, 1992–2017. Very rare. (Breeds Canary Islands, NW Africa & W Mediterranean N through Spain to S France, also Sicily, Sardinia, Cyprus, Israel & W Jordan. Some populations resident, others migrate to winter N Africa, mostly N of Sahara.)

Rose-coloured Starling *Pastor roseus*

BBRC to 2001; HOSRP to date

A rare visitor. Four old records and 27 since 1954.

The first for the county was reported by George Corbin in *The Zoologist* for October 1876, who stated:

It is a pleasure to be able to record the occurrence of a female of this rare species at Woodgreen, on the borders of the forest, near Fordingbridge. I have not seen the bird, but my friend Mr. H. W. Aubrey, of the Rectory, Hale, who is now having it preserved, has kindly sent me the following note: 'It was shot by Mr. Hinxman's gardener, and was apparently feeding on cherries when first observed. It was killed about the middle of July.

Another was reported in *The Zoologist* for October 1896 by John Bucknill, who went on to write *The Birds of Surrey*, published in 1900. He stated that Mr. W. Bradden, the well-known taxidermist of Guildford, informed him that a male specimen of this rare visitor was shot at Greatham, near Liss. It was found in an apple tree. Six more recorded by K & M were outside the present-day county boundary.

One was seen by Dr. C. R. Haines at Petersfield on April 10th 1925 (*HBR*, 1933). Later that year, Mr. A. P. Martin, writing in *The Field* for August 20th, stated that he had watched a *Pastor roseus* in Hampshire on August 9th but did not mention the location. The record was published by *British Birds* (19: 155).

The first modern record was of an adult seen by K. J. J. Monk at Rockbourne on June 27th 1954 (*BNSS report*).

Subsequently, a further 26 have been recorded.

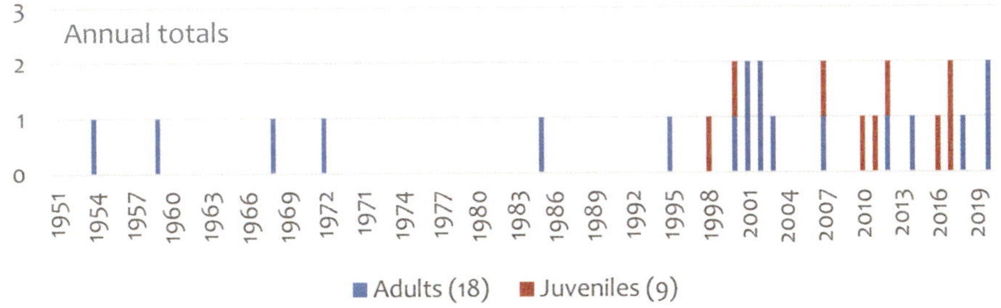

The second was an adult found by Graham Rees at Pilmore Gate Heath, New Forest on March 14th 1959 which was also seen by Johnny Bowers and B. Rozzell. This was followed by further adults, one which regularly visited a bird-table in Noel Elms's garden in Fleet from June 12th to 23rd 1968 and another seen briefly by David Evans with 20 Starlings at Bath Road Recreation Ground, Fareham on April 24th 1972.

There was a long gap before the next, a first-summer which regularly visited a garden in Chetwynd Road, Portsmouth between February 26th (not March 26th as in previous publications) and April 24th 1985. The owners, Mr. and Mrs. R. F. Rae, not knowing what it was, invited David Paradise to identify it. Subsequently, it was confirmed by Pete Potts and seen by other local observers.

Tea, cake and Rose-coloured Starling, Chetwynd Road, Portsmouth. April 17th 1985

Two were found in the 1990s.
An adult was seen and photographed in a garden at Plastow Green, Headley by R. Dobbs on July 30th 1995. A first-winter was present in a housing estate on the eastern outskirts of Andover from November 18th to December 12th 1998. It was first identified by Len Weatherley and regularly observed as it accompanied Starlings in their daily activities. It was easily attracted down to bread and was seen at close quarters by many people.

There was a flurry of records in the 2000s with nine recorded. The first was an adult in Ancells Farm Estate, Fleet on June 25th 2000 which was reported to the author. This was within 500 m of his home and he located it the next evening on a garage roof when it was also seen by other local birders. Unfortunately, it was never seen again. Later that year, Jeff Ross found a first-winter with Starlings at Calshot Spit on October 4th which remained in the area until the 10th and was seen by many visitors.

Tim Parminter found an adult feeding on blackberries with Starlings at Iley Lane, Keyhaven at 1010 hrs on August 26th 2001 but they were flushed by a Peregrine after ten minutes. It was subsequently seen by Martin Cooper, Russell Wynn and Marc Moody during its stay until September 3rd. Another adult was seen and photographed by M. Tate in his garden at Eastoke, Hayling Island between October 23rd and 30th of that year. Two further adults were seen in gardens in 2002: one found by Simon Ingram in Eastleigh on July 14th and a second in Matthew Shaft's garden in Petersfield on July 23rd. The species was recorded for the fourth consecutive year when Mrs. J. Drennan saw an adult in her garden at Crondall on June 12th 2003. The next was an adult in summer plumage seen in flight by R. Hopkins before landing in a small tree at Oxey Marsh at 1100 hrs on June 9th 2007; it was then seen to join a flock of ground feeding Starlings but was not relocated. A juvenile was found by Rob Souter at the old rubbish tip at Lower Pennington Lane at

1100 hrs on September 23rd and remained separate from a nearby flock of Starlings. It was almost certainly present two days earlier but its identity had not been confirmed. It remained in the area and was seen by many observers during its stay until September 28th.

Rose-coloured Starlings: first-summer, Hordle, March 3rd 2012, juvenile, Lower Pennington Lane, October 2nd 2011 and adult, Milford on Sea, June 22nd 2014.
Andy Johnson, Richard Ford and Keith Metcalf

The upsurge in sightings was maintained in the 2010s with nine recorded. A juvenile in conifers at the western end of the car park at Lepe was seen and photographed by Lisle Gwynn on October 20th 2010. It was seen by many observers and stayed until 26th. A juvenile was reported by Andy Davidson perched on brambles with the Starling flock at the old rubbish tip at Lower Pennington Lane on October 2nd 2011. It was relocated in the same area on the 3rd by Marc Moody, David Ryves and Nick and Jackie Hull but flew towards Keyhaven Lagoon around midday and was not seen again. Nearby a first-summer was reported in a garden at Everton Road, Hordle on February 29th 2012 which had been a regular visitor for the previous three weeks. It remained in the area until April 16th at least. Later that year a juvenile was watched by Pete Gammage and others with a large flock of Starlings in a field adjacent to the Hayling Billy footpath north of the Hayling Oyster Beds on September 28th 2012 and was reported to Ralph Hollins *Havant Nature Notes* website as being still present early on 29th. The next was an adult discovered by Keith Metcalf on bird feeders in his garden in Milford on Sea on June 22nd 2014. It was present only for a day but was also seen by Chris Bown murmurating with juvenile Starlings over Fishtail Lagoon, Pennington Marshes from 1910 to 1938 hrs before going to roost. The next was a juvenile photographed by Roger Bushnell at North Gorley on October 25th 2016. Two juveniles were seen in 2017. Steve Keen saw and photographed one in a hedge by the entrance to Hurst View caravan park in Lower Pennington Lane on October 8th and David Ryves saw one in flight with Starlings over Hook Links on November 13th before they flew off towards Fawley. In the following year Bob Marchant and John Faithful saw an adult in flight over Hook Links for 30 seconds on May 29th but unfortunately it was not relocated. In summer 2020 there was a considerable influx into Britain; in Hampshire adults were seen and photographed by Andy Johnson in Treloar Road, Hayling on May 30th and Graham Rogers at Fordingbridge on July 27th.

Analysis of arrival dates shows 18 adults appearing in every month from February to August and October and just nine juveniles from September to November with one staying into December.

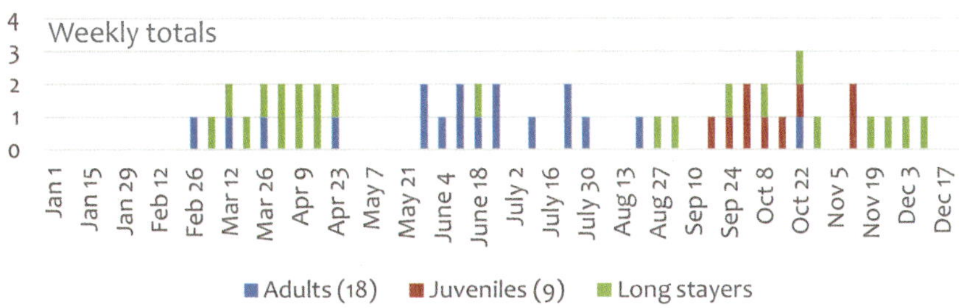

Single records in 2002 and 2020 were not accepted by HOSRP while six sightings in 2003 (3), 2005 and 2016 (2) were reported to bird information services but not submitted to HOSRP.

GB status: 310, 1742–1949; 1,468, 1950–2019. Increasing, occasional influxes becoming more frequent, mean 53 per year in 2010–19. (Breeds locally from Black Sea coasts of Romania and S Ukraine, E through Turkey to steppe region of C Asia. Winters Indian subcontinent S to Sri Lanka, and SE Arabian Peninsula.)

Bluethroat *Luscinia svecica*

HOSRP

A rare passage migrant. 35 records since 1956.

The first for the county was a first-winter female trapped by David Billett and Cliff Henty at Farlington Marshes on September 9th 1956 and still present there the next day. It was in the fields adjacent to North Binness creek. The description of the bird, taken from the Langstone Harbour yearbook, was as follows.

> *Robin sized, olive grey back, whitish underparts striated on sides of upper breast, necklace and moustachial dark markings, white chin, light eye-stripe and second (upper) moustachial stripe. Red sides to uppertail noted. Often flicked tail and the wings were carried below tail level. Trapped in spring-net on 10th (wing 67.5 mm) and ringed.*

Since then, 34 have been recorded.

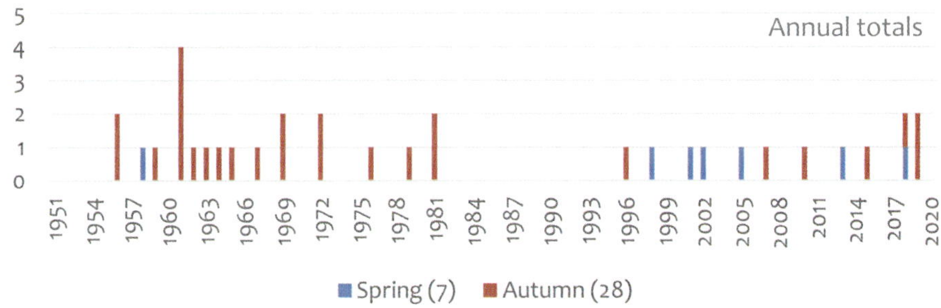

The next was a 'probable' seen by John Ash, a very reliable and experienced observer, at Damerham on October 11th 1956 which was accepted for Cohen (1963). Two further records in the 1950s involved a male of the white-spotted race *cyanecula* seen by Cliff Henty on the lawn in the Rock Garden, Southsea on March 26th 1958 and one found by Norman Orr at Keyhaven on August 30th 1959.

There were 11 records in the 1960s. A male was found at Hurst by Hazel Bidmead and Alan Mitchell on September 2nd 1961, two days later George Clay found one at Farlington Marshes, then one was seen by Tony Heath, Martin Terry, Dave Price and Canning Suffern at Titchfield Haven on September 17th and another was found there by Martin Terry on October 2nd. In the following year, a first-winter male was trapped by Shaun White at Titchfield Haven on September 30th which was also seen by Dave Price. In 1963, David Billett and Graham Rees trapped an adult male at Farlington Marshes on September 22nd and in 1964 Dave Woodbridge saw a first-winter at Keyhaven on September 2nd. The species was recorded for the fifth consecutive year when W. and R. Merritt saw a male at Yateley Gravel Pits on October 3rd 1965. Later in the decade, one was seen at Farlington Marshes by S. Brent, John Barker and Peter Findley on October 27th 1968, another was trapped there by members of the Portsmouth Group on September 20th and presumably a different bird was seen by them on October 4th 1969.

Only four were recorded in the 1970s. A first-winter male was found by members of the Portsmouth Group at Farlington Marshes on August 20th 1972 and later that year a first-winter was seen by Derek Lister, Joan Spurgin and others at Old Winchester Hill between September 15th and 20th. Another first-winter was trapped by Hugh Insley at Lower Test Marshes on August 21st 1976 and a first-winter male was seen by Rev. E. A. Pratt in his garden in Southsea on October 24th 1979.

Occurrences became less frequent thereafter. The only records for the next decade were in 1981 when a first-winter female was trapped by Brian Dudley at Lower Test Marshes on August 25th and Michael Bryant saw one at Farlington Marshes on September 5th.

Bluethroats: South Scrape, Titchfield Haven, March 29th 1998 (left), Keyhaven, April 11th 2013 and Hook-with-Warsash, September 9th 2019. Marc Moody and Andy Collins

The next was not until 1996, when Graham Stephenson located a first-winter male at Black Point, Hayling Island in the morning of November 10th which was seen by many birders during the day. The second of the decade was a male *cyanecula* found by Dan Houghton at Titchfield Haven at 1715 hrs on March 29th 1998 which was still present the next day.

The 2000s produced a slight resurgence of the species. The first for the decade was a male *cyanecula* found by Richard Levett at Upper Titchfield Haven early on the morning of May 30th 2001. Another male *cyanecula* was found by Sue and Peter Morrison at Hook Lake on March 30th 2002 and later that year a first-winter was trapped by Trevor Codlin, Barry Duffin and Pete Carr at Titchfield Haven on September 1st. Mike Wearing raised awareness of an account and photograph in the *Bordon Herald* of a male *cyanecula* which stayed for several days in a garden in Liss Forest in late March 2005. In 2007 one was located by C. Taylor and Steve Piggott at Keyhaven Marshes at 1030 hrs on September 5th and watched intermittently for two hours. What is assumed to be the same individual was relocated by J. Tyler at 1305 hrs on September 11th but not seen again.

The seventh to be trapped in the county was caught by Nigel Jones at Fishlake Meadows on September 30th 2010 and retrapped there on October 2nd. In 2013 an elusive male was watched by Martin Cooper, Marc Moody and Brian Whale at Keyhaven Lagoon from 1445 to 1500 hrs on April 11th 2013 and again briefly by Chris Bown at 1846 hrs. Photographs show that it lacks a white

throat-spot which suggests that it was the blue-throated variant *wolfi* of race *cyanecula* or, less likely, of either the Spanish race *azuricollis* or the eastern Turkish race *magna*. This was the last of an influx of 14 *cyanecula* recorded in southern England from March 23rd (*Birding World* 26: 288–297).

In 2015 Roy Cowley photographed a first-winter male along the Ancient Highway at Keyhaven Marshes at 1200 hrs on October 26th but it was not seen again. In 2018 a male *cyanecula* was found by Andrew Balmer at Hamble Common on April 11th and an adult was photographed by Clare Harrison at Farlington Marshes on September 17th and seen later by Steve Bassett and others. Most recently an elusive bird was found by Kev Roy at Hook-with-Warsash on September 8th 2019 which was still present on 13th. Another showier individual was found by John Hilton at nearby Titchfield Haven on September 19th where it remained until 28th.

Analysis of arrival dates shows six in spring between March 26th and April 11th and another on May 30th, all of which were male *cyanecula*. The remainder in autumn between August 20th and November 10th were either males of the red-spotted race *svecica* or of an indeterminate race.

K & M included a hearsay undated record of one shot by Corporal Strange at Eastney in the 19th century. They did not see the specimen.

GB status: 5,792, 1968–2019. Declining, mean 75 per year in 2010–19. (Red-spotted races breed from mountains of S Norway, E through Fennoscandia & Russia to extreme W Alaska, & S to C Asia. White-spotted races breed C & S Europe E to Caucasus and N Iran. Winters Mediterranean basin, Sahel region of Africa, Arabian Peninsula, Indian subcontinent, S China & SE Asia.)

Red-flanked Bluetail
Tarsiger cyanurus
BBRC to 2016; HOSRP to date

A very rare vagrant. One record in 2010.

The only record for Hampshire is of a first-winter, probably a male, found by Andy Johnson at Sandy Point on 1100 hrs on October 18th 2010. It was seen by many hundreds of observers, often at very close quarters, during its stay until October 23rd. Extracts from Andy's account of the discovery of the bird from the 2010 *HBR* (Johnson 2011) are reproduced below.

Dawn on Monday October 18th 2010 at Sandy Point was cloudy and windless. Having covered the eastern half of the reserve, and with a neap high tide occurring at 0917 hrs, I adjourned to Sparkes Marina to count a distinctly average roost at Black Point which comprised 733 waders of seven species, four Sandwich Terns and 180 Brent Geese. A 'pyoo' at 0935 hrs alerted me to a Snow Bunting over Black Point, which I quickly picked up in the scope – a cracking male bird, it flew round and round calling but unfortunately after a few minutes it left westwards without landing. I mused to myself that the only previous time I had recorded a fly-over Snow Bunting here – on November 13th 2005 – I had gone on to find (with Simon Woolley) a Desert Wheatear...

Shortly after 1100 hrs I approached the open sallow in the northern corner of the reserve. A bird flicked up from the ground on to low branches barely ten yards ahead of me.

Sandy Point, October 19th 2010. Darren Evans and Steve Copsey

October 19th 2010

It sat at an oblique angle, staring at me with a big, beady eye, showing off its bright ORANGE FLANKS and a flicking BLUE TAIL! BAM!!! I practically fainted! One minute it's an ordinary day and a split second later I'm watching a RED-FLANKED BLUETAIL – one of THE birds that patch-watchers hope to stumble across!

The next half-an-hour was a blur of phone calls, camera, expletives, text-messaging and simply trying to watch and enjoy the bird! It soon became apparent that it was favouring this one sallow, and apparently not straying too far away. It continued to show regularly although the intervals between appearances gradually increased to around half an hour. Its habit of continuously flicking its tail meant that its presence was quickly given away, even if it was mostly obscured or in silhouette, and this habit was obvious even to the naked eye. At 1450 hrs it disappeared again for a presumed half hour interlude – but it didn't reappear (much to the dismay of those who arrived after this time).

With Sandy Point's reputation for one day birds, coupled with the fact that it had disappeared in the middle of the afternoon, it was with some surprise and relief that it was still present in the same place early on Tuesday morning – not least for those who had missed it on the previous day! It was then seen every day until the Saturday when it proved more elusive in the morning – possibly a direct result of the large assembled audience. However, in contrast to its stay up until this point, it performed extremely well all afternoon, feeding actively right up to dusk. It seemed obvious that this was in preparation to continue its migration and sure enough, come

Sandy Point, October 23rd 2010. Maxwell Law

Sunday morning, it soon became clear that it had departed. This bird was in 'classic' first-winter (i.e. female-type) plumage. Many of the photographs seem to show a distinct blue tinge to the lesser coverts, possibly indicating that it may have been a first-winter male.

I would like to thank Pete Potts, HCC Senior Countryside Ranger, for allowing news of the exciting discovery (in a restricted access HCC reserve) to be broadcast, very quickly producing signs to direct people to the bird, and wardening the twitch. A huge debt of thanks to Carol Letten and Nick Griffiths at Hayling Island Sailing Club, for allowing the use of their overflow car-park, greatly easing parking pressure on the local roads and the residents!

GB status: 213, 1960–2019. Large increase, mean 15 per year in 2010–19. (Breeds NE Finland E through boreal forests of N Russia & Siberia to Kamchatka, N Japan & NE China. Winters S China, Taiwan & S Japan through SE Asia to N peninsular Thailand.)

Red-breasted Flycatcher *Ficedula parva*

HOSRP

A very rare vagrant. Five records since 1944.

The first for the county was a male seen by Dr. P. Hamilton-Boydon in his garden at Southsea on May 1st 1944. Extracts from an account of the events by D. J. Gunston (*Brit. Birds* 54: 178) are reproduced below.

On the evening of May 1st 1944, a male Red-breasted Flycatcher was identified in his garden in Southsea, by Dr. P. Hamilton-Boyden. The bird was hunting insects in the air with the distinctive flycatcher flight. Its red breast first caught the observer's attention and caused momentary confusion with a Robin, which closer observation dispelled. The bird's slender appearance induced Dr. Boyden to examine it with a glass, and although he unfortunately cannot recall having seen the characteristic tail pattern, he verified the identification at once with a text-book. The garden is surrounded by trees, which, although affording good shelter for birds, made observation somewhat difficult. Dr. Boyden also noticed the unusual call-notes of the bird, sounding distinctively tinkling or bell-like and repeated four or five times. These continued for about half an hour in the vicinity after it had left the garden.

The next record involved two, possibly three, seen by Michael Bryant in and around his garden at Farlington on September 14th 1953. His notes submitted to HFCERC were as follows:

Three birds seen on and off and rapidly moving between trees in several gardens. Very restless. One bird, of which a very close view was obtained, did occasional typical flycatcher flutters after insects. This bird appeared to be a juvenile.

A small flycatcher with typical alert, upright poise, beak short and dark, legs dark, upperparts olive brown, underparts whitish with brownish tinge and distinct dark tips to feathers giving decidedly scaly appearance to breast. Wing feathers not closely observed but tail very conspicuously marked with dark central feathers and outer feather with dark tips and white panels at the base. The male I observed briefly as it perched facing me on a rose arbour; I noted olive-coloured upperparts, orange throat patch and white underparts. While in this position it cocked its tail over its back like a Wren and I observed the same tail pattern as above. It then flicked away and I saw it no more. The third bird was not closely observed at all but seemed to be a female or juvenile, not a male.

Red-breasted Flycatcher, first-summer male, Ampfield, May 29th 2016. Alan Lewis

Arguably both the above records do not stand up to modern scrutiny. In particular, the occurrence in the UK of a juvenile with a scaly breast seems unlikely.

In recent decades there have been only three records. A female or first-winter was found by the author at Broadlands Lake on October 13th 1976, another female or first-winter was watched by Graham Stephenson at Sinah Warren for five minutes on October 3rd 1989 and a first-summer male was found by Nigel Jones at The Straight Mile, Ampfield on the morning of May 29th 2016. News was released and the bird was subsequently enjoyed by many observers during the day.

Records of a male at Pennington Marshes on October 2nd 1968, a female or first-winter at Old Winchester Hill on August 28th 1974 and a male near there on August 18th 1975 which were published in *HBRs* were not included in *BoH* as the available notes were not considered convincing.

GB status: 5,121, 1968–2019. No clear trend, mean 125 per year in 2010–19. (Breeds C Europe from Germany N to S Sweden, C Finland & Russia E to Urals & perhaps beyond, & S to Black Sea & Caucasus. Winters W Indian subcontinent.)

Siberian Stonechat *Saxicola maurus*

BBRC

A very rare vagrant. One record in 2016.

The only record for Hampshire is of a first-summer male found by Amy Robjohns at Titchfield Haven on May 10th 2016 which stayed until the following day. Extracts from Amy's account of the discovery and identification of the bird, taken from the 2016 *HBR* (Robjohns 2017), are reproduced below.

Thick fog at dawn on May 10th prevented any seawatching at Hill Head. Not wanting to admit defeat, I began to wander up the canal path but was soon distracted by the large number of snails on the fence line, so instead began to count and photograph them. After about 350 snails, a stonechat hopped up onto the fence a few feet in front of me. I was immediately struck by two things: firstly, it was a very striking individual, and secondly, what was a stonechat doing here in May? The overwintering stonechats had moved off weeks ago and it certainly looked rather different, so something didn't quite add up. I remember thinking along the lines of 'surely that has to be a Siberian Stonechat' but couldn't yet prove it.

I managed to get a few photos before the bird flew off back into the scrub. As it did so, I noticed the large white rump and cursed myself for not videoing it instead. It didn't reappear for an hour or so and, with coursework to finish, it was time to abandon the patch. The photos were uploaded onto Twitter and Facebook later that morning, causing some debate. I decided to put it out as a possible Siberian Stonechat as I was feeling rather convinced despite the lack of evidence! I also contacted a couple of locals including Dave Stevenson who was able to get useful footage later in the day which confirmed my suspicions.

In hindsight, I regret not returning that evening for another look. However, the final piece of the jigsaw was nailed by Dan Houghton, Simon Ingram, Mark Palmer and Dave Ryves who noticed the extensive white at the base of the tail feathers and managed to get shots confirming it to be a Caspian Stonechat. I couldn't quite believe it when I saw the photos on Facebook! The bird was still present the following day, showing the diagnostic rump and tail features much better than on 10th.

Male Siberian Stonechats have a much blacker head and upperparts, whiter underparts, including flanks, and more restricted orange on the breast than European Stonechats, making their appearance, like the Titchfield bird, much more striking. Other key identification features noted on the photographs include the large, unstreaked white rump, the wider and more prominent white neck patch, the large white area on the inner greater coverts and black axillaries. They are also longer winged, with greater primary projection like Whinchats Saxicola rubetra. In contrast, male European Stonechats have grey axillaries and tend to have streaking on the rump. Male Caspian Stonechats are almost identical to other Siberian Stonechat races but can be separated by the more extensive white in the tail, extending down from the rump to form an inverted 'T', similar to a Wheatear (Svensson et al., 2009; Hellström & Mats Wærn 2011; Clement & Rose, 2015).

This feature was clearly visible in several of the photographs and clinched the identification of this distinctive race.

Caspian Stonechat, Titchfield Haven, May 11th 2016. Simon Colenutt

The excellent photographs helped BBRC to accept the bird as being of the race *hemprichii* known as Caspian Stonechat. There have been only five previous records of this race in Britain; all were of males including two in spring (Stoddart & Hudson 2021).

GB status: 1, 1913; 18, 1985–2020. Status uncertain. (Nominate maurus *breeds Russia & W Siberia, NW Mongolia & N China, S to Tien Shan Mountains, Afghanistan and Turkmenistan. Winters Iran to N India. Race* hemprichii *breeds from W & N Caspian Sea region to lower reaches of Volga & Ural Rivers. Migrates SW to winter SW Asia, Arabian Peninsula & NE Africa. Other races occur in Caucasus region, Himalayas & W China.)*

Siberian/Stejneger's Stonechat *Saxicola maurus/stejnegeri*

BBRC

A very rare vagrant. Three records since 1988.

The first record for the county of a Siberian-type Stonechat was of one found by Barry Duffin at Titchfield Haven on October 31st 1988. Extracts from Barry's description are reproduced below.

At 1545 hrs on 31st October a bird closely resembling a Whinchat was located whilst it perched on dead thistle stems at Titchfield Haven. My attention was drawn first by its overall pale appearance and then by its very distinct white rump as it flew a short distance ahead of me. The bird was watched closely for the next 15 minutes as it flew down from the tops of the thistle heads to feed on the open ground below. I was able to take a detailed description before it flew off to a nearby reed-bed and was lost from view. Head: shape of Common Stonechat. Distinct buffish supercilium from base of bill to well behind eye, noticeably broader behind eye. Lores, ear-coverts and cheek blackish. Crown streaked sandy buff/brown, contrasting with the paler nape and mantle. Upperparts: nape, mantle and back streaked, feathers sandy buff with blackish-brown centres. Wing coverts and primaries dark brown/blackish. Broad white fringes to tertials and inner secondaries, resulting in a noticeable white wing panel. Uppertail-coverts white. Rump extensively white and unstreaked. Tail blackish. Underparts: chin and throat creamy white, being a very distinct feature, contrasting with the uniform pale sandy-buff of the breast. No sign of spots or streaking on underparts. Underwing greyish. Axillaries appeared greyish as rest of underwing although good views were not possible. Soft parts: bill and legs black.

Siberian/Stejneger's Stonechat: Titchfield Haven, October 16th 2014 and Hook-with-Warsash, October 23rd 2015. David Ryves and Daryl Gorman

The next was a first-winter discovered by David Ryves near the Meadow Hide at Titchfield Haven on October 16th 2014. It remained in the area until October 26th, allowing many birders the opportunity to see it well. In the following year another first-winter was found by Bob Marchant and photographed by

Darryl Gorman at Hook-with-Warsash at 0930 hrs on October 23rd. It was in the company of several Stonechats that were quite aggressive towards it. The bird was refound by Mark Palmer at 1420 hrs the next day and seen by several observers before dusk.

Siberian/Stejneger's Stonechat, Titchfield Haven, October 27th 2014

These records were initially accepted as Siberian Stonechats but following the upgrade of races *maurus* and *stejnegeri* to separate species (BOURC 2018) they have been assigned to this wider category. BBRC currently considers that only male Siberian Stonechats of race *hemprichii* to be separable in the field and that other specific identification is only possible with genetic analysis (Stoddart & Hudson 2021). Nonetheless, Paul Leader (*in litt.*), who has much experience with *stejnegeri* in Hong Kong, examined photographs of the 2015 Hook-with-Warsash bird and considered that it was probably a Stejneger's Stonechat.

GB status: 426, 1960–2020. About ten per year; stable.

Desert Wheatear *Oenanthe deserti*

BBRC

A very rare vagrant. Two records in 1961 and 2005.

The first for the county was a female found by George Clay at Farlington Marshes on November 4th 1961 which was seen by many observers during its stay until November 19th. Extracts from George's description are reproduced below.

> *Appearance, very like a pale wheatear with similar behaviour. Generally sandy coloured. Crown, sandy-buff; back, sandy-buff marked greyish; primaries black, light on inner webs; ruff buff-white; tail black. Prominent dark eye; face, throat and breast, sandy-buff marked slightly greyish, a little darker than upperparts; belly and vent white. Legs, feet and bill black. Most of its time was spent around some heaps of rubble on the landward side of the sea wall and it occasionally visited an area of ant hills. It frequently flew up steeply to catch insects as far as 100 yards away.*

Desert Wheatear, Beachlands, Hayling Island, November 13th 2005. photos Richard Ford

The next was not until 2005, when Andy Johnson and Simon Woolley located a first-winter male on the beach just west of Sandy Point at 1010 hrs on November 13th. It disappeared at 1023 hrs but fortunately was relocated by Tim Lawman and Pete Gammage at Beachlands, about two miles to the west, at 1235 hrs. It remained there until dusk and was enjoyed by many observers. Details of the events that day are given by Johnson & Woolley (2007).

GB status: 9, 1880–1949; 155, 1955–2020. About 4–5 per year; stable or increasing. (Breeds desert regions of N Africa from Morocco to Middle East, N to S Caucasus, & C

Asia from C Iran & N Pakistan to Mongolia & N China. Some N African birds resident, but most winter Sahara & Sahel region of N Africa from Mauritania E to Ethiopia & Somalia. Asian breeders winter Arabian Peninsula to NW India.)

Western Black-eared Wheatear *Oenanthe hispanica*

BBRC

A very rare vagrant. One record in 1987.

The only record for Hampshire is of a male found by Geoff Farwell and Terry Jennings at Farlington Marshes at 1415 hrs on June 5th 1987 and subsequently seen by many observers during the day. Extracts from Geoff's description are reproduced below.

> *Initially flushed along the shore and with the naked eye presumed to be a Wheatear but on raising an optic could immediately be identified as a male Black-eared Wheatear of the race* hispanica*. It flew further along the shore and then turned up-shore and settled on a rock allowing good views for a few minutes. Rain started to fall and the bird flew into some bushes. After a lengthy wait we carefully approached the bushes whereupon it flew out allowing good flight views and then settled in some bushes further along the shore. We noted the following details:*
>
> *Size, posture and flight generally similar to Wheatear. General plumage colouration a combination of black, white and rich buff. Bill slender and blackish. Forehead and supercilia whitish – noticeably paler than the rich buff crown, nape and hindneck. Lores, eyes, ear-coverts, chin and throat black forming a neat black facial mask. Upper breast buffish merging into the white, entire underparts. Legs black. Mantle and back uniform rich-buff (extension of hindneck etc); rump and uppertail-coverts white forming a conspicuous white rump in flight. Scapulars, all wing-coverts, tertials, secondaries and primaries black. Tail white with blackish terminal band and fairly conspicuous blackish central bar running vertically along the middle of the tail. The tail sides or outer tail feathers were also blackish but not extending up the tail as far as the central bar. Also worthy of note, the underwings appeared dark grey unlike the black upperwings.*

A previously published record of a male *hispanica* seen by Graham Rees and Peter Le Brocq at Farlington Marshes on September 18th 1954 (*Brit. Birds* 48: 130) is no longer accepted following a review of all major rarities in Britain in 1950–57 (Wallace *et al.* 2006). Other non-accepted records were of one at Farlington Marshes on April 17th 1980 and a male at Pennington Marshes on April 25th 1992, the latter being included in *BoH*. One found dead on board a ship in Southampton Docks on September 25th 1986 was not submitted to BBRC.

GB status: 6, 1892–1949; 12, 1965–2019. Rare, about one per year, stable. 11 pre-1950 and 28, 1950–2019 Western or Eastern Black-eared Wheatears. (Breeds NW Africa & Iberia, E to France & N Italy. Winters western N tropical Africa.)

Eastern Black-eared Wheatear *Oenanthe melanoleuca*

BBRC

A very rare vagrant. One record in 2015.

The only record for Hampshire is of a male discovered by Gary Howard at Acres Down on the morning of June 13th 2015 and subsequently seen by hundreds of birders until dusk. Extracts from Gary's account of its discovery in the 2015 *HBR* (Howard 2016) are reproduced below.

I've been a regular if infrequent visitor to the New Forest for over 30 years and when a planned family weekend away to the Farne Islands got cancelled due to the weather, I suggested we go there instead. We only finalised our plans on the Friday and I booked a place at the campsite on Ocknell Plain and said to my wife, 'That's not far from Acres Down, let's start there and see how we get on'. It was overcast, cool and dry when we arrived just after 0900 hrs on Saturday June 13th 2015 and found the car park empty. The four of us strolled up to the viewpoint in no hurry enjoying views of Tree Pipit and Stonechat en-route.

David Aitken

At about 1030 hrs as I scanned for birds I heard one of the other three birders present say something like 'Is that a shrike?' I looked to my right and about 100 metres away could see a black and white bird on the top of a pine almost at eye-level. I trained my scope on the bird and said, 'No, it's a wheatear', and almost as soon as I said it the bird dropped off the tree and I said something like 'It's not a Wheatear, it's a Black-eared.' I rushed to the edge of the hill but couldn't see anything so I ran a few metres downhill to the base of the pines but still saw nothing. I knew I'd found a 'mega' but it was slipping out of my grip and without even a record shot I feared it would be just another of those ones that got away.

I worked my way slowly and methodically around the base of the hill and after what seemed like an eternity, but was in fact 20 minutes, I saw the bird again. As I reached for my camera it flew away towards a large group of pines at the foot of the hill and out of sight. Jenny, my wife, who was now trying to cope with keeping twin 11-year-old boys in sight and carrying my scope and shadowing me, saw me give the thumbs up and went back up to tell the others (James Burge, Dave Knight and Rob Clements). A few minutes later we were all walking towards the area where I'd seen it fly and I picked it up on open ground amongst the bog cotton. It was a stunning bird and eventually I got some photos, then all too quickly it once again flicked away out of sight. What a bird, I could scarcely believe it!

I was aware of its rarity status and wanted to get the news out as quickly as possible. I photographed the screen image on the back of my camera and attempted to 'tweet' out the news but soon found that I had no phone signal! I dashed off up to the top of the Acres Down viewpoint and finally got the news out just before 1130 hrs. It proved a very popular 'twitch' with hundreds getting to the bird before nightfall. Despite extensive searching the bird was not seen the next day or on any subsequent day.

Comparison of photographs of this bird and one present in Noordwijk, Zuid Holland, the Netherlands three days earlier (*Dutch Birding* 37: 249) shows that they have the same distinctive tail pattern and thus confirms that they were the same individual.

GB status: 4, 1915–1940; 10, 1975–2019. Rare, about one per year, stable. 11 pre-1950 and 28, 1950–2019 Western or Eastern Black-eared Wheatears. (Breeds S Italy to Greece, & SW Asia from Turkey to S Caucasus, S to Israel & SW Iran. Winters eastern N tropical Africa.)

Pied Wheatear *Oenanthe pleshanka*

BBRC

A very rare visitor. One record in 2013.

The only record for Hampshire is of a first-winter female found by Peter Northcote at Rookery Farm, Monk Sherborne on December 7th 2013. Details of the discovery and subsequent identification of the bird were given in a short note in the 2013 *HBR* (Clark 2014), extracts from which are reproduced below.

Peter Northcote first noticed the bird after a day's shooting on December 7th. It was near the beaters' cabin at Rookery Farm. He observed it with binoculars in poor light on the roof of the barn and noticed the white rump and suspected it was a wheatear but could not get better views. He forgot about the bird but when he took part in another shoot on December 12th, he again noticed it hanging around a Land Rover. He took some poor pictures with his phone and sent them to members of the Newbury Ringing Group. It was suspected the bird was a wheatear but they

Peter Northcote

requested better photographs. Peter returned on December 15th and obtained some improved pictures with his Nikon DSLR. Jan Legg, his trainer at the ringing group, suspected it might be a Desert Wheatear. Finally, improved pictures were obtained in good light on December 17th enabling its identity to be confirmed as a first-winter female Pied Wheatear. The County Recorder, Keith Betton, was informed. Peter was understandably anxious about releasing news of the bird on private property but fortunately a public footpath was close by and offered good views over the small area which the bird inhabited. Keith, Peter Hutchins, Barry Stalker and the author went to the area at first light on December 18th to look for the bird and make arrangements to prepare the locals for the inevitable influx of birders which would follow. Unfortunately, as not infrequently happens when the release of news of a bird is delayed, it could not be found. Peter Hutchins continued to check the area over the next few days but it was not seen again.

GB status: 2, 1909–1916; 84, 1954–2019. About 2–3 per year; stable. (Breeds E Romania & Bulgaria, E through S & E Ukraine, S Russia, S Siberia, Kazakhstan & Mongolia to N China, E to Gulf of Bohai. Winters NE & E Africa, & SW Arabian Peninsula.)

Dipper *Cinclus cinclus*

HOSRP

A rare visitor which occasionally breeds, most recently in 1992. Many old records and around 25 since 1953.

The first for the county appears to be that described in Gilbert White's journal for April 20th 1786, when he made the following entry:

> *Young George Tanner shot a water ouzel* Merula aquatica *near James Knight's ponds. This is the first bird of that sort ever observed in this parish (Selborne). This bird, being only pinioned, was caught alive, and put into a cage, to which it soon became reconciled, and is fed with woodlice and small snails. May 12. The water ouzel is living and recovered of its wound.*

K & M listed seven specimens obtained at Emsworth in January 1808, Hayling in 1835, Avington and Worthy in 1878 (3), St. Cross in November 1892 and

Chilbolton in September 1893. The three from 1878 formed part of the Winchester College collection and are on display in the Science School there. Singles were reported on the Test above Longparish on April 26th 1912 and on the Itchen between Twyford and Bishopstoke on May 28th 1933. Dippers evidently extended their breeding range into Hampshire in the 1930s. In the Avon Valley, a pair was near Breamore in 1934 while in the Allen Valley, a pair bred between Fordingbridge and Martin in 1937. In the Test Valley, single pairs were near Andover in 1934, at Andover, Fullerton and Leckford 'as usual' in 1937, at Leckford again in 1938 and at Compton Manor in 1942. In the Itchen Valley, a pair bred at the sluice in front of Kingfisher Lodge, Brambridge in three years just prior to the Second World War. Other records at this time were of singles at Weir House, Alresford on August 2nd and October 18th 1942 and Testwood Mill in August 1942, and two at Longparish on January 15th 1945 (*SEBRs*, Cohen (1963)).

Dipper, Tanners Brook, Southampton, November 27th 1992. Dennis Bright

In 1953, one was at Winchester Water Meadows on June 3rd, and in the following year a pair bred at Breamore Mill (Norman Orr *in litt.*). In the next 25 years the only records were of singles at Brambridge from January to March 1963, Broughton on April 18th 1969, the Itchen Navigation below Winchester on January 3rd 1970, Mislingford from November 19th to 24th 1972, Bere Mill (Whitchurch) on April 17th 1978, Kingsclere in March 1980, Fordingbridge in late December 1983 and early January 1984 and Warnford on January 28th 1989.

A pair was located in the Test Valley at Romsey Water Meadows on November 3rd 1990. They bred successfully nearby at Greatbridge in 1991 and 1992, no doubt aided by the low water level in the river during that period. Further upstream at Kimbridge one was recorded on December 16th 1990, from November 3rd 1991 to January 5th 1992 and on May 17th and June 1st 1992. Further records from the Test Valley were of singles along the Tadburn Stream at Romsey from August 23rd to 30th 1994 and August 24th to 29th 1995 which was still present on September 25th and at Kimbridge on March 3rd, 16th and July 1st 1996 and March 2nd 1997.

Dipper, Greatbridge, Romsey, January 1991

Elsewhere during this period, a first-winter was in the Warnford/Meonstoke area on December 12th and 22nd 1991 and had probably been present since late October, one was at Tanners Brook, Southampton from November 26th 1992 to March 3rd 1993 and another was at Warnford on March 12th 1995.

There have been only two records this century. One was found by Peter Hutchins on the River Test adjacent to Overton Lagoons on December 13th 2003 which stayed until the next day and was seen by many observers. Most recently Simon Woolley saw one at Bickton on December 30th 2012 but it was not relocated.

GB status: Resident W and N England, Wales and Scotland. (Resident much of Europe with other races east to Himalayas, China, Mongolia and Siberia.)

Spanish Sparrow *Passer hispaniolensis*

BBRC

A very rare vagrant. One record in 2011–12.

The only record for Hampshire is of a male first seen by Sue Wilson in her garden at Calshot on December 3rd 2011 which stayed until March 23rd 2012. The story of the discovery and identification of the bird and management of the resultant twitch was given by Sue in a short note in the 2012 *HBR* (Wilson 2013), extracts from which are reproduced below.

While watching the birds feeding in our garden on the morning of December 3rd, I observed a new arrival for our garden bird list. The House Sparrows and Starlings were bickering and feeding on the seed and peanut feeders when suddenly a different male sparrow arrived. This individual was aggressively seeing off another sparrow from a feeding port, his chestnut head most striking from the initial back view. As he settled on the feeder his prominent black chest streaking became clearly visible. It was immediately obvious that this was not a House or Tree Sparrow.

photos previous page
Spanish Sparrow, Calshot,
March 18th 2012
and January 13th 2012.
Steve Gantlett
and Mike Lawrence

I asked my partner Mark to find the bird identification book and grabbed my binoculars for a close-up view. On consulting the book, we realised it was a male Spanish Sparrow. Over the next few days Mark managed to get some photographs through the kitchen window. These were shown to Dave Bishop and Simon Ingram who were watching the Dark-eyed Junco in the New Forest on January 7th 2012 and agreed with the identification. The following day, with the agreement of local residents, Simon Ingram and Bruce Gwynn made arrangements for birders to visit the site. The sparrow remained around the garden and was seen by many hundreds of observers until it was last recorded on March 23rd. Locals opened the Spanish Sparrow Café and collected donations which resulted in £1,800 being raised for local and wildlife charities.

GB status: 10, 1966–2012. (Breeds Cape Verde, Canary Islands & NW Africa, N to Iberian Peninsula, Sardinia, Balkans E to SW Asia, E through C Asia to E Kazakhstan, NW China & S to Afghanistan. European breeders mainly resident, C Asian breeders migratory, wintering NE Africa, Arabian Peninsula & Iran to NW India.)

Tree Sparrow *Passer montanus*

Formerly a scarce resident and erratic but sometimes numerous autumn passage migrant and winter visitor; now a rare visitor not recorded annually.

The Tree Sparrow has always been one of the county's most enigmatic species. It was evidently not known to Gilbert White. K & M stated that, as a breeding species, it is distinctly rare, and even unknown, in many parts and added that the nest had been found in the New Forest and at Bursledon and Laverstoke. They described it as plentiful in winter in most districts of the county. However, there are no dated records for the 19th century.

Records for the first half of the 20th century are almost as sparse. Munn (1919) did not mention the species in his update to K & M. It is mentioned only four times in the county bird reports for 1932–50 which are included in the *HFC Proceedings*. It was noted at Ampfield on September 27th 1938, one was in a garden at Appleslade on April 23rd 1940, it was described as present in fair numbers at Bassett but scarce at Milford on Sea in 1946, and a flock of 20 was at Lyndhurst on January 13th 1947. In addition, one was at Chilcomb, Winchester on November 17th 1949 (Norton *unpub.*). Cohen (1963) knew of only two regular breeding haunts in the 1950s – at Fyfield and Purbrook. The Tree Sparrow was undoubtedly scarce during this period, although many suitable areas in the north of the county were rarely visited by birdwatchers. From 1959 onwards there was a remarkable change in status. Substantial autumn movements and large winter flocks began to be recorded although numbers varied greatly from year to year. The breeding population also increased, particularly in the north-east, where the species became quite widespread.

Tree Sparrows, Titchfield Haven, September 19th 2009

Among the heaviest movements recorded were those of 1,000 east at Gilkicker Point on October 14th 1961, 448 west and 400 east at Titchfield Haven on three dates between October 10th and 25th 1976, 315 mostly west at Fleet Pond between October 9th and November 4th 1976 and 551 west at Hook-with-Warsash October 13th 1983. Large flocks, often feeding in kale fields with Fat Hen *Chenopodium album*, included 500 at Basingstoke Old Sewage Farm in March 1960, 750 at Titchfield Haven in February 1962,

450 at Winchester Sewage Farm in February 1965 and during the 1976/77 winter maxima of 500 at Marsh Court in November, 900 at Tundry Pond in December and 400 at Easton in February. Subsequently, winter numbers declined and the final three-figure flocks were of 180 at Hook Rubbish Tip and 120 at Timsbury during January 1984.

The breeding population peaked at an estimated 500 pairs between the late 1960s and mid-1970s but was not thought to exceed 50 pairs at the end of the HOS Tetrad Atlas Survey in 1991 (*BoH*). The area around Well was the final regular site where breeding was last confirmed in 1993. However, a flock of ten including juveniles at Baker's Bridge, Elvetham on July 5th 1996 indicates local breeding and a pair was at Penn Croft Farm, Crondall on June 14th 1999. Numbers outside the breeding season continued to decline through the 1990s and into this century. Annual totals during this period did not exceed 38 and in the second decade of the new century the maximum was six and none was recorded in 2015–17.

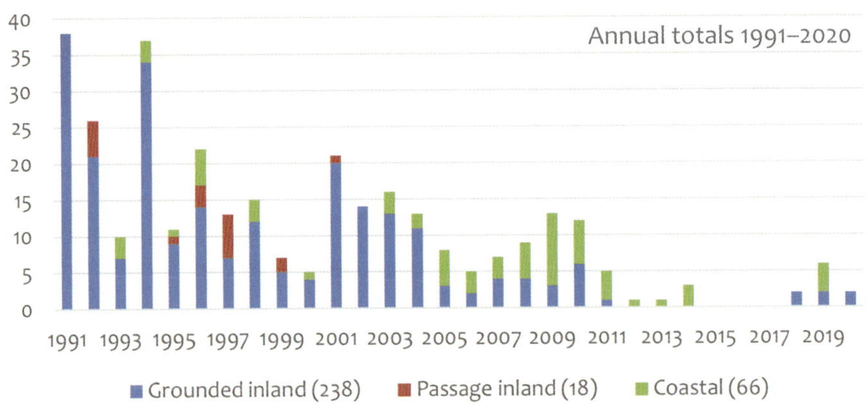

In the 1990s and early 2000s most records involved small groups in weedy fields or adjacent to pheasant feeders, often with gatherings of finches and buntings. The largest flock recorded was of 18 at Well on February 22nd 1992 when the species still bred in the area. Subsequently, a group of seven was found at Hartley Mauditt on February 17th 2001. Small numbers returned to this area in the following four winters with a maximum of 12 on March 10th 2002. Despite the provision of nest boxes there was no indication that they bred locally. A group of four was first noted at a garden seed feeder at Sparsholt on December 24th 2007 and remained until March 29th 2008. One of these had been colour-ringed as a pullus at Beckhampton, Wiltshire in 2007. Up to four were at Bentworth between March 6th and April 23rd 2010 but since then only singles have been recorded at inland localities. Between 1991 and 2001 a total of 18 was noted flying over inland sites (15 at Fleet Pond) on dates between October 3rd and November 22nd but none has been recorded since despite increased coverage at Trigpoint Hill and Tweseldown.

Tree Sparrow, right with House Sparrow, Titchfield Haven, April 29th 2014. Martin Bennett

Most records during the 21st century have been at coastal locations. A group of three coming to feeders at Titchfield Haven from September 17th to 20th 2009 proved very popular with visiting birders, as did another there from April 15th to May 1st 2014. Other records were mostly of birds flying over or briefly grounded. Most were found at Sandy Point where ten have occurred since 2006 including an adult and two juveniles on August 29th 2006; one of the juveniles had been colour-ringed as a pullus near Devizes, Wiltshire in that year.

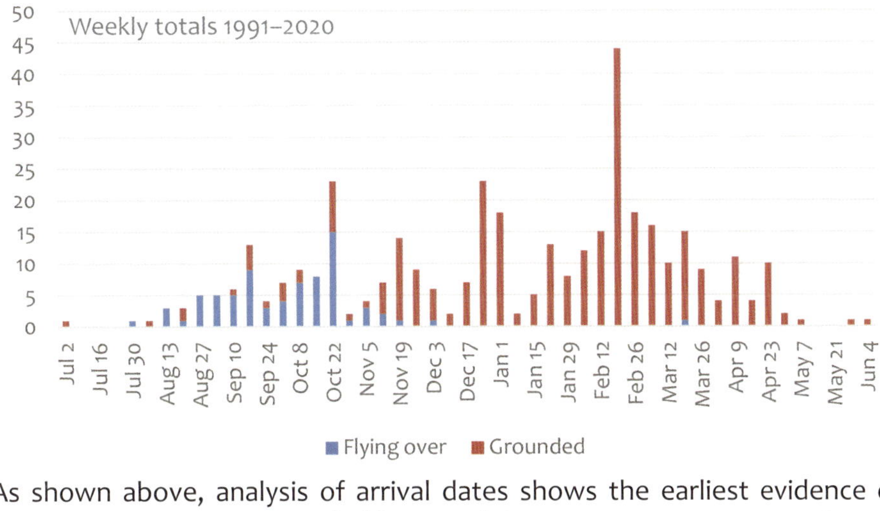

As shown above, analysis of arrival dates shows the earliest evidence of autumn passage was provided by one flying west over Barton on Sea on August 3rd 2004. Very few were seen moving after mid-November with most records subsequently referring to grounded birds. There is some evidence of spring passage with records including the Titchfield Haven bird mentioned above, another there on April 29th 1993, two at Sandy Point on April 21st

1994, one at Sway on June 1st 2004 and one at Sandy Point on June 5th 2013 which was seen again briefly the following morning.

GB status: Resident, passage migrant and winter visitor which has become increasingly rare in S England. (Breeds through most of Europe and Asia.)

Alpine Accentor *Prunella collaris*

BBRC

A very rare vagrant. One record in 2016.

The only record for Hampshire is of one found by George Else at Brownwich Cliffs on April 13th 2016. Extracts from George's account of the bird's discovery in the 2016 *HBR* (Else 2017) are reproduced below.

I visited Brownwich Cliff on the Solent coast on April 13th 2016. It is an area that I have regularly visited for many years for its populations of bees, in particular for one of the mining bees Colletes cunicularius *for which this is the sole known Hampshire site and for another* Andrena vaga *which only occurs here and at Dungeness. I was walking alone along the pebble beach towards Hill Head. Visibility was excellent and the day was warm and sunny with only scattered clouds and a gentle easterly breeze. At about 1050 hrs I was near the north-western end of the cliff when I noticed a bird at the top of the beach only three to four metres away. For five minutes I studied it through the lens of my camera and took a series of 48 images. It gradually ascended the cliff, moving about at times quite actively from ledge to ledge and apparently feeding as it did so. On reaching the summit it disappeared and I did not see it again. It was the size of a Meadow Pipit. The head and breast were unmarked blue/grey. The closed wings were blackish-brown with some conspicuous white spotting. The flanks were a distinct reddish colour. The bill was in part bright yellow. I had no idea what it was though both Alpine Accentor and Rock Thrush* Monticola saxatilis *were at the back of my mind. I immediately dismissed both possibilities as too fanciful but could think of no pipit that remotely resembled it. Later that day, at home, I sent some images to a few friends for their opinions. In the meantime, I studied my field guides when it became obvious, by comparing illustrations with my photographs, that the bird was indeed an Alpine Accentor. This identification was later confirmed by those to whom I had sent the images. The photographs are, alas, of poor quality but sufficient to support the identification. The news was broadcast by Birdguides on Apr 14th but despite further searching that day the bird was not found again.*

George Else

The Hampshire sighting proved to be part of a wider unprecedented northerly movement across western Europe in the previous week which involved 25 at four sites in Germany, two in Holland and one at Scolt Head, Norfolk on 12th.

There is an older record of three seen by Cecil Paddon and two other observers at Beaulieu from January 19th to 21st 1926. Extracts from Cecil's description in *British Birds* (20: 107) are reproduced below.

> *On January 19th 1926, during the second day of a howling blizzard... three strange birds came to food put down at a house in Beaulieu, where they remained until 21st. The points noticeable about the birds were their speckled throats, white wing-bars, deepish red-brown flanks, yellowish beaks and the colour of their legs, which was of a rather pale, but dirty salmon shade. On the ground they had a rather low, sliding movement, almost a crouch, but difficult to describe. They were larger than Chaffinches and their tracks were also larger and quite distinct, with a longish hind claw. Checking the appearance of the birds from Coward's Birds of the British Isles we had little doubt that they were Alpine Accentors.*

The record was included by Witherby *et al.* (1938-41), Cohen (1963) and Naylor (1996) but accidentally overlooked for *BoH*. The record was recently reviewed by a subpanel of HOSRP which decided not to return the species to the county list (Combridge & Clark 2010). It is excluded by Naylor (2021).

GB status: 10, 1817–1932; 20, 1955–2016. Rare, about one every 2-3 years. (Breeds high mountain ranges of NW Africa, Spain & C Europe N to S Germany, S Poland & through Balkan countries to Greece. In Asia, breeds in all major ranges from C Turkey E through C Asia, S to Himalayas & E to Japan & Taiwan. In winter, most European birds descend below snowline near breeding areas but some disperse to lowlands.)

Blue-headed Wagtail *Motacilla flava flava*

HOSRP

A very scarce passage migrant which has bred. Two old records and around 127 since 1951.

The first published records for the county of this race of Yellow Wagtail appear to be those of males seen near Ringwood by C. J. Evans on May 3rd 1937 and at Chilbolton Common by Col. W. A. Payn on May 15th 1942.

Subsequently, around 127 have been recorded as shown on the next page.

Most were recorded in spring with the earliest at Farlington Marshes on April 5th 1980 and April 6th 2011 and peak numbers between mid-April and mid-May. A pure pair bred at Farlington Marshes in 1979 and in other years males have bred with *flavissima* females there with one pair in 1954, three in 1980, one in 1981, four in 1983, three in 1984 and one in 1996. A female summered there in 1977. Other breeding attempts involving *flava* males and *flavissima* females were recorded at Keyhaven in 1966 (five young raised), Fawley in

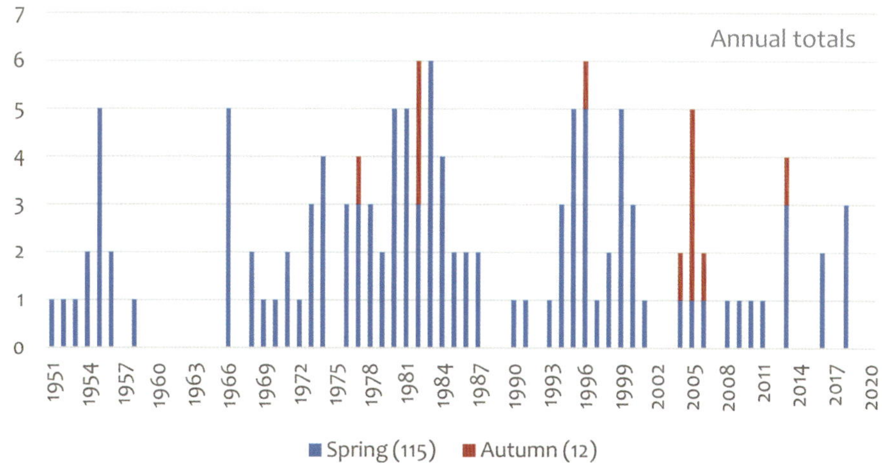

1973, Harbridge in 1974, Winnall Moors in 1976 and 1978 and Southampton Western Docks in 1981. A few have been identified in autumn with the latest a male feeding with *flavissima* Yellow Wagtails among cattle at Farlington Marshes on September 17th 2006.

The weekly totals of new arrivals (excluding some undated summer records) are shown below.

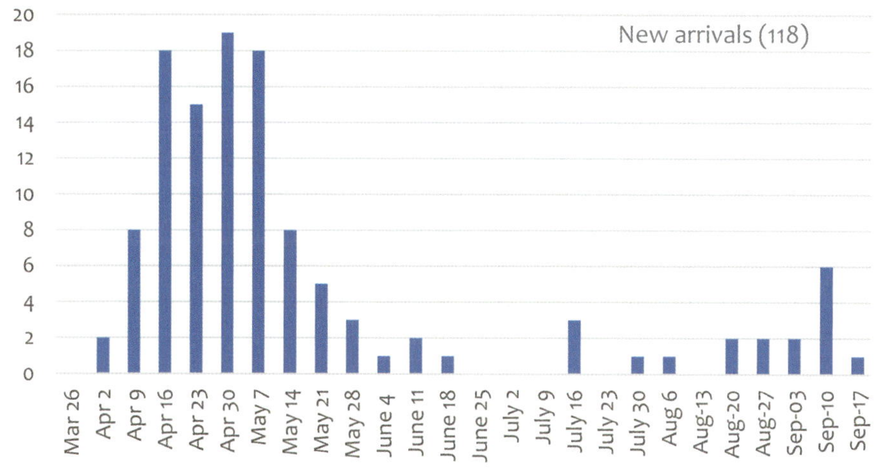

Of the 127 recorded, around 57 have been at the well-watched Farlington Marshes, where the breeding colony of Yellow Wagtails was closely monitored until breeding ceased in 2000. Other records came from 27 locations, mostly at coastal sites and in the main river valleys where Yellow Wagtails bred. Totals include 14 in the Keyhaven area, seven at Titchfield Haven and six at Winnall Moors. Only two have been recorded on Hayling Island, at Sinah on May 4th 1997 and Sandy Point on April 12th 2009. The only records further inland were of males at Anton Lakes on May 2nd 1981, Potbridge on May 13th 1983 and Bidden Water on May 14th 2001.

Blue-headed Wagtail, Lower Pennington Lane, April 28th and 29th 2016. Ian Williamson and David Cuddon

Between 1954 and 1984, records of seven at Farlington Marshes and two at Winnall Moors were described as showing the characteristics of the race *beema*, known as Sykes' Wagtail. Most of these bred with *flavissima* females and are included in the breeding totals given above. Sykes' Wagtail, which is intermediate in appearance between *flava* and *flavissima*, breeds in western Siberia and is unlikely to occur regularly in Britain. It was deleted from the British List in 2009 (BOURC 2010). It has long been known that birds resembling *beema* (and other races) occur in British breeding populations (Smith 1950, Milne 1959). Recently, Yellow Wagtails resembling *beema* and other subspecies have been found nesting in northern France (Alström *et al.* 2003). Such birds have become colloquially known as 'Channel Wagtails' in the popular birding literature (e.g., Dubois 2007, van Duivendijk 2010). Two such birds were photographed at Edenbrook Country Park, Fleet on April 27th 2013 and Pennington Marshes on May 5th 2013. It seems likely that the earlier claims of *beema* related to 'Channel Wagtails', a view supported by Dubois (2007).

GB status: Passage migrant and summer visitor to S and C England. Mixed pairings with flavissima recorded in 19 10 km squares in 2007-11, a pure pair in Cornwall in 2010. (Breeds in western Europe; winters in sub-Saharan Africa.)

Grey-headed Wagtail *Motacilla flava thunbergi*

HOSRP

A very rare vagrant. One record in 1977.

The only record for Hampshire of this race of Yellow Wagtail is of a male seen by David Billett and others at Farlington Marshes on April 8th and 9th 1977. It has not been possible to locate the description.

GB status: 413, 2008–19. Mean 26 per year in 2010–19. (Breeds from C Fennoscandia E across N Russia to NW Siberia; winters in sub-Saharan Africa.)

Eastern Yellow Wagtail *Motacilla tschutschensis*

BBRC

A very rare vagrant. One record in 2020.

The only record for Hampshire is of a first-winter found by Mark Francis in ploughed fields at Brownwich at 1100 hrs on October 5th 2020. It remained in the area until the late afternoon of October 6th and was seen by many Hampshire birders. Extracts from Ashley Howe's submission to BBRC are reproduced below.

As with any normal morning Mark covered his 'patch' which includes Titchfield Haven, the Canal Path and Brownwich Cliffs. The recently ploughed fields adjacent to the clifftop footpath had attracted a good concentration of Meadow Pipits, wagtails, Wheatear and larks, as well as Golden Plover, grazing Canada Geese and a group of active Black-headed Gulls. Conditions that morning were good for scanning the field and especially the smaller passerines for something out of the ordinary. By about 1100 hrs, Mark had picked out a very monochromatic Yellow Wagtail. He posted a few record shots into our local Whatsapp group. Picking up on the message straight away, I felt the bird showed good potential as an Eastern-type Yellow Wagtail. I highlighted the bird to a wider audience in another Whatsapp group for Hampshire Birders. In the meantime, Mark had lost the bird but thankfully I was just arriving on site with Pete Hutchins at 1345 hrs when Mark relocated it. From initial scope views we all agreed that it looked a promising candidate but needed better views to secure the finer details. The birds were very flighty and moving around regularly. Frustratingly we didn't pick up on the bird calling at all. The wagtail continued to show sporadically in the same ploughed field (and the next to the west) up until about 1730 hrs before flying east to roost in reeds at Titchfield Haven.

Brownwich, October 6th 2020. Mark Francis

On October 6th, the weather was much improved in the morning and the bird offered closer views. Mark obtained his best images and Amy Robjohns secured some recordings of its calls. By the afternoon Amy and I had circulated the recordings and back-of-camera shots to various people including Nigel Jones (BBRC) and Josh Jones and Ed Stubbs (Birdguides) who all agreed that the photos and recordings fitted Eastern Yellow Wagtail. We then put the news out nationally.

The bird continued to show well during the second afternoon and it again flew off east to roost. Unfortunately, there was no sign of it the next day.

GB status: 1, 1909; 23, 2009–2020. Status uncertain. (Four races breed in Siberia from upper Yenisey to Kamchatka, S to N Mongolia, NW & NE China & N Japan, also W and N Alaska, USA, and N Yukon, Canada. Winters E and SE Asia, Philippines, Indonesia & N Australia.)

Citrine Wagtail *Motacilla citreola*

BBRC to 2014; HOSRP 2015 to date

A very rare vagrant. Six records since 1983.

The first for the county was a first-winter found and fully described by Bob Chapman at Titchfield Haven on September 22nd 1983. This record was not accepted by BBRC until 1997 and so was not included in *BoH*. Extracts from Bob's description are reproduced below.

In the late afternoon of 22nd September, I was in the South Scrape hide when a bird flew in and landed on a depth marker. As it flew in, I thought it would prove to be a Yellow Wagtail, an idea reinforced by the call which was similar to that species. I did not look at it immediately but when I did, I was struck by its whiteness. The bird was perched in a hunched position on a post within 15 metres from the hide where it remained for about three minutes before flying off. I took a detailed description of the bird's plumage at rest and noted double white wing-bars, grey rump and white outer tail feathers in flight.

Citrine Wagtail, first-summer male, Fleet Pond, May 16th 1993. John Humble

The next was a confiding first-summer male found by John Dixon at Sandy Bay, Fleet Pond at dawn on May 15th 1993 which remained until dusk the next day and was enjoyed by several hundred birders during its stay. Later that year one was heard by Pete Potts and Guy Thompson during an early morning ringing session at Farlington Marshes on August 30th; it was relocated later that day by Keith Betton who aged it as a juvenile. It was last recorded on the morning of September 3rd.

Citrine Wagtail, juvenile, Farlington Marshes, September 1st 1993

In 1996 a first-summer male was found by an unknown birder at Farlington Marshes on July 6th. It stayed until September 6th and was seen by many observers.

The first of the new century was a first-winter identified by Graham Stephenson at Butts Lagoon, Pennington Marshes at 0840 hrs on September 7th 2003; it remained in the area until 11th and was seen by many observers.

The most recent record involved a very vocal and flighty juvenile watched by Dan Houghton at Hook Lake between 1100 and 1120 hrs on August 8th 2020. Despite much searching, it was not relocated.

Another record of one found by Andy Johnson at Sandy Point on September 1st 2008 was harshly considered not proven by BBRC. It flew in from the west, calling repeatedly, at 0812 hrs and landed on the dried-up marsh but flew up and around before it could be viewed on the ground, and left westwards.

GB status: 366, 1954–2019. Increasing, mean 14 per year in 2010–19. (Nominate race breeds Baltic countries, S Finland, Belarus, Ukraine & S Russia, E across N Siberia to Taimyr Peninsula & S to C Siberia, also C & E Turkey E to Kazakhstan, Mongolia & N China. Black-backed race calcarata breeds C Asia to Tibetan Plateau. Winters throughout Indian subcontinent, S China & SE Asia to peninsular Thailand.)

Richard's Pipit *Anthus richardi*

BBRC 1958–70 and 1976–82; HOSRP 1971–75 and 1983 to date.

A rare passage migrant. 43 records of 44 birds since 1955.

The first for the county were two seen by Canning Suffern, Graham Rees and Johnny Bowers at Titchfield Haven on April 16th 1955, an unusual date for this species to occur in Britain. According to Cohen (1963), who quoted a letter from James Ferguson-Lees, the record was accepted by the editors of *British Birds* but not published. Unfortunately, it has not been possible to trace the description.

Since then, a further 42 have been recorded.

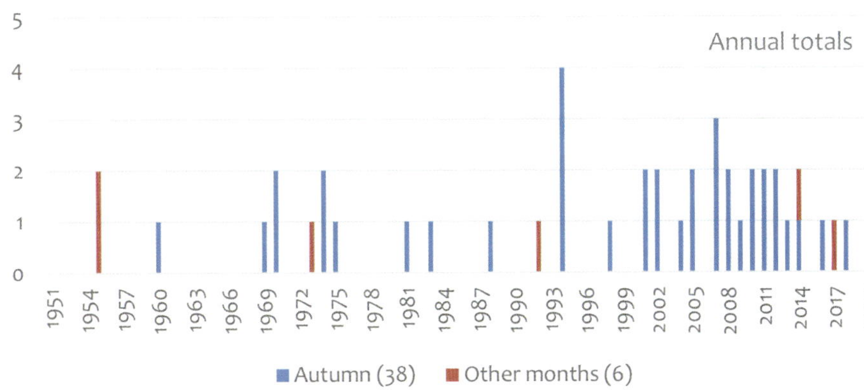

There were two in the 1960s; one seen by Roy Dennis at Needs Ore on October 4th 1960 and the other by George Green at Pennington Marshes on November 1st 1969.

The first of the next decade was one seen well by Barry Duffin and R. E. Jones at Keyhaven Marshes on October 14th 1970; it was seen again by Barry and Dave Wooldridge the next day. A month later on November 15th Barry again located one in the same area which was seen by other observers on 21st and 22nd. David Christie found one at Hook-with-Warsash on July 21st 1973, one of the few British records for that month. In the following year singles were found by Geoffrey Fisher, John Taverner and David Thelwell at Needs Ore on October 6th and Bob Dunn and Ed Wiseman at Keyhaven on November 3rd, and in 1975 another was found by Bob Dunn at Keyhaven on October 1st.

These were followed by an obliging bird found by Keith Wills at Camp Farm Sewage Treatment Works, Aldershot on October 31st 1981 which remained until November 4th and was seen by many local observers. Another long-staying individual was on the reclaimed land at Calshot from October 30th to November 13th 1983. It was discovered by Philip Fawkes and Mike Witherick and enjoyed by many during its two-week stay. The pattern of mostly coastal occurrences was maintained with the next found by Martin Terry and Colin Savage at Gilkicker Point on November 5th 1988.

The first for the 1990s was found by Andy Silcocks by the stream at Brownwich on April 29th 1992 and later seen by others; it was only the second spring record. There was a plethora of sightings in 1994 involving four or more individuals. Barry Duffin saw one fly high to the west over Titchfield Haven at 0840 hrs on October 9th. Later that day, at 1005 hrs, Dave Unsworth saw two fly west over Needs Ore Point, and 20 minutes later he, along with Neil Morris, located one in nearby fields, which later flew off north-west. One was found by Keith M. Grant and Nigel Jones at Needs Ore on October 15th and presumably the same bird was still present the next day when it was located by Alan and Caroline Dawson and Dave Unsworth. The final record of the decade involved a juvenile watched by Glyn Horacek-Davis at Pennington Marshes from 0955 to 1015 hrs on September 19th 1998.

Richard's Pipit, Brownwich Stream, April 29th 1992

The next were in 2001 when Dave Unsworth saw one fly east calling over Needs Ore at 1110 hrs on September 22nd, and Jason Crook and Bob Chapman found one at Farlington Marshes on October 13th. Two occurred in the following year: one was located by Steve Keen at Barton on Sea Golf Course on October 5th and seen by many observers during its stay until October 7th and another was found by Russell Wynn at Pennington Marshes on November 9th. Only the second ever inland was recorded when Lawrie Chappell saw one flying south with Skylarks over Ibsley on October 24th 2004. A year later, Russell Wynn saw and heard one as it flew east over Butts Lagoon, Pennington Marshes at 0830 hrs on September 17th. It was relocated at Oxey Marsh before quickly departing to the north-east. Over two months later, another was located by Simon Woolley in fields adjacent to Lower Pennington Lane at 1015 hrs on November 20th.

Three were recorded in 2007. The first was seen and heard in flight by Jason Crook over Farlington Marshes at 0908 hrs on September 26th. It landed but was not relocated. One seen and heard by Jason as it flew north-east over the site at 1040 hrs on September 30th was assumed to be the same individual. The next was found by Tim Lawman at Mill Rythe, East Hayling on October 2nd. Finally, another was located by Steve Keen as it flew east low over Barton on Sea Golf Course calling frequently at 0913 hrs on November 4th. In 2008 one was seen by Marc Moody three times in flight and briefly grounded with a small flock of Meadow Pipits along Saltgrass Road, Keyhaven between 0829 and 0909 hrs on October 24th and another was recorded by Steve Keen flying west over Barton on Sea on October 28th. The final bird of the decade was found by Steve Keen at Barton on Sea Golf Course at 0820 hrs on September 27th 2009; it was associating with Meadow Pipits and after a few minutes called as it left to the west with them.

Richard's Pipit, Pennington Marshes, March 9th 2014. Roger Murfitt

The run of annual sightings continued in the first half of the 2010s. Jason Crook saw and heard one fly west at very close range over the eastern seawall at Farlington Marshes at 1306 hrs on October 21st 2010. It landed in long grass 150 m away but wasn't seen again until it flew up and left west at 1319 hrs. Steve Piggott found one at Iley Point, Keyhaven on November 11th of that year which had been reported by *Birdguides* the previous day. The next year Mark Edgeller and Tim Parminter independently picked one up on call and saw it flying south with Meadow Pipits at Iley Point at 1055 hrs on September 25th but it was not relocated after it dropped into the saltmarsh. Another was detected by call and then seen well by Andy Johnson as it flew north over Sandy Point at 0825 hrs on October 28th 2011; one which did likewise 13 minutes later was assumed to be the same bird. There were two in 2012 with Steve Piggott recording one in flight over Becton Bunny on October 30th and another found by Andy Rhodes at Calshot on November 17th which was still there the next day.

In the following year Alan Cox watched one at length along the eastern shore of Farlington Marshes on October 19th (not 18th as in the *HBR*). 2014 produced the earliest ever record when Roger Murfitt saw and photographed one at

Richard's Pipit, Oxey Marsh, September 15th 2018. Tommy Saunders and Alan Lewis

Fishtail Lagoon, Pennington Marshes on March 9th. Later that year one was found by Alan Lewis nearby at Keyhaven Lagoon on September 20th which was seen by many observers until 24th. The third inland, although in sight of the sea, was found by Tim Doran at Portsdown Hill on October 9th 2016 where it remained until dusk the next day. Another March bird was found in 2017 when John Pitman watched one for 30 seconds near the yacht clubhouse at Needs Ore on 13th before it flew off. The most recent record was of a flighty bird found by Steve Keen at Pennington Marshes on September 14th 2018 which stayed until the next day and was seen by several birders.

Analysis of arrival dates shows two in March which possibly hints at local wintering, three in April which may indicate passage from further south, one in mid-July and 38 in autumn between September 14th and November 22nd with most in October.

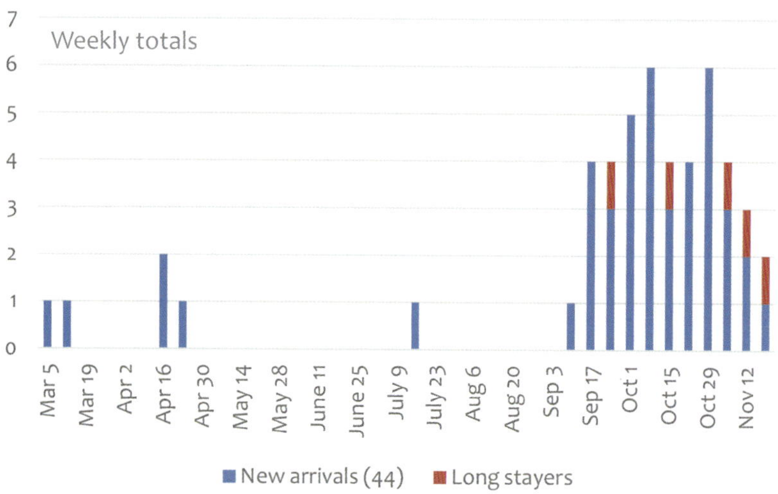

A record of one seen by Bob Dunn at Keyhaven on September 21st 1980 was included in the 1980 *HBR* and *BoH* but evidently not submitted to BBRC; it was also included in Wiseman (2017). One which flew over Titchfield Haven on October 1st 2000 was mentioned by Powell (2002) but not submitted to HOSRP.

Two records in 1959 and 1976 were considered not proven by BBRC. A further 16 records between 1994 and 2018 were adjudged not proven by HOSRP. Most of these involved birds heard and/or seen briefly; there was just one of a grounded bird at Eling Great Marsh on October 28th and 29th 2001 for which the observer could not be traced.

GB status: 108, 1812–1949; 5,142, 1950–2019. No clear trend, mean 105 per year in 2010–19. (Breeds Siberia from Ob River region to Sea of Okhotsk, Mongolia, N & E China. Winters Indian subcontinent, S China & SE Asia, with small numbers locally in NW Africa, Mediterranean basin & Arabian Peninsula.)

Tawny Pipit *Anthus campestris*

BBRC to 1982; HOSRP 1983–2014; BBRC 2015 to date

A rare passage migrant. 18 records since 1963.

The first for the county was an adult found by John Taverner at Needs Ore Point on September 3rd 1963. Extracts from John's description are reproduced below.

> *I found the bird on an area of shingle about 30 yards from the sea. It was thinly covered with low grasses and marine plants. With it were Wheatears, Yellow Wagtails, Meadow Pipits and Linnets. Its size and build were similar to Yellow Wagtail. Plumage. Upperparts: head, neck, back and wing coverts pale sandy brown. At close range it could be seen that the back was lightly streaked darker. The primaries stood out as darker against this sandy brown. Likewise, the tail was darker and the outer tail feathers were dirty white (only visible in flight). There was a noticeable buff superciliary stripe. Underparts: light cream on breast, pale on belly, unstreaked. Soft parts: slender, pale brown bill, legs pale. It moved quickly through the low vegetation and called when flushed as it rose, a loud double 'teep-teep'. I watched it down to ten yards and left it after ten minutes.*

Since then, 17 have been seen. The decline since 1990, with only six found, has been noted elsewhere in Britain and culminated in the species being restored to the BBRC list in 2015.

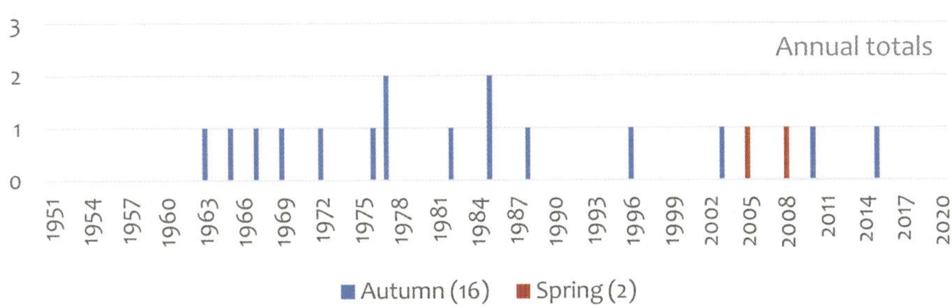

The second was found by Barry Duffin and Martin Terry at Gilkicker Point on September 8th 1965 but not conclusively identified until two days later; it was last seen on September 11th. Gilkicker was the venue for the next found by Martin Terry and Michael Bryant on September 17th 1967. The first inland was a juvenile found by John Lucas at Silchester Common on September 13th 1969 which stayed until 19th and was seen by several observers.

The first of four in the 1970s was found by Derek Lister and Joan Spurgin at Pennington Marshes on September 3rd 1972. This was followed by one, found by George Johnson and later seen by Malcolm Theobald, at Anton Lakes on October 16th 1976. Two were seen in the following year: a juvenile found by Norman Pratt and the author at Dibden Bay on September 4th which was seen by other birders during its stay until 6th and an adult found by George Johnson at Anton Lakes at 0742 hrs on September 11th which was flushed by a Sparrowhawk and flew off east after one minute.

The 1980s also produced four with a juvenile found by John Walters on Sinah Common on October 9th 1982 which was flushed by golfers, another juvenile found by George Clay at Farlington Marshes on August 31st 1985, one heard and seen flying south over Dibden Bay by Mike and Pete Combridge on September 1st of that year and finally one located by Dan Powell at Titchfield Haven on October 2nd 1988.

Tawny Pipit, North Scrape, Titchfield Haven, October 2nd 1988

Gunner Point, Hayling Island, September 5th 2015 (opposite). David Ryves

The next was an adult watched by Dan Houghton and Reg Russell from 1200 to 1230 hrs in fields at Blackwater, Needs Ore on September 17th 1996.

In 2003 Steve Keen watched one at close range on the cliff top just west of Taddiford Gap from 0700 to 0702 hrs on September 27th. It flew over the cliff edge but could not be relocated. Two years later Andy Johnson heard one over Sandy Point at 0625 hrs on May 10th 2005. It landed on the beach where good views were obtained but was then lost to view as it moved off north. The second spring bird was found by Marc Moody in fields adjacent to New Lane, Keyhaven on June 1st 2008 and subsequently seen by many observers.

In 2010 Pete Durnell saw one briefly at Keyhaven Marshes on August 26th but it flew off and could not be relocated. The most recent record was of a first-winter found by Pete Gammage and Keith Turner at Gunner Point, Hayling Island on September 5th 2015. It was discovered while they were looking at the exceptional show of Autumn Lady's-tresses *Spiranthes spiralis*. Despite being flushed by dog walkers on several occasions, it remained long enough for other birders to connect with it.

Analysis of arrival dates shows the two spring records and autumn passage between August 26th and October 16th with a clear peak in the first half of September.

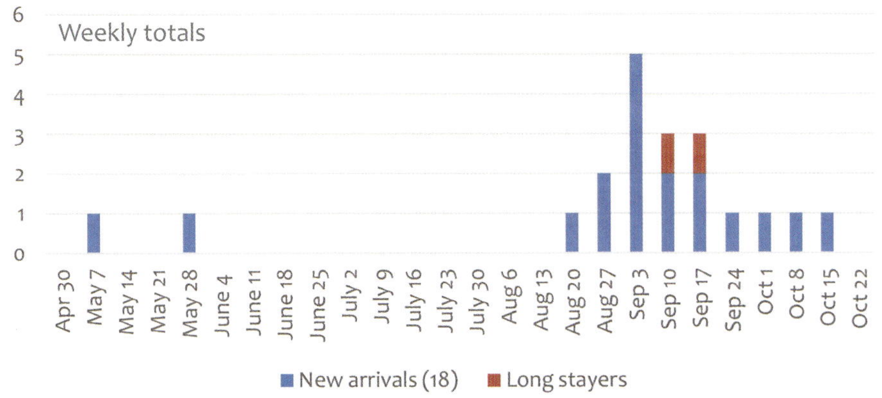

Another record was of one found by John Cloyne at Winchester Sewage Farm on August 25th 1996 which was harshly considered not proven by HOSRP. John remains certain of his identification.

GB status: 22, 1858–1949; 1,283, 1950–2020. About ten per year, recent decrease. (Breeds NW Africa, S & C Europe N to Netherlands & Estonia, E to NW Mongolia, NW China, N Iran & N Afghanistan. Winters S edge of Sahara from Senegal to Sudan, Arabian Peninsula & NW Indian subcontinent.)

Olive-backed Pipit *Anthus hodgsoni*

BBRC to 2012; HOSRP 2013 to date

A very rare vagrant. Five records since 1987.

The first for the county was found by Stuart Sporne on his smallholding at Sway on November 13th 1987, when he flushed a pipit from a young tree belt adjoining his bungalow. It flew into a hedge of sallow and oak and in silhouette the tail was seen to pump as in a wagtail. He assumed it was another Tree Pipit, though very late. On 14th and 15th it was flushed again, flying across a paddock into a line of oaks. No call notes were heard. At 3 pm the wind dropped, two mist nets were erected and the bird was flushed into a net. On taking the bird from the bird bag Stuart's wife, who is not an ornithologist, remarked that it had a green back. It certainly wasn't a Tree Pipit. He took full measurements of the bird and identified it as an Olive-backed Pipit based on the description in the current Svensson which seemed to match the bird perfectly. On 18th the bird was retrapped and Stuart phoned Ed Wiseman to invite him over to see the bird. Ed brought photos of two birds which he had seen trapped on Fair Isle and was able to confirm the identification. It was last seen on November 26th.

The next was found by David Christie at Hook-with-Warsash at 1240 hrs on November 7th 1994 when he flushed one from the shore which briefly paused in long grass before flying into a lakeside tree at 130 metres range. He was able to watch it for a total of 20 minutes. This was followed by one found by Tony Blunden at Hill Coppice, Fareham on October 27th 1999 which he watched at close range. Unfortunately, the news could not be released as it was a private site under development. The first twitchable bird for the county was found by Andy Johnson at Sandy Point at 0955 hrs on October 24th 2009. It showed intermittently in the area for just over three hours in poor weather allowing sufficient time for some birders to enjoy the bird with its finder. The next was also found by Andy Johnson at Sandy Point at 0900 hrs on November 12th 2015. Initially it was very flighty but settled down in the afternoon when several birders saw it well. It was not present the next day.

GB status: 1, 1948; 635, 1964–2019. Increase after decline, mean 30 per year in 2010–19. (Breeds N Urals E across C & E Siberia to N China, Kamchatka, Kuril Islands & Japan.

Nominate race breeds Himalayas & mountains of WC China. Species winters widely throughout Indian subcontinent, S China, Taiwan & N & C parts of SE Asia.)

Red-throated Pipit *Anthus cervinus*

BBRC to 2005; HOSRP 2006–14; BBRC 2015 to date

A very rare vagrant. Three records in 1989, 1992 and 2011.

The first for the county was found by Barry Duffin at Titchfield Haven on October 29th 1989. Extracts from his description are reproduced below.

> *At 1420 hrs I was watching from the South Scrape hide when I heard a call that I immediately recognised as a Red-throated Pipit. Some three to four minutes elapsed before a small pipit with dark upperparts was found feeding amongst vegetation on a shingle island about 40–50 metres away. After viewing the bird through a telescope for a short while, it flew across uttering the same call I had heard earlier and landed in front of the hide at a range of 15–20 metres. I was able to watch it for the next 15 minutes. The distinctive features were its dark streaked upperparts which extended down the back to the rump, the very bold dark streaking on pale underparts, the pale conspicuous 'braces' on the sides of the mantle and not least of all the distinct call, which I described as a noticeable drawn-out, high-pitched piercing 'zeeez' with a trailing off in pitch towards the end of the call. It was far carrying, being distinctly audible for some time as the bird left the area and flew off high to the west when flushed by a Sparrowhawk.*

The next was found by Dave Unsworth at Needs Ore on September 27th 1992. It called several times as it flew low overhead at the point, perching briefly on a barbed wire fence before continuing to the east. The record was submitted to BBRC in 2020 and subsequently accepted although published with the incorrect date of October 15th 1992.

The third record was of one heard calling and seen flying south over Titchfield Haven by Barry Duffin and Trevor Codlin on October 15th 2011.

An earlier record of one seen and heard calling several times by Dave Wooldridge as it flew over Hurst Castle on October 5th 1980 was not submitted to BBRC as he had not seen it grounded. However, Dave, who is experienced with the species, recently confirmed that he was 100% sure about the identification. Another record of one heard by Andy Johnson as it flew over Sandy Point on October 13th 2005 was harshly considered not proven by BBRC.

GB status: 11, 1908–1938; 585, 1950–2020. About ten per year; stable or decreasing. (Breeds N Norway, N Sweden & N Finland then E across N Russia to Chukotskiy Peninsula & S to Kamchatka, occasionally W Alaska. Winters across N & C equatorial Africa, N Indian subcontinent, S China & SE Asia.)

Common Rosefinch *Carpodacus erythrinus*

BBRC to 1982; HOSRP 1983 to date

A very rare vagrant. Two records in 1913 and 2009.

The first for the county was a male seen by Katherine Spear Smith at Titchfield on August 13th 1913 (*Brit. Birds* 7: 179). The finder's account of her sighting is reproduced below.

> *On August 13th, at this place (which is about two miles inland from the Solent), I was startled at seeing a very brilliant scarlet bird perching on a stem where there was a tangled mass of wildflowers and a hedge. I was able to get close to it with field glasses. It was certainly of the finch family, about the size of a Bullfinch, but of a more slender shape; it was scarlet in varying intensity on head, breast and rump; wings and tail dark brown to dark brownish grey; the wing-coverts tipped with buff, making a dull bar across the wing. I thought there was some black through the eye, but it may have been the shadow. It uttered a sort of song, several quick notes in succession. On coming home and looking it up in Saunder's Manual, I could find no bird to answer to the description except the male Scarlet Grosbeak which exactly fitted it, including the song.*

The record was recently reviewed by a sub-committee of *HOSRP* and, despite the reservations of some members, retained on the county list (Combridge & Clark 2010).

The only modern record was of a singing male heard by Simon S. King in an area of gorse and silver birch scrub at Godshill Ridge, New Forest at 1105 hrs on June 1st 2009. He located the bird but it flew off at 1115 hrs and was not relocated. Extracts from Simon's description submitted to HOSRP are reproduced below.

> *At 1105 hrs I heard a male Common Rosefinch singing from an area of mixed silver birch and gorse scrub; I could not at first locate it until it appeared (briefly) on top of gorse, facing away from me and turning its head as it sang. At 1115 hrs it flew away to the west and was lost to sight.* Description: *in size and structure very roughly similar to Greenfinch. Because it was turned away the underparts were not visible, but the upperparts ('back' and tail) were brownish. The head was noticeably reddish, as was the throat (seen when the bird turned its head) and rump (noted when the bird flew off). Bill rather large and stubby; tail seemed longish. In the past all the singing male rosefinches I have observed sang from exposed perches, but this bird was hidden in scrub except for its brief appearance on top of the gorse.* Voice: *song monotonous, to my ear a repetitive three note whistle with the final note down slurred; the song was repeated for several spells in succession.* Prior experience: *in eastern Poland, May 2004.*

GB status: 76, 1906–1949; 5,694, 1950–2019. No clear trend, mean 159 per year in 2010–19. (Breeds N & E Europe, and locally in C Europe and Turkey, E across Russia to NE

Siberia & S to Caucasus & Himalayas. Winters from Iran and Indian subcontinent to S China & Indochina.)*

Twite *Linaria flavirostris*

HOSRP

A declining, now rare, winter visitor and passage migrant, first recorded in 1952.

The first for the county were eight or nine at Farlington Marshes on December 22nd 1952. These were not included in the 1952 *HBR* although it did include a description by Graham Rees of a female there from 23rd to 25th.

No more were seen until November 1956 but in the period from then until 1989/90 there were records in every winter except one with totals in the range 50–331.

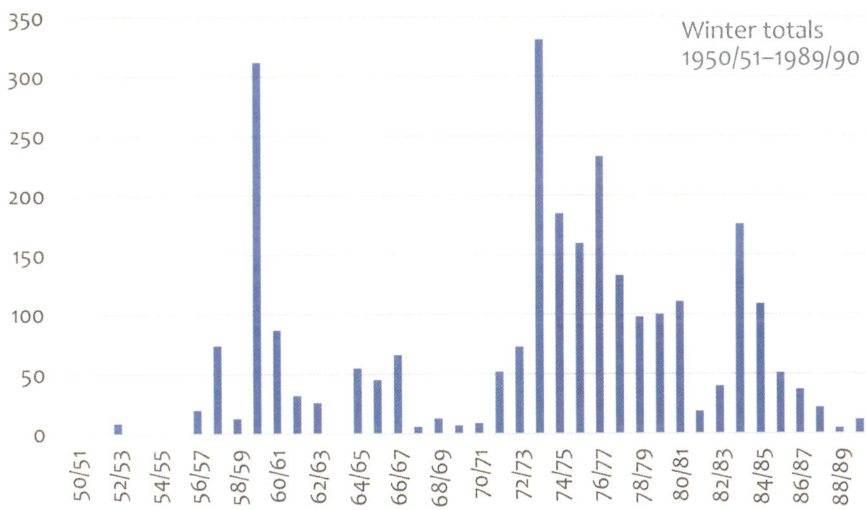

In the late 1950s, Farlington Marshes was the main locality, although numbers there usually showed an autumn peak and were much reduced after the turn of the year. In 1959/60, when a minimum of 312 was recorded, flocks were discovered at Dibden Bay, Needs Ore and Pennington Marshes. After the mid-1960s, Twites became scarce and irregular at Farlington Marshes but during the 1970s sites further west, particularly Dibden Bay and Eling Great Marsh, held substantial flocks. In the late 1980s, numbers were much reduced despite the availability of suitable habitat. Counts at the principal localities during this period are summarised in the table shown overleaf.

Other double-figure counts were made at Needs Ore (23, January 27th 1960), Hook-with-Warsash (22, January 5th 1972; 29, December 22nd 1973; 30, October 15th 1980), Paulsgrove Reclamation (70, February 23rd 1974) and Tournerbury (15, December 12th 1981).

Locality	Range of winters with counts of 10+	No. of winters with counts of 10+	Average max. counts in those winters	Peak counts
Farlington Marshes	1956/57–65/66	9	45	150, October 24th 1959
Eling Great Marsh/Southampton Docks	1976/77–89/90	11	39	84, December 6th 1979
Dibden Bay	1959/60–82/83	14	62	130, December 4th 1976
Fawley/Calshot	1972/73–74/75	3	30	46, February 24th 1974
Normandy/ Pennington Marshes	1959/60–83/84	8	32	60, January 9th 1974

Since then, Twite have become much scarcer with a total of 72 in 16 winters up to 2011/12 but only three since then.

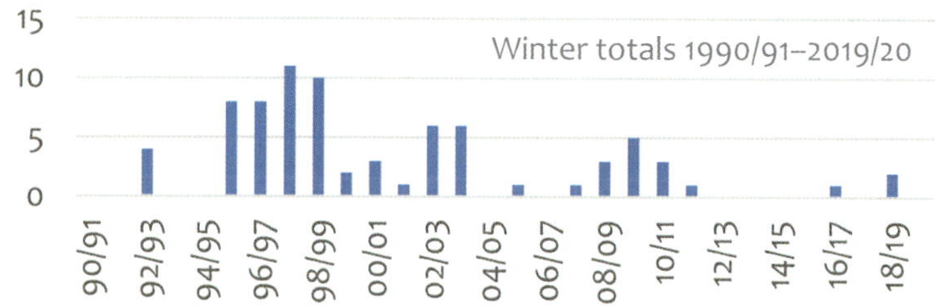

Wintering was recorded at Normandy Marshes in consecutive winters from 1996/97 to 2000/01 with maxima of 7, 6, 10, 2 and 1. At Langstone Harbour there were five in 1997/98, three in 2002/03, four in 2003/04 and one in 2005/06.

Twites normally arrive between mid-October and early November. Earlier records include those of singles at Farlington Marshes on October 3rd 1971, Titchfield Haven on September 30th 1973 and Normandy Marshes on October 3rd 1997. There is evidence of passage through the county in autumn with flocks building up and then declining (as at Farlington Marshes) or staying for one or two days (as at Hook-with-Warsash in 1980). A few have been detected in diurnal passerine movements, e.g. four flying west at Gilkicker Point on November 2nd 1962, six moving west at Hook-with-Warsash on October 20th 1983 and more recently singles west over Browndown on October 21st 2010 and Barton on Sea on November 7th 2010. Peak counts of wintering flocks are usually recorded in December or January with numbers falling from February onwards and the last normally leaving by late March or early April. Later records include those of two at Normandy Marshes on April 9th 1974 and seven at Dibden Bay on April 13th 1975.

Twite: Farlington Marshes, March 16th 2009 and Normandy Lagoon, November 30th 2018. Pete Drury and Alan Lewis

There have been ten inland records, mostly of one to three but including flocks of six at Woolmer Forest on November 24th 1962, flying south-west over Ashley Walk on November 15th 1975 and flying north up the Test Valley at Romsey on February 18th 1982, and four at Athelstan Road allotments (Southampton) on January 18th 1987. The only recent record was of one at Tweseldown on November 13th 2018.

GB status: Declining breeder on uplands and coasts of W and N Scotland, Orkney and Shetland Isles. (Breeds mountainous regions of Europe and Asia.)

Common Redpoll *Acanthis flammea*

HOSRP 2002

A very scarce winter visitor. Around 194 recorded since 1960.

The first record for the county was of two trapped with 59 Lesser Redpolls by John Ash at Sopley on March 12th 1960. Amazingly, one of these, an adult female, was found alive but injured 1,380 km to the north-east in Sweden on September 7th 1962. The next was one found by Dave Wooldridge with a flock of Lesser Redpolls at Bishop's Dyke, New Forest on December 17th 1961.

Awareness of this taxon increased through the 1980s and 1990s. The species Redpoll comprised several races including the nominate, colloquially known as Mealy Redpoll, which is considered here, and Lesser Redpoll *cabaret* which was the common form in southern England. This culminated in the splitting of Lesser Redpoll as a separate species *Acanthis cabaret* (Knox 2001).

The first record of Common Redpoll during this period was of one found by Graham Stephenson among a flock of at least 200 Lesser Redpolls at Fleet Pond on December 28th 1986. Subsequently, there have been sightings in 22 winters up to 2020/21.

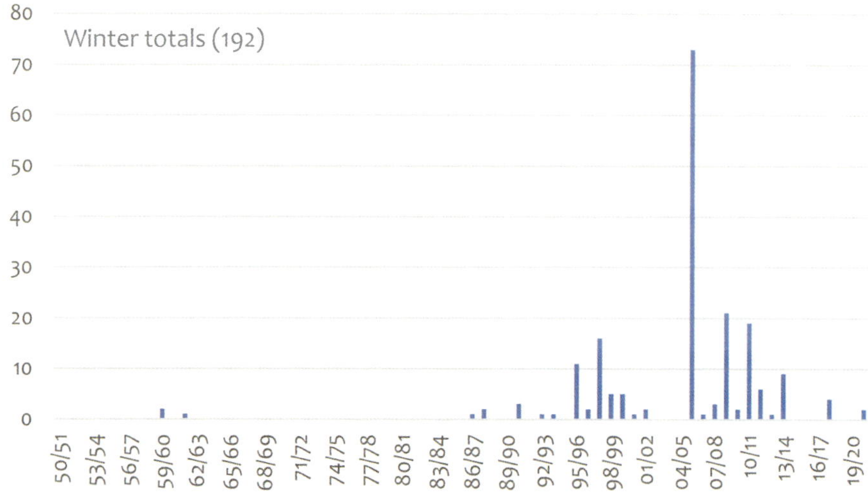

Most records have been of small numbers, usually at feeding stations or detected among flocks of Lesser Redpolls or other species such as Siskins and Reed Buntings. Some large flocks of redpolls have been suspected of being pure Common Redpolls.

Favoured locations have been Fleet Pond, with 30 between 1986/87 and 2005/06 but only five since, Blashford Lakes Centre, with at least 16 between 2007/08 and 2013/14, and The Straight Mile, Ampfield, where Nigel Jones has trapped 16 in his garden including ten in 2008/09 and five in 2013/14. The first sizeable pure flock was of 14 at Bricksbury Hill on February 8th 1998.

Common Redpoll, Blashford Lakes, January 16th 2014 and Greenland Redpoll, Blashford Lakes, April 1st 2010. Martin Bennett and Simon Ingram

There was a large invasion of redpolls into Britain in the 2005/06 winter and several flocks found in the county were thought to include Common Redpolls. A flock of 150 Lessers found at Roke Manor, Romsey on December 29th included at least four Commons. Another group of 29 at Malthouse Farm, Sleaford on January 30th were all assessed as this species although only smaller numbers were claimed subsequently. Another flock of 50 found at Hartley Wintney included at least four Commons but perhaps they all were. Finally, a well-watched mixed finch and bunting flock in game crops at Exbury held at least 120 redpolls of which up to 20 were identified as Commons but clearly more were present. An Arctic Redpoll was also tentatively claimed but identification was not clinched. In the 2008–09 winter, up to seven were present at Blashford Lakes between March 19th and April 13th of which one showed the characteristics of the Greenland race *A. f. rostrata*. Of eight trapped at The Straight Mile between February 27th and April 19th, two showed the characteristics of this race.

Common and Lesser Redpolls, Blashford Lakes, March 8th 2014

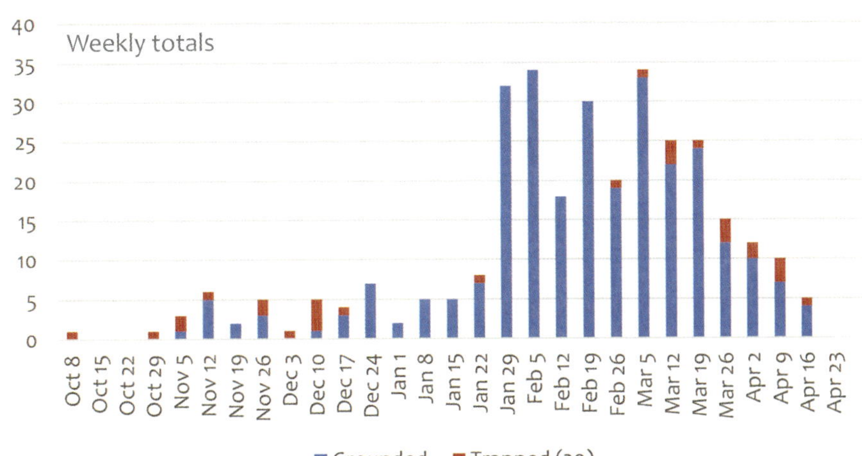

Analysis of arrivals dates shows that few are recorded before late January. The earliest in autumn were a first-winter trapped at Silver Wood near Romsey on October 13th 2012 and an adult male trapped at Castle Bottom on October 29th 2001. Totals for November and December were 16 and 18 respectively. There is a gradual departure from mid-March with the latest recorded being two in Yateley Heath Wood on April 21st 1991 and a male and a female at a feeder in Frensham Lane, Headley between April 12th and 21st 2006. The chart on the previous page shows the weekly distribution of records but no attempt has been made to distinguish new arrivals from long-staying birds.

Two males in the Hart collection, now held at the Horniman Museum, were taken in 1864 and 1911. According to K & M, the 1864 specimen was taken in the New Forest, while two specimens in the Alton Museum were procured in that district by Mr. Pullinger but unfortunately no data were attached to them. These records are too vague to qualify as the first for the county.

GB status: winter visitor in variable numbers; c. 15 pairs breed in N Scotland. (Breeds through N Europe, Asia and N America; winters to S.)

Parrot Crossbil *Loxia pytyopsittacus*

BBRC to 2008; HOSRP 2009 to date

A very rare vagrant. One record in 1991.

The only record for Hampshire is of a male and one or two females found by Kevin Rylands at Bricksbury Hill on March 29th 1991. The authors of *BoH* (1993) were not aware of this record as it was submitted directly to BBRC and was not accepted until after publication. Extracts from Kevin's description are reproduced below.

> *I found a flock of c.40 Common Crossbills and after a few minutes located a female Parrot Crossbill in the top of a larch. It was immediately obvious that this bird was not a 'normal' crossbill. After settling myself I began to look closer at the bird which was soon joined by a male. Nearby a presumed second female was seen although not simultaneously with the pair. The bills totally changed the jizz of the Parrots, giving the head a totally different shape to that of Crossbill. It was very deep, covering the face, giving the bird a flat crown with no forehead. The bill also appeared shorter than the Crossbills. Other differences were that it was less noticeably hooked and the tips were only slightly crossed. The lower mandible had a bulge and not the smooth contours of Crossbill. No calls were distinguished.*

There are several other records of interest. Richard and Spike Millington, Noel Elms and Graham Stephenson saw a male at Fleet Pond on August 28th 1972. It was solitary although over 30 Common Crossbills were around the pond that day. They got very good views of the bird including the head and bill shape and structure. In 2020 Richard told the author that a description of the bird was sent to BBRC but diverted to Roy Dennis. He wrote back saying it sounded fine but Scottish Crossbill could not be eliminated and so the submission was not pursued. Unfortunately, the letter has now been lost. Richard recently commented, 'I remember the bird like it was yesterday. Having seen and found many Parrot Crossbills in subsequent years, I am unwavering in my belief that the Fleet Pond bird was indeed this species ... or the only Scottish Crossbill ever to have left Scotland!'

A putative singing male was at Pitts Wood Inclosure, New Forest from March 5th to 15th 2000 and photographs were published in *Birding World* (13: 94). Dan Powell recently informed the author that he saw the bird but unfortunately the record was not submitted to BBRC. Richard Millington, co-editor of *Birding World* at the time, commented, 'I liked the look of this male ... the almost 'hawfinchy' look, overall chunkiness, the ruffed nape, crown shape and especially bill shape (dumpy, strongly arched upper and lower mandible, almost taller than long, lack of upcross) all – from what can be ascertained – support ID as Parrot'.

Notes made on the 'chunky chap', Pitts Wood Inclosure, March 15th 2000
Photo Paul Gale

Several possible Parrot Crossbills were seen during the 2013/14 winter when a small irruption occurred in Britain. A sub-adult male was located by Mark Palmer and Steve Keen at Slufters Inclosure, New Forest on November 19th 2013, an adult male was seen there by Andy Lester on November 22nd 2013 and a pair was found by Marcus Ward at Bolderwood on January 19th 2014. HOSRP sought advice on these records from BBRC and Richard Millington.

Unfortunately, it was felt that they could not be considered 100% certain due to the brief duration of views, distance from some of the birds and lack of good photographs.

K & M cited two records of Parrot Crossbill, which in the 19th century was considered only as a large stout-billed form of Common Crossbill *L. curvirostra*. Frederick Bond possessed three examples one of which was shot out of a flock of 11 near Lymington in March 1842. K & M quoted this record from Yarrell (1871-85) but in an extensive obituary of Bond published in the *Zoologist* for November 1889 it was given as March 1843. During a large invasion in 1855, Dr. Rake informed Wise that one was killed out of a flock of a dozen at Breamore on November 28th and a few days later several more were killed. In the absence of extant specimens these cannot be considered as genuine records.

GB status: 17, 1818–1928; 677, 1953–2019 including periodic irruptions: 210, 1990/91; 89, 2013/14; 93, 2017/18. Breeds in small numbers in Scottish Highlands. (Circumpolar breeding range at tree line and tundra from N Fennoscandia E to Chukotskiy Peninsula in NE Siberia, N Alaska, N Canada & N Greenland. Disperses erratically to S of breeding range in winter, regularly reaching NW Europe.)

Two-barred Crossbil *Loxia leucoptera*

BBRC

A very rare vagrant. Two records in 1984 and 2013.

The first for the county was a male with Common Crossbills found by Graham and Debbie Barrett at Slufters Inclosure on March 14th 1984. It was subsequently seen by many observers and remained in the area until April 28th. This was a controversial record with some observers convinced it was an aberrant Common Crossbill. Nonetheless, several detailed descriptions supporting its identification as Two-barred were submitted to BBRC and it was eventually accepted on recirculation in 1988.

However, another record of a similarly-plumaged male at Blackwater Arboretum from November 2nd to 14th 1990 was not accepted.

The next was in 2013 when Marcus Ward located one by its distinctive trumpeting call among Common Crossbills as they left a roost at Furzy Lawn Inclosure, New Forest on November 30th. News was released locally and the next morning Steve Keen, Steve Piggott and Mark Halligan got reasonable views of a first-winter female as it left the roost. It was not seen again.

GB status: 70, 1845–1948; 546, 1953–2020. Occasional large influxes: 2008 (59), 2013 (101), 2019 (219). (Eurasian race bifasciata *breeds in larch forests of N Eurasia from N Scandinavia to E Siberia, reaching Sea of Okhotsk & S to Baikal region. Outside breeding season occasionally disperses as far as NW Europe. Nominate* leucoptera *breeds across N North America.)*

Serin *Serinus serinus*

BBRC to 1982; HOSRP 1983 to date

A rare visitor. 45 records since 1961.

The first for the county was one seen by Tim Sharrock at Farlington Marshes on November 4th 1961. Extracts from his description are given below.

> *Small yellow finch flying over from the south at about 40 feet at 1400 hrs, circled and then settled in bushes along seawall and watched for 4 or 5 minutes at 20–30 yards in bright sunlight before it flew off to the north. In flight appeared small and yellow, with short, plump shape and short, forked tail. Flight was undulating. Flight call a melodious twittering. When perched, following notes made: Overall yellow. Two very prominent broad yellow wing-bars. Head and breast bright yellow. Prominent yellow superciliary, extending over forehead. Bill very small and stout – stubby, horn coloured. About same size as Blue Tit in same field of view. Very conspicuous bright yellow rump seen as it flew off.*

Since then, a further 44 have been recorded.

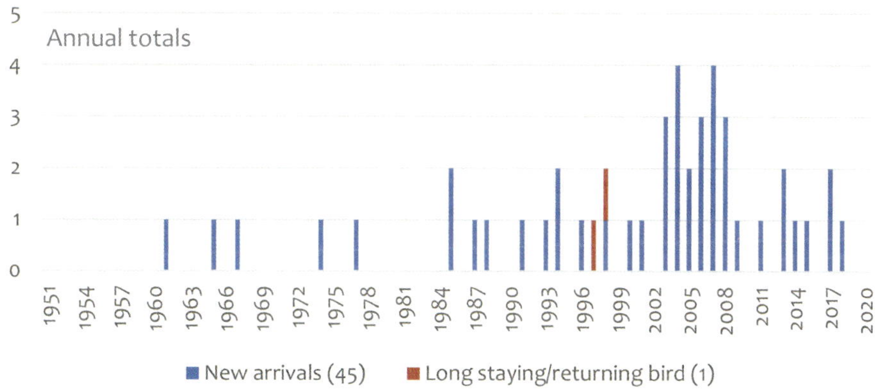

The next record was of a male and two probable females seen by John M. Jones at Culverley Farm, Beaulieu following snowy weather on March 7th 1965. This was followed by one seen by Keith R. Grant and G. Eyers at Farlington Marshes on October 30th 1967.

There were three records in the 1970s but the first of three flying south over Fleet Pond on September 17th 1973 has now been withdrawn by the observer. The other records involved a male seen by Derek Lister at Old Winchester Hill on September 16th 1974 and a singing male located by David Glue at Beaulieu Abbey on March 18th 1977 which was still there the next day.

The next were not until 1985, when Ken Martin found a singing male by the River Itchen at Twyford on May 20th and John Lucas had one calling as it flew south over Tadley on July 9th. In 1987 Chris Wear had a male flying south-west over Farlington Marshes on May 9th and in the following year Andy Parfitt located a male at Needs Ore on April 6th.

The first of the 1990s was a singing male found by Simon Aspinall at Keyhaven on May 2nd 1991; it left east. The next was the first long-staying bird; a male found by Les Oakley and confirmed by Richard Levett which held territory around a large cedar tree in private parkland on the Bossington Estate in the Test Valley from May 28th to June 30th 1993. Two were recorded in 1994. A male was observed at close range by Dave Morrison at Browndown South on May 2nd and one was seen and heard calling by Mark Rowbottom as it flew north-west over The Ridge, Lyndhurst on June 14th. The first ever wintering bird was found by Greg and Lynne Rogers on December 25th 1996 where it was a regular visitor to their bird table in Harestock, Winchester until February 24th. What was presumably the same bird returned in the following winter and was present from February 7th to March 1st 1998. Also in that year, Mick Scott saw a male perch briefly at Fleet Pond at 1030 hrs on April 20th.

Occurrences become more frequent in the first decade of the new century with records in every year but four. Mike Rafter saw one at Romsey on April 26th 2000. The next was one heard and seen flying east over Keyhaven Marshes by Russell Wynn and Marc Moody on October 20th 2001. Simon Wright and Tim Lawman had one at Sinah Warren paddocks from 0830 to 1120 hrs on April 21st 2003 and Jason Crook relocated it at 1400 hrs on April 23rd. One was seen and heard in flight by Russell Wynn over Keyhaven Marshes at 0718 hrs on July 11th of that year and another was recorded by Steve Keen flying west over Beckton Bunny on October 4th. The next was one heard singing by Graham Barrett and watched for two minutes at West End on February 12th 2004. Later that year three were recorded in April. Steve Keen saw a female fly east at Barton on Sea Golf Course on April 25th, Jason Crook flushed one at very close range at Broadmarsh, Langstone Harbour, at 0845 hrs on April 27th which left north-east and Kevin Crisp had one fly north-west over North End, Portsmouth at 1110 hrs on the same day. In 2005 two or possibly three occurred, all found by Andy Johnson at Sandy Point. One called and circled as if going to land but flew off west on April 25th. The next day one was heard calling as it flew east over the beach which is assumed to be the same individual as there was no significant arrival of migrants that day. Finally, one of indeterminate sex was watched feeding in the reserve for over 30 minutes on July 7th.

Three were seen in May 2006. Trevor Carpenter saw one fly west over Stokes Bay at 0907 hrs on 3rd, Tim Walker recorded one flying west over Lee-on-the-Solent promenade at 1400 hrs on 5th and Marc Moody had one fly in off the sea at Hurst Beach on 10th. Spring records monopolised again in 2007. David Thelwell and Marc Moody had one fly north calling over Hurst Beach on April 27th, Marcus and Zoe Ward had another calling over Normandy Marshes at 1950 hrs on May 5th, Andy Johnson watched one in the Sandy Point lifeboat station car park and nearby gardens from 0630 to 0710 hrs on May 16th and

Pete and Mike Oram found a female sitting on an overhead wire at Needs Ore on June 19th. In 2008 Tim Lawman saw a female fly in off the sea at Hayling Bay at 0925 hrs on May 5th. It called and landed to drink by a car park puddle before being disturbed by traffic and flying off inland. In autumn Steve Keen had one over Barton on Sea Golf Course at 0758 hrs on October 25th and Andy Johnson picked up one as it circled over Sandy Point at 0734 hrs on November 9th before departing north-east. The only record in 2009 was of a male found by Simon S. King calling at close range at Castle Hill, Burley Street on May 11th which flew off east at 1152 hrs.

Serin: females, Lee-on-the-Solent, May 17th 2013 and Sandy Point, April 8th 2017. Lynn Kelly and Alan Lewis

Sightings became less frequent in the early years of the 2010s. Simon S. King and Ian Pibworth had one fly low north-west over Testwood Lakes at 1655 hrs on April 24th 2011. The next were in 2013 when Andy Johnson had a male flying around Sandy Point before leaving north-west on April 10th and Lynn Kelly photographed a female on a feeder in her garden on May 17th. A year later Andy Johnson had one flying over Sandy Point on November 15th 2014. Andy also found the next there when a female flew over on May 7th 2015. The domination of Sandy Point continued in 2017 when Andy Johnson found a female on April 5th which showed again to several observers on 7th and 8th. Andy found another female there at 0624 hrs on June 1st which lingered until 0805 hrs before flying off east with three Linnets. The most recent report was of one which Nigel Crook saw fly over Sir George Staunton Country Park, Havant on May 9th 2018.

Analysis of arrival dates shows that there have been records in every month except August. There is a clear peak in spring between April 6th and May 20th with 26 recorded during this period. New arrivals have occurred in all other months except January. Weekly totals are shown on the next page.

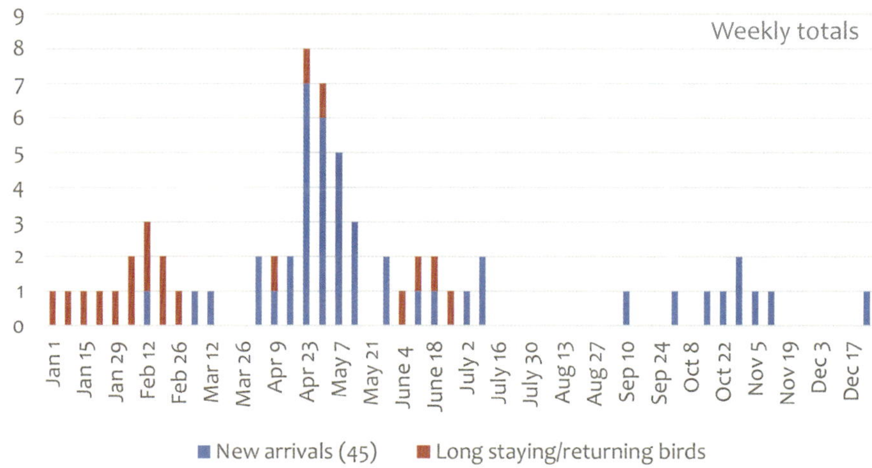

A further 16 claims, mostly of brief flyover encounters, were not accepted by HOSRP.

K & M included the record of one obtained by the Rev. William Hazel near Eastney Fort in April 1852 and reported by him in *The Naturalist* for July 6th 1852. At the time it was believed to be the first for Britain. Alfred Newton in Yarrell (1871-85) says: 'The first recorded example is that by Mr. W. Hazel who said that a male was caught, in April 1852, near Eastney Fort, about a mile from Portsmouth, and having been paired with a hen Canary-bird produced a brood of mules, which unfortunately deceased.' It was not admitted by the BOU in their first list of British Birds (1883: 48) and clearly cannot be considered the first for the Hampshire.

GB status: 27, 1859–1949; 2,289, 1950–2019. Declining, mean 42 per year in 2010–19. (Breeds NW Africa, much of Mediterranean and C Europe N to c. 60°N in Estonia, Turkey, Cyprus and Israel. Northern breeders winter in Mediterranean basin.)

Lapland Bunting *Calcarius lapponicus*

HOSRP

A very scarce passage migrant and winter visitor. Around 213 recorded since 1953.

The first for the county was a male seen by Graham Rees and Colin Tubbs at Black Point on December 23rd 1953. It flew across to East Head on the Sussex side of Chichester Harbour entrance where a second was present; both remained there until January 31st 1954. They noted chestnut on the wing-coverts and nape, yellow bill with darker tip and calls a rather rattling 'ticky-ticky' and 'jip-jip'.

The next was one heard and seen flying over Damerham by John Ash on September 18th 1956 which was followed by the first at Farlington Marshes found by David Billett on October 8th of that year. Winter totals did not exceed seven apart from 2010/11 when at least 101 were recorded including 56 at Keyhaven.

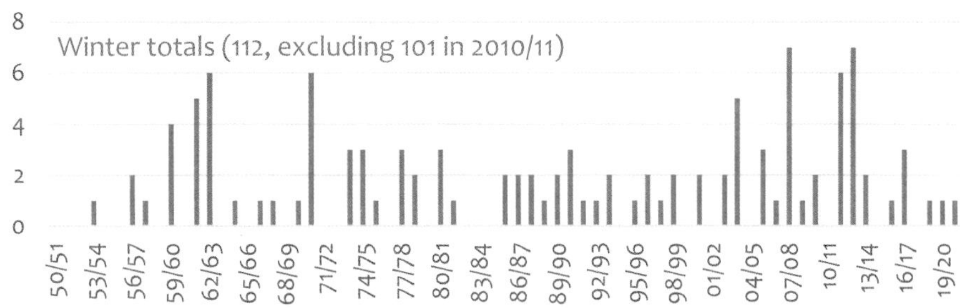

The influx into Britain in 2010/11 started with arrivals in the Western Isles and Cornwall in late August. The first in Hampshire was one which flew into Farlington Marshes from the east on September 9th. This was followed by two singles which flew west over Sandy Point and Brownwich on September 12th. Around 30 were logged in the period up to mid-November; most records were of singles flying over but up to four were grounded at Farlington Marshes between September 17th and 25th, a flock of three circled high over Farlington Marshes between 1630 and 1640 hrs on October 14th before leaving NE, seven were grounded at Brownwich between November 1st and 7th and two flew over Michelmersh in the Test Valley on November 7th. Subsequently, a large gathering was discovered in maize fields at Saltgrass Lane, Keyhaven where Steve Piggott counted at least 15 on November 13th and Bob Chapman saw 13 flying in from the saltmarsh as the tide rose on 21st. Counts of pre-roost gatherings by Marc Moody produced 34 on November 24th, at least 56 on 29th, 44 on December 24th and 21 on January 1st. The final double-figure count was of 14 on January 11th; six were still present on 31st but only singles were recorded in February. Elsewhere, 13 were seen at various sites up to mid-December including five west over Barton on Sea on November 24th. The only evidence of return passage was provided by a male at Ranvilles Lane, Fareham on March 20th, one at Farlington Marshes the next day and one flying north over Iley Point, Keyhaven on 24th.

In other winters most occurred on the coast with the majority in the Lymington/Hurst area (43) and at Farlington Marshes/Langstone Harbour (15). Others were seen at Needs Ore (8, including 3 on January 27th 1960), Hook-with-Warsash (8), Dibden Bay (5, including 3 on September 16th 1970), Barton on Sea, Gilkicker Point, Sandy Point and Titchfield Haven (3 each), Taddiford Gap (2) and Fawley Power Station and Redbridge (1 each).

Lapland Buntings: Saltgrass Lane, Keyhaven, January 3rd and 9th 2011 (left), and Brownwich, October 29th 2005 (above).
Alan Hayden, Trevor Codlin and Richard Ford

Most were flying over or recorded on one or two dates only but longer stays included one at Farlington Marshes from October 30th 1959 to January 2nd 1960, one or two at Pennington Marshes from January 6th to 27th 1963, one at Pennington Marshes from December 13th 1975 to March 14th 1976, two at Hook-with-Warsash from February 9th to 14th 1991, one at Brownwich from October 29th to November 2nd 2005 and up to three at Saltgrass Lane, Keyhaven between November 17th 2007 and March 15th 2008.

Seventeen have been recorded at inland locations on dates between September 18th and December 13th including six in 2010. Records came from Beacon Hill (Highclere), Bishop's Dyke, Butser Hill, Damerham, Godshill (2), Hazeley Heath, Michelmersh (2), Miles Hill, Old Winchester Hill (2), Picket Post, Ranvilles Lane, South Warnborough, Trigpoint Hill and West End (Southampton). All were singles flying over apart from one grounded at Hazeley Heath on December 13th 1980, one at Ranvilles Lane from October 31st to November 7th 2010 and the two over Michelmersh mentioned above. In addition to the bird at Ranvilles Lane in March 2011, two males have been recorded grounded on spring passage at Cheesefoot Head on March 25th 2004 and Over Wallop on April 23rd 2010.

The weekly totals show a marked autumn passage with a few new arrivals through the winter and a hint of return passage in spring.

Brownwich, November 1st 2005.
Rosemary Powell

The earliest were at Pennington Marshes from September 9th to 15th 1970 and Farlington Marshes on September 9th 2010. Some inland birds on spring passage in late March have been mentioned above but the latest coastal record was of a singing male at Keyhaven on March 27th 2004. By far the latest ever was the male feeding with Linnets at Over Wallop on April 23rd 2010.

Three records pre-date the accepted first for Hampshire above. Meinertzhagen recorded one on plough at Mottisfont on January 4th 1898 but all his sightings lacking independent verification have now been discredited. Miss Sheila Young *in litt.* to Ken Rooke reported that she and Brian Vesey-Fitzgerald saw one settle in full view on the seawall near the mouth of the Lymington River in the company of five Wheatears on March 31st 1949. The same observers watched another at the edge of a small duck pond by Warren Farm, Beaulieu, on April 9th 1949, having it under observation for some time at about seven yards range. Ken sought details of the identifications but none was forthcoming and so the records were not accepted.

GB status: regular passage migrant mostly in small numbers; occasional breeder in Scotland from 1977. (Breeds throughout tundra of Eurasia and North America; winters to south.)

Ortolan Bunting *Emberiza hortulana*

HOSRP

A rare autumn passage migrant. Twelve records since 1956.

The first for the county was one seen by John Ash at Damerham on September 17th 1956. The 1956 *HBR* states, 'One, and probably a second. This bird was rather wild, but fair details were obtained and the observer has had several opportunities of watching this species recently at Portland'. Cohen (1963) accepted two.

The next was one seen by Tim Sharrock at Needs Ore on October 15th 1961. Tim's description, accepted by HFCERC was as follows:

> *Flight call 'ptwick' heard as flying over at 1600 hrs. Settled in bramble and observed at 40 yards through x25 telescope and x10 binoculars for five minutes. Pinkish bunting-shaped bill, pinkish legs and white eye-ring confirmed original identification. Faint wing-bars, generally brownish, streaked breast. Dark moustachial streaks offset by creamy moustachial stripe. It was also seen through the telescope by my wife and Mr. A Bailey.*

The third was recorded by David Christie when he flushed a female or juvenile at Hook-with-Warsash at 1000 hrs on September 22nd 1983 which flew off across Southampton Water towards Calshot. In the following year one was picked up on call by Andy Hanby as it flew up from a grassy area in the then grounds of Sandy Point Hospital on September 2nd; it perched in a gorse bush for about a minute enabling Dave Bishop, Trevor Codlin, Keith Turner and John Walters to scope it before it flew off high to the north-west with a small group of Meadow Pipits.

Ortolan Bunting, Warsash, August 29th 2015

The only bird in the next decade was seen by Mike and Pete Combridge along the Ancient Highway at Keyhaven on August 31st 1992.

The first of the 21st century was one watched for a few minutes by Richard Carpenter on a treetop in his garden at Hill Head on August 23rd 2000. In the following year, John Norton discovered one at Brownwich on October 12th which was seen by two other observers but could not be found the next day.

The next was not until 2013 when Andy Johnson flushed one from a treetop on the eastern edge of Sandy Point on September 21st. It spiralled up, calling, and left north-east. Dan Powell saw one fly up into a hedge near Workman's Lane, Warsash on August 29th 2015 and with Rosemary Powell he was able to watch it for five minutes before it flew back into the field. It was a record year in 2018 when three were recorded in quick succession. One was discovered by Marc Moody near Bishop's Dyke, New Forest on the afternoon of September 13th which was seen by those able to respond to the news quickly. A day later, Pete Gammage found one on farmland along Pigeon House Lane, Portsdown Hill. This provided good views to many observers during the afternoon and throughout the following day enabling them to add this tricky species to their county lists. Also on September 14th, Andy Lester saw one flying high to the south over Emer Bog at 1840 hrs.

Ortolan Bunting, Portsdown Hill, September 14th 2018. Lee Fuller

GB status: 270, 1776–1949; 2,927, 1950–2019. Declining, mean 31 per year in 2010–19. (Breeds Europe from C Spain to C Fennoscandia, E across C Russia to Baikal region, and Turkey E to N Kazakhstan. Winters across N & C equatorial Africa.)

Cirl Bunting *Emberiza cirlus*

HOSRP

Formerly a moderately common but locally distributed resident; now extinct as a breeder and not recorded since 1985.

The Cirl Bunting may not have colonised England until the late 18th century, for it was not until 1800 that it was first discovered by Montagu in Devon. It was unknown to White, who lived at Selborne until his death in 1793. However, Blyth, writing in the second volume of *The Naturalist* in 1837, described the Cirl as 'nowhere more plentiful than in the vicinity of Alton, whence it is strange that Gilbert White should have overlooked it'. K & M described it as 'a local resident .... highly characteristic of the southern parts of the county .... and found commonly in most inland localities, but not in all'.

Numbers were possibly at their highest level during the first half of the 20th century although only anecdotal evidence is available for that period. Between the late 1930s and the 1950s, breeding season records were traced for 160 tetrads and the population was estimated to approach or even exceed 1,000 pairs (*BoH*). However, a steep decline was underway in the 1960s and few breeding pairs remained by the late 1970s. Small numbers continued to breed in the Avon Valley at this time with the final confirmed breeding at Rockford in 1982 and probable breeding there a year later. Elsewhere, there were reports from only six areas between 1977 and 1982 and apart from two instances of pairs seen all referred to single singing males.

Cirl Bunting, male

Since then, the only records have been of a singing male seen by D. W. Lewis at Longwater Lawn, New Forest on six dates between March 27th and May 8th 1983, a male with Yellowhammers seen by Christopher Wear at West Hayling on October 25th 1983, at least two found by Owen Haisell, Len Mummery and Ed Wiseman at Pennington Marshes on September 23rd 1984 and a male scoped by Kevin Haggar among 15 Yellowhammers feeding at Anton Lakes on January 27th 1985.

During the 1960s, coverage of most areas where the Cirl Bunting was known to occur was poor, e.g. around Andover and Odiham. Thus we will never know the full picture of its disappearance, but the population was probably well below 50 pairs at the time of the national 1968–72 Atlas, when it was recorded in sixteen 10 km squares. There are several factors which may have contributed to its decline in Hampshire and most other English counties. In coastal areas especially, much habitat was destroyed by house building. The use of herbicides and insecticides reduced summer survival and breeding success while the autumn ploughing of stubbles affected winter food supplies. This was a major factor contributing to the decline (Evans & Smith 1994). Since the late 1980s, the species has been confined to south Devon, where it has been the subject of intensive research and targeted conservation measures, led by the RSPB and Natural England (Balmer *et al.* 2013). The last full survey in 2016 produced 896 territories with an estimated total of 1,079 (Holling *et al.* 2018).

GB status: 1,000 + breeding pairs, SW England. (Resident N W Africa, W and S Europe and W Turkey.)

Little Bunting *Emberiza pusilla*

BBRC to 1993; HOSRP 1994 to date

A very rare vagrant. Seven records since 1986.

The first for the county was a male trapped by Charles Cuthbert at Winnall Moors on March 18th 1986. The record was not submitted to BBRC until 2015. Extracts from Charles's description are reproduced below:

I was catching Reed Buntings at a winter roost at Winnall Moors when the Little Bunting was caught in a mist net. I recognised it as being quite different at that moment, and probably a Little Bunting, so after ringing the bird I immediately stopped all further ringing activity and took the bunting home in a bird bag. That evening I telephoned a few friends and work colleagues whom I thought would be interested. Everyone who saw it agreed with my identification based on size, wing coverts and bill shape. It was well marked and close to the top of the size range for Little Bunting which led me to consider that it was probably a male. The following morning at dawn I returned to Winnall Moors with Wilf Simcox where the bird was safely released at the ringing site but not seen again.

Little Bunting, Winnall Moors, March 18th 1986. Barry Duffin

The next was a female trapped by Pete Potts at Titchfield Haven on February 16th 1992. It remained in the area until April 17th 1992. It regularly came to seed put out for small passerines and was seen by many observers.

The first of the 21st century was a first-winter which frequented John Parson's garden in Stubbington between February 19th and March 17th 2000, feeding on food put out for Reed Buntings. Its identity was confirmed by Barry Duffin and Dan Powell and it was trapped on at least one occasion. The next was one seen by James Andrews in a pre-roost gathering with other buntings before dropping into a reedbed at Barton's Mill, Old Basing on October 24th and 25th 2008; further views were obtained of presumably the same bird nearby at noon on November 1st.

In the 2010s three were recorded. The first was a first-winter trapped by Nigel Jones at Fishlake Meadows on October 26th 2016. This was followed by a first-winter found by Pete Gammage with 50 Reed Buntings in game cover in a field just west of Fort Widley, Portsdown Hill on January 25th 2017. It was elusive but eventually seen by patient observers and last recorded on February 9th. Finally, one was seen and photographed by Carol Thorpe in her garden at North Baddesley between March 17th and April 10th 2018.

Little Bunting, North Baddesley, March 23rd 2018. Andy Thorpe

An older record of a male found by George Clay at Bedhampton Creek on November 8th 1953 was included in the Langstone Harbour yearbook but not accepted for publication in *British Birds*. This was attributed to the submission of an inconclusive description of a bird seen at a later date by another observer (Graham Rees *in litt.*).

GB status: 119, 1902–1949; 1,786, 1950–2019. Increasing, mean 73 per year in 2010–19. (Breeds N Eurasia from N Fennoscandia E to the Chukotskiy Peninsula & Sea of Okhotsk in E Siberia. Winters NE India to S China, S to N Thailand & N Indochina.)

Rustic Bunting *Emberiza rustica*

BBRC to 2005; HOSRP 2006-14; BBRC 2015 to date

A very rare vagrant. One record in 2010.

The only record for Hampshire is of one, probably a female, identified by Russell Wynn in his garden at Brockenhurst on March 20th 2010. It was still present the next day. Extracts from Russell's account of events in the 2010 *HBR* (Wynn 2011) are reproduced below.

On the morning of March 20th 2010 I was settling down to do some work in our conservatory when my girlfriend mentioned that she could see a sparrow-like bird feeding on birdseed at the bottom of the garden. As sparrows are quite scarce in our rural location, I assumed it would be a Dunnock, but a quick glance with compact binoculars revealed it was a bunting. My first thought was Reed Bunting (a potential garden tick) but I thought it looked a bit contrasting and compact for that species. I dashed to get my Leicas to get a better look at the bird and was soon staring at a stunning Rustic Bunting! I then set up a scope and tripod and got a few digiscoped record shots. The bird showed intermittently at the feeding station but frequently disappeared into the adjacent hedge, presumably seeking shelter from the heavy showers. However, it looked fairly settled, so I called local birders

Marcus Ward, Marc Moody, Pete Durnell and Tim Parminter (accompanied by Alan Lewis) over to support the sighting. We continued to watch the bird on and off for a few hours from within the conservatory and agreed that the plumage features indicated it was probably a female. The night of March 20th was cloudy and damp, and unsurprisingly the bird was still present the next morning. It remained around the feeding station all day despite us working in the garden for long periods. I managed to get some better shots with the DSLR, and a couple of video clips. The bird survived an attack by a female Sparrowhawk and was seen to dig a worm out of the manure heap on one occasion. I occasionally heard the bird call; an unobtrusive 'sic'. A clear night on March 21st presumably saw the bird continue its northwards migration as it was not seen again.

Rustic Bunting, Brockenhurst, March 21st 2010. Russell Wynn

GB status: 19, 1881–1949; 582, 1953–2020. About ten per year; stable or decreasing. (Breeds N Eurasia from Sweden & Finland, E across N Russia to E Siberia & Kamchatka, & S to Baikal region. Migrates SE to winter E China, Korean Peninsula & S Japan.)

Black-headed Bunting *Emberiza melanocephala*

BBRC

A very rare vagrant. One record in 1993.

The only record for Hampshire is of a male discovered by Mrs. P. Gawthorpe in her garden at Morelands Road, Purbrook on May 28th 1993. She telephoned Gary Wearn who lived nearby; he immediately ran round to her bungalow. He saw it after a few moments and immediately confirmed its identity. Extracts from his description taken from the 1993 *HBR* (Wearn 1994) are reproduced below.

The head was entirely black with a distinct yellow neck collar, forming a neat line between the chestnut shoulders and black hood. The eye was dark. The nape, mantle and rump were chestnut brown with maybe a hint of olive on the rump. The wings were dark brown/black but heavily fringed and tipped buff. The chin, breast, belly and flanks were entirely yellow and unstreaked. The bill was quite long and pointed and silver/grey in colour. The legs were flesh-coloured. It was in good condition, showing no signs of previous captivity and was unringed. The bird was loosely associating with juvenile Starlings and sparrows and during the 15 minutes I watched it spent time feeding on the lawn and sitting quite high up in the surrounding bushes. It eventually left the garden and was not seen again.

Another record was of a female feeding on bird seed in gardens at Cholderton Road, Grateley from mid-March until April 4th 1993 when it was found dead by Margaret Cooper. The skin was subsequently identified at the BTO by John Marchant. The identification of this individual was accepted by BBRC but, in view of the date, it was considered most likely to have been an escape.

GB status: 5, 1884–1949; 227, 1951–2020. About five per year; decreasing. (Breeds C Italy to Greece, Turkey, N Iraq, W Iran, & N through Caucasus to Ukraine & S Russia. Winters W & C India.)

Dark-eyed Junco *Junco hyemalis*

BBRC

A very rare vagrant. Four records since 1987.

The first for the county was a singing first-summer male discovered by Peter Maynard at Long Valley near Fleet on May 30th 1987. It stayed into July of that year and was seen again in the same area on May 20th 1988. It was seen but not identified by David Dimmock in his garden at nearby Church Crookham on February 9th 1989. In the following winter, David refound the bird on December 26th and alerted the author, who was able to confirm its identity. It was then seen by many observers until March 9th 1990. Extracts from Peter's account of his initial discovery of the bird are reproduced below.

On May 30th at 5.30 am while out on patrol as a warden on the Aldershot Military Lands, I heard a bird song which was new to me. The bird was perched on top of a tall Scots pine. I could see it was chocolate brown above with a pale underside which was formed by a crescent shape across its chest and wing-shoulder edges. The eyes were dark but the bill was light straw coloured. As it sang it tipped its head back and I could see its feathers extended. The song was very repetitive like a bunting's and sung in a key very like the lower notes of a male Redstart. The tail was long with white outer tail feathers which showed as it flew down to feed. The bird was in this area for over a week but because of secret testing of tanks

Dark-eyed Junco, Church Crookham, January 1990. David Dimmock

etc I was unable to allow other people to see it. On June 7th I saw it outside this location with two other birdwatching friends and we watched it for over an hour before it flew off towards Fleet.

Church Crookham, January 1st 1990

The second was one seen and photographed by John Applin and David Twigg which stayed all day in gardens at Picket Piece, Andover on May 5th 1996. It was not seen subsequently.

The next was one discovered by Karen Bennie while walking at Hawkhill Inclosure, New Forest on December 24th 2011. A photo sent to County Recorder Keith Betton confirmed the sighting as a first-winter male of this species. The news was released but the bird was not seen again until December 30th. It was faithful to a small area where seed was provided and it was seen by many birders until last recorded on March 11th 2012.

The fourth record involved one seen and photographed by Duncan Burchell in his garden in Petersfield on May 2nd 2017. It remained all day but was not seen subsequently.

An earlier record of one carrying a blue ring seen by Steve Keen in a garden in Fareham between February 3rd and 14th 1982 was accepted by BBRC in 2020 but it was regarded as an escape.

See Appendix IV for an additional record.

GB status: 50, 1960–2019. Rare, about one per year; stable or increasing. (Breeds North America from N Alaska & N Canada, S to S California, N Texas & N Georgia. All British records belong to nominate 'Slate-coloured Junco', breeding throughout N & E of range, S to Georgia. Northern populations migratory, wintering to S of breeding range.)

Dark-eyed Junco,
Hawkhill Inclosure, New Forest,
February 20th 2012. Gordon Small

White-throated Sparrow,
Old Winchester Hill, April 11th
and 14th 2009.
Darren Evans and Chris Holt

White-throated Sparrow *Zonotrichia albicollis*

BBRC

A very rare vagrant. Three records in 1961, 2007 and 2008–11.

The first for the county was a bird glimpsed by Tim Sharrock at Needs Ore on May 5th 1961 and subsequently identified on May 19th. It was almost certainly the same individual which was on board the RMS Queen Elizabeth before it docked at Southampton on May 2nd 1961 (Durand 1961, Sharrock 1961).

The next was one photographed by R. Rothwell on board the RMS Queen Elizabeth 2 en-route from New York to Southampton which was seen in the dockland area after the ship's arrival on May 12th and 13th 2007 but not subsequently.

The third was a first-winter male identified by John Rowe at Old Winchester Hill on November 12th 2008, having been heard on November 5th. It was seen again on November 21st and remained into 2009. News of the bird was not released but it was discovered by visiting birders Kevin McCoy, David Mead and Heather Rowe on April 11th and seen by many visitors up to April 20th. Subsequently, it became elusive but remained in the area, being seen daily until June 11th, then on June 22nd and finally on July 17th. It was heard again by Dave Compton on August 7th and September 13th 2010 and then again from November 14th until April 4th 2011 and then on eight dates until last recorded on July 17th.

See Appendix IV for additional records

GB status: 2, 1893–1909; 48, 1961–2018. About 1-2 per year, stable. (Breeds North America from SE Yukon E to Newfoundland, S to Great Lakes & N USA to New Jersey. Winters SE USA, from Massachusetts S to Florida, Texas & into N Mexico & California.)

American Redstart *Setophaga ruticilla*

BBRC

A very rare vagrant. One record in 1985.

The only record for Hampshire is of a first-winter male found by Simon Woolley at Winchester Water Meadows on October 4th 1985. It remained in the area until October 6th. Extracts from his description are reproduced below.

> *On October 4th 1985 I found a completely unfamiliar bird that instantly flew up to a bush and was not seen again that day. Noticeable features included large yellow patches on the wings and tail, and fluttery wing-beats. Despite reference to books, I found nothing resembling the bird. Next morning, I found the bird again on a sprawling thicket near the river and was able to make full notes and a sketch. Later, with still no idea as to its identity, a birdwatcher, D. Jackman, ran up and asked*

if I had seen the American Redstart. On comparing notes, I realised we had seen the same bird, but despite contacting several observers, no one saw the bird that afternoon. On the 6th at 9 am, A. Baldock saw a bird with large yellow patches in the wing and tail dive into a bush, but it was not seen again.

Size comparable to Blue Tit, but shape reminiscent of a small flycatcher, with a rounded rather bulbous head and shortish wings and tail. Upperparts, greeny-brown/olive except for darker primaries, a bright yellow wing flash, a faint pale yellow eye-stripe behind eye and a white eye-ring. Tail pattern similar to Wheatear but brilliant yellow instead of white patches. Underparts, off white/grey except for pure white throat and undertail-coverts, rather darker across breast. Bright primrose yellow patches on sides of breast, tinged orange towards front edge, reminiscent of a Firecrest's 'cape' but more noticeable. Bill thin and short, dark.

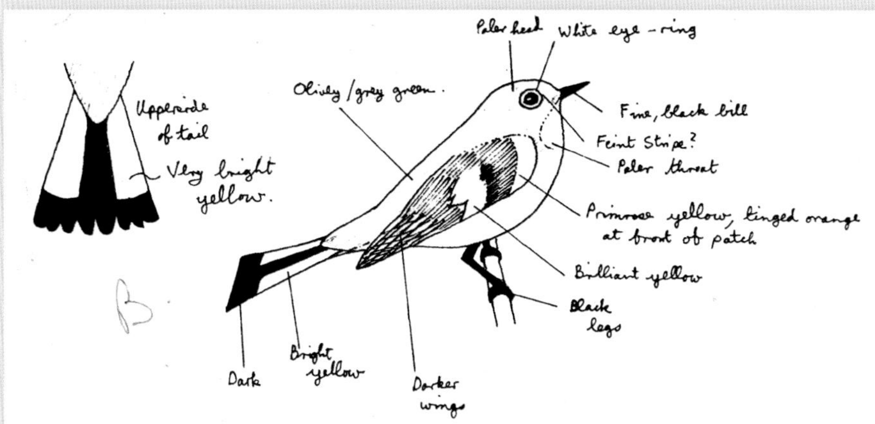

Very distinctive habit of fanning tail, even at rest, and constantly flicked tail, deliberately and unlike Redstart's nervous quiver. General actions resembled a Phylloscopus *warbler but included agile flycatching sorties. Called several times; a firm 'chitt' and much softer 'swit'.*

GB status: 6, 1967–2017. (Breeds Canada from S Yukon E to Newfoundland & S to USA mostly E of Rockies from Washington to E coast & S to SE Texas & NW Florida. Migrates through E USA to winter S USA, S to C & NE South America & Greater & Lesser Antilles.)

Stilt Sandpiper with Lapwing, Pennington Marshes, July 22nd 2002 (above right)

Appendices

I Records of selected species not in category A, B or C.

Falcated Duck *Mareca falcata*
BBRC

A very rare vagrant or escape. Two records in 2001 and 2013.

A male was seen with a flock of Wigeon by Pete Combridge during a WeBS count at Bury Marsh on February 9th 2013. The species was recently accepted into category A of British List by BOURC on the strength of a male at WWT Welney in December 1986 (Stoddart and McInerny 2020). BBRC employed a criteria-based approach to decide whether subsequent records involved wild birds. As the Bury Marsh bird was not seen to be unringed, despite meeting the other criteria, it was not accepted (Holt *et al.* 2020).

There was also an earlier record of a male, with a metal ring on one leg, seen by Tim Parminter and Russell Wynn at Keyhaven Marshes from April 27th to 29th 2001; it had previously been seen at Yarmouth, IOW on April 15th.

GB status: 7, 1988-2008. (Breeds E Siberia from Yenisey River & Baikal region E to Sea of Okhotsk & S to NE China & Hokkaido, Japan. Winters from S Japan to SE China, locally W to Nepal.)

Laughing Dove *Streptopelia senegalensis*
BBRC

Presumed escape. One record in 1984.

One was at Eastleigh from April 15th to 18th 1984. The identification was accepted by BBRC but it was regarded as an escape.

GB status: Not on British List. (Breeds Africa, Malta, Middle East, C Asia and Indian sub-continent.)

Greater Flamingo *Phoenicopterus roseus*

BBRC

One 19th century record.

An adult was shot by Lord Montagu's keeper Mr. Goff at the mouth of the Beaulieu River on November 26th 1883. It is preserved in the Palace House, Beaulieu. Montagu, writing in *The Zoologist* for August 1884, describes how the bird arrived following a south-westerly gale two weeks previously. It could fly well and evaded initial attempts to secure it. He concluded that the bird was a wild one and had been blown to England by stress of weather.

GB status: Not on British List. (Partial migrant breeding SE Spain, S France, NW, E and S Africa, Madagascar, C Asia; in winter in Indian Subcontinent.)

Demoiselle Crane *Grus virgo*

BBRC

Presumed escapes. Three records in 1927, 1963 and 2008.

One was seen by Cecil Paddon in flight over Needs Ore at dusk on October 23rd 1927 and on the next two days was in the saltmarsh there accompanying three Spoonbills. It was still present on November 9th (*Brit. Birds* 21: 182). Another was seen by Colin Tubbs, Ed Wiseman and others at Stoney Cross Airfield on August 17th 1963. A colour-ringed bird was found by Steve Mansfield at Lode Farm Sand Pit, Kingsley on August 27th 2008 and was still present the next day.

GB status: Not on British List. (Breeds from E Europe to Mongolia and NE China; winters Lake Chad to Ethiopia and Indian Subcontinent.)

Band-rumped (Madeiran) Storm Petrel *Oceanodroma castro*

BBRC

One early 20th century record, no longer accepted by BOURC.

A freshly dead storm petrel was found on the beach at Milford on Sea by Roland Follett, a 17-year-old apprentice butcher, on November 19th 1911. The corpse was given to the Rev. John Kelsall, who in turn passed it to Philip Munn. It was subsequently identified by William Ogilvie-Grant as Britain's third (what was then called) Madeiran Fork-tailed Petrel (Munn 1912, Combridge & Wiseman 2009). The first two British records were later rejected because of the suspicion of fraud (Nicholson & Ferguson-Lees 1962). Later, the Milford petrel was also removed from the British List, on the grounds that identification was uncertain because the specimen was lost and thus unavailable for re-examination, and that there was no proof that the bird had died in British waters (BOURC 2009).

This removal was questioned by Combridge & Wiseman (2009), who provided additional information unknown to the reviewing committee. This questioning was identified by Harrop *et al.* (2012) as an example of 'denial' that specimen fraud had taken place in the past, thus implying that fear of sharp practice (of which BOURC provided no evidence) played a significant part in the decision to delete the record (see also Wiseman & Combridge 2012).

That Band-rumped Storm Petrel occurs, albeit rarely, in British waters was demonstrated when one, later accepted by BBRC, was photographed at sea on July 28th 2007 (Hudson *et al.* 2011, plate 295). However, field identification away from breeding colonies is complicated by the fact that Band-rumped, mainly on vocal and genetic grounds, is no longer considered to be a monotypic species but rather a group of perhaps ten cryptic species, six of which occur in the Atlantic (Howell & Zufelt 2019, and references therein).

GB status: 4, 2007-2020. (North Atlantic population breeds on islands off coast of Portugal, Azores, Madeira, Canary Islands & Cape Verde group. Elsewhere, breeds Ascension & St Helena, South Atlantic, & Galapagos, Hawaii & islands of N Japan. Distribution outside breeding season poorly understood.)

Contributed by Pete Combridge

Scopoli's Shearwater *Calonectris diomedea*
BBRC

One 19th century record.

K & M stated that two Great Shearwaters were obtained by Edward Hart in 1878 and 1894. Hart's catalogue states that he shot two from a boat in Christchurch Bay between the Shingles and the Needles on May 29th 1878. These specimens passed to Leicester Museum for restoration along with the rest of the Hart collection in the 1950s. A copy of Hart's catalogue annotated at Leicester states that specimen number 1, an adult female, had gone to pieces due to fat burn, and that number 2, an adult male, was relaxed and made into a cabinet skin in October 1959. This was examined by the author and Marc Moody at Leicester in October 2021 and appears to be of this species. There are only three accepted British records of this taxon, which was recently split from Cory's Shearwater, in 2004, 2019 and 2020. Given the somewhat confused original details and the lack of references in contemporary journals such as *The Zoologist* and *The Field*, the record is unlikely to satisfy the criteria for acceptance as a first for Britain. It has been passed to BOURC for consideration.

Scopoli's Shearwater, Leicester Museum. Marc Moody

GB status: 3, 2014–2020. (Breeds widely throughout W & C Mediterranean but local in E. Disperses widely outside breeding season where distribution poorly known.)

Dalmatian Pelican *Pelecanus crispus*
BBRC

Presumed escape. One record in 1967.

A first-winter was found in Langstone Harbour on November 29th 1967 where it remained until December 3rd. On the final date it was watched by John Smith swimming off the southern tip of Farlington Marshes between 1100 and 1300 hrs. He described it as follows:

The general shape, enormous size and massive bill distinguished it as a pelican. When swimming it appeared a dirty white colour. In flight the primary tips were black, this colouring extending down the near edge of the upper wing surface and tapering off at the secondaries. The underwing surface was uniform dirty white. At one stage it was swimming among a large flock of Brent Geese. It made frequent short flights but at 1200 hrs flew off south down Langstone Harbour but returned 20 minutes later. At 1300 hrs it again flew south and was watched as it disappeared above Eastney.

This individual was first seen in Essex on October 29th 1967 and was also recorded in Kent, Dorset and Cornwall before taking up residence at St. Agnes Pool, Scilly from January 10th to 25th 1968. It was known to have escaped from Colchester Zoo in October 1967 (Penhallurick 1969).

GB status: 1, 2016. (Localised colonial breeder in S & E Europe including Montenegro, Albania, Greece, Bulgaria, Romania, Ukraine & S Russia, E to Azerbaijan, Iran, Turkmenistan, Uzbekistan & Kazakhstan. European birds winter E Mediterranean, Asian breeders winter S to Iraq, Iran & Indian subcontinent. Presumed now extinct in C China & Mongolia, formerly wintered E China S to Hong Kong.)

Egyptian Vulture *Neophron percnopterus*

BBRC

A very rare vagrant or escape. Two records in 1968 and 1969.

An adult was seen by Ken Nalder and Ted Green near Denny Wood, New Forest on June 16th 1968. Extracts from Ken's description submitted to BBRC are reproduced below.

While walking on the B3056 road towards Beaulieu Road Station, Ted Green and myself heard a Curlew alarming as it flew out of a low-lying bog and to our amazement a black and white raptor, slightly larger than the Buzzards that were in the vicinity, came low over the heathland towards us with a shallow glide and an occasional single wingbeat. As the bird flew past at about 100 yards, I noticed the unusual yellowish face and long bill. The white area on the front of the wings had several small dark patches but on the underside these were not apparent. The remaining area of the long, large, finger-edged wings was black and for this reason the white wedge-shaped medium tail stood out noticeably and because of its unusual shape and colour for a raptor. As it reached the edge of the heathland it started to soar to a height of about 200 feet when the very short head and exceptionally long, very straight and wide wings were noted as it spiralled. It passed within 300 yards of us and then flew in a north-easterly direction until out of sight.

The same bird was seen again by Phil Agland at nearby Bishop's Dyke on June 18th. Another adult was at Farlington Marshes on October 31st and November 1st 1969. It was subsequently caught in Sussex and showed signs of having had its primaries clipped, thus denoting previous captivity.

In a review of all records carried out by BOURC in 1999, the New Forest bird was admitted into category D. This indicates that natural occurrence could not be excluded although the species is an unlikely vagrant and there is sufficient evidence to suggest that escape from captivity was possible. However, it was not stated what the evidence was! A record of an adult on the Isles of Scilly in June 2021 will be considered by BOURC for admission to category A; acceptance would presumably lead to a further review of the 1968 record.

GB status: 2, 1825 and 1868. (Summer visitor NW Africa and Spain E to C Asia; resident W Africa across Sahel zone to Arabia and Indian Subcontinent.)

Red-headed Bunting *Emberiza bruniceps*

BBRC

Presumed escape. One record in 1961.

An unringed male was found by Bill Oddie, D. A. Whitehouse and R. W. Butler on the banks of a muddy creek at Keyhaven on July 3rd 1961. The identification was accepted by BBRC but it was regarded as an escape.

GB status: Not on British List. (Breeds C Asia, winters Indian sub-continent.)

II 'the ones that got away'. Selected not proven and non-submitted records.

Black Scoter *Melanitta americana*

BBRC

Two possible modern records.

Male scoters seen by Marc Moody off Hurst Beach on February 6th 1989 and Mick Hay and Keith Maycock off Pennington Marshes on July 22nd 1989 showed extensive yellow on their upper mandibles. This is characteristic of the race *americana* of Common Scoter which has since been split as Black Scoter. Detailed notes of the second sighting were submitted to BBRC but the record was eventually considered not proven (Rogers *et al.* 1995). See Dean *et al.* (1989) for details of the criteria for identification of *americana* and Garner (1989) for notes on a male Common Scoter with extensive yellow on its bill.

GB status: 15, 1987–2020. (Breeds Siberian tundra from Yana River E to Alaska, & N Canada E to Newfoundland. In North Atlantic, winters along coasts of E USA, N to South Carolina, & inland on Great Lakes. Elsewhere, winters ice-free seas along both coasts of N Pacific Ocean, S to N Japan & California.)

Sharp-tailed Sandpiper *Calidris acuminata*

BBRC

One possible modern record.

Bryan Renyard found a wader at Farlington Marshes on October 25th 1969 which he identified as a Pectoral Sandpiper. It remained until October 27th. His description in the Langstone Harbour yearbook does not specifically rule out either species although David Billett has added, 'With hindsight this was probably a Sharp-tailed Sandpiper'. David excluded the record from his account of Pectoral Sandpiper in Billett (1987). Steve Gantlett confirmed in 2020 that he saw the bird. In an e-mail exchange with the author he said, 'This was the first 'rarity' I ever saw, on October 26th 1969. I scoped it at close range from 'the humps' and watched it for ages. I was told it was a Pectoral Sandpiper and just accepted it at the time (remember this was before most field guides were published and some books lumped Sharp-tailed). I was too inexperienced to take proper notes, but I did note that it definitely had green legs (I didn't know the significance, but not ochre like all the other Pecs I have seen since) and a noticeable rufous cap. Also, although I watched it for ages (on my own) at close range, it didn't have a noticeable breast band and I was surprised by that as I was looking hard for it and wondering where it was....'.

GB status: 4, 1848–1892; 28, 1956–2012. (Breeds Siberian tundra from Yana River to Kolyma River delta, possibly further E. Migrant through coastal Alaska, China & Japan to winter New Guinea, Australia & New Zealand.)

Ross's Gull *Rhodostethia rosea*

BBRC

One possible modern record.

Martin Cooper saw an adult among a large gathering of Black-headed Gulls off Hurst Castle which worked its way west during a strong south-westerly gale at 1500 hrs on January 25th 2002. He noted a small gull with a pink flush on the underparts, grey underwings and a distinctive flight. The record has not been submitted to BBRC. An adult was discovered at Blackstone Meadow, Plymouth, Devon on January 28th where it remained until March 5th. Martin saw photographs of the Devon individual and believes that the bird seen in Hampshire was the same individual.

GB status: 1, 1936; 101, 1960–2019. About two per year; stable or decreasing. (Locally common on tundra of NE Siberia from Lena River E to at least Kolyma River. In Canada, rare & local breeder W Hudson Bay region, perhaps elsewhere. Siberian birds migrate E past Point Barrow, Alaska in September to unknown wintering area assumed to lie near edge of pack ice in Bering Sea or N Pacific, S to N Japan.)

Masked Booby *Sula dactylatra*

BBRC

One possible modern record.

A juvenile was found by Marc Moody feeding with Gannets off Hurst Beach on July 29th 2007. It was later seen by Tim Parminter. Marc submitted detailed notes to BBRC with additional information supplied by Tim but the record was considered not proven. Marc's annotated sketches and extracts from his notes are reproduced below.

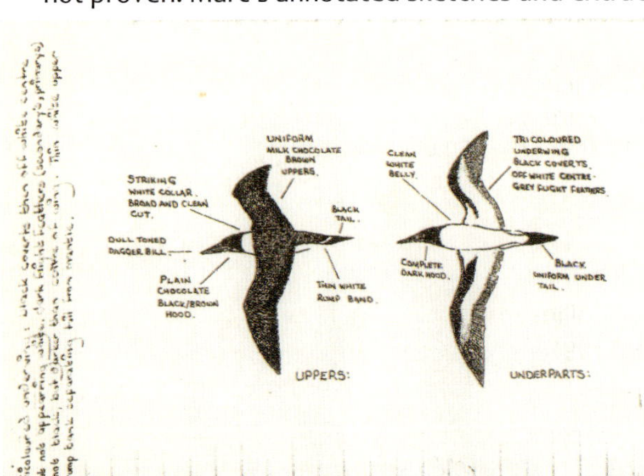

On July 29th 2007 I found an unusual juvenile Morus/Sula sp among a large gathering of Gannets. At 0830 hrs, while looking through the flock of 300+ Gannets fishing below the Isle of Wight cliffs, a very striking bird was observed. Its black hood with broad white collar was so prominent. It could easily be picked out among the mêlée of diving Gannets. After 12 minutes of continuous observation the bird headed east with Gannets into the Solent and was lost from

view. TP arrived soon after and at 0900 hrs he relocated the bird in the original area. We watched it for the next 15 minutes, both in flight and sat on the sea. It then left the main gathering, flying away to the west on its own, which Gannets also did.

It seems likely that this was the same bird seen by Dave Wooldridge, Derek Hale and Andrew Rann off St. Catherine's Point, Isle of Wight on May 19th 2007 (Wooldridge 2008). There were also reports of a possible immature Masked Booby seen from a yacht in the English Channel off Portland, Dorset on May 19th and off Porthgwarra, Cornwall on May 25th.

GB status: Not on British list. (Breeds small islands throughout tropics.)

Lessser Spotted Eagle *Aquila pomarina*

BBRC

One possible modern record.

One was observed independently by two observers over Langstone Harbour between 1242 and 1307 hrs on October 2nd 2001. Detailed notes were supplied to BBRC but the record was considered not proven.

GB status: Not on British list. (Breeds E Europe, winters E & S Africa.)

Crag Martin *Ptyonoprogne rupestris*

BBRC

One possible modern record.

One was seen by Martin Cooper at Pennington Marshes at 1130 hrs on May 3rd 1998. There was a cold northerly blowing with clear blue skies and many hirundines were flying low over the marsh and offshore. Martin noted a brownish hirundine which lacked the breast band of a Sand Martin, was paler underneath with black axillaries. At close range he was able to see it spread its tail to reveal a row of white spots. No other observers were present and by the time Martin had returned with Ed Wiseman it was nowhere to be found. The record has not been submitted to BBRC although Martin remains certain of his identification.

GB status: 12, 1988–2019. Very rare, last in 2018. (Breeds NW Africa & Iberian Peninsula N to S Germany & E through Mediterranean & C Asia, N to Baikal region of S Siberia, S to Tibetan Plateau & E to NE China. S European population mostly resident but Asian populations migratory, wintering NE Africa, & NW India to NC China.)

American Robin *Turdus migratorius*

BBRC

One possible modern record.

One was seen briefly by Margaret Collins in her garden at Leigh Park, Havant in very cold weather on February 10th 1991. Extracts from Margaret's description are reproduced below.

I looked out of the kitchen window into our garden, where we feed the birds, and an unusual bird caught my eye which I realised was an American Robin. I called my husband Barry over to see it but it flew to the other end of the garden and disappeared under the privet hedge. Then a cat came through the hedge. Description Upperparts: head black with a very obvious white eye-ring; rest of upperparts dark grey. Underparts: breast, belly and flanks brick red, vent and undertail-coverts white. Bare parts: bill yellow, legs dark. Size: about the same as Blackbird. We searched for the bird over the next week but were unable to relocate it.

BBRC harshly considered the record not proven, but Margaret remains certain of its identity.

GB status: 28, 1952–2018. Rare, about one every 2-3 years. (Breeds North America from tree line of Alaska & N Canada, S to S Mexico. Winters S Canada to S USA & C America, S to Guatemala.)

Isabelline Wheatear *Oenanthe isabellina*

BBRC

One possible modern record.

One was found by Mike and Pete Combridge at Pennington Marshes on October 24th 1987. Extracts from their submission to BBRC are reproduced below.

Weather dry, sunny, light north-east wind. Isabelline observed down to 20 yards (8 x 40, 10 x 40 bins, two 25 x 60 scopes). Watched 0930–1130 (approx.). Found on gravel track behind seawall, then moved to top of old rubbish tip. Pale underwings and tail pattern (broad, black lower half) constant features.

Upperparts: crown, nape, mantle and wing-coverts pale sandy brown, the greater coverts pale tipped, forming a narrow wing-bar. Ear-coverts slightly warmer sandy buff. Primaries and tertials darker brown with pale edges, secondaries sandy brown. Alula blackish, isolated in surrounding pale area of wing. Rump, uppertail-coverts and upper half of tail white, lower half of tail black with a very short projecting T stem. Supercilium dull white, extending from above bill to above eye, better defined when viewed from the front at an angle than in direct side views. A black line extended from bill to eye.

Underparts: chin and throat white, breast and flanks buff, rest of underparts white. Underwings noticeably pale in flight, whitish.

Bare parts: bill, eye and legs black. Bill noticeably deep at base, in length 1.5 to 2 times loral distance, with a small hook to tip of upper mandible.

General appearance and actions: noticeably pale, looking leggy. Ran extensively and perched freely on low vegetation and piles of earth. Most noticeable action a constant tail-wag, at times with bursts so vigorous that it recalled a wagtail. Noticeably broad-winged in flight and rather full-tailed.

The record was harshly considered not proven by BBRC but Pete confirmed in 2020 that he remains certain that it was an Isabelline Wheatear.

GB status: 1, 1887; 50, 1977–2019. About 1-2 per year, stable. (Breeds Black Sea coast from E Greece N to Ukraine & SW Russia, & E from Turkey through Kazakhstan & Mongolia to N China, S to Iran & N Pakistan. Winters Africa from N Sahel zone to E Africa, & Middle East to S Iran, Pakistan & NW India.)

III Suspect old records of interest

Gyr Falcon *Falco rusticolus*

BBRC

One suspect old record.

A pale morph was claimed by Richard Meinertzhagen at Mottisfont in January 1896. This record was documented long after the event by Meinertzhagen in his book *Pirates and Predators*, published in 1959. It was considered unsafe in a review of records of category B species undertaken by HOSRP (Combridge & Clark 2010). An extract from the panel's report is reproduced below.

> *According to Meinertzhagen's account, it caught and ate an Aylesbury Duck and was painted by the famous bird artist George Lodge who 'was staying with us at the time' (the Meinertzhagen family then owned the Mottisfont Estate). The painting, which shows a Gyr clutching an Aylesbury Duck on snowy stubbles, is reproduced below (plate 37 in Meinertzhagen (1959)).*
>
> *The record entered Hampshire's ornithological literature when it was published by Cohen (1963), who in his introduction stated that 'Colonel Meinertzhagen... allowed me to extract whatever I wanted from his notes relating to the Mottisfont estate'. Cohen dated the Gyr Falcon to January 18th, but Meinertzhagen's published account simply mentions the month and implies that it was present for two days.*

Reproduced by permission of the George Lodge Trust

> *Unfortunately, Richard Meinertzhagen's reputation, not only as an ornithologist but also in the wider sphere, has been tarnished recently. It is now widely acknowledged that he falsified not only ornithological data but also other aspects of his life, to such an extent that none of his claims can be taken at face value unless independently verified (Rasmussen & Prys-Jones 2003, Garfield 2007). With this in mind, the panel contacted Brian Bird of the George Edward Lodge Trust; Bird informed us that according to Lodge's published memoirs (1946) he had seen just three wild Gyr Falcons: two in Sweden and one in Norway. It seems inconceivable*

that Lodge, according to Bird 'a stickler for getting things just right', would not have mentioned seeing a Gyr in Britain, let alone have forgotten such an encounter. It may also be significant that Lodge died in 1954, five years before publication of the Gyr account in Meinertzhagen's book. With no independent verification of the record, the panel voted to delete it from the Hampshire List.

GB status: 187, 1835–1949; 185, 1950–2019. About three per year; stable. (In Europe, most numerous Iceland & Norway, smaller populations breeding N Sweden, Finland & Arctic Russia. To E, breeds Arctic Siberia, Alaska, N Canada & Greenland. European birds mostly resident but high-Arctic breeders from N Canada & Greenland migratory, occasionally wintering S to NW Europe.)

White's Thrush *Zoothera dauma*

BBRC

One suspect old record.

One was said to have been shot near Southampton in December 1893 (*The Field*, 24 February 1894). It is also mentioned in the second edition of Harting's *Handbook of British Birds* which says 'seen in the possession of Mr. Hill at Bulford, Amesbury, Wilts Sept. 6, 1894'. This suggests that it was seen by Harting as no other name is mentioned. The report is not mentioned by Saunders in his *Manual* (1899, 2nd edition), so perhaps he had some doubts over the thrush's true identity (he says that *lunulatus* 'has not infrequently been passed off as White's Thrush'). *Zoothera lunulata* is a native of eastern Australia. In a review of records of category B species undertaken by HOSRP, it was stated that, as the record had never been accepted by any national authority, without good evidence the panel was disinclined to accept it in retrospect. The panel also noted the curious fact that it was reported by a certain '*G. White*'; the Heron Court bird of 1828 (outside present-day Hampshire – see Appendix V) was named by Eyton in 1836 as *Turdus whitei*, White's Thrush, in honour of Gilbert White of Selborne, raising the possibility that the Southampton report was a hoax (Combridge & Clark 2010). The 1828 bird is on display at the Bournemouth Natural Science Society Museum but the origin of another specimen held by the Hampshire Cultural Trust at Chilcomb House, Winchester is uncertain (Ross Turle *in litt.*).

GB status: 24, 1828-1949; 70, 1952–2020. 1 or 2 per year, increasing. (Race aurea breeds European Urals & E across Siberia from Yenisey River to Ussuriland, S to N Mongolia, extreme NE China, Korean Peninsula & Japan. Winters widely across S China, Taiwan & S Japan to Indochina & C Thailand. Nominate race resident or altitudinal migrant in Himalayas, SW China & Taiwan.)

Pine Grosbeak *Pinicola enucleator*

BBRC

One suspect old record.

Henry Reeks, writing in the Zoologist for 1864, stated: 'Pine Grosbeak at Thruxton. On Saturday the 30th of January, my brother killed a fine male specimen of the Pine Grosbeak which was preening its feathers on the branch of a fir tree. Upon skinning the bird, I found the crop filled with small seeds .... which appear to be of Pinus sylvestris. I believe this to be the bird's first appearance in this part of Hampshire'.

K & M evidently overlooked Reeks' note but referred to an adult male purchased by Edward Hart in April 1876 which was 'formerly in the possession of the late H. Treasure Jenkins and killed in the neighbourhood of the New Forest many years ago'. Although Thruxton is near Andover, some distance from the New Forest, it is assumed that only one bird was involved. Saunders (1899), citing both J. H. Gurney and Professor Newton in support, wrote that most of the 'about 40' British reports were 'unworthy of belief, although the identification of the specimens still existing is correct'. Saunders stated that individuals were often imported as cage birds or as frozen specimens and, indeed, K & M themselves dismissed a report of four Pine Grosbeaks, said to have been shot at Woolmer Forest 'about March 1st, 1889', due to a suspicion that they had been 'sent in ice from Russia'. In a review of records of category B species undertaken by HOSRP, the panel felt that, because contemporary authorities doubted the authenticity of most 19th century records, Pine Grosbeak could not safely be added to the Hampshire List (Combridge & Clark 2010).

The taxidermy case containing the Thruxton specimen is held at the Horniman Museum, London.

© Horniman Museum

GB status: 4, 1831–1890; 8, 1954–2012. (Resident or dispersive across N Scandinavia, E to E Siberia & Kamchatka, & S to Hokkaido, Japan. Widespread across much of N North America, E to Newfoundland & S through mountains to C California & N New Mexico.)

IV Records of ship-assisted birds

These records relate to North American landbirds which were aboard trans-Atlantic liners when they arrived at Southampton. Most were fed on board and/or did not leave the ship when it docked and thus do not qualify for acceptance as category A records by BBRC.

Grey Catbird *Dumetella carolinensis*

One was seen by Malcolm Draper on board the RMS Queen Elizabeth 2 when it docked at Southampton on October 21st 1998. It did not leave the ship which left the next day (Dave Unsworth *in litt.*).

Song Sparrow *Melospiza melodia*

One was on board the RMS Mauretania when it docked at Southampton on October 14th 1962; it was still present the next day (Durand 1963). Another was seen by Malcolm Draper with the Catbird on board the RMS Queen Elizabeth 2 which docked at Southampton on October 21st 1998 (Dave Unsworth *in litt.*).

Dark-eyed Junco *Junco hyemalis*

One was on board the RMS Mauretania when it docked at Southampton on October 14th 1962 and was still present the next day (Durand 1963).

White-crowned Sparrow *Zonotrichia leucophrys*

One was found on the RMS Queen Elizabeth 2 on September 21st 1988, soon after leaving New York. It was still on board as the ship arrived in Southampton on September 26th, having fed on crumbs during the voyage (Frankland 1989).

White-throated Sparrow *Zonotrichia albicollis*

Four which arrived in Southampton on a Cunard ship in October or November 1958 were placed in East Park aviary in the city, where the last died in January 1964 (Sharrock 1965). Two were on board the RMS Mauretania when it docked at Southampton on October 14th 1962; both were still present on the following day but on 16th at least one was seen in a small park outside the dock gates (Durand 1963).

Common Yellowthroat *Geothlypis trichas*

One was on board the MV Aurora en route from St. John's, Newfoundland from September 19th 2008. It was fed on scraps by passengers and was still on board on the evening of September 22nd prior to the ship docking at Southampton (S. L. Gelder *in litt.*). There were no further reports.

Northern Parula *Setophaga americana*

One was found on the RMS Mauretania during an eastbound crossing of the North Atlantic on September 19th 1962. It was kept in semi-captivity and fed on soft fruit, crumbs and water. It died after the ship docked in Southampton (Durand 1972). The specimen is held at Liverpool Museum.

Blackpoll Warbler *Setophaga striata*

One was present on the RMS Queen Elizabeth in Southampton on October 12th 1961. Despite attempts to get it to fly ashore, it remained on board and subsequently died during the crossing to New York (Durand 1972).

Yellow-rumped Warbler *Setophaga coronata*

Mark Smith found a corpse of one in Southampton Western Docks on November 14th 2013. Identification was accepted by BBRC.

Short-billed Dowitcher!!

Adult male, disused saltpans, Mudeford, Christchurch, September 1872; juvenile female, Stanpit Marsh, October 7th 1902. According to K & M, Edward Hart discovered two skins of this species among some 'Sabine's Snipes' in his collection. **Now at Chilcomb House, Winchester.** Not accepted by BOURC because dates of collection do not match plumage state of birds. Identification not disputed.

Great Snipe§

Adult male shot by E. Budden at Cross Meadows, Christchurch, October 1849; adult male shot by R. Laidlawe at Gunter Meadows, Christchurch, August 9th 1876; adult male shot by Edward Hart at Gunter Meadows, September 16th 1878; adult female shot by R. Keynes at Chipmead, Christchurch, September 14th 1880; adult female shot by Edward Hart at Gunter Meadows, August 27th 1883. All except the 1849 specimen mounted by Edward Hart. **All now at Horniman Museum.**

Juvenile shot near Christchurch, September 1st 1868, mounted by J. H. Gurney, Jnr. (*Zool.* 1868: 1422) – not in K & M.

One Christchurch, shot by Edward Hart, September 19th 1868 (*Zool.* 1868: 1461; 1869: 1511; *The Field,* October 10th 1868) – not in K & M; incorrectly given as September 29th in Naylor (1996).

One, Duntish Common near Christchurch, shot September 4th 1875 (*The Field,* September 18th 1875).

One, Christchurch, early November 1882 (*The Field,* November 11th 1882) – not in K & M.

Red-necked Phalarope

Two immature males and an immature female shot by Edward Hart at Christchurch Meadows, September 29th 1916. **Now at Horniman Museum.** Overlooked by Cohen (1963). One seen by Miss C. Popham in Christchurch Harbour on October 4th and 5th 1937 (*SEBR*).

Grey Phalarope

Five adult males and six adult females shot by Edward Hart on flooded meadows around Christchurch in 1865, 1866, 1870, 1872, 1875, 1886 and 1891, when especially numerous. **Now at Horniman Museum** (one specimen missing).

Two seen by Miss C. Popham at Christchurch Harbour, December 29th 1935 (*Proc. BNSS* 28: 33) and one seen there by C. Osman on November 7th 1937 (*SEBR*).

Sabine's Gull§

Adult found dead in field near Bournemouth, October 13th 1891 (*Zool.* 1892: 22).

Immature shot by Edward Hart near Christchurch, September 26th 1896. **Now at Chilcomb House, Winchester.**

Immature shot by Mr. E. A. Dobree, Christchurch, November 1903 (*Bull. BOC.* 15: March 16th 1904).

Immature shot by Mr. A. Ponchaud, Christchurch Harbour, August 31st 1921 (*Proc. BNSS* 23: 44).

Glaucous Gull§

Second- or third-winter, shot by Edward Hart, Christchurch, February 1877. **Now at Horniman Museum.**

Iceland Gull§

Second- or third-winter, shot by Edward Hart, Mudeford, 31 December 1874. **Now at Chilcomb House, Winchester.**

Gull-billed Tern§

Adult shot by Edward Hart near Christchurch, May 14th 1872. The specimen passed to Mr. A. von Hügel (*Zool.* 1872:3149, *The Field,* July 20th 1872). **Now at Natural History Museum.**

Caspian Tern§

Adult male shot by the Hon. Grantley F. Berkeley at Highcliffe, early autumn 1852. Not known by whom mounted. Hart collection. **Now at Horniman Museum.**

Whiskered Tern§

Adult male shot by Edward Hart at Old Stage, Warren-bite, Christchurch Harbour, June 1875. **Now at Horniman Museum.**

White-winged Black Tern§

Singles shot by Edward Hart on the River Stour on May 2nd 1883 and at Barling Lake, Christchurch the next day.
Now at Chilcomb House, Winchester.

Long-tailed Skua§

Immature female shot near Christchurch, December 2nd 1889; adult male shot on the Sand Dunes, Christchurch Harbour, October 22nd 1891. **Now at Horniman Museum.** K & M incorrectly gives date of 1889 specimen as September 25th.

Hart noted, 'During October 1891 we had quite an invasion to our bays and harbour of this elegant species. Wherever a party of gulls were feeding four or five of these Skuas could be seen mobbing and taking their food. Unfortunately, the majority had shed their long tail feathers. I found they were not difficult to snare, and succeeded in capturing seven or eight, sending some to the zoo. They were the first of this species ever shown there. They are easily kept in confinement, feeding and thriving on any animal food, as well as fish. Unfortunately, I underestimated the cutting power if their bills, for after about 2 months they cut the brails I had placed on their wings, and so escaped from my open aviary.

At that date so little was known of the progressive stages of plumage which this species passes through, or of the age at which it commences to breed, that their escape was a great loss.

Black Guillemot§

Adult female winter plumage, December 16th 1866; adult male changing plumage November 5th 1875; adult male summer plumage, April 1877; adult male winter plumage December 1891; adult female changing plumage, October 1895. All shot in Christchurch Bay by Edward Hart except the 1877 bird shot by Mr. T. M. Pike. **All now at Horniman Museum.**
Dates given in K & M are January 18th 1862, December 16th 1866, November 13th 1875 and December 1891.

Puffin§

Adult male winter plumage January 1880; adult female summer plumage, May 1883; adult male winter plumage January 29th 1886; adult female winter plumage, January 4th 1890; adult male, October 1891; adult female, November 4th 1894; two young swimming on the sea having left their nesting holes near The Needles, June 15th 1895; immature male, November 1896. All mounted by Edward Hart. **Now at Horniman Museum.**

© Horniman Museum

Storm Petrel

Adults procured near Christchurch, November 1870, 1871 and 1883. Hart collection. **Now at Leicester Museum.**
Two seen by Roy Pitman off Hengistbury Head on November 17th 1934 (*SEBR*).

Leach's Petrel

Two procured near Christchurch, November 1881. Hart collection.
Now at Leicester Museum.

White Stork

Adult male shot by Edward Hart at Party Meade Rails, Christchurch Harbour, July 29th 1881. **Now at Leicester Museum.**
Two at Poole Harbour, Dorset in April 1884, later shot at Christchurch Harbour (Prendergast & Boys 1983). Whereabouts of specimens unknown.

Glossy Ibis§

One shot Christchurch Harbour, October 1853 (*The Naturalist,* November 12th 1853) said to have been preserved by Mr. Hart but not mentioned by K & M or retained in Hart collection.

Immature pair shot by Edward Hart near Christchurch Harbour, September 1859 (K & M). These were possibly from a flock of six or seven in Poole Harbour of which five were shot in September 1859 (*The Field*, October 8th 1859, *Zool.* 1878: 296, Prendergast & Boys 1983). Current whereabouts unknown.

A pair shot by J. Derham in Christchurch Harbour, August 1870. Hart collection. **Now at Leicester Museum.**

An immature pair shot by Edward Hart in Christchurch Harbour, September 1876. **Now at Horniman Museum.** K & M only gives one for September 1876.

One shot in Christchurch Harbour, October 26th 1929 (*Brit. Birds* 24: 168)

Little Bittern

One shot near Christchurch, 1773 (Latham 1785).

One shot at Heron Court on River Stour, April 26th 1862 (*The Field,* May 3rd 1862). **Malmesbury collection.**

One shot near Christchurch, late September 1866 (*The Field,* October 7th 1866). Not in K & M.

Adult female shot by Edward Hart at Parley Court, Christchurch, May 1869. **Now at Horniman Museum.** Given as May 1868 in K & M and Naylor (1996).

Pair and five young obtained, Claypool, Christchurch Harbour, June 5th 1869. Presented by Mr. T. M. Pike.
Now at Chilcomb House, Winchester.

One, Christchurch, September 8th 1869, reported by William Hart (*Zool.* 1869: 1917). Not in K & M.

Adult male shot by Edward Hart at Gascoigne Meadow, Christchurch, April 22nd 1878. **Now at Horniman Museum.**

Adult male shot at Heron Court, River Stour, June 9th 1893 (K & M). Malmesbury collection. **Now at Bournemouth Natural Science Society museum.**

Night-heron§

Immature shot, Christchurch, 1836 (Yarrell 1845; Harting 1872). Preserved by Mr. Mott but not included in Hart collection or mentioned by K & M.

Adult male shot, Christchurch, August 7th 1879 (*The Field,* August 16th 1879). Hart collection. Specimen destroyed at Leicester Museum due to moth damage.

First-winter shot, Christchurch, November 5th 1884 (K & M). Hart collection. Specimen destroyed at Leicester Museum due to moth damage.

A female shot by the river keeper on the River Stour at Christchurch, November 6th 1891 (*Zool.* 1891: 471–472). The specimen was seen but not retained by Edward Hart.

Squacco Heron§

Adult male shot by Mr. Saunders near Christchurch, June 8th 1832. Hart collection. **Now at Leicester Museum.**

One seen several times by Edward Hart near the Blackwater Ferry, River Stour and shot, June 1893 (Cohen 1963). **Malmesbury collection.**

Little Egret§

Adult shot by William Lockyer at Christchurch, beginning of July 1822. Preserved by Barrow and passed to Hart collection. **Now at Horniman Museum.** Incorrectly given as 1882 in Cohen (1963) and Naylor (1996). This would be the first for Britain but is not accepted by Naylor (2021). It is currently being reassessed by BOURC.

Red Kite
Adult female shot by T. Footner on Stanpit Marsh in 1851. Mounted by Barron and re-mounted by Edward Hart. **Now at Horniman Museum.** A sight record of one mentioned by Edward Hart as flying over Christchurch on October 26th 1884 – 'the last that was seen, or ever may be seen in this district'. Little did he know!

White-tailed Eagle
One caught near Christchurch, November 3rd 1862 (*The Field,* November 15th 1862), not in K & M.
Adult male shot by Edward Hart, Hengistbury Head, December 4th 1871. **Now at Leicester Museum.**
Adult shot Heron Court, January 9th 1893 (K & M). **Malmesbury collection.**

Hoopoe
Two eggs taken by Edward Hart, 1886. Specimens shot near Christchurch on August 14th and September 2nd 1878, July 12th 1892 and September 9th 1903. Mounted by Edward Hart. **Now at Horniman Museum.** Sight records from Hengistbury Head on August 28th 1932, Chewton Common on March 31st 1944, Hengistbury Head on Apr 13th 1952 and nearby Wick on April 17th 1952, Highcliffe on May 5th 1952 and Hengistbury Head on June 1st 1953 and May 3rd 1955.

Bee-eater§
Adult male shot by Mr. T. Critchell in his garden near Sheepwash, Iford on May 21st 1888. Mounted by Edward Hart. **Now at Horniman Museum.**
One reported by Edward Hart as seen at Holdenhurst, June 1903 (K & M).

Wryneck
Five specimens shot by Edward Hart near Christchurch between 1865 and 1868. **Now at Horniman Museum.** Sight records at Mudeford on September 26th 1937, Bournemouth on April 18th 1937, Mudeford on May 1st 1941, Hengistbury Head on May 24th 1944 and Mudeford on September 19th 1945.

Red-footed Falcon§
Adult male shot at Parley, 1854, no date. Hart collection.
Now at Chilcomb House, Winchester.
Adult female shot at Tuckton by H. Moser's keeper, Mr. Brewer, May 16th 1882 (*The Field,* May 20th 1882). Mounted by Edward Hart.
Now at Chilcomb House, Winchester.

Lesser Grey Shrike§
Adult shot at Heron Court, September 1842. **Malmesbury collection.** Identified in 1894 by Rev. R. A. Mathew (*Zool.* 1894: 345). First British record.

Woodchat Shrike
Adult male shot by Edward Hart at Somerford near Christchurch, May 29th 1880. **Now at Horniman Museum.**

Golden Oriole
Adult male shot by Edward Hart at Blackwater, Christchurch, May 1869; adult female shot by Edward Hart at Somerford, Christchurch, May 15th 1884. Female **now at Horniman Museum.** Male specimen destroyed due to moth damage.

Nutcracker
Male shot by gamekeeper S. Plowman at Pike's Hill, St. Catherine's Hill, November 6th 1868 (*Zool.* 1868: 1481, 1869: 1511). Mounted by Edward Hart.
Now at Horniman Museum.

Chough§
Adult male and female shot by Edward Hart and adult male shot by Robert Keynes, Christchurch Harbour, January 3rd 1871. All mounted by Edward Hart. **Now at Horniman Museum.** The only specimens Edward Hart ever came across, although said to have been formerly resident in Hampshire and Dorset.

Hooded Crow
Singles seen by Ken Rooke at Stanpit Marsh on April 7th 1934 (*Proc. BNSS* 26: 36) and Ken Smith at Hurn on December 1st 1935, said to have been present for some weeks (*Proc. BNSS* 28: 31).

Waxwing
Adult male shot at Hurn, 1866; adult female shot at Avon, 1872; pair shot at Mudeford, January 1895. All mounted by Edward Hart.
Now at Horniman Museum.

Crested Tit!!
Adult male shot by Mr. Footner from a clump of fir trees near Stanpit Marsh, 1846. Mounted by Barron and re-mounted by Edward Hart. **Now at Horniman Museum.** Not accepted by BOURC.

Shore Lark§
Adult male shot October 2nd 1875; two adult males and one female shot October 21st 1875. All obtained by Edward Hart on the Sand Dunes, Mudeford.
Now at Horniman Museum.

Great Reed Warbler
Adult male shot by Edward Hart on the Mill Stream, Christchurch, May 10th 1900. **Now at Horniman Museum.**

Aquatic Warbler§
Adult male shot by Edward Hart on Smuggler's Island, Christchurch Harbour, September 1876.
Now at Horniman Museum.

Marsh Warbler§
Adult male and female and nest with five eggs, obtained by Edward Hart on the railway embankment running through a small damp osier bed on Burton Farm, Conleys, near Christchurch, May 1869.
Now at Horniman Museum.

Rose-coloured Starling§
Adult male shot by F. Hart, Purewell, 1841, mounted by Barron; adult male, shot by J. Cull, Creedy Gardens, Christchurch, July 8th 1862, mounted by Edward Hart.
Now at Horniman Museum.

White's Thrush!!
Female, shot by the Earl of Malmesbury, Heron Court, January 24th 1828. (Eyton 1836). Described as a male in *The Zoologist* (1874: 4045). **Now at Bournemouth Natural Science Society Museum.** First British record.

Alpine Accentor§
Adult male shot by W. Humby at Castle Gardens, Christchurch, no date, 1885. Mounted by Edward Hart.
Now at Chilcomb House, Winchester.

Blue-headed Wagtail§
Adult males shot by Edward Hart at Burton Meadows in May 1876 and May 1883 (2) and an adult female shot there by Hart in May 1890. **Now at Horniman Museum.** Sight records of males at Mudeford on May 15th 1936, Stanpit on April 14th 1942 and April 10th 1945 (Miss C. Popham) and a female at Stanpit on July 2nd 1945 (E. Cohen).

Tawny Pipit§
One shot by Edward Hart at Stanpit Marsh, August 1879.
Now at Leicester Museum.

Twite§
Adult male shot October 29th 1877 and adult female shot November 1st 1877 by Edward Hart on the Sand Dunes, Mudeford. **Now at Horniman Museum.**

Parrot Crossbill§
Two males procured in March 1862, a pair on November 28th 1874 and two males on December 5th 1901. All shot by Edward Hart in Grove Wood, Christchurch. **Now at Horniman Museum.** Examination of a high resolution photograph of the display case (which is in storage) by the author in 2021 suggests that at least two of these are genuine Parrot Crossbills.

Lapland Bunting§

Adult male in changing plumage March 13th 1891 and adult male in winter plumage October 2nd 1893 both shot by Edward Hart at Portfields, Christchurch. **Now at Horniman Museum.**

Records for Christchurch Harbour, 1956–March 1974

§ indicates pre-dates first record for present-day Hampshire.

Spotted Crake

Singles Stanpit, September 11th 1964, September 9th 1966, September 6th 1967, July 30th and August 6th 1968; trapped Wick, August 11th 1968; trapped Stanpit, September 6th 1969 and August 8th 1970; Wick, August 12th and 22nd (heard only) and October 23rd 1970; Stanpit, August 19th 1971; 2, Wick, August 25th 1971 (heard only); Wick, September 2nd 1971.

Little Crake

Female, Wick Hams, March 21st, 25th and 27th 1970 (R. J. Booth, G. E. Fordham, Chris Husband).

Crane

26 SE, Hengistbury Head, November 3rd 1963 (Toddy Cooper).

Kentish Plover

Singles, Stanpit Marsh, September 1st 1956, May 18th 1958, September 11th-14th 1959, September 2nd 1961 and July 28th 1962.

Dotterel

Hengistbury Head, August 31st and September 2nd 1960; Christchurch Harbour, May 18th–27th 1972.

Temminck's Stint

Singles, Stanpit Marsh, September 20th 1967 (G. Richards) and September 7th 1968 (Chris Husband).

Pectoral Sandpiper

Stanpit Marsh, October 9th 1966 (Tony Wise *et al.*).

Dowitcher *sp.*

2, Stanpit Marsh, September 23rd–27th 1968 (J. Jenkins and S. F. W. Paul); Stanpit Marsh, December 2nd and 3rd 1973 (Mike Arnold and Chris Husband).

Grey Phalarope

2, Stanpit, October 9th and 10th 1958, 1 to 11th when killed; 6 off Hengistbury, October 18th 1959, 2 there on 22nd and 24th and 1 off Mudeford, 25th; 2 in a car park near Hengistbury, October 8th 1960 and 20 on sea, October 9th; 2, Stanpit, October 28th–31st 1961; 2, Hengistbury, November 24th 1963; 1 trapped, Stanpit, November 4th and 5th 1966; 1 off Hengistbury, September 28th 1968; 1 off Hengistbury, November 27th 1971.

Collared Pratincole

Stanpit Marsh, September 14th 1957 (Johnny Bowers and Eric Jones). Now considered unacceptable as bird was seen swimming. See Wallace, Bradshaw & Rogers (2006), where date is erroneously given as September 13th 1951.

Sabine's Gull

A juvenile off Hengistbury Head, September 19th 1970 was probably the bird at Pennington Marshes from September 12th to October 11th.

Gull-billed Tern

Stanpit Marsh, September 2nd 1959 (H. J. Clase).

Caspian Tern

Mudeford, September 7th 1958 (Mike Adams); Stanpit Marsh, July 20th 1968 (Chris Husband, Michael Webber).

Whiskered Tern

One in winter plumage, Stanpit Marsh, August 13th–16th 1967 (Chris Husband, Tony Wise *et al.*).

White-winged Black Tern

First-winter, Hengistbury, November 2nd and 3rd 1967 (Phil Agland, Chris Husband, Dave Smith).

Long-tailed Skua

Adult east, Hengistbury, June 17th 1972 (R. G. Mayall).

Glossy Ibis
Christchurch Harbour, August 10th 1972 (Chris Husband).

Little Bittern
A male picked up on the beach at Boscombe, April 21st 1963, later died (Miss H. J. Brotherton, Frank Clafton).

Night-heron
First-winter, Christchurch, from early January until January 28th 1962 (P. S. Bulson, E. Cohen). Probably not seen in *CHOG* recording area.

Purple Heron
An adult flew west off Hengistbury Head at 0730 hrs on September 13th 1970 (Chris Husband).

Little Egret§
Hengistbury, May 8th 1956 (Frank Clafton).

Hoopoe
Singles at Hengistbury Head on April 2nd 1959, April 20th 1968 and May 5th and 6th 1969.

Wryneck
Singles at Hengistbury Head on September 15th and 21st 1963, May 15th 1965, September 2nd and 24th 1966 and August 25th 1971 and at Stanpit on September 19th 1965.

Woodchat Shrike
Hengistbury, May 19th 1968 (Denis Hewitson, Dr. R. G. Mayall, Kevin Standring *et al.*); Hengistbury, first-winter trapped, October 14th 1970 (John Morgan, Mike Stewart); Wick Hams, May 12th 1971 (not submitted to BBRC).

Red-rumped Swallow§
A juvenile east Hengistbury, September 12th 1973 (Chris Husband).

Aquatic Warbler
Stanpit, August 9th and 12th 1959; singles trapped Stanpit, September 6th, 8th and 19th 1959; singles trapped Stanpit, August 14th and 24th 1965; Wick, trapped August 22nd 1967; Stanpit, September 7th 1968; Stanpit, trapped, August 6th 1969; Stanpit, September 13th 1969; Stanpit, trapped, August 9th 1970; Wick, August 9th 1970; Wick, trapped, August 20th 1970; 2, Stanpit, September 20th 1970 (not submitted to BBRC); Stanpit, August 16th 1971 (not submitted to BBRC); trapped, Wick, August 26th 1971; adult trapped, Wick, August 16th 1972; singles trapped Wick, September 1st and 7th 1972; adult trapped, Wick, August 13th 1973; trapped, Wick, August 20th 1973.

Melodious Warbler
Hengistbury, September 12th 1973 (Chris Husband).

Icterine Warbler§
Hengistbury, September 10th 1966 (Chris Husband and Tony Wise).

Bluethroat
First-winter female trapped Stanpit, September 7th 1958 and found dead in Spain, November 9th 1958; first-winter trapped Stanpit, September 12th 1959; first-winter male *cyanecula* trapped Stanpit, September 19th 1959; Stanpit, September 23rd 1959; first-winter male Stanpit, September 4th 1966; adult male, Hengistbury, October 14th 1970; first-winter male trapped Wick, September 7th 1971; first-winter male trapped Wick, September 14th 1972; first-winter male Wick, September 23rd 1973.

Blue-headed Wagtail
Hengistbury, September 22nd 1962. Wick, May 4th 1970; 3, Stanpit Marsh April 20th 1971 and 1 on July 23rd and August 23rd 1971; 1, Stanpit, April 13th and May 7th 1972; 3 trapped Stanpit, July 23rd and Aug 1st 1972.

Citrine Wagtail§
2 first-winters, Stanpit, October 15th 1966 (Chris Husband, Mike Stewart, Tony Wise).

Serin
2, Hengistbury, October 15th 1966 (Chris Husband); male, Wick Hams, October 27th and November 5th 1969 (Mike Arnold, Peter Sparkes, R. G. Booth *et al.*); male, Stanpit, December 13th 1970 (Chris Husband).

VI Significant records in the Christchurch Harbour area, April 1974–December 2020.

§ indicates pre-dates first record in present-day Hampshire
!! indicates not recorded in present-day Hampshire.

Little Swift§
Hengistbury, June 5th 1997 (Eric Lloyd).

Great Spotted Cuckoo§
Sub-adult, Hengistbury Head, April 20th 1994 (Dave Smith).

Pacific Golden Plover§
Adult, Stanpit, July 25th–27th 1990 (Graham Armstrong, Eric King *et al.*).

American Golden Plover§
Adult, Stanpit, May 22nd–24th 1992 (Lawrie Chappell *et al.*).

Spotted Sandpiper§
Summer plumage adult, Stanpit, May 7th–14th 1976 (Chris Husband, Dave Smith, Paul Prior *et al.*).

Bonaparte's Gull§
Adult, Stanpit, April 9th–12th 1975 (R. G. Mayall, Mike Arnold, Chris Husband *et al.*).

Ross's Gull !!
First-summer, Stanpit Marsh, June 16th–August 20th 1974 (Frank Clafton, K. L. Hainsworth, R. Sutton *et al.*).

Lesser Crested Tern !!
West over Hengistbury, April 25th 1995 (Graham Armstrong).

Elegant Tern§
Stanpit, May 10th 2005 (Lawrie Chappell).

Black Guillemot§
Off Hengistbury, January 2nd 1980.

Barolo/Boyd's/ Audubon's Shearwater (group)!!
West off Hengistbury, June 13th 1984 (Dave Smith). Under review by BBRC.

Great White Egret§
Stanpit and Hengistbury, June 12th and 13th 1974 (Norman Meader *et al.*).

Red-eyed Vireo!!
Hengistbury, October 12th 1987 (Dave Smith and Paul Morrison).

Penduline Tit§
Juvenile, Stanpit, October 21st and 22nd 1988 (Mark Hallett, Paul Morrison *et al.*).

Western Bonelli's Warbler!!
Singing male May 24th and 25th 1984 (Bruce Caswell and Maurice Opie).

Bonelli's Warbler *sp.*!!
Singing male, Hengistbury, May 31st 1975 (Tim Christmas, Chris Husband and Geoff Sherwood).

Radde's Warbler§
Hengistbury, October 21st 2003 (Lawrie Chappell).

Dusky Warbler§
Hengistbury, October 29th 1991 (Dave Smith).

Greenish Warbler!!
Hengistbury, October 1st–3rd 1985, trapped on 2nd (Ed Brett, P. J. Holmes and Dave Smith *et al.*).

Fan-tailed Warbler!!
Hengistbury, May 20th–31st 2000 (Dave Smith *et al.*).

Red-throated Pipit§
Hengistbury, May 25th 1987 (Dave Smith).

Little Bunting§
Female trapped, Wick, May 1st 1976 (Chris Husband, John Morgan and Paul Prior).

Rustic Bunting§
Male, Crouch Hill, Stanpit, March 20th 1993 (Paul Stancliffe).

Bobolink!!
Wick, November 1st–23rd 2002 (Steve Simmonds, Graham Armstrong, Dave Masser and Ian Southworth *et al.*).

Northern Parula!!
First-winter female, Hengistbury, September 30th–October 12th 1985, trapped on 9th (Dave Smith *et al.*).

VII Records previously accepted but now considered not proven following a review of 1950–57 British rarities.

Moustached Warbler
2, Eling Great Marsh, August 13th 1951 (*Brit. Birds* 45: 219–220).

(Eastern) Black-eared Wheatear
Male, Farlington Marshes, September 18th 1954 (*Brit. Birds* 48: 130). See Wallace, Bradshaw & Rogers (2006).

VIII BBRC not proven records 1958–2020.

Black Brant
Hook-with-Warsash, February 10th 2002–19th March 2002.

Lesser White-fronted Goose
Portsmouth Harbour, November 21st–29th 1991.

American Wigeon
Boarhunt, March 12th 1997.

Ring-necked Duck
Lymington River, December 18th 1977; Alresford Pond, May 14th 1979.

Harlequin Duck
Fareham, January 19th 1965; Pitts Deep, November 22nd 1981.

Black Scoter
Pennington Marshes, July 22nd 1989.

Common Nighthawk
Pitts Deep, September 17th 1973.

Alpine Swift
Toyd Down, June 30th 1988; Lymington, March 21st 1990; Farnborough, May 12th 2001; Alresford, May 1st 2004.

Pallid Swift
Winchester, April 30th 2010; Titchfield Haven, July 12th 2012.

Little Swift
Southampton University, February 6th 2004; Titchfield Haven, September 17th to 18th 2015.

Great Spotted Cuckoo
Winchester, July 15th 1994.

Yellow-billed Cuckoo
Bishop's Dyke, October 22nd 1974; Broomy Bottom, November 12th 2020.

Sandgrouse *sp.*
Keyhaven, November 17th 1973.

Oriental Turtle Dove
Pennington Marshes, June 6th 1977.

Little Crake
New Forest, April 16th 1972; Fawley, November 3rd & 6th 1983.

Crane
Upper Hamble Country Park, April 26th 1984; 2, Chawton, April 12th 1987.

Black-winged Stilt
Blashford Lakes, August 2nd 2014.

Sociable Plover
Brockhampton, July 27th 1994.

Pacific Golden Plover
Warsash, August 21st 1992.

American Golden Plover
Browndown, September 3rd 1999.

Broad-billed Sandpiper
Pennington Marshes, December 26th 1968; Fareham, February 12th–16th 1991.

Stilt Sandpiper
Needs Ore Point, December 5th 1997.

Baird's Sandpiper
Titchfield Haven, January 7th 1984; Ibsley Water, October 11th 2010; Pennington Marshes, July 23rd 2014.

Least Sandpiper
Keyhaven, September 9th 2005.

White-rumped Sandpiper
Pennington Marshes, September 4th–6th 1970; Langstone Harbour, September 8th 1979; Keyhaven Marshes, August 4th 2004.

Semipalmated Sandpiper
Keyhaven Marshes, September 19th–20th 2014; Winchester SF, September 12th 2016.

Western Sandpiper
Warsash, October 27th 1973.

Dowitcher *sp.*
Oxey Marsh, November 6th 1970; Fawley Reclamation, October 1st 1978.

Great Snipe
Farlington Marshes, April 4th 1959; Romsey, February 8th 1968; near Fordingbridge, January 28th 1976; Cholderton, February 13th–March 13th 1976; Quarley, February 3rd-March 1976; Bishop's Dyke, November 20th 1984; Ashurst, January 20th 1985.

Terek Sandpiper
2, Farlington Marshes, September 24th 1967; Titchfield Haven, May 8th 1981; Keyhaven Marshes, October 27th 1989.

Spotted Sandpiper
Curbridge, August 3rd 1987–September 3rd 1987.

Solitary Sandpiper
Lymington, August 12th 1967.

Lesser Yellowlegs
Keyhaven, September 25th 1970; Pennington Marshes, September 16th 1977; Milford on Sea, February 17th 1998.

Marsh Sandpiper
Lepe, March 22nd 1981; Eling Great Marsh, August 30th 2010.

Greater Yellowlegs
Pennington Marshes, February 26th 1980; 2, Keyhaven Marshes, March 31st 1985.

Slender-billed Gull
Gilkicker Point, May 24th 1968; Eling Great Marsh, August 7th 1990.

Bonaparte's Gull
Titchfield Haven, September 5th 1960.

Laughing Gull
Lymington River, January 27th 1986; Pennington Marshes, May 16th 1999; Milford on Sea, September 23rd 1999; Gosport, November 10th 2005.

Franklin's Gull
Southampton, February 18th 2014.

Mediterranean Gull
Langstone Harbour, September 20th 1959.

Great Black-headed Gull
Keyhaven Marshes, October 30th 1976.

Ring-billed Gull
Southampton Water, February 2nd 1982; Netley, April 28th 1983; Eling Great Marsh, February 9th 1984; Fareham Creek, October 13th 1984; Farlington Marshes, December 2nd 1984; River Itchen, Southampton, January 14th 1985; Titchfield Haven, January 19th 1985; Southsea, January 20th 1985; Southsea, October 17th–18th 1987.

Caspian Gull
Eling Great Marsh, February 2nd 1998–April 30th 1998; Eling Great Marsh, April 18th 1998–September 4th 1998; Eling Great Marsh, August 28th 1998–September 2nd 1998; Eling Great Marsh, December 18th–27th 1998.

'Azorean' Yellow-legged Gull
Ibsley Water, November 21st 2015–24th March 2016; Ibsley Water, January 25th 2020.

Gull-billed Tern
Milford on Sea, May 5th 1970; Pennington Marshes, September 16th 1970; Calshot, October 25th 1987; Farlington Marshes, July 5th 1988; Titchfield Haven, July 15th 1988.

Caspian Tern
Keyhaven Marshes, August 28th 1977; Lepe, June 11th 1978; Farlington Marshes, July 8th 1996; Normandy, October 10th 2003; Hook-with-Warsash, June 27th 2005.

Sooty Tern
Lymington, August 21st 1973; Solent, May 11th 1980.

Forster's Tern
Hill Head, September 5th 2008.

Whiskered Tern
Keyhaven Marshes, September 8th 1990; 2, Needs Ore Point, June 1st 1993; Titchfield Haven, August 2nd 1995; Hayling Bay, June 8th 1996; Gosport, April 14th 2003; Hamble-le-Rice, October 2nd 2007.

White-winged Black Tern
Farlington Marshes, August 20th 1959; Gilkicker Point, July 18th 1965; Pennington

Marshes, September 22nd 1983; Titchfield Haven, September 26th 2004.

Pacific Diver
Weston Shore, February 13th–14th 2020.

Bulwer's Petrel
Keyhaven, May 30th 1991.

Cory's Shearwater
Hayling Bay, October 19th 1981.

Black Stork
2, Southsea, July 30th 2015.

White Stork
Fareham, June 13th 1960; Titchfield Haven, August 27th 1967; 2, Bentley, March 1971; 2, Isington, April 1973; North Camp GP, June 21st 1974; 12, Bentley, March 8th 1977; Regent's Park, Southampton, September 17th 1978.

Frigatebird *sp.*
Pennington Marshes, March 21st 1971.

Glossy Ibis
Titchfield Haven, September 26th 2010.

Night-heron
Arlebury Lake, March 25th 1990; Keyhaven, October 9th 1998; 2, Southampton, May 13th 2001.

Squacco Heron
North Warnborough, June 3rd 2007.

Cattle Egret
Colden Common, June 17th 1960; Stoney Cross, July 21st 2005; M3 near Cove, March 26th 2006.

Great White Egret
Pylewell Point, April 15th 1995; Beaulieu Estuary, August 17th 2003; West Hayling, May 10th 2005.

Short-toed Eagle
Totton, April 27th 1990; Clanfield, June 21st 2014; Totton, June 24th 2014; Braishfield, September 14th 2015.

Lesser Spotted Eagle
Farlington Marshes, October 9th 2001.

Red Kite
Adhurst St. Mary, July 14th 1960.

Black Kite
Chilbolton, June 21st 1987; Portchester, May 26th 1988; New Forest, April 21st 1989; Well, May 7th 1990; Fleet, July 3rd 1993; Romsey, June 2nd 1996; Eastleigh, September 16th 1996; East Stratton, May 4th 1997; Lower Test Marshes, April 18th 1998; Hill Head, April 30th 1998; Warnford area, July 5th 1998; Cheesefoot Head, January 19th–5th March 2003; near Wilverley Plain, June 25th 2003; Longwood Warren, June 17th 2005.

White-tailed Eagle
New Milton, October 22nd 1990.

'Steppe' Buzzard
Burghclere, June 5th 2006.

Bee-eater
Titchfield Haven, May 11th 1960.

Yellow-bellied Sapsucker
Andover, December 25th 1988.

Black Woodpecker
Ashurst, early July 1984.

Lesser Kestrel
Winchester, January 24th 1979.

Red-footed Falcon
Female, Verely Hill, New Forest, May 17th 1964; Lord's Wood, July 1st 1967; Farlington Marshes, October 28th 1968; Bishop's Dyke, May 23rd 1969; female, Hasley Inclosure, May 20th 1976; female, Lower Test Marshes, June 13th 1978; New Forest, May 20th 1979; Bishop's Dyke, July 26th 1979; male, Farlington Marshes, May 12th 1980; New Forest, May 30th 1980; male, New Forest, July 12th 1980; Portsmouth, July 29th 1981; New Forest, May 27th 1982; female, Bishop's Dyke, June 11th 1983; Sway, December 1st 1985; New Forest, May 25th 1987; Basingstoke, June 10th 1996; male, Hayling Bay, May 5th 1997; Titchfield Haven, September 26th 2004.

Gyr Falcon
Matley Wood, October 10th 2002.

Nutcracker
4, Cheriton, August 13th 1972; Ringwood, October 15th 1989; Lymington, October 28th 2001.

Penduline Tit
Burley, December 10th 1998; Farnborough, October 12th 2000.

Oriental Skylark
Hurst Beach, December 12th 1988.

Short-toed Lark
Southampton, January 27th 1979; Beaulieu Road, May 6th 1989; Warnford, September 10th 1989.

Red-rumped Swallow
Black Gutter Bottom, April 24th 1987; Titchfield Haven, September 19th 1988.

American Cliff Swallow
Needs Ore Point, June 3rd 2016.

Bonelli's Warbler *sp.*
Fareham, July 27th–28th 1960; 2, Nursling, October 31st–November 7th 1976; one to 13th; New Forest, May 17th 1980; Alresford, September 12th 1984.

Western Bonelli's Warbler
New Forest, May 15th 2000; Chalton, August 23rd 2007; Bordon, August 10th 2013.

Iberian Chiffchaff
Elvetham Heath, April 6th 2012.

Two-barred Greenish Warbler
Titchfield Haven, October 26th 2017.

Greenish Warbler
Cowplain, October 2nd 1960.

Great Reed Warbler
Broughton, April 3rd 1966; Titchfield Haven, May 12th 2001.

Aquatic Warbler
Titchfield Haven, September 4th 1960; Keyhaven, August 31st 1969; Warsash, September 7th 1972; Keyhaven Marshes, September 20th 1981; Needs Ore Point, September 20th 2016; Farlington Marshes, August 21st 2017.

Blyth's Reed Warbler
Keyhaven Marshes, June 14th 2014.

Moustached Warbler
Southampton, August 24th 1978.

Booted Warbler
Keyhaven Marshes, June 30th 2014–1st July 2014.

Olivaceous Warbler *sp.*
Lower Itchen Meadows, September 10th 1988.

Savi's Warbler
Farlington Marshes, September 29th 2001.

Fan-tailed Warbler
Damerham, May 28th 1979.

Sardinian Warbler
Farlington Marshes, June 17th 2005.

Subalpine Warbler *sp.*
Pennington Marshes, September 3rd 2019.

Spectacled Warbler
Lepe, April 15th 1988.

White's Thrush
Winchester, November 21st 1996.

Naumann's Thrush
Black Gutter Bottom, October 21st 1998.

Dusky Thrush
3, Liss Forest, January 14th 1982.

American Robin
Netley, December 2nd 1975; Leigh Park, February 10th 1991; Blashford Lakes, March 12th 2010.

Collared Flycatcher
Male, Milford on Sea, June 8th 1976.

Siberian/Stejneger's Stonechat
Titchfield Haven, February 14th/15th 1983; Farlington Marshes, March 25th 1986; Pennington Marshes, October 9th 1993.

Isabelline Wheatear
Pennington Marshes, October 24th 1987.

Black-eared Wheatear *sp.*
Farlington Marshes, April 17th 1980; Keyhaven, April 25th 1992.

Black Wheatear
Farlington Marshes, October 30th 1966.

Citrine Wagtail
Lymington, October 12th 2000; Sandy Point, September 1st 2008.

Richard's Pipit
Farlington Marshes, October 27th 1959; Pennington Marshes, October 21st 1976.

Tawny Pipit
2, Gilkicker Point, September 5th 1961; South Moor, Langstone Harbour, October 14th 2016.

Pechora Pipit
Near Burley, February 11th 1985.

Red-throated Pipit
Titchfield Haven, October 14th 1998; Wilverley Plain, March 17th 2003; Sandy Point, October 13th 2005.

Pine Grosbeak
Near Beaulieu, January 22nd 1981.

Trumpeter Finch
Needs Ore Point, May 10th 2008; Gosport, September 7th 2014.

Two-barred Crossbill
Rhinefield, November 2nd–14th 1990; Pennington Marshes, November 22nd 1998.

Little Bunting
Keyhaven, October 24th 1976; Ringwood, January 3rd 1987; Keyhaven Marshes, November 3rd 1990.

Rustic Bunting
Ringwood, March 11th 1977.

Black-headed Bunting
2, Compton, May 1979.

Dark-eyed Junco
Bordon, April 3rd 2017.

Scarlet Tanager
Mouth of Meon River, August 16th 1969.

References

Glossy Ibis, Bickerley Common, December 8th 2012

Alström, P., Mild, K. & Zetterström, B. 2003. *Pipits and Wagtails of Europe, Asia and North America.* Helm, London.

Anon. 2001. Rarities Committee news. Request for records of nominate *cachinnans*, 'Caspian Gull'. *Brit. Birds* 93: 303.

Antliff-Clark, E & Bridge, D. 2019. The Great Crane Project 2019 Report. RSPB. http://www.thegreatcraneproject.org.uk/sites/default/files/Public%20Report%20Compressed.pdf

Balmer, D. E., Gillings, S., Caffrey, B. J., Swann, R. L., Downie, I. S. & Fuller, R. J. 2013. *Bird Atlas 2007–11: the breeding and wintering birds of Britain and Ireland.* BTO Books. Thetford.

Bibby, C. J. 1973. The Red-backed Shrike: A vanishing British species. *Bird Study* 20: 103-110.

Billett, D. F. 1987. The Birds of Langstone Harbour and Farlington Marshes 1950–86. Unpub. MS in J. M. Clark library.

Billett, D. F. 1993. The Birds of Langstone Harbour and Farlington Marshes 1987–92. Unpub. MS in J. M. Clark library.

Billett, D. F. & Grant, P. J. 1971. Franklin's Gull in Hampshire: a species new to Britain and Ireland. *Brit. Birds* 64: 310-313.

British Ornithologists' Union (BOU). 2018. The British List: a Checklist of Birds of Britain (9th edition). *Ibis* 160: 190–240.

Boyd, H. 1954. The "wreck" of Leach's Petrels in the autumn of 1952. *Brit. Birds* 47: 137-163.

Boyes, R. 1774. Alresford Pond. Manuscript notes in the Hampshire County Record Office, Winchester. Mentioned in A History of Alresford. A.J. Robertson. 1937. Winchester (Warren).

British Ornithologists' Union. 2017. The British List: A Checklist of Birds of Britain (9th edition). *Ibis* 160: 190-240.

British Ornithologists' Union Records Committee (BOURC) 2009. 37th report. *Ibis* 151: 224–230.

British Ornithologists' Union Records Committee (BOURC) 2010. 38th report. *Ibis* 152: 192–204.

British Ornithologists' Union Records Committee (BOURC) 2014. 42nd report. *Ibis* 157: 413.

British Ornithologists' Union Records Committee (BOURC) 2018. 48th report. *Ibis* 160: 241–248.

British Ornithologists' Union Records Committee (BOURC) 2021. 52nd report. *Ibis* 163: 305–308.

British Ornithologists' Union Records Committee (BOURC) 2022. 54th report. *Ibis.* In press.

Bucknill, J. A. 1900. *The Birds of Surrey*. R. H. Porter. London.

Burn, D. M. & Mather, J. R. 1974. The White-billed Diver in Britain. *Brit Birds* 67: 257–296.

Chafin, W. 1818. *Anecdotes and History of Cranborne Chase.* The Dovecote Press. Stambridge.

Chappell, Lawrence. 2004. Mongolian Plover – the mongolus form of Lesser Sand Plover. A new species for Hampshire and subspecies for Britain. *HBR* 2003. 186–191.

Cheke, A. S. 1962. *Birds of the Winchester District.* P. & G. Wells. Winchester.

Clark, J. M. 1981. The Seasonal Occurrences of Birds in Hampshire, 1950–80. *HBR* 1980: 53–60.

Clark, J. M. 1984. *Birds of the Hants/Surrey Border*. Hobby Books. Fleet.

Clark, J. M. 2002. Species New to Hampshire. *HBR* 2002: 161–162.

Clark. J. M. 2007. Invasion of Waxwings *Bombycilla garrulus* into Hampshire, December 2004–May 2005. *HBR* 2005: 211–215.

Clark. J. M. 2013. Invasion of Waxwings *Bombycilla garrulus* into Hampshire, November 2010–April 2011. *HBR* 2011: 211–219.

Clark, J. M. & Eyre, J. A. (eds). 1993. *Birds of Hampshire*. Hampshire Ornithological Society.

Clark, John. 2014. First for Hampshire – Pied Wheatear, Monk Sherborne, December 7th–17th 2013. *HBR* 2013: 214–215.

Cloyne, John & Pearson, Paul. 2016. High numbers of post-breeding Mediterranean Gulls around Winchester. *HBR* 2015: 232-233.

Cohen, E. 1963. *Birds of Hampshire and the Isle of Wight*. Oliver & Boyd, Edinburgh & London.

Cohen, E & Taverner, J. H. 1972. *A Revised List of Hampshire and Isle of Wight Birds.* Oxford Illustrated Press. Oxford.

Colenutt, S. 2015. First for Hampshire – Short-toed Eagle in the New Forest, June 8th to July 5th 2014. *HBR* 2014: 208–210.

Combridge, P & Clark, J. M. 2010. Category B Species on the Hampshire List: a review. *HBR* 2009: 209–219.

Combridge, P & Wiseman, E. 2009. The curious case of the disappearing storm-petrel. *Brit. Birds:* 102: 213–215.

Cox, Alan. 2007. Fifty years of the Hampshire Bird Report. *HBR* 2006: 245-251.

Cox, Alan. 2017. First for Hampshire and Great Britain: Chinese Pond Heron at East Dean in 2004. *HBR* 2016: 237–242.

Cox, Alan, Clark, John & Eyre, John. 2019. *Birds of Hampshire. A Checklist. Third Edition.* HOS.

Crook, J. 2002. Letter on unusual Brent Geese in Norfolk and Hampshire. *Brit. Birds* 95: 466–467.

Crook, J. 2008. First for Hampshire – Greater Yellowlegs. *HBR* 2007: 203–206.

Dean A. R. and the Rarities Committee. 1989. Distinguishing characters of American/ East Asian race of Common Scoter. *Brit. Birds* 82: 616–616.

Dean, A., Bradshaw, C., Martin, J., Stoddart, A. & Walbridge, G. 2010. The status in Britain of 'Siberian Chiffchaff'. *Brit. Birds* 103: 320–338.

Dewar, G 1899. *Wildlife in Hampshire Highlands*. Dent. London.

Doran, T. M. J. 2015. Greenland White-fronted Geese at Farlington Marshes October 2014. *HBR* 2015: 214–215.

Dubois, Philippe. 2007. Yellow, Blue-headed, 'Channel' and extra-limital wagtails: from myth to reality. *Birding World* 20: 104–112.

Duffin, Barry. 1984. *The Birds of Titchfield Haven. An annotated checklist.* Privately published.

Duffin, Barry. 1991. *The Birds of Titchfield Haven*. Pekkari Books. Hill Head.

Duffin, Barry. 2003. Savi's Warblers Breeding in Hampshire – a Historical Perspective. *HBR* 2002: 205–209.

Duffin, Barry. 2009. *Titchfield Haven Birds Day by Day.* Hampshire County Council. Winchester.

Duffin, Barry. 2015. First for Hampshire – Pallas's Grasshopper Warbler at Titchfield Haven. *HBR* 2014: 211–213.

Durand, A. L. 1961. White-throated Sparrow and American Robin crossing Atlantic on board ship. *Brit. Birds* 54: 439–440.

Durand, A. L. 1963. A remarkable fall of American land-birds on the Mauretania, New York to Southampton, October 1962. *Brit. Birds* 56: 157–164.

Durand, A. L. 1972. Landbirds over the North Atlantic: unpublished records 1961–65 and thoughts a decade later. *Brit. Birds* 65: 428–442.

Else, George. 2017. First for Hampshire: Alpine Accentor at Brownwich. *HBR* 2016: 233–234.

Evans, A. D. & Smith, K. W. 1994. Habitat selection of Cirl Buntings *Emberiza cirlus* wintering in Britain. *Bird Study* 41: 81–87.

Eyton, T. C. 1836. *A History of Rarer British Birds.* Longmans. London.

Fearon, W. A. (Ed). 1900. *Winchester College Memorial Buildings: Collection of Hampshire Birds 1863–1900.* Warren & Son. Winchester.

Frankland, J. B. 1989. North American landbirds on the QE2. *Brit. Birds* 82: 568–569.

Fraser, P. A., Rogers, M. J. and the Rarities Committee. 2007. Report on Rare Birds in Great Britain in 2005, part 2. *Brit. Birds* 100: 72–104.

French, Paul. 2021. Rarities Committee News. *Brit. Birds* 114: 292.

Gammage, Peter. 2020. First for Hampshire: Eastern Olivaceous Warbler at Farlington Marshes. *HBR* 2019: 238–241.

Garfield, B. 2007. *The Meinertzhagen Mystery: The Life and Legend of a Colossal Fraud.* Potomac Books, Virginia.

Garner, M. S. 1989. Common Scoter of nominate race with extensive yellow on bill. *Brit. Birds.* 82: 616–618.

Garr, J. J. 1994. Species new to Hampshire – White-billed Diver at Lymington. *HBR* 1993: 111-112.

Gibb, John. 1948. Report on the Immigration of Waxwings, winter, 1946–7. *Brit. Birds* 41: 2–9, 34–40.

Gibbons, D. W., Reid J. B. & Chapman R. A. (eds). 1993. *The New Atlas of Breeding Birds in Britain & Ireland 1988–91.* Poyser, London.

Green, George. 2004. *The Birds of Dorset.* Christopher Helm. London.

Gurney, J. H. Jr. 1867. *A Summary of the Occurrences of the Grey Phalarope in Great Britain during the Autumn of 1866.* John Van Voorst. London.

Hack, Philip & Jones, Nigel. 2013. First for Hampshire – Spectacled Warbler. *HBR* 2011: 179–182.

Hagemeijer W. J. M. & Blair, M. J. (eds). 1997. *The EBCC Atlas of European Breeding Birds.* Poyser, London.

Harber, D. D. 1964. The influx of Cranes in October 1963. *Brit. Birds* 57: 502–508.

Harrop, A. H. J., Collinson, J. M. & Melling, T. 2012. What the eye doesn't see: the prevalence of fraud in ornithology. *Brit. Birds* 105: 236–257.

Harrop, A. H. J. 2002. The Ruddy Shelduck in Britain. A review. *Brit. Birds* 95: 123–128.

Harrop, A. H. J. & McGowan, R. Y. 2009. Britain's first Baikal Teal. *Brit. Birds* 102: 691–696.

Hart, Edward. 1913. *Annotated List of the Birds and Animals.* Sidney J. Mate. Bournemouth.

Hartert, E., Jourdain, F. C. R., Ticehurst, N. F. & Witherby, H. F. 1912. *A Hand-List of British Birds.* Witherby, London.

Harting, J. E. 1872. *A Handbook of British Birds.* John Van Voorst. London.

Hellström, M. & Mats Wærn, M. 2011. Field identification and ageing of Siberian Stonechats in spring and summer. *Brit. Birds* 104: 236-254.

Holling, M and the Rare Breeding Birds Panel. 2012. Rare breeding birds in the UK in 2010. *Brit. Birds* 105: 352–416.

Holling, M and the Rare Breeding Birds Panel. 2018. Rare breeding birds in the UK in 2016. *Brit. Birds* 111: 644–694.

Hollyer, J. N. 1970. The invasion of Nutcrackers in autumn 1968. *Brit. Birds* 70: 353–373.

Holt, C and the Rarities Committee. 2018. Report on Rare Birds in Great Britain in 2017. *Brit. Birds* 111: 557–627.

Holt, C and the Rarities Committee. 2019. Report on Rare Birds in Great Britain in 2018. *Brit. Birds* 112: 556–626.

Holt, C., French, P. and the Rarities Committee. 2020. Report on Rare Birds in Great Britain in 2019. *Brit. Birds* 113: 585–655.

Holt, C., French, P. and the Rarities Committee. 2021. Report on Rare Birds in Great Britain in 2020. *Brit. Birds* 114: 570–628.

Houghton, D. & Jones, N. 2011. First for Hampshire – Iberian Chiffchaff in the New Forest. *HBR* 2010: 176–179.

Howard, Gary. 2016. Black-eared Wheatear at Acres Down: first record of eastern race in Hampshire. *HBR* 2015: 229–231.

Howell, S. N. G. & Zufelt, K. 2019. *Oceanic Birds of the World.* Princeton University Press. Princeton & Oxford.

Hudson, N. and the Rarities Committee. 2009. Rare Birds in Great Britain 2008. *Brit. Birds* 102: 528–601.

Hudson, N. and the Rarities Committee. 2011. Rare Birds in Great Britain 2010. *Brit. Birds* 104: 557–629.

Hudson, N. and the Rarities Committee. 2013. Rare Birds in Great Britain 2012. *Brit. Birds* 106: 570–641.

Hudson, N. and the Rarities Committee. 2014. Rare Birds in Great Britain 2013. *Brit. Birds* 107: 579–653.

Johnson, Andy. 2011. First for Hampshire – Red-flanked Bluetail at Sandy Point. *HBR* 2010: 192–194.

Johnson, Andy. 2014. First for Hampshire – Semipalmated Plover, Black Point, October 17th–29th 2013. *HBR* 2013: 209–214

Johnson, Andy. 2018. First for Hampshire: Elegant Tern on Hayling Island. *HBR* 2017: 242–245.

Johnson, Andy & Woolley, Simon. 2007. Desert Wheatear, Hayling Seafront, November 2005. *HBR* 2005: 218–223.

Jones, Nigel. 2017. First for Hampshire: Pallid Harrier at Needs Ore. *HBR* 2016: 235–236.

Knox, A. G., Helbig, A. J., Parkin, D. T. & Sangster, G. 2001. The taxonomic status of Lesser Redpoll. *Brit. Birds* 94: 260–267.

Kelsall, J. E. & Coles, R. E. 1913. The Birds of Milford. *Milford-on-Sea Record Soc.* 1 (6): August.

Kelsall, J. E. & Munn, P.W. 1905. *The Birds of Hampshire and the Isle of Wight*. Witherby, London.

Kelso, J. E. H. 1912. *Notes on Some Common & Rare British Birds.* Century Press. London.

Lewis, Alan. 2018. First for Hampshire: Lesser Scaup at Blashford Lakes. *HBR* 2017: 246–249.

Lewis, Alan. 2019. First for Hampshire: Thayer's Gull at Ibsley Water and Blackheath Down. *HBR* 2018. 233–236.

Lodge, G. 1946. *Memoirs of an Artist Naturalist*. Gurney & Jackson, London.

Longcroft, Charles John. 1856. *The Hundred of Bosmere.* John Russell Smith. Southampton.

Longstaff, T. G. 1926. Local Changes in Distribution. *Ibis Series XII*. 2: 637–656.

McInerney, C. J. & Stoddart, A. 2017. Recent BOURC decisions: Mugimaki Flycatcher and Chinese Pond Heron. *Brit. Birds* 110: 345–354.

Mansel-Pleydell, J. C. 1888. *The Birds of Dorsetshire.* Ballantyne, Hanson & Co. Edinburgh & London.

Marchant, Bob. 2014. First for Hampshire – Brown Shrike, Chilling, September 20th 2013. *HBR* 2013: 207–208.

Martin, J. 2002. From the Rarities Committee's files: Unusual Brent Geese in Norfolk and Hampshire. *Brit. Birds* 95: 129–136.

Meinertzhagen, R. 1959. *Pirates and Predators*. Oliver & Boyd, Edinburgh & London.

Milinets-Raby, P. 2020. First confirmed breeding of Cattle Egrets in Hampshire. *HBR* 2019: 215–219.

Millington, Richard. 1981. *A Twitcher's Diary.* Blandford Press. Poole.

Milne, B. S. Variation in a population of Yellow Wagtails. *Brit. Birds* 52: 281–295.

Monk, J. F. 1963. The past and present status of the Wryneck in the British Isles. *Bird Study* 10: 112–132.

Munn, P. W. 1912. Madeiran Fork-tailed Petrel in Hampshire. *Brit. Birds* 5: 252–253.

Munn, P. W. 1919. Notes on the Birds of Hampshire and the Isle of Wight. Part 1. *Proc. Hants Field Club Archaeol. Soc.* 8 (3): 277-288.

Munn, P. W. 1920. Notes on the Birds of Hampshire and the Isle of Wight. Part 2. *Proc. Hants Field Club Archaeol. Soc.* 9 (1): 23-36.

Naylor, K. A. 1996. *A Reference Manual of Rare Birds in Great Britain and Ireland. Volume 1.* Privately published. Nottingham.

Naylor, K. A. 2021. *Historical Rare Birds.* https://www.historicalrarebirds.info

Newman, E. 1849. *The Letters of Rusticus on the Natural History of Godalming.* John Van Voorst. London.

Nicholson, E. M. & Ferguson-Lees, I. J. 1962. The Hastings Rarities. *Brit. Birds* 55: 299–384.

Norton, John. 2011. First for Hampshire – Isabelline Shrike at Forton Lake, Gosport. *HBR* 2010: 183–186.

Norton, John A. & Raby, Peter N. 2007. First

for Hampshire – Laughing Gull. *HBR* 2005 224–226.

Norton, W. J. E. Unpub. Bird records 1949–89 for Winchester, especially Morestead area. Manuscript held in J. M. Cloyne collection.

Ogilvie, F. M. 1892. On the recent occurrence in the British Islands of the Ruddy Sheldrake. *The Zoologist* 1892: 392–398.

Paddon, C. 1926. Probable Alpine Accentors in Hampshire. Brit. Birds 20: 107.

Payne-Gallwey, Sir R. 1893. *The Diary of Colonel Peter Hawker*. Longmans. London.

Peakall, D. B. 1962. The past and present status of the Red-backed Shrike in Great Britain. *Bird Study* 9: 198-216.

Penhallurick, R. D. 1969. *Birds of the Cornish Coast.* Bradford Barton. Truro.

Pierce, G. G. 1951. Walking in Hampshire: Winnall – Compton. *Hampshire Review* 9: 16–24.

Pinchen, Bryan. 2015. First for Hampshire – Hume's Warbler at Pennington. *HBR* 2014: 213–214.

Powell, D. 2002. October 1st 2000 – One of Those Days. *HBR* 2000: 162–164.

Prendergast, E. D. V. & Boys, J. V. 1983. *The Birds of Dorset.* David & Charles. Newton Abbot.

Rasmussen, P.C. & Prys-Jones, R.P. 2003. History vs Mystery: the reliability of Museum Specimen Data. In: Collar, N.J., Fisher, C.T., & Feare, C.J. (eds). Why Museums Matter. *Bull. Brit. Orn.* Cl. Supplement, 123A: 1–360.

Riddington, R. 2019. Female Green-winged Teal in Shetland. *Brit. Birds* 112: 35–39.

Robjohns, Amy. 2017. Caspian Stonechat at Titchfield Haven: First record of distinctive race in Hampshire. *HBR* 2016: 243–244.

Rogers, M. J. and the Rarities Committee. 1979. Report on rare birds in Great Britain in 1978. *Brit. Birds* 72: 503–549.

Rogers, M. J. and the Rarities Committee. 1981. Report on rare birds in Great Britain in 1980. *Brit. Birds* 74: 493–495.

Rogers, M. J. and the Rarities Committee. 1991. Report on rare birds in Great Britain in 1990. *Brit. Birds* 84: 449–505.

Rogers, M. J. and the Rarities Committee. 1995. Report on rare birds in Great Britain in 1994. *Brit. Birds* 88: 493–558.

Rogers, M. J. and the Rarities Committee. 1997. Report on rare birds in Great Britain in 1996. *Brit. Birds* 90: 453–522.

Rogers, M. J. and the Rarities Committee. 2002. Report on rare birds in Great Britain in 2002. *Brit. Birds* 96: 542–609.

Rogers, M. J. and the Rarities Committee. 1997. Report on rare birds in Great Britain in 2003. *Brit. Birds* 97: 558–625.

Rolfe, Mark. 2016. Second for Hampshire: Greater Yellowlegs at Titchfield. *HBR* 2015: 227–228.

Saunders, H. 1899. *An Illustrated Manual of British Birds*. 2nd edn. Gurney & Jackson, London.

Sharrock, J. T. R. 1961. White-throated Sparrow in Hampshire. *Brit. Birds* 54: 366–367.

Sharrock, J. T. R. 1965. White-throated Sparrows in Hampshire. *Brit. Birds* 58: 230.

Smith, G. W. 1898. Ornithological Notes from mid-Hants: Autumn and Winter 1897. *The Zoologist* 1898: 126–130.

Smith, S. 1950. *The Yellow Wagtail.* Collins, London.

Spear Smith, K. 1913. Probable Scarlet Grosbeak in Hampshire. Brit. Birds 7: 179.

Stancliffe, Paul. 1995. Species new to Hampshire – Semipalmated Sandpiper at Keyhaven Marsh. *HBR* 1994: 120–122.

Stoddart, A. M. 2008. Rarities Committee news. Siberian Chiffchaffs in Britain. *Brit. Birds.* 101: 165–166.

Stoddart, A. & Hudson, N. 2021. From the rarities committee's files: BBRC and newly split species. *Brit. Birds* 114: 8-17.

Stoddart, A. & McInerny, C. 2020. Falcated Duck in Norfolk: New to Britain. *Brit. Birds* 113: 44–53.

Suffern, C & Ferguson-Lees, I. J. 1964. Cetti's Warbler in Hampshire. *Brit. Birds* 57: 365-366.

Summers-Smith, J. D. 1950. A History of the Birds of the Newbury District. *Trans. Newbury District Field Club* 9: 59-86.

Svennson, L., Mullarney, K., Zetterström, D & Grant, P.J. 2010. *Collins Bird Guide.* 2nd Edition. Harper Collins. London

Taverner, J. H. 1956. Hoopoes breeding in Hampshire. *Brit. Birds* 49: 454.

Taverner, J. H. 1970. Mediterranean Gulls nesting in Hampshire. *Brit. Birds* 63: 67–79.

Taverner, J. H. 1972. Mediterranean Gulls in Hampshire 1970-71. *Brit. Birds* 65: 185–186.

Taylor, M. 2002. *Guardian Spirit of the East Bank: A celebration of the life of R. A. Richardson*. Wren Publishing, Norfolk.

Tubbs, C. R. 1968. *The New Forest: An ecological history*. David & Charles. Newton Abbot.

Unsworth, Dave. 2011. First for Hampshire – Booted Warbler at Needs Ore. *HBR* 2010: 187–191.

Van Duivendijk, N. 2010. *Advanced Bird ID Guide*. New Holland, London.

Vesey-Fitzgerald, B. 1946. *British Game*. Collins. London.

Wallace, D. I. M. 1976. Distinguishing Great Snipe from Snipe. Brit. Birds 69: 377–383.

Wallace, D. I. M.1977. Further definition of Great Snipe characters. Brit. Birds 70: 283–289.

Wallace, D. I. M. 1999. History of the Common Rosefinch in Britain and Ireland, 1869–1996. Brit. Birds 92: 445–471.

Wallace, D. I. M., Bradshaw, C. & Rogers, M. J. 2006. A review of the 1950–57 British rarities. *Brit. Birds* 99: 460–464

Waters, E. & Waters, D. 2005. The former status of Great Bustard in Britain. *Brit. Birds* 98: 295-305.

Wearn, Gary. 1994. Species new to Hampshire – Black-headed Bunting at Purbrook. *HBR* 1993: 106-107.

Wilson, Sue. 2013. First for Hampshire – Spanish Sparrow. *HBR* 2013: 204–205.

Wiltshire Ornithological Society. 2007. *Birds of Wiltshire*. WOS. Devizes.

Wise, J. R. 1862. *The New Forest: its History and Scenery*. Sotheran. London.

Wiseman, E. J. 2009. The Portsmouth Group. *Brit. Birds* 102: 132–136.

Wiseman, E. J. 2017. *One Man's Marshes. The birds of Lymington and Keyhaven*. Powellwildlifeart.com

Wiseman, E. J. & Combridge, P. 2012. What the eye does see: misrepresenting the views of others, and a Freudian slip? *Brit. Birds* 105: 483.

Witherby, H. F. 1922. On the White-billed Northern Diver as a British Bird. *Brit. Birds* 16: 9-12.

Witherby, H. F., Jourdain F. R. C., Ticehurst N. F. & Tucker B. W. 1938–41. *The Handbook of British Birds*. Witherby, London.

Wooldridge, D. B. 2008. Report of a Masked Booby *Sula dactylatra* at St. Catherine's Point, 19th May 2007. *The Isle of Wight Bird Report* 2007: 145–146.

Woolley, S. 2005. An unusual kite at Cheesefoot Head in early 2003. *HBR* 2004: 215–219.

Wynn, Russell B. 2003a. A Record Breaking Year for Seabirds. *HBR* 2001: 181–183.

Wynn, Russell B. 2003b. Stilt Sandpiper. A new species for Hampshire. *HBR* 2002: 198–201.

Wynn, Russell. 2003c. A dark morph Montagu's Harrier in Hampshire and Norfolk. *Birding World* 16: 257.

Wynn, Russell. 2011. First for Hampshire – Rustic Bunting at Brockenhurst. *HBR* 2010: 180–182.

Yarrell, W. 1843. *A History of British Birds*. John Van Voorst. London.

Yarrell, W. 1845. *A History of British Birds*. 2nd Edition. John Van Voorst. London.

Yarrell, W. 1871-85. *A History of British Birds*. 4th edition, revised by A. Newton and H. Saunders. John Van Voorst, London.

Roller, Pig Bush, New Forest, May 31st 1987

Index

Wryneck, Hook-with Warsash, September 5th 2015

Lapland Bunting, male, Saltgrass Lane, Keyhaven, January 22nd 2011

First published 2022 by
Hampshire Ornithological Society
www.hos.org.uk

ISBN 978-1-9993092-3-7

Printed by Henry Ling
www.henryling.co.uk

Cover photographs
front. *Wilson's Phalarope, Keyhaven, October 11th 2021. Ian Wells*
Short-toed Eagle, New Forest, June 30th 2014. Alan Lewis
spine. *Franklin's Gull, Farlington Marshes, March 19th 1970. J.B. & S. Bottomley*
Woodchat Shrike, near Hambledon, May 8th 2006. Richard Ford
back. *Red-breasted Goose, Milford on Sea, October 18th 2013. Chris Holt*
Scops Owl, Dummer, May 1980. Pete Wheeler
Penduline Tit, Titchfield Haven, December 11th 2015. Lee Fuller
Waxwing, Romsey, January 9th 2011. Ian Williamson

End page illustrations
half title. *Cliff Swallow, Titchfield Haven, October 1st 2000*
frontispiece. *Lesser Sand Plover, Keyhaven, July 22nd 2003*
contents. *Desert Wheatear, South Hayling, November 13th 2005*
back page. *Little Swift, Shatterford Bottom, June 4th 2000. Dan Powell*